Matrix Analysis of Framed Structures

Second Edition

Matrix Analysis of Framed Structures

Second Edition

William Weaver, Jr.
Stanford University

James M. Gere
Stanford University

CBS Publishers & Distributors Pvt. Ltd.

New Delhi • Bengaluru • Chennai • Kochi • Kolkata • Mumbai
Hyderabad • Nagpur • Patna • Pune • Vijayawada

ISBN: 81-239-1151-3

First Indian Edition: 1986
Reprint: 1998, 2000, 2002, 2004

This edition has been published in India by arrangement with
Van Nostrand Reinhold Company, Inc., USA

Published by:
Satish Kumar Jain for CBS Publishers & Distributors Pvt. Ltd.,
4819/XI Prahlad Street, 24 Ansari Road, Daryaganj, New Delhi - 110002
delhi@cbspd.com, cbspubs@airtelmail.in • www.cbspd.com
Ph.: 23289259, 23266861, 23266867 • Fax: 011-23243014

Corporate Office: 204 FIE, Industrial Area, Patparganj, Delhi - 110 092
Ph: 49344934 • Fax: 011-49344935
E-mail: publishing@cbspd.com • publicity@cbspd.com

Branches:
- ***Bengaluru:*** 2975, 17th Cross, K.R. Road, Bansankari 2nd Stage, Bengaluru - 70 • Ph: +91-80-26771678/79 • Fax: +91-80-26771680 E-mail: cbsbng@gmail.com, bangalore@cbspd.com
- ***Chennai:*** No. 7, Subbaraya Street, Shenoy Nagar, Chennai - 600030 Ph: +91-44-26681266, 26680620 • Fax: +91-44-42032115 E-mail: chennai@cbspd.com
- ***Kochi:*** Ashana House, 39/1904, A.M. Thomas Road, Valanjambalam, Ernakulum, Kochi • Ph: +91-484-4059061-65 Fax: +91-484-4059065 • E-mail: cochin@cbspd.com
- ***Kolkata:*** 6-B, Ground Floor, Rameshwar Shaw Road, Kolkata - 700014 Ph: +91-33-22891126/7/8 • E-mail: kolkata@cbspd.com
- ***Mumbai:*** 83-C, Dr. E. Moses Road, Worli, Mumbai - 400018 Ph: +91-9833017933, 022-24902340/41 • E-mail: mumbai@cbspd.com

Representatives:

- Hyderabad: 0-9885175004
- Nagpur: 0-9021734563
- Patna: 0-9334159340
- Pune: 0-9623451994
- Vijayawada: 0-9000660880

Printed at:
Neekunj Print Process, Delhi (India)

Preface

The advent of the digital computer made it necessary to reorganize the theory of structures into matrix form, and the first edition of this book was written for that purpose. It covered the analysis of all types of framed structures by the flexibility and stiffness methods, with emphasis on the latter approach. At that time, it was evident that the stiffness method was superior for digital computation, but for completeness both methods were extensively discussed. Now the flexibility method should play a less important role and be characterized as a supplementary approach instead of a complementary method. The flexibility method cannot be discarded altogether, however, because it is often necessary to obtain stiffnesses through flexibility techniques.

This book was written as a text for college students on the subject of the analysis of framed structures by matrix methods. The preparation needed to study the subject is normally gained from the first portion of an undergraduate engineering program; specifically, the reader should be familiar with statics and mechanics of materials, as well as algebra and introductory calculus. A prior course in elementary structural analysis would naturally be beneficial, although it is not a prerequisite for the subject matter of the book. Elementary matrix algebra is used throughout the book, and the reader must be familiar with this subject. Since the topics needed from matrix algebra are of an elementary nature, the reader can acquire the necessary knowledge through self-study during a period of two or three weeks. A separate mathematics course in matrix algebra is not necessary, although some students will wish to take such a course in preparation for more advanced work. To assist those who need only an introduction to matrix algebra, without benefit of a formal course, the authors have written a supplementary book on the subject.*

There are several reasons why matrix analysis of structures is vital to the structural analyst. One of the most important is that it makes possible a comprehensive approach to the subject that is valid for structures of all types. A second reason is that it provides an efficient means of describing the various steps in the analysis, so that these steps can be more easily programmed for a digital computer. The use of matrices is natural when performing calculations with a computer, because they permit large groups of numbers to be manipulated in a simple and effective manner. The reader will find that the methods of analysis developed in this book are highly organized and that the same basic procedures can be followed in the analysis of all types of framed structures.

In writing this book, the authors have assumed that readers will be interested in preparing their own programs for structural analysis or will at least like to see how that is done. While the details of coding a program in a particular computer language are not covered completely, the em-

* *Matrix Algebra for Engineers*, D. Van Nostrand, New York, 1965.

phasis is upon FORTRAN, which is the language most frequently used. Chapter 5 contains FORTRAN-oriented flow charts of programs for analyzing the six basic types of framed structures. These charts will help readers code their own programs should they desire to do so.

The first chapter of this book covers the fundamental concepts of structural analysis that are needed for the remaining chapters. Those who have previously studied theory of structures will find that this material is mostly review. Anyone who is encountering structural theory for the first time, however, will find it necessary to become thoroughly familiar with these basic matters.

The flexibility and stiffness methods are introduced in Chapters 2 and 3, respectively. In Chapter 4, the stiffness method is further developed in a computer-oriented manner in preparation for programming. Flow charts of computer programs for analyzing all types of framed structures are presented in Chapter 5. In order to maintain the earlier emphasis on fundamental matters, many special topics are postponed until Chapter 6, where such topics as nonprismatic members, temperature effects, and elastic connections are considered. All of these subjects can be viewed as modifications of the basic procedures described in earlier chapters.

Problems for hand solution are given at the ends of the first four chapters, and these problems are generally placed in order of increasing difficulty. All examples and problems in Chapters 1, 2, and 3 (and the Appendixes) are in literal form, while the numerical examples and problems in Chapters 4 and 5 are in both US and SI units.

References for further study, a list of notation, and appendixes containing tables of useful information are at the end of the book. Notable among the appendix topics are the unit-load method for calculating displacements (Appendix A) and computer routines for solving equations (Appendix D). Answers to all of the problems are also given at the end of the book.

In this edition, several important topics have been added to make the book more up to date and complete. The following sections and appendixes are new: 1.13 Energy Concepts, 1.14 Virtual Work, 2.7 Formalization of the Flexibility Method, 3.6 Formalization of the Stiffness Method, 5.12 Combined Program for Framed Structures, 6.2 Rectangular Framing, 6.3 Symmetric Structures, 6.5 Automatic Dead Load Analysis, 6.10 Translation of Axes, 6.11 Member Stiffnesses and Fixed-End Actions from Flexibilities, 6.13 Curved Members, 6.17 Offset Connections, Appendix D: Computer Routines for Solving Equations, and Appendix E: Solution Without Rearrangement. The word Matrix now appears in the title to clarify the fact that the book deals with the analysis of framed structures by matrix methods.

The authors wish to express their appreciation to graduate students Paul R. Johnston and Moshe Eisenberger for doing an excellent job of coding and improving the programs. Thanks are also due Laura J. Selby for outstanding work in typing the manuscript. Finally, the authors are much indebted to Dr. Winfred O. Carter of Utah State University for his helpful advice during the preparation of this edition.

William Weaver, Jr.
James M. Gere

Contents

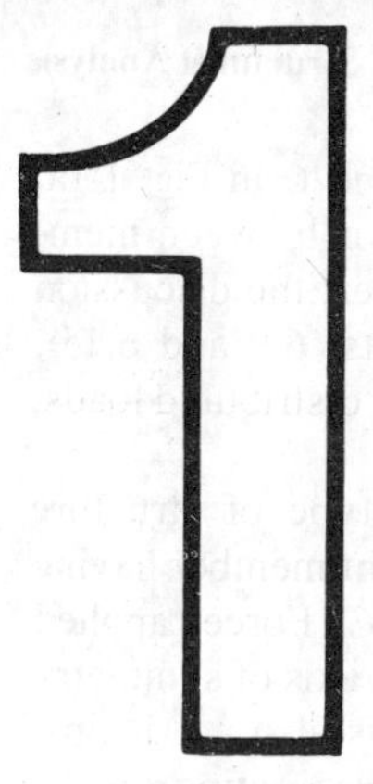

Basic Concepts of Structural Analysis

1.1 Introduction. This book describes matrix methods for the analysis of framed structures with the aid of a digital computer. Both the flexibility and stiffness methods of structural analysis are covered, but emphasis is placed upon the latter because it is more suitable for computer programming. While these methods are applicable to discretized structures of all types, only framed structures will be discussed. After mastering the analysis of framed structures, the reader will be prepared to study the finite element method for analyzing discretized continua (see Refs. C-1 through C-8).

In this chapter various preliminary matters are considered in preparation for the matrix methods of later chapters. These subjects include descriptions of the types of framed structures to be analyzed and their deformations due to loads and other causes. Also discussed are the basic concepts of equilibrium, compatibility, determinacy, mobility, superposition, flexibility and stiffness coefficients, equivalent joint loads, energy, and virtual work.

1.2 Types of Framed Structures. All of the structures that are analyzed in later chapters are called *framed structures* and can be divided into six categories: beams, plane trusses, space trusses, plane frames, grids, and space frames. These types of structures are illustrated in Fig. 1-1 and described later in detail. These categories are selected because each represents a class of structures having special characteristics. Furthermore, while the basic principles of the flexibility and stiffness methods are the same for all types of structures, the analyses for these six categories are sufficiently different in the details to warrant separate discussions of them.

Every framed structure consists of members that are long in comparison to their cross-sectional dimensions. The *joints* of a framed structure are points of intersection of the members, as well as points of support and free ends of members. Examples of joints are points *A, B, C,* and *D* in Figs. 1-1a and 1-1d. Supports may be *fixed,* as at support *A* in the beam of Fig. 1-1a, or *pinned,* as shown for support *A* in the plane frame of Fig. 1-1d, or there may be *roller supports,* illustrated by supports *B* and *C* in Fig. 1-1a. In special instances the connections between members or between members and supports may be elastic (or semi-rigid). However, the discussion of this possibility will be postponed until later (see Arts. 6.9 and 6.15). *Loads* on a framed structure may be concentrated forces, distributed loads, or couples.

Consider now the distinguishing features of each type of structure shown in Fig. 1-1. A *beam* (Fig. 1-1a) consists of a straight member having one or more points of support, such as points *A, B,* and *C*. Forces applied to a beam are assumed to act in a plane which contains an axis of symmetry of the cross section of the beam (an axis of symmetry is also a principal axis of the cross section). Moreover, all external couples acting on the beam have their moment vectors normal to this plane, and the beam deflects in the same plane (the *plane of bending*) and does not twist. (The case of a beam which does not fulfill these criteria is discussed in Art. 6.17.) Internal stress resultants may exist at any cross section of the beam and, in the general case, may include an axial force, a shearing force, and a bending moment.

A *plane truss* (Fig. 1-1b) is idealized as a system of members lying in a plane and interconnected at hinged joints. All applied forces are assumed to act in the plane of the structure, and all external couples have their moment vectors normal to the plane, just as in the case of a beam. The loads may consist of concentrated forces applied at the joints, as well as loads that act on the members themselves. For purposes of analysis, the latter loads may be replaced by statically equivalent loads acting at the joints. Then the analysis of a truss subjected only to joint loads will result in axial forces of tension or compression in the members. In addition to these axial forces, there will be bending moments and shear forces in those members having loads that act directly upon them.

A *space truss* (see Fig. 1-1c) is similar to a plane truss except that the members may have any directions in space. The forces acting on a space truss may be in arbitrary directions, but any couple acting on a member must have its moment vector perpendicular to the axis of the member. The reason for this requirement is that a truss member is incapable of supporting a twisting moment.

A *plane frame* (Fig. 1-1d) is composed of members lying in a single plane and having axes of symmetry in that plane (as in the case of a beam). The joints between members (such as joints *B* and *C*) are rigid connections. The forces acting on a frame and the translations of the frame are in the plane of the structure; all couples acting on the frame have their moment vectors normal to the plane. The internal stress resultants acting at any cross section of a plane frame member may consist in general of a bending moment, a shearing force, and an axial force.

A *grid* is a plane structure composed of continuous members that either intersect or cross one another (see Fig. 1-1e). In the latter case the connec-

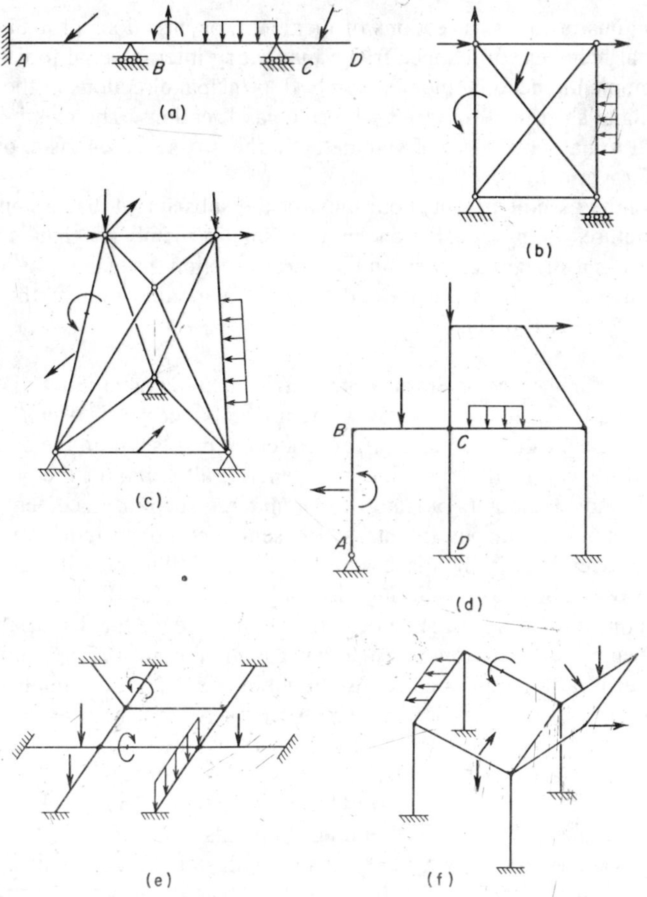

Fig. 1-1. Types of framed structures: (a) beam, (b) plane truss, (c) ce truss, (d) plane frame, (e) grid, and (f) space frame.

tions between members are often considered to be hinged, whereas in the former case the connections are assumed to be rigid. While in a plane frame the applied forces all lie in the plane of the structure, those applied to a grid are normal to the plane of the structure; and all couples have their vectors in the plane of the grid. This orientation of loading may result in torsion as well as bending in some of the members. Each member is assumed to have two axes of symmetry in the cross section, so that bending and torsion occur independently of one another (see Art. 6.17 for a discussion of non-symmetric members).

The final type of structure is a *space frame* (Fig. 1-1f). This is the most general type of framed structure, inasmuch as there are no restrictions on the locations of joints, directions of members, or directions of loads. The individual members of a space frame may carry internal axial forces, torsional moments, bending moments in both principal directions of the cross section, and shearing forces in both principal directions. The members are assumed to have two axes of symmetry in the cross section, as explained above for a grid.

It will be assumed throughout most of the subsequent discussions that the structures being considered have prismatic members; that is, each member has a straight axis and uniform cross section throughout its length. Nonprismatic members are treated later by a modification of the basic approach (see Art. 6.12).

1.3 Deformations in Framed Structures. When a structure is acted upon by loads, the members of the structure will undergo *deformations* (or small changes in shape) and, as a consequence, points within the structure will be displaced to new positions. In general, all points of the structure except immovable points of support will undergo such displacements. The calculation of these displacements is an essential part of structural analysis, as will be seen later in the discussions of the flexibility and stiffness methods. However, before considering the displacements, it is first necessary to have an understanding of the deformations that produce the displacements.

To begin the discussion, consider a segment of arbitrary length cut from a member of a framed structure, as shown in Fig. 1-2a. For simplicity the bar is assumed to have a circular cross section. At any cross section, such as at the right-hand end of the segment, there will be stress resultants that in the general case consist of three forces and three couples. The forces are the axial force N_x and the shearing forces V_y and V; the couples are the twisting moment T and the bending moments M_y and M_z. Note that moment vectors are shown in the figure with double-headed arrows, in order to distinguish them from force vectors. The deformations of the bar can be analyzed by taking separately each stress resultant and determining its effect on an element of the bar. The element is obtained by isolating a portion of the bar between two cross sections a small distance dx apart (see Fig. 1-2a).

The effect of the axial force N_x on the element is shown in Fig. 1-2b. Assuming that the force acts through the centroid of the cross-sectional area, it is found that the element is uniformly extended, the significant strains in the element being normal strains in the x direction. In the case of a shear force (Fig. 1-2c), one cross section of the bar is displaced laterally with respect to the other. There may also be some distortion of the cross sections, but this has a negligible effect on the determination of displacements and can be ignored. A bending moment (Fig. 1-2d) causes relative rotation of the two cross sections so that they are no longer parallel to one another. The resulting strains in the element are in the longitudinal direction of the bar, and they consist of contraction on the compression side and

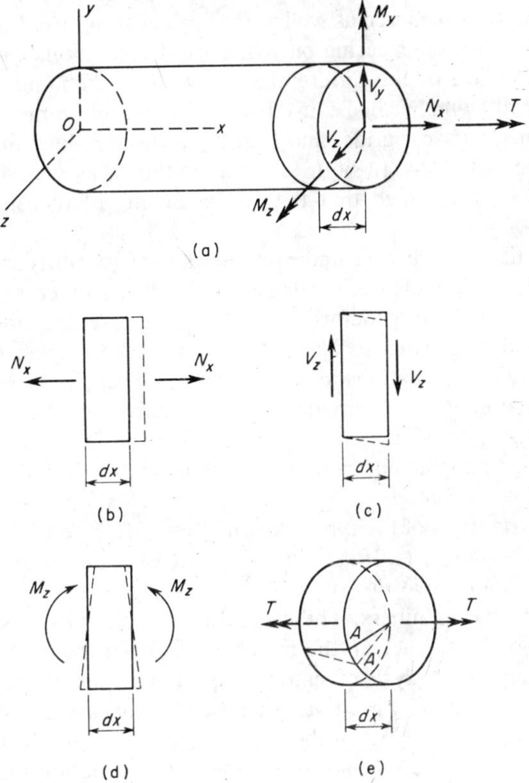

Fig. 1-2. Types of deformations: (b) axial, (c) shearing, (d) flexural, and (e) torsional.

extension on the tension side. Finally, the twisting moment T causes a relative rotation of the two cross sections about the x axis (see Fig. 1-2e) and, for example, point A is displaced to A'. In the case of a circular bar, twisting produces only shearing strains; and the cross sections remain plane. For other cross-sectional shapes, distortion of the cross sections will occur.

The deformations shown in Figs. 1-2b, 1-2c, 1-2d, and 1-2e are called axial, shearing, flexural, and torsional deformations, respectively. Their evaluation is dependent upon the cross-sectional shape of the bar and the mechanical properties of the material. This book is concerned only with materials that are linearly elastic, that is, materials that follow Hooke's law. For such materials the various formulas for the deformations, as well as those for the stresses and strains in the element, are given for reference purposes in Appendix A, Art. A.1.

The *displacements* in a structure are caused by the cumulative effects of the deformations of all the elements. There are several ways of calculating these displacements in framed structures, depending upon the type of deformation being considered as well as the type of structure. For example, deflections of beams considering only flexural deformations can be found by direct integration of the differential equation for bending of a beam. Another method, which can be used for all types of framed structures including beams, trusses, grids, and frames, is the unit-load method, discussed later in Art. 1.14. In both of these methods, as well as others in common use, it is assumed that the displacements of the structure are small.

In any particular structure under investigation, not all types of deformations will be of significance in calculating the displacements. For example, in beams the flexural deformations normally are the only ones of importance, and it is usual to ignore the axial deformations. Of course, there are exceptional situations in which beams are required to carry large axial forces, and under such circumstances the axial deformation must be included in the analysis. It is also possible for axial forces to produce a beam-column action which has a nonlinear effect on the displacements (see Art. 6.18).

For truss structures of the types shown in Figs. 1-1b and 1-1c, the analyses are made in two parts. If the joints of the truss are idealized as hinges and if all loads act only at the joints, then the analysis involves only axial deformations of the members. The second part of the analysis is for the effects of the loads that act on the members between the joints, and this part is essentially the analysis of simply supported beams. If the joints of a truss-like structure actually are rigid, then bending occurs in the members even though all loads may act at the joints. In such a case, flexural deformations could be important, and in this event the structure may be analyzed as a plane or space frame.

In plane frames (see Fig. 1-1d) the significant deformations are flexural and axial. If the members are slender and are not triangulated in the fashion of a truss, the flexural deformations are much more important than the axial ones. However, the axial contributions should be included in the analysis of a plane frame if there is any doubt about their relative importance.

In grid structures (Fig. 1-1e) the flexural deformations are always important, but the cross-sectional properties of the members and the method of fabricating joints will determine whether or not torsional deformations must be considered. If the members are thin-walled open sections, such as I-beams, they are likely to be very flexible in torsion, and large twisting moments will not develop in the members. Also, if the members of a grid are not rigidly connected at crossing points, there will be no interaction between the flexural and torsional moments. In either of these cases, only flexural deformations need be taken into account in the analysis. On the other hand, if the members of a grid are torsionally stiff and rigidly interconnected at crossing points, the analysis must include both torsional and flexural deformations. Usually, there are no axial forces present in a grid because the forces are normal to the plane of the grid. This situation is analogous to that in a beam having all its loads perpendicular to the axis of the beam, in which case there are no axial forces in the beam. Thus, axial deformations are not included in a grid analysis.

Space frames (Fig. 1-1f) represent the most general type of framed structure, both with respect to geometry and with respect to loads. Therefore, it follows that axial, flexural, and torsional deformations all may enter into the analysis of a space frame, depending upon the particular structure and loads.

Shearing deformations are usually very small in framed structures and hence are seldom considered in the analysis. However, their effects may always be included if necessary in the analysis of a beam, plane frame, grid, or space frame (see Art. 6.16).

There are other effects, such as temperature changes and prestrains, that may be of importance in analyzing a structure. These subjects are discussed in later chapters in conjunction with the flexibility and stiffness methods of analysis.

1.4 Actions and Displacements. The terms "action" and "displacement" are used to describe certain fundamental concepts in structural analysis. An *action* (sometimes called a generalized force) is most commonly a single force or a couple. However, an action may also be a combination of forces and couples, a distributed loading, or a combination of these actions. In such combined cases, however, it is necessary that all the forces, couples, and distributed loads be related to one another in some definite manner so that the entire combination can be denoted by a single symbol. For example, if the loading on the simply supported beam AB shown in Fig. 1-3 consists of two equal forces P, it is possible to consider the combination of the two loads as a single action and to denote it by one symbol, such as F. It is also possible to think of the combination of the two loads plus the two reactions R_A and R_B at the supports as a single action, since all four forces have a unique relationship to one another. In a more general situation, it is possible for a very complicated loading system on a structure to be treated as a single action if all components of the load are related to one another in a definite manner.

In addition to actions that are external to a structure, it is necessary to deal also with internal actions. These actions are the resultants of internal stress distributions, and they include bending moments, shearing forces,

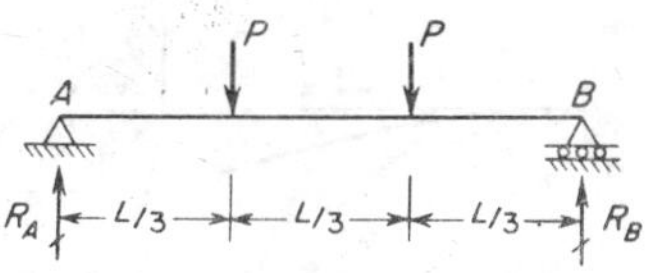

Fig. 1-3.

axial forces, and twisting moments. Depending upon the particular analysis being made, such actions may appear as one force, one couple, two forces, or two couples. For example, in making static equilibrium analyses of structures these actions normally appear as single forces and couples, as illustrated in Fig. 1-4a. The cantilever beam shown in the figure is subjected at end B to loads in the form of actions P_1 and M_1. At the fixed end A the reactive force and reactive couple are denoted R_A and M_A, respectively. In order to distinguish these reactions from the loads on the structure, they are drawn with a slanted line (or slash) across the arrow. This convention for identifying reactions will be followed throughout the text (see also Fig. 1-3 for an illustration of the use of the convention). In calculating the axial force N, bending moment M, and shearing force V at any section of the beam in Fig. 1-4a, such as at the midpoint, it is necessary to consider the static equilibrium of a portion of the beam. One possibility is to construct a free-body diagram of the right-hand half of the beam, as shown in Fig. 1-4b. In so doing, it is evident that each of the internal actions appears in the diagram as a single force or couple.

There are situations, however, in which the internal actions appear as two forces or couples. This case occurs most commonly in structural analysis when a "release" is made at some point in a structure, as shown in Fig. 1-5 for a continuous beam. If the bending moment is released at joint B of the beam, the result is the same as if a hinge were placed in the beam at that joint (see Fig. 1-5b). Then, in order to take account of the bending moment M_B in the beam, it must be considered as consisting of two equal and opposite couples M_B that act on the left- and right-hand portions of the beam with the hinge, as shown in Fig. 1-5c. In this illustration the moment

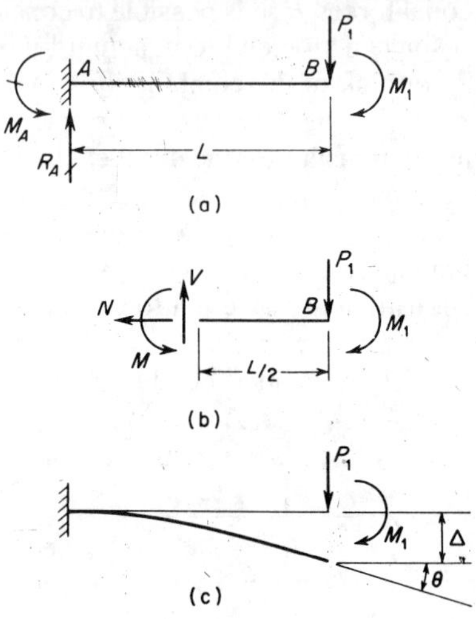

Fig. 1-4.

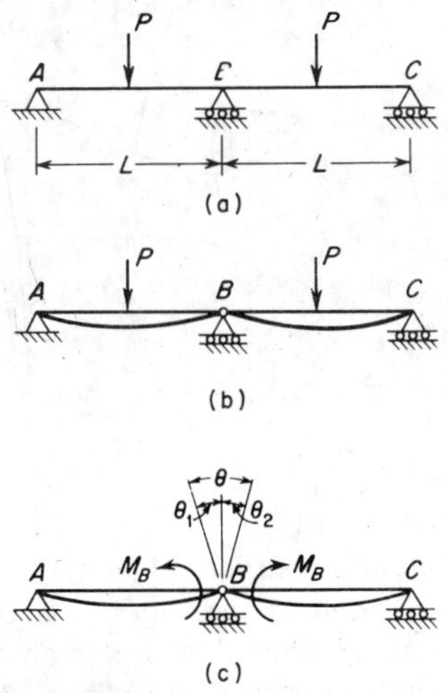

Fig. 1-5.

M_B is assumed positive in the directions shown in the figure, signifying that the couple acting on the left-hand beam is counterclockwise and the couple acting on the right-hand beam is clockwise. Thus, for the purpose of analyzing the beam in Fig. 1-5c, the bending moment at point B may be treated as a single action consisting of two couples. Similar situations are encountered with axial forces, shearing forces, and twisting moments, as illustrated later in the discussion of the flexibility method of analysis.

A second basic concept is that of a *displacement,* which is most commonly a translation or a rotation at some point in a structure. A translation refers to the distance moved by a point in the structure, and a rotation means the angle of rotation of the tangent to the elastic curve (or its normal) at a point. For example, in the cantilever beam of Fig. 1-4c, the translation Δ of the end of the beam and the rotation θ at the end are both considered as displacements. Moreover, as in the case of an action, a displacement may also be regarded in a generalized sense as a combination of translations and rotations. As an example, consider the rotations at the hinge at point B in the two-span beam in Fig. 1-5c. The rotation of the right-hand end of the member AB is denoted θ_1, while the rotation of the left-hand end of member BC is denoted θ_2. Each of these rotations is considered as a displacement. Furthermore, the sum of the two rotations, denoted as θ, is also a displacement. The angle θ can be considered as the relative rotation at point B between the ends of members AB and BC.

Another illustration of displacements is shown in Fig. 1-6, in which a plane frame is subjected to several loads. The horizontal translations Δ_A, Δ_B, and Δ_C of joints A, B, and C, respectively, are displacements, as also are the rotations θ_A, θ_B, and θ_C of these joints. Joint displacements of these types play an important role in the analysis of framed structures.

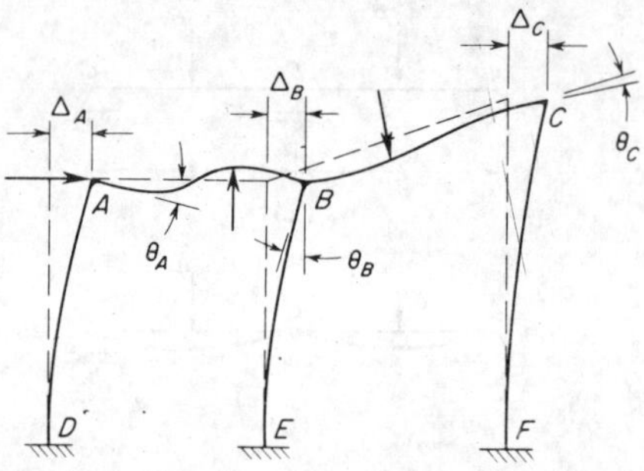

Fig. 1-6.

It is frequently necessary in structural analysis to deal with actions and displacements that *correspond* to one another. Actions and displacements are said to be corresponding when they are of an analogous type and are located at the same point on a structure. Thus, the displacement corresponding to a concentrated force is a translation of the structure at the point where the force acts, although the displacement is not necessarily caused by the force. Furthermore, the corresponding displacement must be taken along the line of action of the force and must have the same positive direction as the force. In the case of a couple, the corresponding displacement is a rotation at the point where the couple is applied and is taken positive in the same sense as the couple.

As an illustration, consider again the cantilever beam shown in Fig. 1-4a. The action P_1 is a concentrated force acting downward at the end of the beam, and the downward translation Δ at the end of the beam (see Fig. 1-4c) is the displacement that corresponds to this action. Similarly, the couple M_1 and the rotation θ are a corresponding action and displacement. It should be noted, however, that the displacement Δ corresponding to the load P_1 is not caused solely by the force P_1, nor is the displacement θ corresponding to M_1 caused by M_1 alone. Instead, in this example, both Δ and θ are displacements due to P_1 and M_1 acting simultaneously on the beam. In general, if a particular action is given, the concept of a corresponding displacement refers only to the definition of the displacement, without regard to the actual cause of that displacement. Similarly, if a displacement is given, the concept of a corresponding action will describe a particular kind of action on the structure, but the displacement need not be caused by that action.

As another example of corresponding actions and displacements, refer to the actions shown in Fig. 1-5c. The beam in the figure has a hinge at the middle support and is acted upon by the two couples M_B, which are considered as a single action. The displacement corresponding to the action M_B consists in general of the sum of the counterclockwise rotation θ_1 of the left-hand beam and the clockwise rotation θ_2 of the right-hand beam. Therefore, the angle θ (equal to the sum of θ_1 and θ_2) is the displacement corresponding to the action M_B. This displacement is the relative rotation between the two beams at the hinge and has the same positive sense as M_B. If the angle θ is caused only by the couples M_B, then it is described as the displacement corresponding to M_B and caused by M_B. This displacement can be found with the aid of the table of beam displacements given in Appendix A (see Table A-3, Case 5), and is equal to

$$\theta = \theta_1 + \theta_2 = \frac{M_B L}{3EI} + \frac{M_B L}{3EI} = \frac{2M_B L}{3EI}$$

in which L is the length of each span and EI is the flexural rigidity of the beam.

There are other situations, however, in which it is necessary to deal with a displacement that corresponds to a particular action but is caused by some other action. As an example, consider the beam in Fig. 1-5b, which is the same as the beam in Fig. 1-5c except that it is acted upon by two forces P instead of the couples M_B. The displacement in this beam corresponding to M_B consists of the relative rotation at joint B between the two beams, positive in the same sense as M_B, but due to the loads P only. Again using the table of beam displacements (Table A-3, Case 2), and also assuming that the forces P act at the midpoints of the members, it is found that the displacement θ corresponding to M_B and caused by the loads P is

$$\theta = \theta_1 + \theta_2 = \frac{PL^2}{16EI} + \frac{PL^2}{16EI} = \frac{PL^2}{8EI}$$

The concept of correspondence between actions and displacements will become more familiar to the reader as additional examples are encountered in subsequent work. Also, it should be noted that the concept can be extended to include distributed actions, as well as combinations of actions of all types. However, these more general ideas have no particular usefulness in the work to follow.

In order to simplify the notation for actions and displacements, it is desirable in many cases to use the symbol A for actions, including both concentrated forces and couples, and the symbol D for displacements, including both translations and rotations. Subscripts can be used to distinguish between the various actions and displacements that may be of interest in a particular analysis. The use of this type of notation is shown in Fig. 1-7, which portrays a cantilever beam subjected to actions A_1, A_2, and A_3. The displacement corresponding to A_1 and due to all loads acting simultaneously is denoted by D_1 in Fig. 1-7a; similarly, the displacements corresponding to A_2 and A_3 are denoted by D_2 and D_3.

Now consider the cantilever beam subjected to action A_1 only (see Fig. 1-7b). The displacement corresponding to A_1 in this beam is denoted by D_{11}. The significance of the two subscripts is as follows. The first subscript

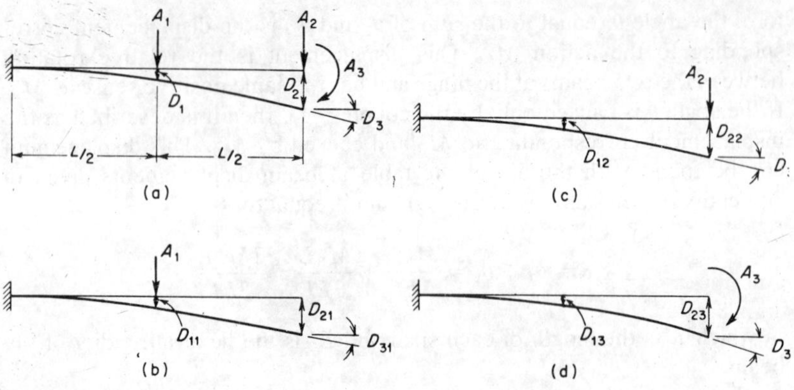

Fig. 1-7.

indicates that the displacement corresponds to action A_1, and the second indicates that the cause of the displacement is action A_1. In a similar manner, the displacement corresponding to A_2 in this beam is denoted by D_{21}, where the first subscript shows that the displacement corresponds to A_2 and the second shows that it is caused by A_1. Also shown in Fig. 1-7b is the displacement D_{31}, corresponding to the couple A_3 and caused by A_1.

The displacements caused by action A_2 acting alone are shown in Fig. 1-7c, and those caused by A_3 alone are shown in Fig. 1-7d. In each case the subscripts for the displacement symbols follow the general rule that the first subscript identifies the displacement and the second gives the cause of the displacement. In general, the cause may be a single force, a couple, or an entire loading system. Unless specifically stated otherwise, this convention for subscripts will always be used in later discussions.

For the beams pictured in Fig. 1-7 it is not difficult to determine the various displacements (see Table A-3, Cases 7 and 8). Assuming that the beam has flexural rigidity EI and length L, it is found that the displacements for the beam in Fig. 1-7b are

$$D_{11} = \frac{A_1 L^3}{24EI} \qquad D_{21} = \frac{5A_1 L^3}{48EI} \qquad D_{31} = \frac{A_1 L^2}{8EI}$$

In a similar manner the remaining six displacements in Figs. 1-7c and d (D_{12}, D_{22}, . . . , D_{33}) can be found. Then the displacements in the beam under all loads acting simultaneously (see Fig. 1-7a) are determined by summation:

$$D_1 = D_{11} + D_{12} + D_{13}$$
$$D_2 = D_{21} + D_{22} + D_{23}$$
$$D_3 = D_{31} + D_{32} + D_{33}$$

These summations are expressions of the principle of superposition, which is discussed more fully in Art. 1.9.

1.5 Equilibrium. One of the objectives of any structural analysis is to determine various actions pertaining to the structure, such as reactions at the supports and internal stress resultants (bending moments, shearing forces, etc.). A correct solution for any of these quantities must satisfy all conditions of static equilibrium, not only for the entire structure, but also for any part of the structure taken as a free body.

Consider now any free body subjected to several actions. The resultant of all the actions may be a force, a couple, or both. If the free body is in static equilibrium, the resultant vanishes; that is, the resultant force vector and the resultant moment vector are both zero. A vector in three-dimensional space may always be resolved into three components in mutually orthogonal directions, such as the x, y, and z directions. If the resultant force vector equals zero, then its components also must be equal to zero, and therefore the following equations of static equilibrium are obtained:

$$\sum F_x = 0 \qquad \sum F_y = 0 \qquad \sum F_z = 0 \tag{1-1a}$$

In these equations the expressions ΣF_x, ΣF_y, and ΣF_z are the algebraic sums of the x, y, and z components, respectively, of all the force vectors acting on the free body. Similarly, if the resultant moment vector equals zero, the moment equations of static equilibrium are

$$\sum M_x = 0 \qquad \sum M_y = 0 \qquad \sum M_z = 0 \tag{1-1b}$$

in which ΣM_x, ΣM_y, and ΣM_z are the algebraic sums of the moments about the x, y, and z axes, respectively, of all the couples and forces acting on the free body. The six relations in Eqs. (1-1) represent the static equilibrium equations for actions in three dimensions. They may be applied to any free body such as an entire structure, a portion of a structure, a single member, or a joint of a structure.

When all forces acting on a free body are in one plane and all couples have their vectors normal to that plane, only three of the six equilibrium equations will be useful. Assuming that the forces are in the x-y plane, it is apparent that the equations $\Sigma F_z = 0$, $\Sigma M_x = 0$, and $\Sigma M_y = 0$ will be satisfied automatically. The remaining equations are

$$\sum F_x = 0 \qquad \sum F_y = 0 \qquad \sum M_z = 0 \tag{1-2}$$

and these equations become the static equilibrium conditions for actions in the x-y plane.

In the stiffness method of analysis, the basic equations to be solved are those which express the equilibrium conditions at the joints of the structure, as described later in Chapter 3.

1.6 Compatibility. In addition to the static equilibrium conditions, it is necessary in any structural analysis that all conditions of compatibility be satisfied. These conditions refer to continuity of the displacements throughout the structure and are sometimes referred to as conditions of geometry. As an example, compatibility conditions must be satisfied at all points of support, where it is necessary that the displacements of the structure be consistent with the support conditions. For instance, at a fixed support there can be no translation and no rotation of the axis of the member.

Compatibility conditions must also be satisfied at all points throughout the interior of a structure. Usually, it is compatibility conditions at the joints of the structure that are of interest. For example, at a rigid connection between two members the displacements (translations and rotations) of both members must be the same.

In the flexibility method of analysis the basic equations to be solved are equations that express the compatibility of the displacements, as will be described in Chapter 2.

1.7 Static and Kinematic Indeterminacy. There are two types of indeterminacy that must be considered in structural analysis, depending upon whether actions or displacements are of interest. When actions are the unknowns in the analysis, as in the flexibility method, then *static indeterminacy* must be considered. In this case, indeterminacy refers to an excess of unknown actions as compared to the number of equations of static equilibrium that are available. The equations of equilibrium, when applied to the entire structure and to its various parts, may be used for the calculation of reactions and internal stress resultants. If these equations are sufficient for finding all actions, both external and internal, then the structure is statically determinate. If there are more unknown actions than equations, the structure is statically indeterminate. The simply supported beam shown in Fig. 1-3 and the cantilever beam of Fig. 1-4 are examples of statically determinate structures, since in both cases all reactions and stress resultants can be found from equilibrium equations alone. On the other hand, the continuous beam of Fig. 1-5a is statically indeterminate.

The unknown actions in excess of those that can be found by static equilibrium are known as *static redundants,* and the number of such redundants represents the *degree* of static indeterminacy of the structure. Thus, the two-span beam of Fig. 1-5a is statically indeterminate to the first degree, since there is one redundant action. For instance, it can be seen that it is impossible to calculate all of the reactions for the beam by static equilibrium alone. However, after the value of one reaction is obtained (by one means or another), the remaining reactions and all internal stress resultants can be found by statics alone.

Other examples of statically indeterminate structures are shown in Fig. 1-8. The propped cantilever beam in Fig. 1-8a is statically indeterminate to the first degree, since there are four reactive actions (H_A, M_A, R_A, and R_B) whereas only three equations of equilibrium are available for the calculation of reactions (see Eqs. 1-2).

The fixed-end beam of Fig. 1-8b is statically indeterminate to the third degree because there are six reactions to be found in the general case. In

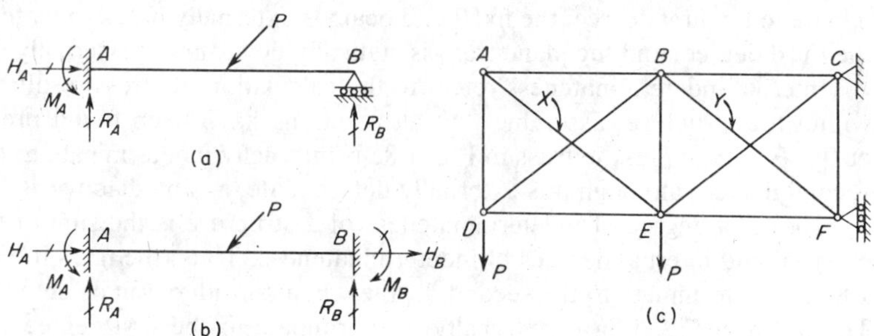

Fig. 1-8. Examples of statically indeterminate structures.

the special case when all the concentrated forces on a fixed-end beam act in a direction which is perpendicular to the axis of the beam, there will be no axial forces at the ends of the beam. In such a case the beam can be analyzed as if it were statically indeterminate to the second degree.

The plane truss in Fig. 1-8c is statically indeterminate to the second degree. This conclusion can be reached by cutting two bars, such as X and Y, thereby releasing the forces in those bars. The truss with the cut bars is then statically determinate, since all reactions and bar forces can be found by a direct application of equations of equilibrium. Each bar that is cut represents one action, namely, the force in the bar, that is released from the truss. The number of actions that must be released in order to reduce the statically indeterminate structure to a determinate structure will be equal to the degree of indeterminacy. This method of ascertaining the degree of static indeterminacy is quite general and can be used with many types of structures.

As another example of this method for determining the degree of indeterminateness of a structure, consider the plane frame shown in Fig. 1-6. The object is to make cuts, or releases, in the frame until the structure has become statically determinate. If bars AB and BC are cut, the structure that remains consists of three cantilever portions (the supports of the cantilevers are at D, E, and F), each of which is statically determinate. Each bar that is cut represents the removal (or release) of three actions (axial force, shearing force, and bending moment) from the original structure. Because a total of six actions was released, the degree of indeterminacy of the frame is six.

A distinction may also be made between external and internal indeterminateness. External indeterminateness refers to the calculation of the reactions for the structure. Normally, there are six equilibrium equations available for the determination of reactions in a space structure, and three for a plane structure. Therefore, a space structure with more than six reactive actions, and a plane structure with more than three reactions, will usually be externally indeterminate. Examples of external indeterminateness can be seen in Fig. 1-8. The propped cantilever beam is externally indeterminate to the first degree, the fixed-end beam is externally indeterminate to the third degree, and the plane truss is statically determinate externally.

Internal indeterminateness refers to the calculation of stress resultants within the structure, assuming that all reactions have been found previously. For example, the truss in Fig. 1-8c is internally indeterminate to the second degree, although it is externally determinate, as noted above.

The total degree of indeterminateness of a structure is the sum of the external and internal degrees of indeterminateness. Thus, the truss in Fig. 1-8c is indeterminate to the second degree when considered in its entirety. The beam in Fig. 1-8a is externally indeterminate to the first degree and internally determinate, inasmuch as all stress resultants in the beam can be readily found after all the reactions are known. The plane frame in Fig. 1-6 has nine reactive actions, and therefore it is externally indeterminate to the sixth degree. Internally the frame is determinate since all stress resultants can be found if the reactions are known. Therefore, the frame has a total indeterminateness of six, as previously observed.

Occasionally, there are special conditions of construction that affect the degree of indeterminacy of a structure. The three-hinged arched truss shown in Fig. 1-9 has a central hinge at joint B which makes it possible to calculate all four reactions by statics. For the truss shown, the bar forces in all members can be found after the reactions are known, and therefore the structure is statically determinate overall.

Several additional examples of statically indeterminate structures are given at the end of this article. These examples illustrate how the degree of indeterminacy can be obtained for many structures by intuitive reasoning. Other examples will be encountered in Chapter 2 in connection with the flexibility method of analysis. However, for large structures it is desirable to have more formalized methods of establishing static indeterminacy; such methods are discussed elsewhere and are not repeated here.*

In the stiffness method of analysis the displacements of the joints of the structure become the unknown quantities. Therefore, the second type of indeterminacy, known as *kinematic indeterminacy*, becomes important. In

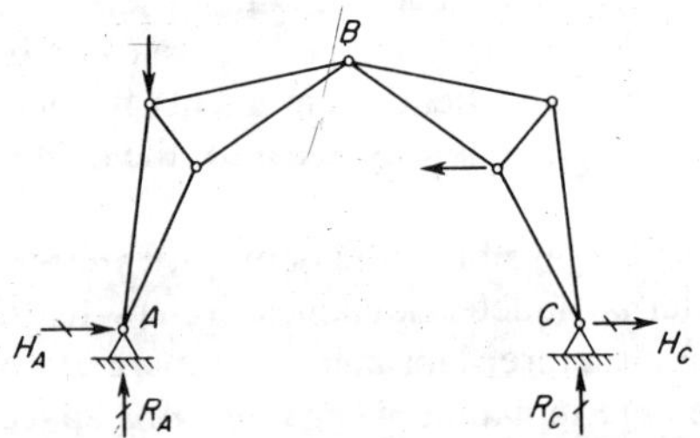

Fig. 1-9. Three-hinged arched truss.

*For a more complete discussion of static indeterminacy, see *Elementary Structural Analysis*, 3rd ed., by C. H. Norris, J. B. Wilbur, and S. Utku, McGraw-Hill, New York, 1976.

order to understand this type of indeterminacy, it should be recalled that joints in framed structures are defined to be located at all points where two or more members intersect, at points of support, and at free ends. When the structure is subjected to loads, each joint will undergo displacements in the form of translations and rotations, depending upon the configuration of the structure. In some cases the joint displacements will be known because of restraint conditions that are imposed upon the structure. At a fixed support, for instance, there can be no displacements of any kind. However, there will be other joint displacements that are not known in advance, and which can be obtained only by making a complete analysis of the structure. These unknown joint displacements are the kinematically indeterminate quantities, and are sometimes called kinematic redundants. Their number represents the degree of kinematic indeterminacy of the structure, or the number of *degrees of freedom* for joint displacement.

To illustrate the concepts of kinematic indeterminacy, it is useful to consider again the examples of Fig. 1-8. Beginning with the beam in Fig. 1-8a, it is seen that end *A* is fixed and cannot undergo any displacement. On the other hand, joint *B* has two degrees of freedom for joint displacement, since it may translate in the horizontal direction and may rotate. Thus, the beam is kinematically indeterminate to the second degree, and there are two unknown joint displacements to be calculated in a complete analysis of this beam. In many practical analyses it would be permissible to neglect axial deformations of the beam; in such a case, joint *B* would have only one degree of freedom (rotation), and the structure would be analyzed as if it were kinematically indeterminate to the first degree.

The second example of Fig. 1-8 is a fixed-end beam. Such a beam has no unknown joint displacements, and therefore is kinematically determinate. By comparison, the same beam was statically indeterminate to the third degree.

The third example in Fig. 1-8 is the plane truss that was previously shown to be statically indeterminate to the second degree. Joint *A* of this truss may undergo two independent components of displacement (such as translations in two perpendicular directions) and hence has two degrees of freedom. Rotation of a joint of a truss has no physical significance because, under the assumption of hinged joints, rotation of a joint produces no effects in the members of the truss. Thus, the degree of kinematic indeterminacy of a truss is always found as if the truss were subjected to loads at the joints only. This philosophy is the same as in the case of static indeterminacy, wherein only axial forces in the members are considered as unknowns. The joints *B, D,* and *E* of the truss in Fig. 1-8 also have two degrees of freedom each, while the restrained joints *C* and *F* have zero and one degree of freedom, respectively. Thus, the truss has a total of nine degrees of freedom for joint translation and is kinematically indeterminate to the ninth degree.

The rigid frame shown in Fig. 1-6 offers another example of a kinematically indeterminate structure. Since the supports at D, E, and F of this frame are fixed, there can be no displacements at these joints. However, joints A, B, and C each possess three degrees of freedom, since each joint may undergo horizontal and vertical translations and a rotation. Thus, the total number of degrees of kinematic indeterminacy for this frame is nine. If the effects of axial deformations are omitted from the analysis, the degree of kinematic indeterminacy is reduced. There would be no possibility for vertical displacement of any of the joints because the columns would not change length. Furthermore, the horizontal translations of joints A and B would be equal, and the horizontal translation of C would have a known relationship to that of joint B. In other words, if axial deformations are neglected the only independent joint displacements are the rotations of joints A, B, and C and one horizontal displacement (such as that of joint B). Therefore, the structure would be considered to be kinematically indeterminate to the fourth degree.

In summary, two simple rules can always be used to find the static and kinematic indeterminacy of a framed structure. First, to find the number of redundant actions, count the number of releases necessary to obtain a statically determinate structure. This can be done indirectly by finding the number of unknown actions in excess of those that can be found from static equilibrium equations. Second, to find the number of degrees of freedom for joint displacement, count the number of joint restraints that must be added to obtain a kinematically determinate structure (no joint displacements). Several examples involving both static and kinematic indeterminacy will now be given.

Example 1. The space truss shown in Fig. 1-10 has pin supports at A, B, and C. The degrees of static and kinematic indeterminacy for the truss are to be obtained.

In determining the degree of static indeterminacy, it can be noted that there are three equations of equilibrium available at every joint of the truss for the purpose of calculating bar forces or reactions. Thus, a total of 18 equations of statics is available. The number of unknown actions is 21, since there are 12 bar forces and 9 reactions (three at each support) to be found. The truss is, therefore, statically inde-

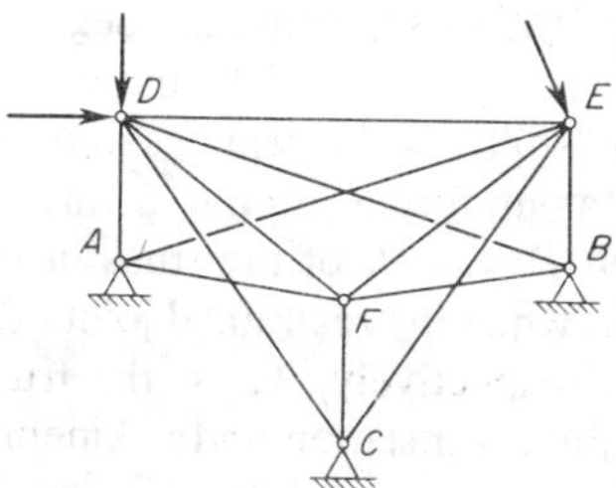

Fig. 1-10. Example 1.

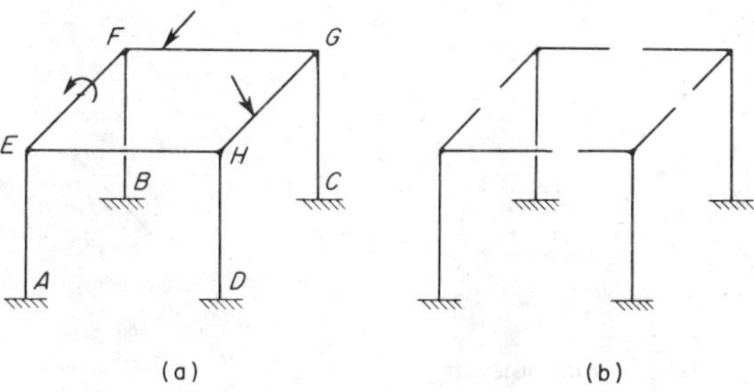

Fig. 1-11. Example 2.

terminate to the third degree. More specifically, the truss is externally indeterminate to the third degree, because there are nine reactions but only six equations for the equilibrium of the truss as a whole. The truss is internally determinate since all bar forces can be found by statics after the reactions are determined.

Each of the joints D, E, and F has three degrees of freedom for joint displacement, because each joint can translate in three mutually orthogonal directions. Therefore, the truss is kinematically indeterminate to the ninth degree.

Example 2. The degrees of static and kinematic indeterminacy are to be found for the space frame shown in Fig. 1-11a.

There are various ways in which the frame can be cut in order to reduce it to a statically determinate structure. One possibility is to cut all four of the bars EF, FG, GH, and EH, thereby giving the released structure shown in Fig. 1-11b. Since each release represents the removal of six actions (axial force, two shearing forces, twisting couple, and two bending moments) the original frame is statically indeterminate to the 24th degree.

The number of possible joint displacements at E, F, G, and H is six at each joint (three translations and three rotations); therefore, the frame is kinematically indeterminate to the 24th degree.

Now consider the effect of omitting axial deformations from the analysis. The degree of static indeterminacy is not affected, because the same number of actions will still exist in the structure. On the other hand, there will be fewer degrees of freedom for joint displacement. The columns will not change in length, thereby eliminating four joint translations (one each at E, F, G, and H). In addition, the four horizontal members will not change in length, thereby eliminating four more translations. Thus, it is finally concluded that the degree of kinematic indeterminacy is 16 when axial deformations are excluded from consideration.

Consider next the effect of replacing the fixed supports at A, B, C, and D by immovable pinned supports. The effect of the pinned supports is to reduce the number of reactions at each support from six to three. Therefore, the degree of static indeterminacy becomes 12 less than with fixed supports, or a total of 12 degrees. At the same time, three additional degrees of freedom for rotation have been added at each support, so that the degree of kinematic indeterminacy has been increased by 12 when compared to the frame with fixed supports. It can be seen that removing restraints at the supports of a structure serves to decrease the degree of statical indeterminacy, while increasing the degree of kinematic indeterminacy.

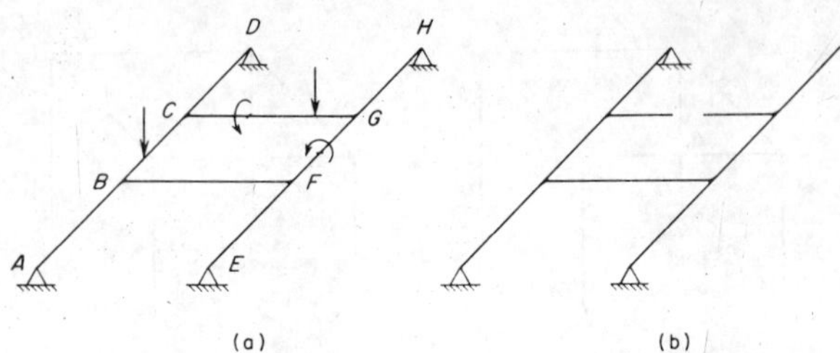

Fig. 1-12. Example 3.

Example 3. The grid shown in Fig. 1-12a lies in a horizontal plane and is supported at *A*, *D*, *E*, and *H* by simple supports. The joints at *B*, *C*, *F*, and *G* are rigid connections. What are the degrees of static and kinematic indeterminacy?

Because there are no axial forces in the members of a grid, only vertical reactions are developed at the supports of this structure. Therefore, the grid is externally indeterminate to the first degree, because only three equilibrium equations are available for the structure in its entirety, but there are four reactions. After removing one reaction, the grid can be made statically determinate by cutting one member, such as *CG* (see Fig. 1-12b). The release in member *CG* removes three actions (shearing force in the vertical direction, twisting moment, and bending moment). Thus, the grid can be seen to be internally indeterminate to the third degree and statically indeterminate overall to the fourth degree.

In general there are three degrees of freedom for displacement at each joint of a grid (one translation and two rotations). Such is the case at joints *B*, *C*, *F*, and *G* of the grid shown in Fig. 1-12a. However, at joints *A*, *D*, *E*, and *H* only two joint displacements are possible, inasmuch as joint translation is prevented. Therefore, the grid shown in the figure is kinematically indeterminate to the 20th degree.

1.8 Structural Mobilities. In the preceding discussion of external static indeterminacy, the number of reactions for a structure was compared with the number of equations of static equilibrium for the entire structure taken as a free body. If the number of reactions exceeds the number of equations, the structure is externally statically indeterminate; if they are equal, the structure is externally determinate. However, it was tacitly assumed in the discussion that the geometrical arrangement of the reactions was such as to prevent the structure from moving when loads act on it. For instance, the beam shown in Fig. 1-13a has three reactions, which is the same as the number of static equilibrium equations for forces in a plane. It is apparent, however, that the beam will move to the left when the inclined load *P* is applied. A structure of this type is said to be *mobile* (or *kinematically unstable*). Other examples of mobile structures are the frame of Fig. 1-13b and the truss of Fig. 1-13c. In the structure of Fig. 1-13b the three reactive forces are concurrent (their lines of action intersect at point *O*). Therefore, the frame is mobile since it cannot support a general load, such as the force *P*, which does not act through point *O*. In the truss of Fig. 1-13c there are two bars which are collinear at joint *A*, and there is no other

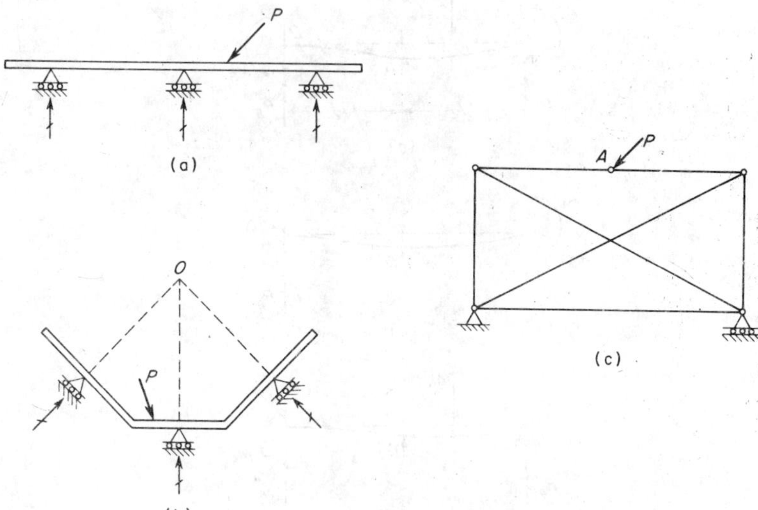

Fig. 1-13. Mobile structures.

bar meeting at that joint. Again, the structure is mobile since it is incapable of supporting the load P in its initial configuration.

From the examples of mobile structures given in Fig. 1-13, it is apparent that both the supports and the members of any structure must be adequate in number and in geometrical arrangement to insure that the structure is not movable. Only structures meeting these conditions will be considered for analysis in subsequent chapters.

1.9 Principle of Superposition. The principle of superposition is one of the most important concepts in structural analysis. It may be used whenever linear relationships exist between actions and displacements (the conditions under which this assumption is valid are described later in this article). In using the principle of superposition it is assumed that certain actions and displacements are imposed upon a structure. These actions and displacements cause other actions and displacements to be developed in the structure. Thus, the former actions and displacements have the nature of causes, while the latter are effects. In general terms the principle states that the effects produced by several causes can be obtained by combining the effects due to the individual causes.

In order to illustrate the use of the principle of superposition when actions are the cause, consider the beam in Fig. 1-14a. This beam is subjected to loads A_1 and A_2, which produce various actions and displacements throughout the structure. For instance, reactions R_A, R_B, and M_B are developed at the supports, and a displacement D is produced at the midpoint of the beam. The effects of the actions A_1 and A_2 acting sepa-

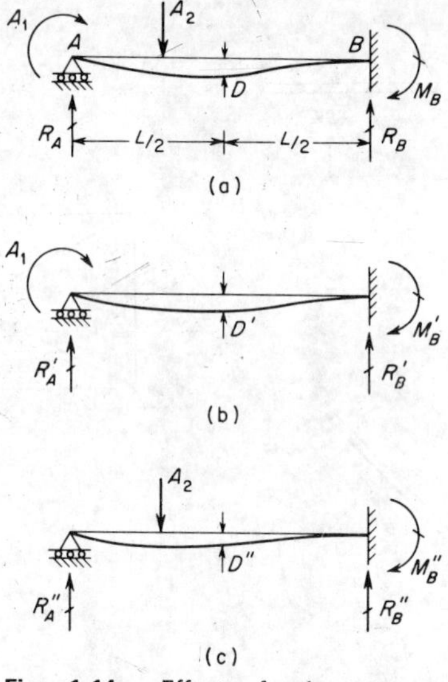

Fig. 1-14. Effects of actions.

rately are shown in Figs. 1-14b and 1-14c. In each case there is a displacement at the midpoint of the beam and reactions at the ends. A single prime is used to denote quantities associated with the action A_1, and a double prime is used for quantities associated with A_2.

According to the principle of superposition, the actions and displacements caused by A_1 and A_2 acting separately (Figs. 1-14b and 1-14c) can be combined in order to obtain the actions and displacements caused by A_1 and A_2 acting simultaneously (Fig. 1-14a). Thus, the following *equations of superposition* can be written for the beam in Fig. 1-14:

$$\begin{aligned} R_A &= R'_A + R''_A \qquad & R_B &= R'_B + R''_B \\ M_B &= M'_B + M''_B \qquad & D &= D' + D'' \end{aligned} \tag{1-3}$$

Of course, similar equations of superposition can be written for other actions and displacements in the beam, such as stress resultants at any cross section of the beam and displacements (translations and rotations) at any point along the axis of the beam. This manner of using superposition was illustrated previously in Art. 1.4.

A second example of the principle of superposition, in which displacements are the cause, is given in Fig. 1-15. The figure portrays again the beam AB with one end simply supported and the other fixed. When end B of the beam is displaced downward through a distance Δ and, at the same time, is caused to be rotated through an angle θ (see Fig. 1-15a), various actions and displacements in the beam will be developed. For example, the reactions at each end and the displacement at the center are shown in Fig. 1-15a. The next two sketches (Figs. 1-15b and 1-15c) show the beam with

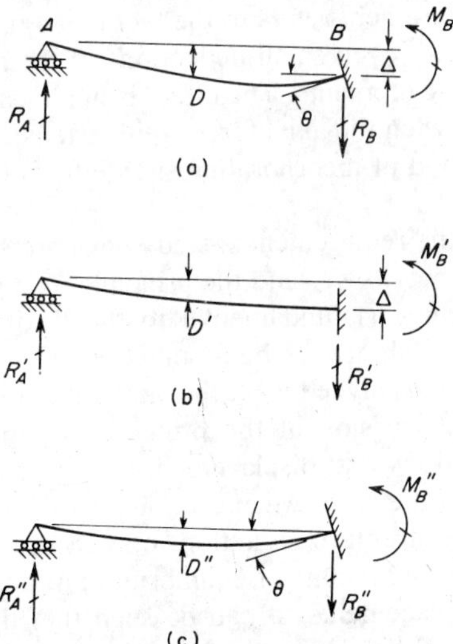

Fig. 1-15. Effects of displacements.

the displacements Δ and θ occurring separately. The reactions at the ends and the displacement at the center are again denoted by the use of primes; a single prime is used to denote quantities caused by the displacement Δ, and double primes are used for quantities caused by the rotation θ. When the principle of superposition is applied to the reactions and the displacement at the midpoint, the superposition equations again take the form of Eqs. (1-3). This example illustrates how actions and displacements caused by displacements can be superimposed. The same principle applies to any other actions and displacements in the beam.

The principle of superposition may be used also if the causes consist of both actions and displacements. For example, the beam in Fig. 1-15a could be subjected to various loads as well as to the displacements Δ and θ. Then the actions and displacements in the beam can be obtained by combining those due to each load and displacement separately.

As mentioned earlier, the principle of superposition will be valid whenever linear relations exist between actions and displacements of the structure. This occurs whenever the following three requirements are satisfied: (1) the material of the structure follows Hooke's law; (2) the displacements of the structure are small; and (3) there is no interaction between axial and flexural effects in the members. The first of these requirements means that the material is perfectly elastic and has a linear relationship between stress and strain. The second requirement means that all calculations involving the over-all dimensions of the structure can be based upon the original dimensions of the structure; it is also a basic assumption for calculating displacements by the methods described in Appendix A. The third requirement implies that the effect of axial forces on the bending of the members is neglected. This requirement refers to the fact that axial forces in a member, in combination with even small deflections of the member, will have an effect on the bending moments. The effect is nonlinear and can be omitted from the analysis when the axial forces (either tension or compression) are not large. (A method of incorporating such effects into the analysis is described in Art. 6.18.)

When all three of the requirements listed above are satisfied, the structure is said to be *linearly elastic*, and the principle of superposition can be used. Since this principle is fundamental to the flexibility and stiffness methods of analysis, it will always be assumed in subsequent discussions that the structures being analyzed meet the stated requirements.

In the preceding discussion of the principle of superposition it was assumed that both actions and displacements were of importance in the analysis, as is generally the case. An exception, however, is the analysis of a statically determinate structure for actions only. Since an analysis of this kind requires the use of equations of equilibrium but does not require the calculation of any displacements, it can be seen that the requirement of linear elasticity is superfluous. An example is the determination of the reactions for a simply supported beam under several loads. The reactions are linear functions of the loads regardless of the characteristics of the material of the beam. It is still necessary, however, that the deflections of the beam be small, since otherwise the positions of the loads and reactions would be altered.

1.10 Action and Displacement Equations. The relationships that exist between actions and displacements play an important role in structural analysis and are used extensively in both the flexibility and stiffness methods. A convenient way to express the relationship between the actions on a structure and the displacements of the structure is by means of action and displacement equations. A simple illustration of such equations is obtained by considering the linearly elastic spring shown in Fig. 1-16. The action A will compress the spring, thereby producing a displacement D of the end of the spring. The relationship between A and D can be expressed by a *displacement equation*, as follows:

$$D = FA \tag{1-4}$$

In this equation F is the *flexibility* of the spring and is defined as the displacement produced by a unit value of the action A.

The relationship between the action A and the displacement D for the

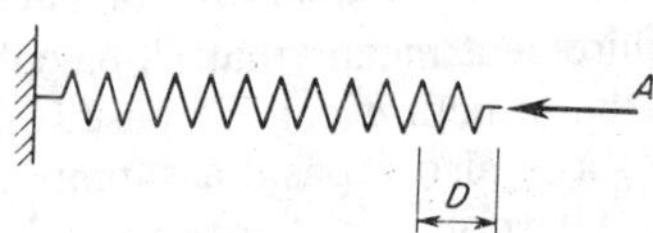

Fig. 1-16. Linearly elastic spring.

spring in Fig. 1-16 can also be expressed by an *action equation* which gives A in terms of D:

$$A = SD \tag{1-5}$$

In this equation S is the *stiffness* of the spring, which is defined as the action required to produce a unit displacement. It can be seen from Eqs. (1-4) and (1-5) that the flexibility and stiffness of the spring are inverse to one another, as follows:

$$F = \frac{1}{S} = S^{-1} \qquad S = \frac{1}{F} = F^{-1} \tag{1-6}$$

The flexibility of the spring has units of length divided by force, while the stiffness has units of force divided by length.

The same general relationships (Eqs. 1-4 to 1-6) that apply to the spring will hold also for any linearly elastic structure that is subjected to a single action. An example is given in Fig. 1-17a, which shows a simply supported beam acted upon by a concentrated force A at the midpoint. The displacement D shown in the figure is the vertical, downward deflection of the beam at the point where A acts upon the beam. Hence, in this example the displacement D not only corresponds to A but also is caused by A. The action and displacement equations given previously (Eqs. 1-5 and 1-4, respectively) are valid for the beam of Fig. 1-17a, provided that the flexibility F and the stiffness S are determined appropriately. In this case the flexibility F is seen to be the displacement produced by a unit value of the load, as shown in Fig. 1-17b (see Case 2 of Table A-3 in Appendix A):

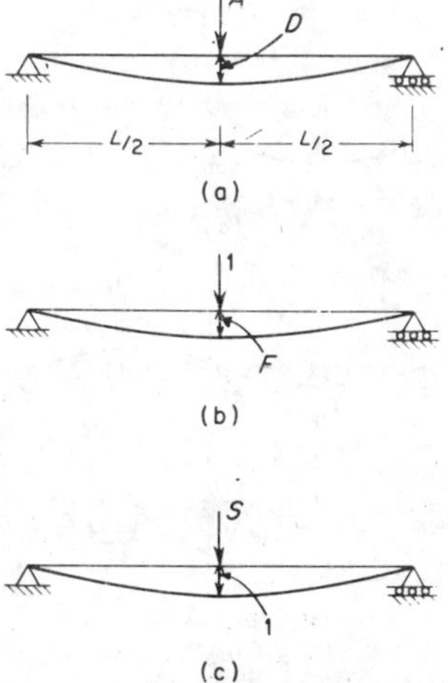

Fig. 1-17. Flexibility and stiffness of a beam subjected to a single load.

$$F = \frac{L^3}{48EI}$$

in which L is the length of the beam and EI is the flexural rigidity. The stiffness S, equal to the inverse of the flexibility, is the action required to produce a unit value of the displacement (see Fig. 1-17c):

$$S = \frac{48EI}{L^3}$$

Note again that S has units of force divided by length. Also, it should be emphasized that the simple relationships expressed by Eq. (1-6) are valid only when the structure is subjected to a single load.

Now consider a more general example in which a structure is acted upon by three loads (Fig. 1-18a). The loads on the beam are denoted by A_1, A_2, and A_3, and are taken positive in the directions (or senses) shown in the figure. The deflected shape produced by the loads acting on the beam is shown in Fig. 1-18b. In this figure the displacements in the beam corresponding to A_1, A_2, and A_3, and caused by all three loads acting simulta-

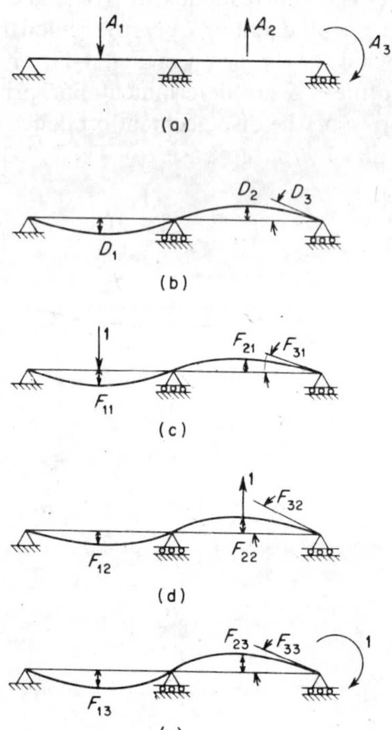

Fig. 1-18. Illustration of flexibility coefficients.

neously, are denoted by D_1, D_2, and D_3, respectively, and are assumed positive in the same directions (or senses) as the corresponding actions.

By using the principle of superposition, each of the displacements in Fig. 1-18b can be expressed as the sum of the displacements due to the loads A_1, A_2, and A_3 acting separately. For example, the displacement D_1 is given by the expression

$$D_1 = D_{11} + D_{12} + D_{13}$$

in which D_{11} is the displacement corresponding to A_1 and caused by A_1, D_{12} is the displacement corresponding to A_1 and caused by A_2, and D_{13} is the displacement corresponding to A_1 and caused by A_3. In a similar manner two additional equations can be written for D_2 and D_3. Each of the displacements appearing on the right-hand sides of such equations is a linear function of one of the loads; that is, each displacement is directly proportional to one of the loads. For example, D_{12} is a displacement caused by A_2 alone, and it is equal to A_2 times a certain coefficient. Denoting such coefficients by the symbol F, it is possible to write equations for the displacements D_1, D_2, and D_3 explicitly in terms of the loads, as follows:

$$\begin{aligned} D_1 &= F_{11}A_1 + F_{12}A_2 + F_{13}A_3 \\ D_2 &= F_{21}A_1 + F_{22}A_2 + F_{23}A_3 \\ D_3 &= F_{31}A_1 + F_{32}A_2 + F_{33}A_3 \end{aligned} \tag{1-7}$$

In the first of these equations the expression $F_{11}A_1$ represents the displacement D_{11}, the expression $F_{12}A_2$ represents the displacement D_{12}, and so forth. Each term on the right-hand sides of the above equations is a displacement that is written in the form of a coefficient times the action that produces the displacement. The coefficients are called *flexibility coefficients*, or more simply, flexibilities.

Each flexibility coefficient F represents a displacement caused by a unit value of a load. Thus, the coefficient F_{11} represents the displacement corresponding to action A_1 and caused by a unit value of A_1; the coefficient F_{12} is the displacement corresponding to A_1 and caused by a unit value of A_2; and so forth. The physical significance of the flexibility coefficients is shown in Figs. 1-18c, 1-18d, and 1-18e. The displacements of the beam caused by a unit value of the action A_1 are shown in Fig. 1-18c. All of the flexibility coefficients in this figure have a second subscript equal to one, thereby denoting the cause of the displacements. The first subscript in each case identifies the displacement by denoting the action that corresponds to it. Similar comments apply also to the displacements pictured in Figs. 1-18d and 1-18e. The calculation of flexibilities may be a simple or a difficult task, depending upon the particular structure being investigated. An example involving a very simple structure is given at the end of Art. 1.11. The more general use of flexibility coefficients in structural analysis, as well as methods for calculating them, will be shown in Chapter 2.

Instead of expressing the displacements in terms of the actions, as was done in Eqs. (1-7), it is also possible to write action equations expressing the actions in terms of the displacements. Such equations can be obtained, for instance, by solving simultaneously the displacement equations. Thus, if Eqs. (1-7) are solved for the actions in terms of the displacements, the resulting action equations have the form:

$$\begin{aligned} A_1 &= S_{11}D_1 + S_{12}D_2 + S_{13}D_3 \\ A_2 &= S_{21}D_1 + S_{22}D_2 + S_{23}D_3 \\ A_3 &= S_{31}D_1 + S_{32}D_2 + S_{33}D_3 \end{aligned} \tag{1-8}$$

in which each S is a *stiffness coefficient* or, more simply, a stiffness. As stated previously, a stiffness represents an action due to a unit displacement. Thus, the stiffness coefficient S_{11} represents the action corresponding to A_1 when a unit displacement of type D_1 is introduced while the other displacements, namely, D_2 and D_3, are kept equal to zero. Similarly, the stiffness coefficient S_{12} is the action corresponding to A_1 caused by a unit displacement of type D_2 while D_1 and D_3 are equal to zero. By continuing in this manner, all of the stiffness coefficients can be defined as actions produced by unit displacements.

Physical interpretations of the stiffness coefficients are shown in Fig. 1-19. The first two parts of the figure (Figs. 1-19a and 1-19b) are repeated from Fig. 1-18 in order to show the actions and displacements in the original beam. In Fig. 1-19c the beam is shown with a unit displacement corresponding to A_1 induced in the structure while the displacements corresponding to A_2 and A_3 are made equal to zero. To impose these displacements on the beam requires that appropriate artificial restraints be provided. These are shown in the figure by the simple supports corresponding to A_1 and A_2 and the rotational restraint corresponding to A_3. The restraining actions developed by these artificial supports are the stiffness coefficients. For example, it can be seen from the figure that S_{11} is the action corresponding to A_1 and caused by a unit displacement corresponding to A_1 while the displacements corresponding to A_2 and A_3 are retained at zero. The stiffness coefficient S_{21} is the action corresponding to A_2 caused by a unit displacement corresponding to A_1 while the displacements corresponding to A_2 and A_3 are made equal to zero, and so on, for the other stiffnesses. Note that each stiffness coefficient is a reaction for the restrained structure, and therefore a slanted line is used across the vector in order to distinguish it from a load vector. Each stiffness coefficient is shown acting in its assumed positive direction, which is automatically the same direction as the corresponding action. If the actual direction of one of the stiffnesses is opposite to that assumed, then the coefficient will have a negative value when it is calculated. The stiffness coefficients caused by unit displacements corresponding to A_2 and A_3 are shown in Figs. 1-19d and 1-19e.

The calculation of the stiffness coefficients for the continuous beam of Fig. 1-19 would be a lengthy task. However, in analyzing a structure by the

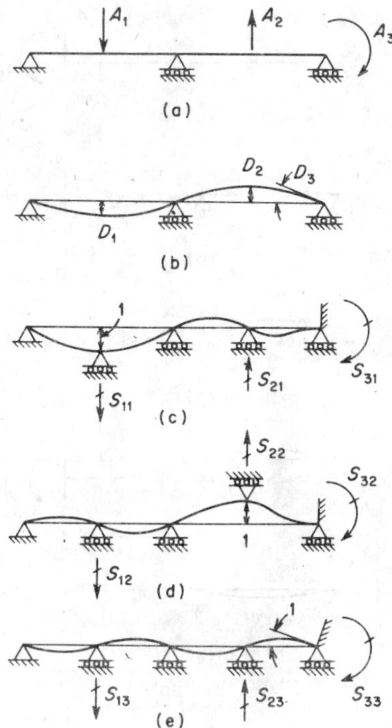

Fig. 1-19. Illustration of stiffness coefficients.

stiffness method (as is done in Chapters 3 and 4), this difficulty is avoided by limiting the calculation of stiffnesses to very special structures that are obtained by completely restraining all the joints of the actual structure. The primary purpose of the preceding discussion and the following two examples is to aid the reader in visualizing the physical significance of stiffness and flexibility coefficients, without regard to matters of practical calculation.

Example 1. The beam shown in Fig. 1-20 is subjected to loads A_1 and A_2 at the free end. The physical significance of the flexibility and stiffness coefficients corresponding to these actions is to be portrayed by means of sketches.

Unit loads corresponding to the actions A_1 and A_2 are shown in Figs. 1-20b and 1-20c, respectively. The displacements produced by these unit loads, and which correspond to the actions A_1 and A_2, are the flexibility coefficients. These coefficients (F_{11}, F_{21}, F_{12}, and F_{22}) are identified in the figures.

The stiffness coefficients (see Figs. 1-20d and 1-20e) are found by imposing unit displacements corresponding to A_1 and A_2, respectively, while at the same time maintaining the other corresponding displacement equal to zero. To accomplish this result requires the introduction of suitable restraints against translation and rotation at the free end of the beam. The restraint actions corresponding to A_1 and A_2 are the stiffness coefficients, and they are labeled in the figures.

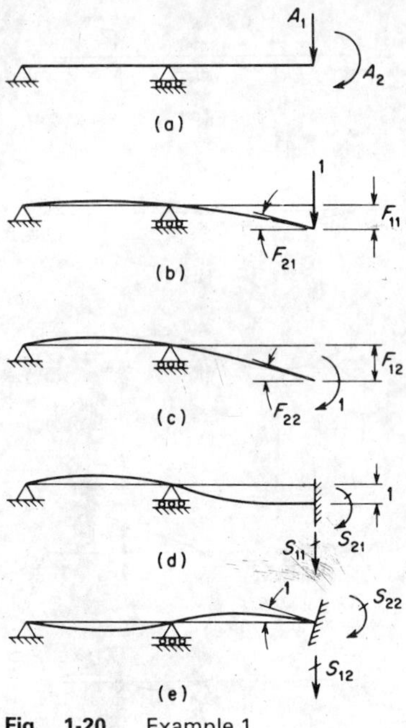

Fig. 1-20. Example 1.

Example 2. The plane truss illustrated in Fig. 1-21a is subjected to two loads A_1 and A_2. The sketches in Figs. 1-21b, 1-21c, 1-21d, and 1-21e show the physical significance of the flexibility and stiffness coefficients corresponding to A_1 and A_2. Note that the stiffnesses shown in Figs. 1-21d and 1-21e are the restraint actions required when the loaded joint of the truss is displaced a unit distance in the directions of A_1 and A_2, respectively. A restraint of this type is provided by a pinned support.

1.11 Flexibility and Stiffness Matrices. In the preceding article the meaning of action and displacement equations was discussed with reference to particular examples. It is easy to generalize from that discussion and thereby to obtain the equations for a structure subjected to any number of corresponding actions and displacements. Thus, if the number of actions applied to the structure is n, the equations that give the n corresponding displacements are (compare with Eqs. 1-7):

$$\begin{aligned} D_1 &= F_{11}A_1 + F_{12}A_2 + \cdots + F_{1n}A_n \\ D_2 &= F_{21}A_1 + F_{22}A_2 + \cdots + F_{2n}A_n \\ \cdots & \quad \cdots \quad \cdots \quad \cdots \quad \cdots \\ D_n &= F_{n1}A_1 + F_{n2}A_2 + \cdots + F_{nn}A_n \end{aligned} \tag{1-9}$$

In these equations each displacement D corresponds to one of the actions A and is caused by all of the actions acting simultaneously on the structure.

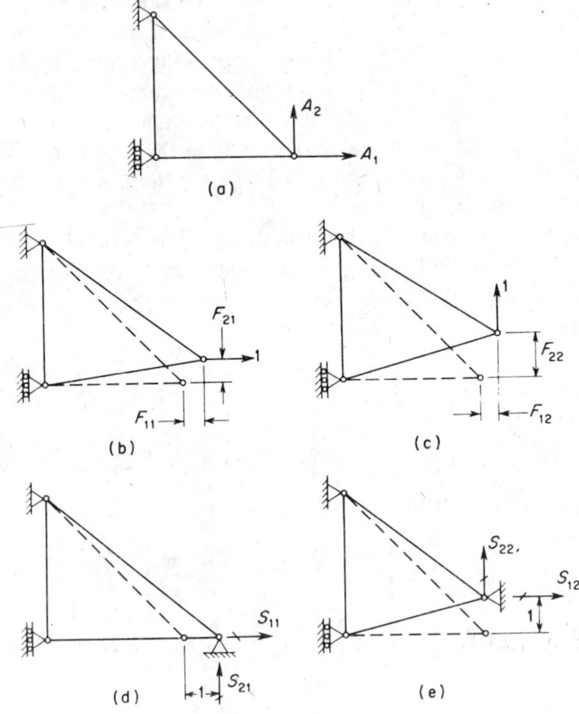

Fig. 1-21. Example 2.

For example, D_1 is the displacement corresponding to A_1 and caused by all of the actions $A_1, A_2, \ldots, A_n$. Each flexibility coefficient F represents a displacement caused by a unit value of one of the actions, while the other actions are zero. For example, F_{21} is the displacement corresponding to action A_2 and caused by a unit value of the action A_1. In general, the flexibility coefficient F_{ij} is the i-th displacement (that is, the displacement corresponding to the i-th action) due to a unit value of the j-th action. The coefficient is taken as positive when it is in the positive direction (or sense) of the i-th action.

In matrix form the displacement equations (Eqs. 1-9) become

$$\begin{bmatrix} D_1 \\ D_2 \\ \cdots \\ D_n \end{bmatrix} = \begin{bmatrix} F_{11} & F_{12} & \cdots & F_{1n} \\ F_{21} & F_{22} & \cdots & F_{2n} \\ \cdots & \cdots & \cdots & \cdots \\ F_{n1} & F_{n2} & \cdots & F_{nn} \end{bmatrix} \begin{bmatrix} A_1 \\ A_2 \\ \cdots \\ A_n \end{bmatrix}$$

or

$$\mathbf{D} = \mathbf{F}\mathbf{A} \tag{1-10}$$

in which **D** is a displacement matrix of order $n \times 1$; **F** is a square *flexibility* *matrix* of order $n \times n$; and **A** is an action or load matrix of order $n \times 1$.* The flexibility coefficients F_{ij} that appear on the principal diagonal of **F** are called *direct flexibility coefficients* and represent displacements caused by unit values of the corresponding actions. The remaining flexibility coefficients are called *cross flexibility coefficients*, and each represents a displacement caused by a unit value of an action that does not correspond to the displacement. It is apparent that $i = j$ for the direct flexibilities and $i \neq j$ for the cross flexibilities.

The action equations for the structure with n actions A acting upon it can be obtained by solving simultaneously Eqs. (1-9) for the actions in terms of the displacements. This operation gives the following action equations:

$$\begin{aligned} A_1 &= S_{11}D_1 + S_{12}D_2 + \cdots + S_{1n}D_n \\ A_2 &= S_{21}D_1 + S_{22}D_2 + \cdots + S_{2n}D_n \\ \cdots &\quad \cdots \quad\quad \cdots \quad\quad \cdots \quad\quad \cdots \\ A_n &= S_{n1}D_1 + S_{n2}D_2 + \cdots + S_{nn}D_n \end{aligned} \tag{1-11}$$

These action equations in matrix form are

$$\begin{bmatrix} A_1 \\ A_2 \\ \cdots \\ A_n \end{bmatrix} = \begin{bmatrix} S_{11} & S_{12} & \cdots & S_{1n} \\ S_{21} & S_{22} & \cdots & S_{2n} \\ \cdots & \cdots & \cdots & \cdots \\ S_{n1} & S_{n2} & \cdots & S_{nn} \end{bmatrix} \begin{bmatrix} D_1 \\ D_2 \\ \cdots \\ D_n \end{bmatrix}$$

or

$$\mathbf{A} = \mathbf{SD} \tag{1-12}$$

As described previously, the matrices **A** and **D** represent the action and displacement matrices of order $n \times 1$. The matrix **S** is a square *stiffness matrix* of order $n \times n$. Each stiffness coefficient S_{ij} can be defined as the i-th action due to a unit value of the j-th displacement, assuming that the remaining displacements are equal to zero. If $i = j$ the coefficient is a *direct stiffness coefficient*; if $i \neq j$, it is a *cross stiffness coefficient*.

Since Eqs. (1-11) were obtained from Eqs. (1-9) and the actions A and displacements D appearing in these equations are corresponding, it follows that the flexibility and stiffness matrices are related in a special manner. This relationship can be seen by solving Eq. (1-10) for **A**, giving the expression

$$\mathbf{A} = \mathbf{F}^{-1}\mathbf{D} \tag{1-13}$$

in which $\mathbf{F}^{-1}$ denotes the inverse of the flexibility matrix **F**. The vectors† **A** and **D** in this equation are the same as those in Eq. (1-12) and, therefore, it is apparent that

*Identifiers for matrices are printed in bold-face type to distinguish them from scalars.

†The term *vector* is frequently used for a matrix of one column or one row.

$$\mathbf{S} = \mathbf{F}^{-1} \qquad \text{and} \qquad \mathbf{F} = \mathbf{S}^{-1} \tag{1-14}$$

This relationship shows that the stiffness matrix is the inverse of the flexibility matrix and vice versa, provided that the same set of actions and corresponding displacements is being considered in both the action and displacement equations.

A somewhat different situation that occurs in structural analysis is the following. A set of displacement equations relating actions $\mathbf{A}_1$ and corresponding displacements $\mathbf{D}_1$ is obtained for a particular structure, as follows:

$$\mathbf{D}_1 = \mathbf{F}_1\mathbf{A}_1$$

In this equation $\mathbf{F}_1$ is the flexibility matrix relating the displacements $\mathbf{D}_1$ to the actions $\mathbf{A}_1$. Independently, a set of action equations relating another set of actions $\mathbf{A}_2$ to the corresponding displacements $\mathbf{D}_2$ may be written for the same structure; thus:

$$\mathbf{A}_2 = \mathbf{S}_2\mathbf{D}_2$$

It is, of course, not true that the flexibility and stiffness matrices $\mathbf{F}_1$ and $\mathbf{S}_2$ are the inverse of one another. However, it is always possible to obtain the inverse of $\mathbf{F}_1$, and this inverse can correctly be called a stiffness matrix. Specifically, it is the stiffness matrix $\mathbf{S}_1$ that is associated with the action equation relating $\mathbf{A}_1$ and $\mathbf{D}_1$. That is,

$$\mathbf{A}_1 = \mathbf{F}_1^{-1}\mathbf{D}_1 = \mathbf{S}_1\mathbf{D}_1$$

Similarly, the inverse of the matrix $\mathbf{S}_2$ is a flexibility matrix relating $\mathbf{D}_2$ and $\mathbf{A}_2$, as follows:

$$\mathbf{D}_2 = \mathbf{S}_2^{-1}\mathbf{A}_2 = \mathbf{F}_2\mathbf{A}_2$$

This discussion shows that a flexibility or stiffness matrix is not an array that is determined only by the nature of the structure; it is also related directly to the set of actions and displacements that is under consideration. The important fact to be remembered is that the flexibility matrix $\mathbf{F}$ obtained for a structure being analyzed by the flexibility method (see Chapter 2) is not the inverse of the stiffness matrix $\mathbf{S}$ for the same structure being analyzed by the stiffness method (see Chapter 3). The reason is that different sets of actions and corresponding displacements are utilized in the two methods.

Example. The cantilever beam shown in Fig. 1-22a is subjected to actions A_1 and A_2 at the free end. The corresponding displacements are denoted by D_1 and D_2 in the figure.

The flexibility coefficients are identified in Figs. 1-22b and 1-22c and can be evaluated without difficulty (see Table A-3, Appendix A):

$$F_{11} = \frac{L^3}{3EI} \qquad F_{12} = F_{21} = \frac{L^2}{2EI} \qquad F_{22} = \frac{L}{EI}$$

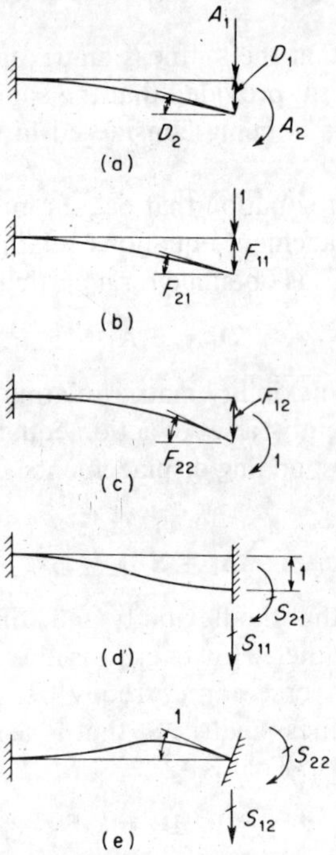

Fig. 1-22. Example of flexibility and stiffness coefficients.

Therefore, the displacement equations are

$$D_1 = \frac{L^3}{3EI}A_1 + \frac{L^2}{2EI}A_2$$

$$D_2 = \frac{L^2}{2EI}A_1 + \frac{L}{EI}A_2$$

and it is seen that the flexibility matrix is

$$\mathbf{F} = \begin{bmatrix} \dfrac{L^3}{3EI} & \dfrac{L^2}{2EI} \\ \dfrac{L^2}{2EI} & \dfrac{L}{EI} \end{bmatrix}$$

The stiffness coefficients are the restraint actions shown in Figs. 1-22d and 1-22e. In this particular example, the coefficients are reactions for a fixed-end beam, and their expressions can be obtained from a table of fixed-end actions. Such a table is given in Appendix B (see Table B-4), from which the following expressions are obtained:

$$S_{11} = \frac{12EI}{L^3} \qquad S_{12} = S_{21} = -\frac{6EI}{L^2} \qquad S_{22} = \frac{4EI}{L}$$

Thus, the action equations are

$$A_1 = \frac{12EI}{L^3}D_1 - \frac{6EI}{L^2}D_2$$

$$A_2 = -\frac{6EI}{L^2}D_1 + \frac{4EI}{L}D_2$$

and the stiffness matrix is

$$\mathbf{S} = \begin{bmatrix} \frac{12EI}{L^3} & -\frac{6EI}{L^2} \\ -\frac{6EI}{L^2} & \frac{4EI}{L} \end{bmatrix}$$

When the flexibility and stiffness matrices are multiplied, the identity matrix **I** is obtained, as follows:

$$\mathbf{FS} = \mathbf{SF} = \begin{bmatrix} 1 & 0 \\ 0 & 1 \end{bmatrix} = \mathbf{I}$$

Thus, the matrices **F** and **S** are the inverse of one another because the same set of actions and corresponding displacements was considered in both cases.

1.12 Equivalent Joint Loads. The analysis of structures by formalized versions of the flexibility and stiffness methods (given at the ends of Chapters 2 and 3) requires that the structure be subjected to loads acting only at the joints. In general, however, the actual loads on a structure do not meet this requirement. Instead, the loads may be divided into two types: loads acting at the joints and loads acting on the members. Loads of the latter type must be replaced by equivalent loads acting at the joints if the requirement stated above is to be fulfilled. The joint loads that are determined from the loads on the members are called *equivalent joint loads*. When these loads are added to the actual joint loads, the total loads which result are called *combined joint loads*. Thereafter, the structure can be analyzed by matrix methods for the effects of the combined joint loads.

It is advantageous in the analysis if the combined joint loads are evaluated in such a manner that the resulting displacements of the structure are the same as the displacements produced by the actual loads. This result can be achieved if the equivalent loads are obtained through the use of fixed-end actions, as demonstrated by the example in Fig. 1-23. Part (a) of this figure shows a beam *ABC* supported at joints *A* and *B* and subjected to several loads. Some of these loads are actual joint loads (see Fig. 1-23b) while the remaining loads act on the members (see Fig. 1-23c). To accomplish the replacement of the member loads by equivalent joint loads, the joints of the structure are restrained against all displacements. For the beam in the figure, this procedure results in two fixed-end beams (Fig. 1-23d). When these fixed-end beams are subjected to the member loads, a set of fixed-end actions is produced. The end-actions can be obtained by means of the formulas in Appendix B and are shown in Fig. 1-23d for the particular loads in this example. The same fixed-end actions are also shown in Fig. 1-23e, where they are represented as restraint actions for the

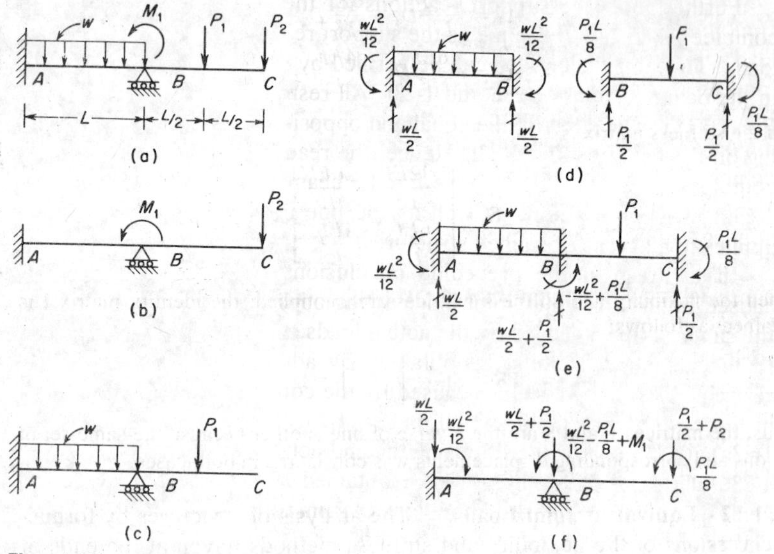

Fig. 1-23. Combined joint loads.

restrained structure. If these restraint actions are reversed in direction, they constitute a set of forces and couples that is equivalent to the member loads. Such equivalent joint loads, when added to the actual joint loads (Fig. 1-23b), produce the combined joint loads shown in Fig. 1-23f. Then the combined loads are used in carrying out the structural analysis, which is described later.

In general, the combined joint loads for any type of structure can be found by the procedure illustrated in Fig. 1-23. The first step is to separate the actual joint loads from the member loads. Then the structure is restrained against joint displacement by introducing appropriate joint restraints. Next, the restraint actions produced by the member loads on the restrained structure are calculated, using the formulas given in Table B-1. The negatives of these actions are the equivalent joint loads. Such loads are added to the actual joint loads to give the combined loads.

It was stated earlier that the displacements of the structure under the action of the combined loads should be the same as those produced by the actual loads. In order to observe that this requirement is satisfied, consider again the beam pictured in Fig. 1-23. It is apparent from the figure that the superposition of the combined loads (Fig. 1-23f) and the actions on the restrained structure (Fig. 1-23e) will give the actual loads on the beam (Fig. 1-23a). It follows, therefore, that the superposition of the joint displacements in the beams of Figs. 1-23e and 1-23f must produce the joint displacements in the actual beam. But, since all joint displacements for the restrained beam are zero, it can be concluded that the joint displacements in the beam under the actual loads and the combined loads are the same.

Furthermore, the support reactions for the structure subjected to the combined loads are the same as the support reactions caused by the actual loads. This conclusion also can be verified by superposition of the actions in the beams of Figs. 1-23e and 1-23f. All restraint actions in the beam of Fig. 1-23e are negated by the equal and opposite equivalent joint loads acting on the beam of Fig. 1-23f. Hence, the reactions for the beam with the combined loads are the same as for the beam with the actual loads (Fig. 1-23a). This conclusion, as well as the one in the preceding paragraph, applies to all types of framed structures.

In contrast to the preceding conclusions, the member end-actions caused by the combined joint loads acting on the structure are not usually the same as those caused by the actual loads. Instead, the end-actions due to the actual loads must be obtained by adding the end-actions in the restrained structure to those caused by the combined loads. For example, in the case of the beam shown in Fig. 1-23, the actual end-actions (Fig. 1-23a) are found by superimposing the end-actions from the beams in Figs. 1-23e and 1-23f. The latter will be obtained as a result of the structural analysis itself, and the former are known from the calculations for the fixed-end actions.

The remaining quantities of interest in a structural analysis performed by the flexibility method are the redundant actions themselves. Whether or not the redundant actions are the same in the actual structure and the structure with the combined joint loads depends upon the particular situation. If the redundant action is a reaction for the structure, it will be the same in both cases. If it is a member end-action, it must be treated in the manner described above for end-actions.

1.13 Energy Concepts. Most of the equations in this book are derived from principles of equilibrium, compatibility, and superposition. However, energy methods also play a vital role in structural theory, especially for the analysis of discretized continua by finite elements. In this article the concepts of strain energy and complementary strain energy will be developed and related to stiffnesses and flexibilities. The discussion pertains only to structures for which strains and displacements are small and no energy is dissipated in the static loading process. That is, the external work of gradually applied loads is equal to the energy stored in the structure, and the system is said to be *conservative*.

When a structure is loaded statically, various types of internal stresses and strains occur, depending upon the nature of the problem. Figure 1-24 shows the most general case of an infinitesimal element within a three-dimensional solid.* Appearing in the figure are three components of *normal stresses* (σ_x, σ_y, and σ_z) and six components of *shear stresses* (τ_{xy}, τ_{yz},

*See S. P. Timoshenko and J. N. Goodier, *Theory of Elasticity,* 3rd ed., McGraw-Hill, New York, 1970, p. 4.

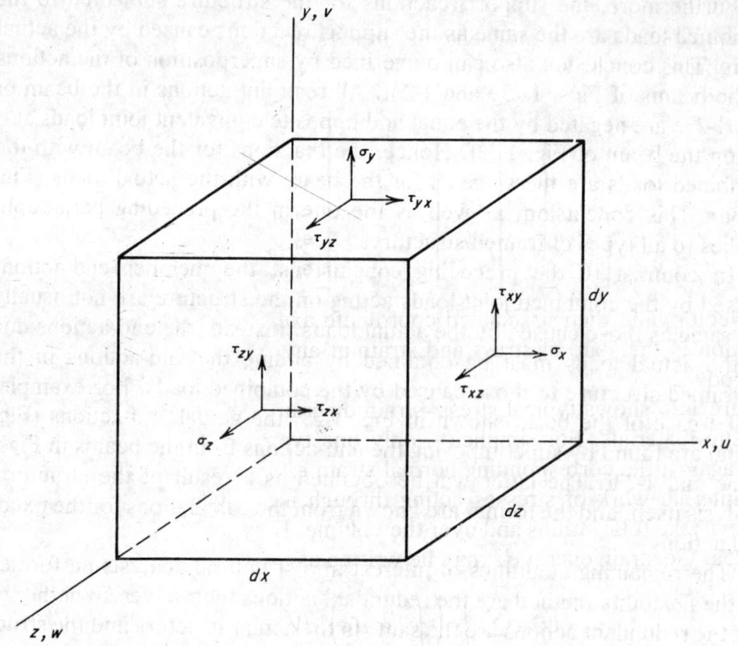

Fig. 1-24. Stresses on an infinitesimal element.

etc.). The shear stresses are not all independent; from equilibrium of the element the following relationships are known:

$$\tau_{xy} = \tau_{yx} \qquad \tau_{yz} = \tau_{zy} \qquad \tau_{zx} = \tau_{xz} \tag{a}$$

Consequently, only six independent components of stress need be considered in the most general case.

Corresponding to the normal stresses are three types of *normal strains* (ϵ_x, ϵ_y, and ϵ_z), defined as

$$\epsilon_x = \frac{\partial u}{\partial x} \qquad \epsilon_y = \frac{\partial v}{\partial y} \qquad \epsilon_z = \frac{\partial w}{\partial z} \tag{b}$$

where u, v, and w are displacements (translations) in the x, y, and z directions, respectively. Similarly, six *shear strains* (γ_{xy}, γ_{yz}, etc.) correspond to the six shear stresses. However, only three are independent, as can be seen from the following definitions of shear strains:

$$\gamma_{xy} = \frac{\partial u}{\partial y} + \frac{\partial v}{\partial x} = \gamma_{yx} \qquad \gamma_{yz} = \frac{\partial v}{\partial z} + \frac{\partial w}{\partial y} = \gamma_{zy} \qquad \gamma_{zx} = \frac{\partial w}{\partial x} + \frac{\partial u}{\partial z} = \gamma_{xz}$$

(c)

For convenience, the six independent stresses and corresponding strains will be represented as column vectors. Thus,

$$\boldsymbol{\sigma} = \begin{bmatrix} \sigma_1 \\ \sigma_2 \\ \sigma_3 \\ \sigma_4 \\ \sigma_5 \\ \sigma_6 \end{bmatrix} = \begin{bmatrix} \sigma_x \\ \sigma_y \\ \sigma_z \\ \tau_{xy} \\ \tau_{yz} \\ \tau_{zx} \end{bmatrix} \qquad \boldsymbol{\epsilon} = \begin{bmatrix} \epsilon_1 \\ \epsilon_2 \\ \epsilon_3 \\ \epsilon_4 \\ \epsilon_5 \\ \epsilon_6 \end{bmatrix} = \begin{bmatrix} \epsilon_x \\ \epsilon_y \\ \epsilon_z \\ \gamma_{xy} \\ \gamma_{yz} \\ \gamma_{zx} \end{bmatrix} \tag{d}$$

These vectors and their associated coordinate axes constitute a complete description of the state of stress and strain at any point in a three-dimensional body.

Figure 1-25 shows typical stress-strain diagrams for nonlinear and linear elastic materials. For simplicity, only one type of normal stress σ is plotted against the corresponding normal strain ϵ. *Strain energy* is defined as the internal work of stresses acting through incremental strains, integrated over the total strains and over the volume. For the nonlinear case in Fig. 1-25a, the strain energy U may be written as

$$U = \sum_{i=1}^{n_s} \int_V \left(\int_0^{\epsilon_i} \sigma \, d\epsilon \right) dV \tag{1-15}$$

in which V denotes volume and n_s is the number of strain components ϵ (see Eq. d) involved in a particular problem. The expression in parentheses represents the area between the curve and the strain axis in Fig. 1-25a. Evaluation of this integral for the linear case in Fig. 1-25b (where $\sigma = \epsilon\sigma_i/\epsilon_i$) produces the result

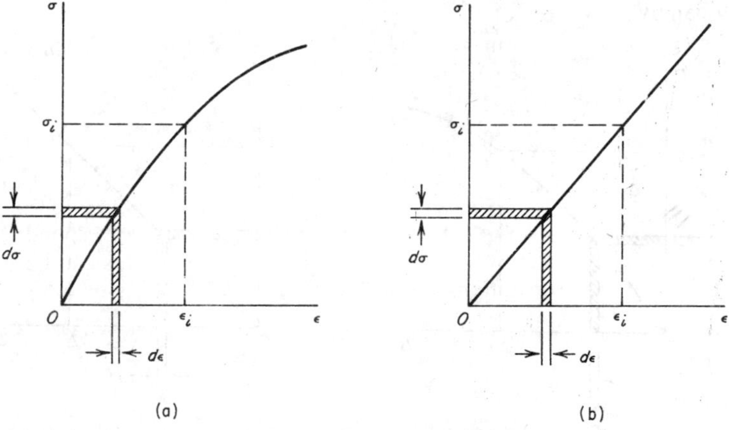

Fig. 1-25. Stress-strain diagrams: (a) nonlinear and (b) linear.

$$U = \frac{1}{2}\sum_{i=1}^{n_s}\int_V \sigma_i\epsilon_i\, dV = \frac{1}{2}\int_V \boldsymbol{\sigma}^{\mathrm{T}}\boldsymbol{\epsilon}\, dV \tag{1-16}$$

in which $\boldsymbol{\sigma}^{\mathrm{T}}$ is the transpose of the column matrix $\boldsymbol{\sigma}$. Note that the indicial-summation form of the expression for U is equivalent to the matrix-multiplication form, but the latter is more convenient.

Complementary strain energy is defined as the internal work of strains multiplied by incremental stresses, integrated over the total stresses and over the volume. For the nonlinear case in Fig. 1-25a, the expression for complementary strain energy U^* is

$$U^* = \sum_{i=1}^{n_s}\int_V \left(\int_0^{\sigma_i} \epsilon\, d\sigma\right) dV \tag{1-17}$$

Here the term in parentheses represents the area between the curve and the stress axis in Fig. 1-25a. Evaluating this integral for the linear case in Fig. 1-25b (where $\epsilon = \sigma\epsilon_i/\sigma_i$) gives the result

$$U^* = \frac{1}{2}\sum_{i=1}^{n_s}\int_V \epsilon_i\sigma_i\, dV = \frac{1}{2}\int_V \boldsymbol{\epsilon}^{\mathrm{T}}\boldsymbol{\sigma}\, dV \tag{1-18}$$

in which $\boldsymbol{\epsilon}^{\mathrm{T}}$ is the transpose of the column matrix $\boldsymbol{\epsilon}$. Comparison of Eqs. (1-16) and (1-18) shows that the strain energy is equal to the complementary strain energy for linearly elastic materials. Of course, this fact can also be observed from Fig. 1-25b.

The *external work of loads* may be written in a manner similar to that for strain energy, but for concentrated surface loads no integration over the volume is required. For this purpose, Fig. 1-26 shows nonlinear and linear load-displacement diagrams for a single load P and the corresponding displacement Δ. The external work W of all such loads in the nonlinear case is given by

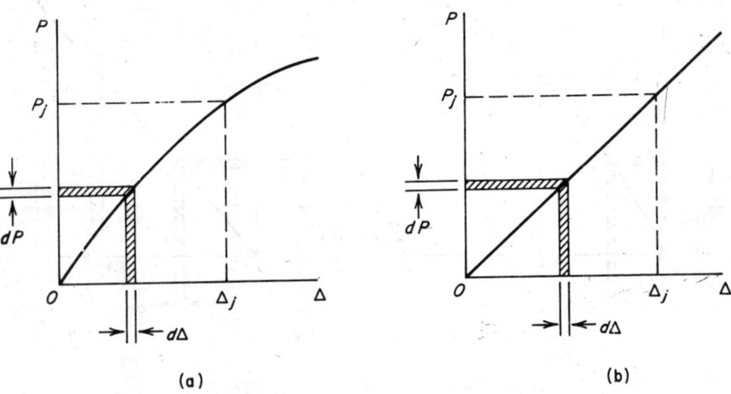

Fig. 1-26. Load-displacement diagrams: (a) nonlinear and (b) linear.

$$W = \sum_{j=1}^{n_p} \int_0^{\Delta_j} P\, d\Delta \tag{1-19}$$

in which n_p is the number of loads. The integral in Eq. (1-19) represents the area between the curve and the displacement axis in Fig. 1-26a. Performing this integration for the linear case in Fig. 1-26b yields

$$W = \frac{1}{2}\sum_{j=1}^{n_p} P_j\Delta_j = \frac{1}{2}\mathbf{A}^{\mathrm{T}}\mathbf{D} \tag{1-20}$$

The symbol $\mathbf{A}^{\mathrm{T}}$ in the latter form of this expression denotes the transpose of a column vector containing all of the applied actions, and **D** is a column vector of the corresponding displacements.

In addition, the *complementary work of loads* takes a form similar to that for complementary strain energy. For the nonlinear case in Fig. 1-26a, the complementary work of loads W^* is defined as

$$W^* = \sum_{j=1}^{n_p} \int_0^{P_j} \Delta\, dP \tag{1-21}$$

In this instance the integral is seen to be the area between the curve and the load axis in Fig. 1-26a. Integration for the linear case in Fig. 1-26b gives

$$W^* = \frac{1}{2}\sum_{j=1}^{n_p} \Delta_j P_j = \frac{1}{2}\mathbf{D}^{\mathrm{T}}\mathbf{A} \tag{1-22}$$

For the linear case, Eqs. (1-20) and (1-22) show that the work of the loads is equal to the complementary work of the loads, as can be observed from Fig. 1-26b.

For linearly elastic structures, it is possible to express strain energy and complementary strain energy in terms of stiffnesses and flexibilities instead of stresses and strains. The first expression of this type is obtained by using the *principle of conservation of energy,* which states that the work of the loads W equals the strain energy U stored in the structure. Furthermore, the applied actions are known to be related to the corresponding displacements by either flexibilities (Eq. 1-10) or stiffnesses (Eq. 1-12). Substitution of the latter relationship into the transposed form of Eq. (1-20) produces

$$U = W = \frac{1}{2}\mathbf{D}^{\mathrm{T}}\mathbf{S}\mathbf{D} \tag{1-23}$$

Similarly, the *principle of conservation of complementary energy* leads to the second expression

$$U^* = W^* = \frac{1}{2}\mathbf{A}^{\mathrm{T}}\mathbf{F}\mathbf{A} \tag{1-24}$$

which results from substituting Eq. (1-10) into the transpose of Eq. (1-22).

Of course, it is also true that $U = U^*$ and $W = W^*$, as has been shown before.

Energy theorems developed by Castigliano, Crotti, and Engesser* are widely used by structural engineers for establishing equilibrium and compatibility equations. *Castigliano's first theorem* states that if the strain energy of an elastic body is expressed as a function of a set of displacements, the first partial derivative of that function with respect to a particular displacement is equal to the corresponding action. This theorem is valid for either nonlinear or linear elasticity, and application to the linear case of Eq. (1-23) gives

$$\frac{\partial U}{\partial D_j} = \sum_{k=1}^{n} S_{jk} D_k = A_j \qquad (j = 1, 2, \ldots, n) \qquad \text{(1-25)}$$

which may be verified by performing the detailed operations of matrix multiplication and differentiation. Equation (1-25) represents a set of n equilibrium conditions for a linearly elastic structure.

The *Crotti–Engesser theorem* states that if the complementary strain energy is written in terms of a set of applied actions, the first partial derivative of such a function with respect to a particular action is equal to the corresponding displacement. Differentiation of Eq. (1-24) in this manner yields

$$\frac{\partial U^*}{\partial A_j} = \sum_{k=1}^{n} F_{jk} A_k = D_j \qquad (j = 1, 2, \ldots, n) \qquad \text{(1-26)}$$

Equation (1-26) represents a set of n compatibility conditions for a linearly elastic structure.

Although the Crotti-Engesser theorem is valid for either nonlinear or linear systems, it is applied here only to the linear case. Thus, the complementary strain energy U^* in Eq. (1-26) could be replaced by the strain energy U. This specialized form of the Crotti–Engesser theorem is known as *Castigliano's second theorem,* which is valid only for linearly elastic structures.

By further differentiation of Eqs. (1-25) and (1-26), it is possible to obtain stiffnesses and flexibilities from strain energy and complementary strain energy. On the one hand, differentiating Eq. (1-25) with respect to D_k yields a typical stiffness term S_{jk}, as follows:

$$\frac{\partial^2 U}{\partial D_j \partial D_k} = \frac{\partial A_j}{\partial D_k} = S_{jk} \qquad \begin{array}{l} j = 1, 2, \ldots, n \\ k = 1, 2, \ldots, n \end{array} \qquad \text{(1-27)}$$

Thus, it is seen that S_{jk} is equal to the second partial derivative of U (expressed in terms of displacements) with respect to D_j and D_k.

*For a more thorough explanation of these energy theorems, accompanied by historical references, see S. P. Timoshenko and J. M. Gere, *Mechanics of Materials,* D. Van Nostrand, New York, 1972, Chapter 11.

On the other hand, differentiation of Eq. (1-26) with respect to A_k results in a typical flexibility coefficient F_{jk}. That is,

$$\frac{\partial^2 U^*}{\partial A_j \partial A_k} = \frac{\partial D_j}{\partial A_k} = F_{jk} \qquad \begin{matrix} j = 1, 2, \ldots, n \\ k = 1, 2, \ldots, n \end{matrix} \tag{1-28}$$

Hence, F_{jk} is equal to the second partial derivative of U^* (expressed as a function of applied actions) with respect to A_j and A_k.

At this point in the discussion it is convenient to observe *reciprocal relations* for stiffnesses and flexibilities. If the sequence of differentiation in Eq. (1-27) is reversed, the result must be the same, showing that

$$S_{jk} = S_{kj} \tag{1-29}$$

Since all pairs of cross stiffnesses are equal, the matrix **S** must be symmetric. Thus, it is equal to its own transpose:

$$\mathbf{S} = \mathbf{S}^{\mathrm{T}} \tag{1-30}$$

This equality is also evident from Eq. (1-23), because U is a scalar and can be transposed.

Similarly, reversal of the sequence of differentiation in Eq. (1-28) shows that

$$F_{jk} = F_{kj} \tag{1-31}$$

Therefore, all pairs of cross flexibilities are equal,* and the matrix **F** is symmetric and equal to its own transpose:

$$\mathbf{F} = \mathbf{F}^{\mathrm{T}} \tag{1-32}$$

This equality may also be deduced from Eq. (1-24) by observing that $U^* = U^{*\mathrm{T}}$, which is a scalar.

Illustrations of the reciprocal relations discussed here can be seen by referring to the example at the end of Art. 1.11 (see p. 33). In that example the cross flexibilities F_{12} and F_{21} are equal, and the flexibility matrix is symmetric. Similarly, the cross stiffnesses S_{12} and S_{21} are equal, and the stiffness matrix is also symmetric. Many additional examples of the reciprocal relations will be encountered later in connection with the flexibility and stiffness methods of analysis.

1.14 Virtual Work.

This article will be devoted to concepts of virtual work and energy, as distinguished from the real work and energy discussed in the preceding article. Virtual work methods are more versatile than those involving real work because the displacements utilized may be due to influences other than the applied loads. When such loads are subjected to virtual displacements, they act at constant values instead of increasing gradually from zero. Use of this rather subtle idea provides

*The reciprocal relation for flexibilities is also called *Maxwell's theorem,* because it was first presented by J. C. Maxwell in 1864.

additional techniques for establishing equilibrium or compatibility conditions for structures. Furthermore, any desired actions, displacements, stiffnesses, or flexibilities may be determined without having to formulate strain energy or take derivatives.

Consider a conservative structure that is in a state of equilibrium under the influence of a set of applied actions **A** that cause internal stresses $\boldsymbol{\sigma}$. Then let the structure be subjected to any pattern of *compatible virtual deformations* that may be of interest. Such deformations must be continuous within the structure and consistent with geometric boundary conditions, but otherwise the selection is arbitrary. Included in the pattern will be external *virtual displacements* $\delta\mathbf{D}$, corresponding to the real actions **A**; and there also will be internal *virtual strains* $\delta\boldsymbol{\epsilon}$, corresponding to the real stresses $\boldsymbol{\sigma}$. The *principle of virtual work* states that the external virtual work of the real actions **A** multiplied by the virtual displacements $\delta\mathbf{D}$ is equal to the internal virtual work of the real stresses $\boldsymbol{\sigma}$ multiplied by the virtual strains $\delta\boldsymbol{\epsilon}$, integrated over the volume.

The *external virtual work* δW is given by

$$\delta W = \mathbf{A}^{\mathrm{T}}\,\delta\mathbf{D} \tag{1-33}$$

and the *internal virtual work* δU may be written as

$$\delta U = \int_V \boldsymbol{\sigma}^{\mathrm{T}}\delta\boldsymbol{\epsilon}\; dV \tag{1-34}$$

A more appropriate name for this quantity is *virtual strain energy*. Applying the principle of virtual work gives the equality

$$\delta W = \delta U \tag{1-35}$$

and substituting Eqs. (1-33) and (1-34) into Eq. (1-35) produces

$$\mathbf{A}^{\mathrm{T}}\delta\mathbf{D} = \int_V \boldsymbol{\sigma}^{\mathrm{T}}\delta\boldsymbol{\epsilon}\; dV \tag{1-36}$$

To understand this virtual work equality, one should think of the real actions doing external work as they act at constant values through the virtual displacements. Simultaneously, the real stresses do internal work as they act at constant values during the virtual strains.

Equation (1-36) is true only if the structure is in equilibrium when it is also subjected to compatible virtual deformations. Conversely, if Eq. (1-36) is true, the structure is proven to be in equilibrium. In fact, equations of equilibrium may be derived from Eq. (1-36) by introducing appropriate stress–strain and strain–displacement relationships for the structure. This approach is used in the method of finite elements but is not suitable for the analysis of framed structures.

Instead, a method for calculating actions and stiffnesses will be developed by specializing Eq. (1-36) as follows. If an action A_j is unknown (such as a reaction), it can be found by letting $\delta D_j = 1$ while all other elements of $\delta\mathbf{D}$ are zero. Then Eq. (1-36) yields

$$A_j(1) = \int_V \boldsymbol{\sigma}^{\mathrm{T}} \delta\boldsymbol{\epsilon}_j \, dV \tag{a}$$

where $\delta\boldsymbol{\epsilon}_j$ is a vector of internal virtual strains for the condition $\delta D_j = 1$. The equivalent transposed form

$$(1)A_j = \int_V \delta\boldsymbol{\epsilon}_j^{\mathrm{T}} \boldsymbol{\sigma} \, dV \tag{1-37}$$

will be more convenient in the ensuing discussion.

Equation (1-37) represents a technique known as the *unit-displacement method,* which can also be used to calculate stiffness coefficients. For this purpose, define a vector **D** of real displacements corresponding to the actions **A**, and let $D_k = 1$ while all other elements of **D** are zero. In this case the quantity A_j in Eq. (1-37) becomes the action of type j due to a unit displacement of type k, which is the definition for S_{jk}. Thus, Eq. (1-37) becomes

$$(1)S_{jk} = \int_V \delta\boldsymbol{\epsilon}_j^{\mathrm{T}} \boldsymbol{\sigma}_k \, dV \tag{1-38}$$

in which $\boldsymbol{\sigma}_k$ is a vector of internal stresses for the condition $D_k = 1$. This integral equation for S_{jk}, derived by the unit-displacement method, is used more frequently than the differential equation (Eq. 1-27) developed from strain energy in the preceding article.

To see the physical meaning of Eq. (1-38), it is helpful to refer to an example, such as that at the end of Art. 1.11 (see Fig. 1-22). For the stiffness term S_{12} as an illustration, Eq. (1-38) gives

$$(1)S_{12} = \int_V \delta\boldsymbol{\epsilon}_1^{\mathrm{T}} \boldsymbol{\sigma}_2 \, dV \tag{b}$$

In this case the external virtual work is equal to the product of S_{12} in Fig. 1-22c and the unit displacement (of type 1) in Fig. 1-22d. The internal virtual work is calculated as the product of the stresses of type 2 (Fig. 1-22e) and the virtual strains of type 1 (Fig. 1-22d). Because such calculations are likely to be confusing, the subscripts on both sides of Eq. (1-38) have been arranged in the same sequence (by earlier transposition) to serve as a memory guide.

All of the concepts discussed so far in this article have complementary counterparts that may be examined by introducing the notion of *virtual loads* instead of virtual displacements. Such an approach leads to the unit-load method, which is unexcelled for the purpose of calculating deflections and flexibilities for framed structures. To begin, suppose that a conservative structure is in a state of compatible deformation with external displacements **D** and internal strains $\boldsymbol{\epsilon}$. Then subject the structure to any desired set of virtual loads that are in equilibrium. Included in the set will be *virtual actions* $\delta\mathbf{A}$, corresponding to the real displacements **D**; and there also will be internal *virtual stresses* $\delta\boldsymbol{\sigma}$, corresponding to the real strains $\boldsymbol{\epsilon}$. The *principle of complementary virtual work* states that the external work of the virtual actions $\delta \mathbf{A}$ multiplied by the real displacements $\mathbf{D}$ is equal to the internal work of the virtual stresses $\delta \boldsymbol{\sigma}$ multiplied by the real strains $\boldsymbol{\epsilon}$ integrated over the volume.

The *external complementary virtual work* δW^* will be written as

$$\delta W^* = \delta \mathbf{A}^{\mathrm{T}} \mathbf{D} \tag{1-39}$$

and the *internal complementary virtual work* δU^* is

$$\delta U^* = \int_V \delta \boldsymbol{\sigma}^{\mathrm{T}} \boldsymbol{\epsilon} \, dV \tag{1-40}$$

This quantity is called the *complementary virtual strain energy*. Application of the principle of complementary virtual work yields

$$\delta W^* = \delta U^* \tag{1-41}$$

Substitution of Eqs. (1-39) and (1-40) into Eq. (1-41) gives

$$\delta \mathbf{A}^{\mathrm{T}} \mathbf{D} = \int_V \delta \boldsymbol{\sigma}^{\mathrm{T}} \boldsymbol{\epsilon} \, dV \tag{1-42}$$

To visualize the nature of this complementary virtual work equality, it is helpful to imagine the virtual loads acting at constant values through the real displacements while the virtual stresses act at constant values during the real strains. Thus, one may think of the virtual loads as existing before the real displacements occur, instead of the sequence first described.

Equation (1-42) applies only to a structure in a compatible state of deformation when it is also subjected to virtual loads in equilibrium. Conversely, if Eq. (1-42) is true, the structure is proven to be in a compatible state. It follows that compatibility equations may be derived from Eq. (1-42) by bringing in appropriate strain-stress and stress-action relationships for the structure. However, this approach will not be pursued here because it is not suitable for the analysis of framed structures.

Instead, a method for calculating displacements and flexibilities will be developed by specializing Eq. (1-42) as follows. If the value of a particular displacement D_j is desired, it can be found by introducing a unit load $\delta A_j = 1$ while all other elements of $\delta \mathbf{A}$ are zero. (Typically, equilibrium is maintained by virtual reactions that do no work.) Then Eq. (1-42) becomes

$$(1) D_j = \int_V \delta \boldsymbol{\sigma}_j^{\mathrm{T}} \boldsymbol{\epsilon} \, dV \tag{1-43}$$

where $\delta \boldsymbol{\sigma}_j$ is a vector of internal virtual stresses for the condition $\delta A_j = 1$.

This *unit-load method* can also be used to calculate flexibility coefficients. To do this, let the vector $\mathbf{A}$ contain real actions corresponding to the displacements $\mathbf{D}$, and let $A_k = 1$ while all other elements of $\mathbf{A}$ are zero. In this instance the quantity D_j in Eq. (1-43) becomes the displacement of type j due to a unit action of type k, which is the definition for F_{jk}. Thus,

$$(1)F_{jk} = \int_V \delta\sigma_j^T \epsilon_k \, dV \tag{1-44}$$

in which ϵ_k is a vector of internal strains for the condition $A_k = 1$. This integral equation for F_{jk}, derived by the unit-load method, is much more useful than the differential equation (Eq. 1-28) developed from complementary strain energy in the preceding article.

As an illustration, consider the example at the end of Art. 1.11 (see Fig. 1-22). Application of Eq. (1-44) to the calculation of the flexibility coefficient F_{21} gives

$$(1)F_{21} = \int_V \delta\sigma_2^T \epsilon_1 \, dV \tag{c}$$

Here the external complementary virtual work is equal to the product of the unit load (of type 2) in Fig. 1-22c and the flexibility F_{21} in Fig. 1-22b. The internal complementary virtual work is equal to the product of the virtual stresses of type 2 (Fig. 1-22c) and the strains of type 1 (Fig. 1-22b). Note that subscripts on both sides of Eq. (1-44) have been deliberately arranged in the same sequence to serve as a memory guide.

Detailed examples of the unit-load method applied to various types of framed structures appear in Appendix A.2 for convenient reference throughout the book. In studying that appendix, the reader will see that for the slender members of framed structures it is simpler to work with stress resultants and corresponding displacements than with actual stresses and strains.

Problems

1.4-1. A prismatic member is subjected to equal and opposite axial forces P at the ends, thereby producing uniform tension in the bar. Define the displacement that corresponds to the two forces P.

1.4-2. Define the actions that correspond to the displacements Δ_c and θ_c in the plane frame shown in Fig. 1-6.

1.4-3. The simply supported beam shown in the figure has constant flexural rigidity EI and length L. The loads on the beam are two actions A_1 and A_2 as illustrated. Using Table A-3, obtain expressions in terms of A_1, A_2, E, I, and L for each of the following: (a) the displacement D_{11} corresponding to action A_1 and caused by action A_1 acting alone; (b) the displacement D_{12} corresponding to action A_1 and caused by A_2 acting alone; and (c) the displacement D_1 corresponding to A_1 and caused by both actions acting simultaneously.

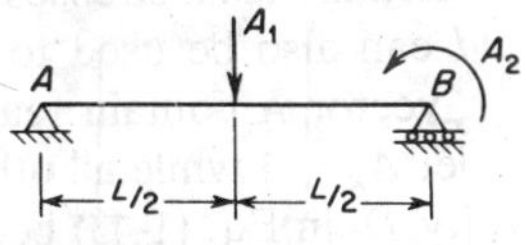

Prob. 1.4-3.

1.4-4. For the cantilever beam shown in the figure, determine the displacements D_{11}, D_{21}, and D_{31} using Table A-3. Assume that the beam has constant flexural rigidity EI.

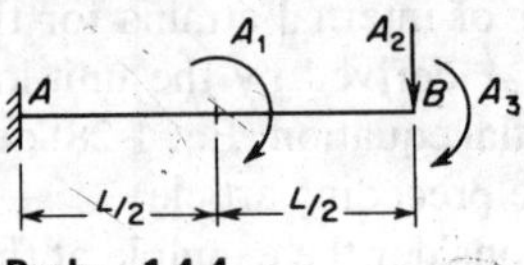

Prob. 1.4-4.

1.4-5. The overhanging beam shown in the figure is subjected to loads A_1, A_2, and A_3. Assuming constant flexural rigidity EI for the beam, determine the displacements D_{11}, D_{23}, and D_{33} using Table A-3.

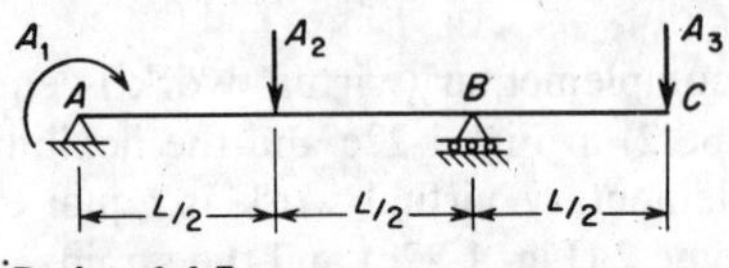

Prob. 1.4-5.

1.7-1. (a) What is the degree of kinematic indeterminacy for a simply supported beam (see figure)? (b) If the effects of axial deformations are neglected, what is the degree of kinematic indeterminacy? (c) and (d) Repeat questions (a) and (b) for a cantilever beam that is fixed at one end and free at the other.

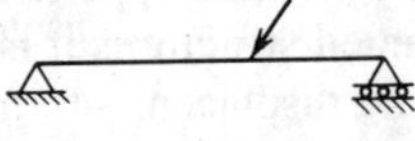

Prob. 1.7-1.

1.7-2. (a) What is the degree of static indeterminacy for the continuous beam shown in Fig. 1-1a? (b) What is the degree of kinematic indeterminacy? (c) If the effects of axial deformations are neglected, what is the degree of kinematic indeterminacy?

1.7-3. (a) Determine the number of degrees of static indeterminateness for the plane truss shown in Fig. 1-1b. (b) Determine the degree of kinematic indeterminacy.

1.7-4. What is the degree of kinematic indeterminacy for the plane truss shown in Fig. 1-9?

1.7-5. Determine the degrees of (a) static indeterminacy and (b) kinematic indeterminacy for the space truss in Fig. 1-1c.

1.7-6. Repeat Prob. 1.7-2 for the plane frame shown in the figure.

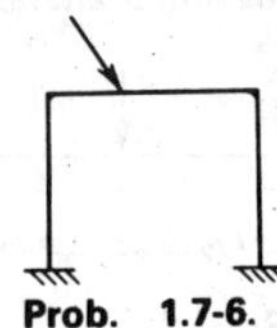

Prob. 1.7-6.

1.7-7. Repeat Prob. 1.7-2 for the plane frame with pinned supports (see figure).

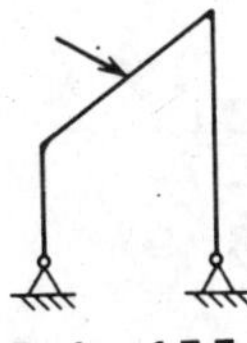

Prob. 1.7-7.

1.7-8. Repeat Prob. 1.7-2 for the plane frame shown in Fig. 1-1d.

1.7-9. For the grid shown in the figure, find (a) the degree of static indeterminacy and (b) the degree of kinematic indeterminacy.

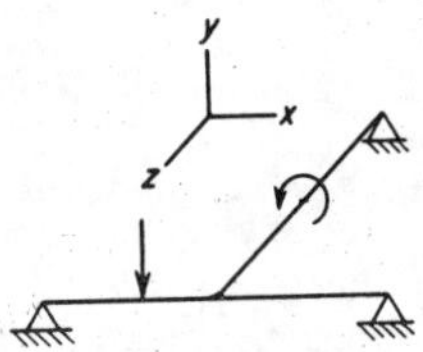

Prob. 1.7-9.

1.7-10. Solve the preceding problem for the grid shown in the figure.

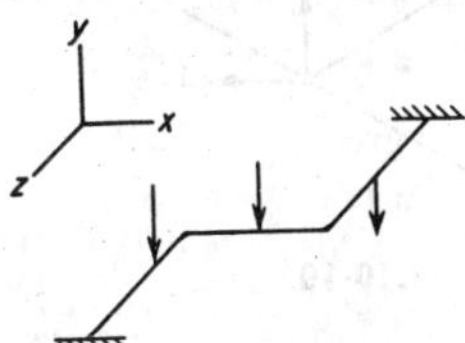

Prob. 1.7-10.

1.7-11. Repeat Prob. 1.7-9 for the grid shown in Fig. 1-1e.

1.7-12. Repeat Prob. 1.7-2 for the space frame pictured in Fig. 1-1f.

1.10-1 to 1.10-10. Illustrate by means of sketches the physical significance of the flexibility and stiffness coefficients corresponding to the actions shown in the figures.

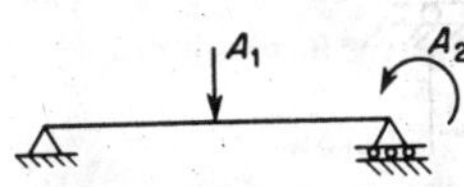

Prob. 1.10-1.

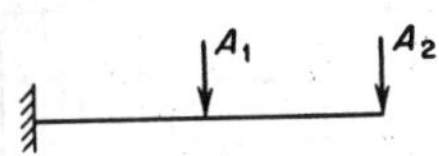

Prob. 1.10-3.

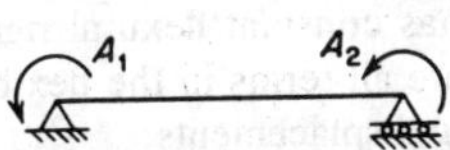

Prob. 1.10-2.

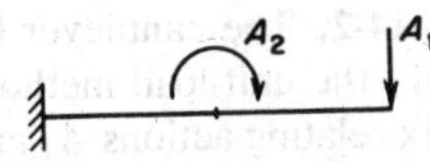

Prob. 1.10-4.

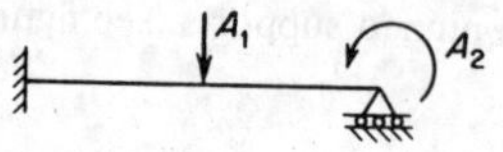

Prob. 1.10-5.

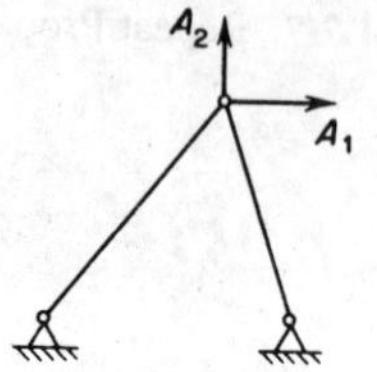

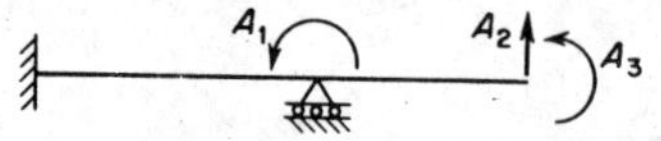

Prob. 1.10-6.

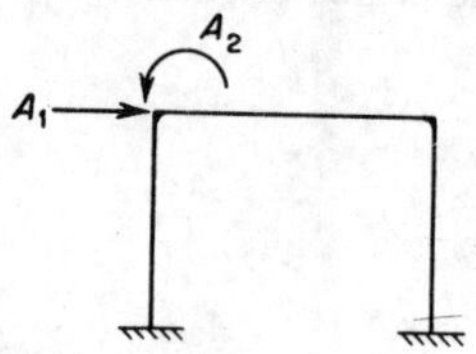

Prob. 1.10-7.

Prob. 1.10-8.

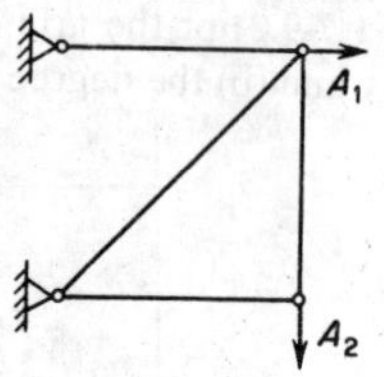

Prob. 1.10-9.

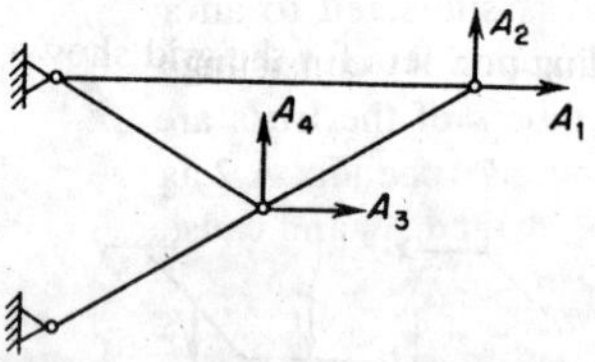

Prob. 1.10-10.

1.14-1. The simply supported beam shown in the figure has constant flexural rigidity EI. By the unit-load method of Appendix A.2, determine the following displacements due to actions A_1 and A_2 applied simultaneously: (a) rotation at point A (positive counterclockwise); (b) translation at mid-span (positive upward); and (c) rotation at mid-span (positive counterclockwise).

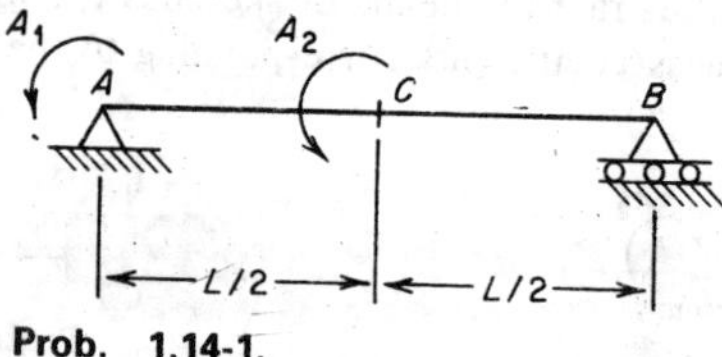

Prob. 1.14-1.

1.14-2. The cantilever beam shown in the figure has constant flexural rigidity EI. By the unit-load method of Appendix A.2, calculate all terms in the flexibility matrix relating actions A_1 and A_2 to their corresponding displacements.

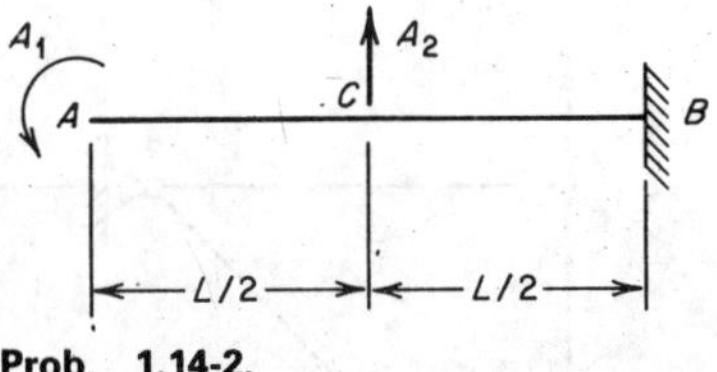

Prob. 1.14-2.

1.14-3. For the overhanging beam in the figure, determine all coefficients in the flexibility matrix relating actions A_1 and A_2 to their corresponding displacements. Use the unit-load method of Appendix A.2, assuming that the beam has constant flexural rigidity EI.

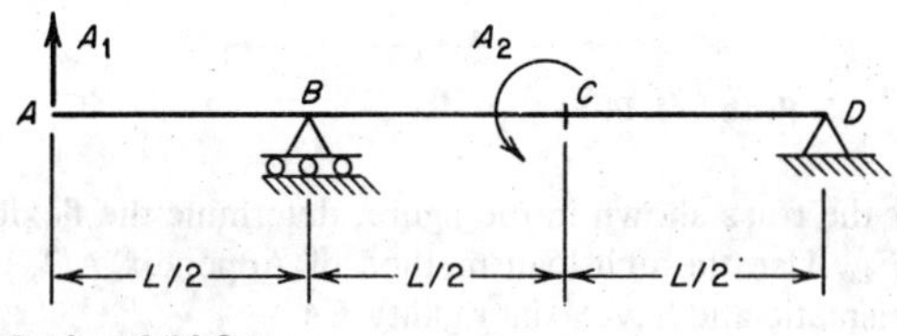

Prob. 1.14-3.

1.14-4. The figure shows a truss subjected to an action A_1 in the form of a horizontal force at joint A and an action A_2 consisting of two equal and opposite forces at joints A and D. All members of the truss are prismatic and have axial rigidity EA. Using the truss example in Appendix A.2 as a guide, find the displacements D_1 and D_2 corresponding to A_1 and A_2 and caused by these actions applied simultaneously.

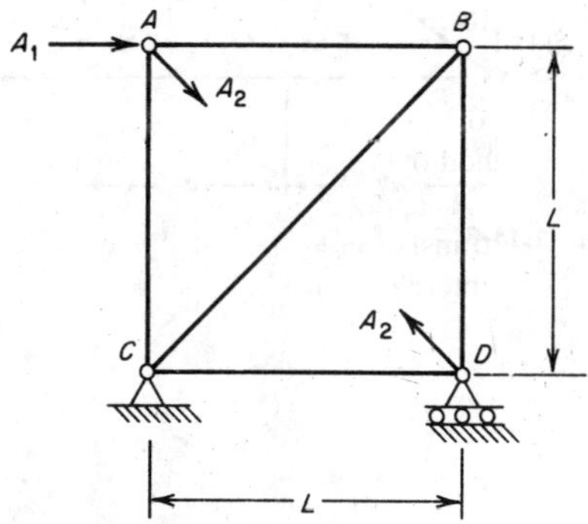

Prob. 1.14-4.

1.14-5. The truss shown in the figure has the same axial rigidity EA for all of its members, which are prismatic. By the unit-load method of Appendix A.2, calculate all coefficients in the flexibility matrix relating actions A_1 and A_2 to their corresponding displacements.

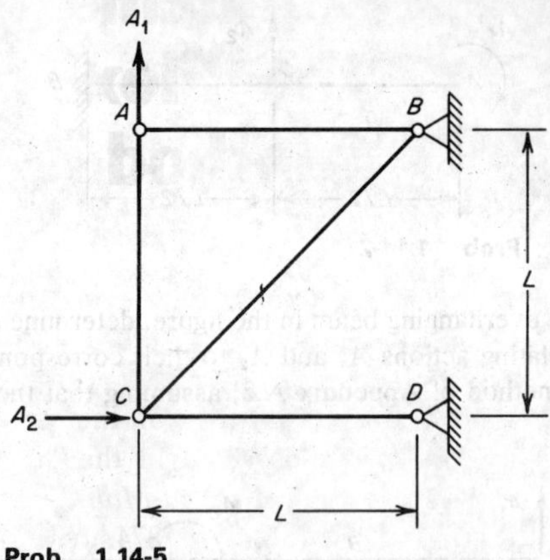

Prob. 1.14-5.

1.14-6. For the truss shown in the figure, determine the flexibility coefficients F_{12}, F_{22}, and F_{23}. Use the unit-load method of Appendix A.2, assuming that all members are prismatic and have axial rigidity EA.

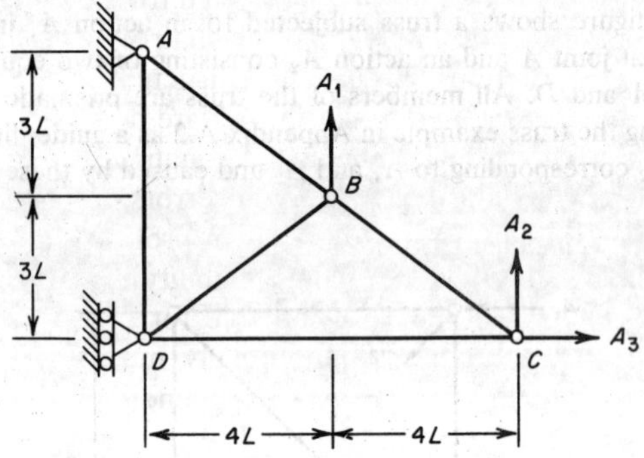

Prob. 1.14-6.

Fundamentals of the Flexibility Method

2.1 Introduction. Basic concepts of the *flexibility method* (also called the *force method*) are described in this chapter. The flexibility method is a generalization of the *Maxwell-Mohr method* developed by J. C. Maxwell in 1864 and O. C. Mohr in 1874. In this approach statically indeterminate structures are analyzed by writing compatibility equations in terms of flexibility coefficients and selected redundants. Also involved in such equations are displacements calculated for a statically determinate version of the original structure (with the redundants released). Because the flexibility method requires extensive use of calculated displacements, the material in Appendix A will be referenced frequently.

The flexibility method is presented in matrix form in order to permit immediate generalization to complicated structures, even though the first problems to be solved are very simple and are devised solely to illustrate the basic concepts. A primary advantage of matrix notation is that it serves to organize the work and helps to avoid errors in the solution process. Moreover, the solution process for a given problem is carried out in this chapter by finding the inverse of the flexibility matrix. This approach is symbolically convenient and is acceptable for small problems with few redundants. However, it is computationally more efficient to solve the simultaneous equations (as explained in Appendix D) than to calculate the inverse.

It should be realized at the outset that the flexibility method is not conducive to computer programming because the choice of redundants is not unique. In spite of this drawback, the student is urged to learn the method because it is useful for hand calculations. It also provides a good background for learning the stiffness method, which is mathematically similar and much more suitable for computer programming.

In this chapter the flexibility method is first developed for the basic purpose of finding redundants in structures that are subjected to applied loads only. Subsequently, the effects of temperature changes, prestrains, and support displacements are considered; and formulas for calculating joint displacements, member end-actions, and support reactions are developed as well. Then flexibility matrices for individual members are presented and used in assembling the over-all flexibility matrix for the structure by a formal procedure.

2.2 Flexibility Method. In order to illustrate the fundamental ideas of the flexibility method, consider the example shown in Fig. 2-1a. The beam *ABC* in the figure has two spans of equal length and is subjected to a uniform load of intensity w. The beam is statically indeterminate to the first degree, since there are four possible reactions (two reactions at A, one at B, and one at C) and three equations of static equilibrium for actions in a plane. Because the loads are perpendicular to the beam, however, there is no need to consider a horizontal reaction at point A. The reaction R_B at the middle support will be taken as the statical redundant, although other possibilities also exist. If this redundant is released (or removed), a statically determinate *released structure* is obtained. In this case, the released structure is the simply supported beam shown in Fig. 2-1b.

Under the action of the uniform load w, the released structure will deflect as illustrated in Fig. 2-1c. The displacement of the beam at point B is denoted as Δ_B and is given by the expression (see Table A-3 of Appendix A)

$$\Delta_B = \frac{5wL^4}{384EI}$$

in which EI is the flexural rigidity of the beam. However, the actual beam is assumed to have no translation (that is, no settlement) at point B. Therefore, the redundant reaction R_B must be such as to produce in the released structure an upward displacement equal to Δ_B (see Fig. 2-1d). According to the principle of superposition, the final displacement at point B in the

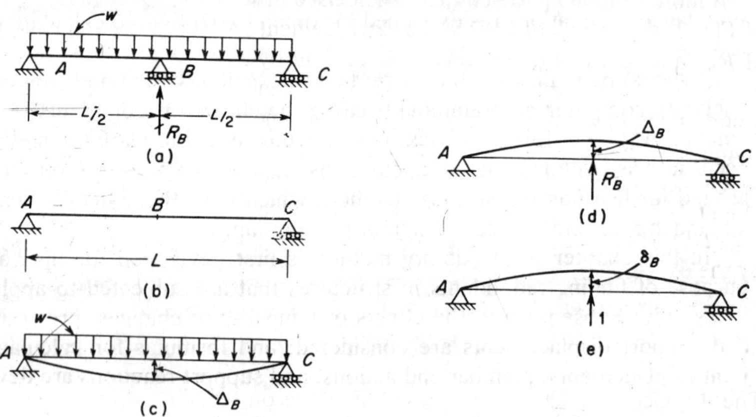

Fig. 2-1. Illustration of flexibility method.

released structure is the resultant of the displacements caused by the load w and the redundant R_B. The upward displacement due to R_B is

$$\Delta_B = \frac{R_B L^3}{48EI}$$

as obtained from Table A-3 of the Appendix. Equating the two expressions obtained above for Δ_B gives

$$\frac{5wL^4}{384EI} = \frac{R_B L^3}{48EI} \tag{2-1}$$

The unknown reaction R_B may be obtained by solving Eq. (2-1):

$$R_B = \frac{5wL}{8}$$

After R_B has been found, the remaining reactions for the two-span beam can be found from static equilibrium equations.

In the preceding example, the procedure was to calculate displacements in the released structure caused by both the loads and the redundant action and then to formulate an equation pertaining to these displacements. Equation (2-1) expressed the fact that the downward displacement due to the load was equal to the upward displacement due to the redundant. In general, an equation of this type can be called an *equation of compatibility,* because it expresses a condition relating the displacements of the structure. It may also be called an *equation of superposition,* since it is based upon the superposition of displacements caused by more than one action. Still another name would be *equation of geometry,* because the equation expresses a condition that applies to the geometry of the structure.

A more general approach that can be used in solving the two-span beam of Fig. 2-1a consists of finding the displacement produced by a unit value of R_B and then multiplying this displacement by R_B in order to obtain the displacement caused by R_B. Also, it is more general and systematic to use a consistent sign convention for the actions and displacements at B. For example, it may be assumed that both the displacement at B and the reaction at B are positive when in the upward direction. Then the application of a unit force (corresponding to R_B) to the released structure, as shown in Fig. 2-1e, results in a positive displacement δ_B. This displacement is given by the expression

$$\delta_B = \frac{L^3}{48EI} \tag{2-2}$$

The displacement caused by R_B acting alone on the released structure is $\delta_B R_B$. The displacement caused by the uniform load w acting alone on the released structure is

$$\Delta_B = -\frac{5wL^4}{384EI} \tag{2-3}$$

This displacement is negative because Δ_B is assumed to be positive when in the upward direction. Superposition of the displacements due to the load w and the reaction R_B must produce zero displacement of the beam at point B. Thus, the compatibility equation is

$$\Delta_B + \delta_B R_B = 0 \tag{2-4}$$

from which

$$R_B = -\frac{\Delta_B}{\delta_B} \tag{2-5}$$

When the expressions given above for δ_B and Δ_B (see Eqs. 2-2 and 2-3) are substituted into Eq. (2-5), the result is

$$R_B = \frac{5wL}{8}$$

which is the same as that obtained before. The positive sign in this result denotes the fact that R_B is in the upward direction.

An important part of the preceding solution consists of writing the superposition equation (Eq. 2-4) which expresses the geometrical fact that the beam does not deflect at the support B. Included in this equation are the effects of the load and the redundant reaction. The displacement caused by the reaction has been conveniently expressed as the product of the reaction itself and the displacement caused by a unit value of the reaction. The latter is a flexibility coefficient, since it is the displacement due to a unit action. If all terms in the equation are expressed with the same sign convention, then the sign of the final result will denote the true direction of the redundant action.

If a structure is statically indeterminate to more than one degree, the approach used in the preceding example must be further organized, and a more generalized notation must be introduced. To illustrate these features, another example of a prismatic beam will be considered (see Fig. 2-2a). The beam shown in the figure is statically indeterminate to the second degree; hence, a statically determinate released structure may be obtained by removing two redundant actions from the beam. Several choices for the redundants and the corresponding released structures are possible. Four such possibilities for the released structure are shown in Figs. 2-2b, 2-2c, 2-2d, and 2-2e. In the first of these figures, the reactive moment at A and the force at B are taken as the redundants; thus, the rotational restraint at A and the translational restraint at B are removed from the original beam in order to obtain the released structure. In the next case (Fig. 2-2c) the reactive moment at A and the internal bending moment at B are released. Therefore, the released structure has no rotational restraint at A and no

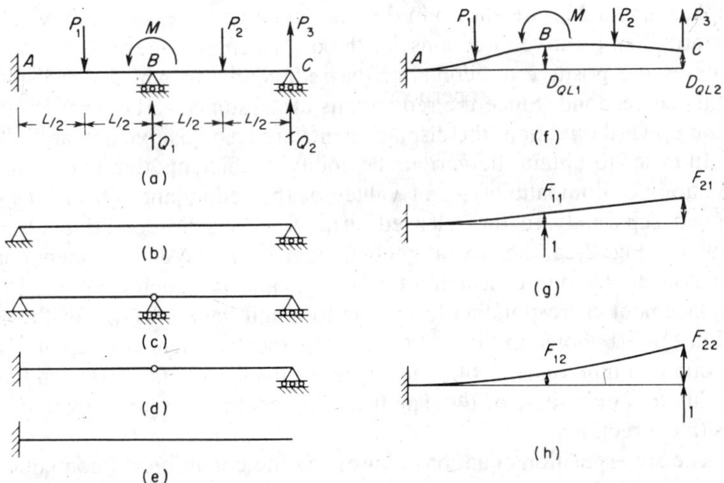

Fig. 2-2. Illustration of flexibility method.

restraint against bending moment at B. The latter condition is represented by the presence of a hinge in the beam at point B. The released structure shown in Fig. 2-2d is obtained by releasing both the reaction and bending moment at point B. Lastly, the released structure shown in Fig. 2-2e is obtained by selecting the reactions at joints B and C as the redundants. This particular released structure is selected for the analysis which follows, although any of the others would be suitable. All of the released structures shown in Fig. 2-2 are statically determinate and immobile. In general, only structures of this type will be used in the discussions of the flexibility method.

The redundant actions that are selected for the analysis are denoted as Q_1 and Q_2 in Fig. 2-2a. These actions are the reactive forces at joints B and C. In Fig. 2-2f is shown the released structure acted upon by the various loads on the original beam, which in this example are assumed to be three concentrated forces P_1, P_2, and P_3 and a couple M. These loads produce displacements in the released structure, and, in particular, displacements that correspond to Q_1 and Q_2 are produced at joints B and C. These displacements are denoted D_{QL1} and D_{QL2} in the figure. In this notation the symbol D_{QL} is used to represent a displacement corresponding to a redundant Q and caused by the loads on the structure. The numerical subscripts that follow the symbol denote the redundant to which the displacement corresponds. Thus, it is convenient to envisage D_{QL} as one symbol, while the numbers that follow the symbol are the subscripts.* In Fig. 2-2f the

*In computer programming, the symbol D_{QL} would be written DQL because of the necessity that all characters, whether alphabetic or numeric, be on the same line. When subscripts are present, they can be added to the basic symbol. For example, in a computer program $DQL(1)$ is the usual way of writing D_{QL1}. A similar style can be used for the other symbols that are encountered in subsequent work.

displacements D_{QL} are shown in their assumed positive direction, which is upward. The positive directions for the displacements must always be the same as the positive directions of the redundants to which the displacements correspond. Since the redundants are assumed to be positive when in the upward direction, the displacements are also positive upward.

In order to obtain the various flexibility coefficients that appear in the equations of compatibility, unit values of the redundants Q_1 and Q_2 are applied separately to the released structure. For the condition $Q_1 = 1$ shown in Fig. 2-2g, the flexibility coefficient F_{11} is the displacement corresponding to Q_1 due to a unit value of Q_1, and the coefficient F_{21} is the displacement corresponding to Q_2 due to a unit value of Q_1. For the condition $Q_2 = 1$ shown in Fig. 2-2h, F_{12} is the displacement corresponding to Q_1 due to a unit value of Q_2, and F_{22} is the displacement corresponding to Q_2 due to a unit value of Q_2. The flexibility coefficients are shown in their positive directions.

The superposition equations expressing the conditions of compatibility at joints B and C of the actual beam may now be written. Since the translational displacements at supports B and C are zero, the equations become

$$\begin{aligned} D_{QL1} + F_{11}Q_1 + F_{12}Q_2 &= 0 \\ D_{QL2} + F_{21}Q_1 + F_{22}Q_2 &= 0 \end{aligned} \tag{2-6}$$

The first of these equations represents the total displacement at B, which consists of three parts: the displacement due to loads, the displacement due to Q_1, and the displacement due to Q_2. The superposition of all three displacements gives the total displacement, which is zero. Similar comments apply to the second equation. The two equations can be solved simultaneously for Q_1 and Q_2, after which all other actions in the beam can be found by statics.

It is desirable at this stage of the discussion to write the superposition equations in a slightly more general form. It is always possible that support movements corresponding to the redundants will occur in the original beam, and these displacements can be included readily in the analysis. Assume that D_{Q1} and D_{Q2} represent the actual displacements in the beam corresponding to Q_1 and Q_2. Thus, $\mathbf{D}_{Q1}$ represents the support displacement at B, with the upward direction being positive. Similarly, D_{Q2} is the support displacement at C. The superposition equations express the fact that the final displacements corresponding to Q_1 and Q_2 are equal to the sums of the displacements caused by the loads and the redundants; thus, the equations are:

$$\begin{aligned} D_{Q1} &= D_{QL1} + F_{11}Q_1 + F_{12}Q_2 \\ D_{Q2} &= D_{QL2} + F_{21}Q_1 + F_{22}Q_2 \end{aligned} \tag{2-7}$$

If there are no support displacements, as assumed in this problem, then D_{Q1} and D_{Q2} are both zero, and Eqs. (2-7) reduce to Eqs. (2-6). If there are support displacements that do not correspond to a redundant, such as a displacement at joint A, they must be handled by the methods described later in Art. 2.4.

The superposition equations (2-7) can be written in matrix form as

$$\mathbf{D}_{Q} = \mathbf{D}_{QL} + \mathbf{FQ} \tag{2-8}$$

in which $\mathbf{D}_{Q}$ is the matrix of actual displacements corresponding to the redundants, $\mathbf{D}_{QL}$ is the matrix of displacements in the released structure corresponding to the redundant actions Q and due to the loads, and $\mathbf{F}$ is the flexibility matrix for the released structure corresponding to the redundant actions Q. For the equations given above, these matrices are:

$$\mathbf{D}_{Q} = \begin{bmatrix} D_{Q1} \\ D_{Q2} \end{bmatrix} \qquad \mathbf{D}_{QL} = \begin{bmatrix} D_{QL1} \\ D_{QL2} \end{bmatrix} \qquad \mathbf{F} = \begin{bmatrix} F_{11} & F_{12} \\ F_{21} & F_{22} \end{bmatrix} \qquad \mathbf{Q} = \begin{bmatrix} Q_1 \\ Q_2 \end{bmatrix}$$

It should be noted that an alternate (and more consistent) symbol for a flexibility coefficient F would be D_{QQ}, which represents a displacement in the released structure corresponding to a redundant Q and caused by a unit value of a redundant Q. Thus, the matrix $\mathbf{F}$ could also be denoted as $\mathbf{D}_{QQ}$ for consistency in notation, but the use of $\mathbf{F}$ for "flexibility" is more convenient.

The vector $\mathbf{Q}$ of redundants can be obtained by solving Eq. (2-8). Symbolically, the result is

$$\mathbf{Q} = \mathbf{F}^{-1}(\mathbf{D}_{Q} - \mathbf{D}_{QL}) \tag{2-9}$$

in which $\mathbf{F}^{-1}$ denotes the inverse of the flexibility matrix. From this equation the redundants can be calculated after first obtaining the matrices $\mathbf{D}_{Q}$, $\mathbf{D}_{QL}$, and $\mathbf{F}$. The first of these matrices will be known from the support conditions that exist in the original structure, while the last two are calculated from the properties of the released structure. The problem can be considered as solved when the matrix $\mathbf{Q}$ is known, since all other actions then can be found by static equilibrium. When the actions throughout the structure have been found, the displacements at any point can also be found. A method for incorporating into the analysis the calculation of actions and displacements at various points in the structure will be given in Art. 2.5.

The matrix $\mathbf{D}_{Q}$ normally will be a null matrix $\mathbf{0}$ (that is, a matrix with all elements equal to zero), except when one or more of the redundants is a support reaction that has a support displacement corresponding to it. If the matrix $\mathbf{D}_{Q}$ is null, Eq. (2-9) for the redundant actions $\mathbf{Q}$ becomes

$$\mathbf{Q} = \mathbf{F}^{-1}(\mathbf{0} - \mathbf{D}_{QL}) = -\mathbf{F}^{-1}\mathbf{D}_{QL} \tag{2-10}$$

This equation may be used instead of Eq. (2-9) whenever the displacements D_{Q} are zero.

To show the use of the matrix equations given above, consider again the beam in Fig. 2-2a. In order to have a specific example, assume that the beam has constant flexural rigidity EI in both spans and that the actions on the beam are as follows:

$$P_1 = 2P \qquad M = PL \qquad P_2 = P \qquad P_3 = P$$

Also, assume that there are no support displacements at any of the supports of the structure.

The matrices to be found first in the analysis are $\mathbf{D_Q}$, $\mathbf{D_{QL}}$, and $\mathbf{F}$, as mentioned previously. Since in the original beam there are no displacements corresponding to Q_1 and Q_2, the matrix $\mathbf{D_Q}$ is a null matrix. The matrix $\mathbf{D_{QL}}$ represents the displacements in the released structure corresponding to the redundants and caused by the loads. These displacements are found by considering Fig. 2-2f, which shows the released structure under the action of the loads. The displacements in this beam corresponding to Q_1 and Q_2 can be found by the methods described in Appendix A (see Example 3, Art. A.2), and the results are:

$$D_{QL1} = \frac{13PL^3}{24EI} \qquad D_{QL2} = \frac{97PL^3}{48EI}$$

The positive signs in these expressions show that both displacements are upward. From the results given above, the vector $\mathbf{D_{QL}}$ is obtained:

$$\mathbf{D_{QL}} = \frac{PL^3}{48EI}\begin{bmatrix} 26 \\ 97 \end{bmatrix}$$

The flexibility matrix $\mathbf{F}$ is found by referring to the beams pictured in Figs. 2-2g and 2-2h. The beam in Fig. 2-2g, which is subjected to a unit load corresponding to Q_1, has displacements given by the expressions

$$F_{11} = \frac{L^3}{3EI} \qquad F_{21} = \frac{5L^3}{6EI}$$

Similarly, the displacements in the beam of Fig. 2-2h are

$$F_{12} = \frac{5L^3}{6EI} \qquad F_{22} = \frac{8L^3}{3EI}$$

From the results listed above, the flexibility matrix can be formed:

$$\mathbf{F} = \frac{L^3}{6EI}\begin{bmatrix} 2 & 5 \\ 5 & 16 \end{bmatrix}$$

The inverse of the flexibility matrix can be found by any one of several standard methods,* the result being

*See, for instance, J. M. Gere and W. Weaver, Jr., *Matrix Algebra for Engineers,* D. Van Nostrand, New York, 1965

$$\mathbf{F}^{-1} = \frac{6EI}{7L^3}\begin{bmatrix} 16 & -5 \\ -5 & 2 \end{bmatrix}$$

It should be noted that both the flexibility matrix and its inverse are symmetric matrices.

As the final step in the analysis, Eq. (2-10) may be used to obtain the redundant actions **Q**, as follows:

$$\mathbf{Q} = \begin{bmatrix} Q_1 \\ Q_2 \end{bmatrix} = -\frac{6EI}{7L^3}\begin{bmatrix} 16 & -5 \\ -5 & 2 \end{bmatrix}\frac{PL^3}{48EI}\begin{bmatrix} 26 \\ 97 \end{bmatrix} = \frac{P}{56}\begin{bmatrix} 69 \\ -64 \end{bmatrix}$$

From this expression it is seen that the vertical reactions at supports B and C of the beam in Fig. 2-2a are

$$Q_1 = \frac{69P}{56} \qquad Q_2 = -\frac{8P}{7}$$

The negative sign for Q_2 indicates that this reaction is downward.

The redundant actions having been obtained in the manner shown in this example, the remaining actions in the beam can be found from static equilibrium equations. Also, the displacements at any point in the beam can now be obtained without difficulty, inasmuch as all actions can be considered as known. For example, one method of finding displacements is to subdivide the beam into two simple beams. Then each simple beam is acted upon by known moments at the ends as well as by the loads, and the displacements can be calculated.

While the matrix equations of the flexibility method (Eqs. 2-8, 2-9, and 2-10) were derived by discussion of the two-span beam of Fig. 2-2, they are actually completely general. They can be applied to any statically indeterminate framed structure having any number of degrees of indeterminacy. Of course, in this event the matrices appearing in the equations will be of a different order than for the two-span beam. In general, if there are n degrees of statical indeterminacy, the order of the flexibility matrix **F** will be $n \times n$, and the order of all the other matrices will be $n \times 1$. In the next article several examples of finding redundant actions by the flexibility method will be given.

2.3 Examples. In order to illustrate the application of the flexibility method to structures of various types, several examples are given in this article. In each example it is assumed that the object of the analysis is to calculate the values of certain selected redundants; hence, the problem is considered to be solved when the matrix **Q** is determined. The redundants are selected in each example primarily for illustrative purposes; however, many other choices of redundants are possible.

Example 1. The beam AB shown in Fig. 2-3a is fixed at both ends and is subjected to a concentrated load P and a couple M at the midpoint. It is assumed that the beam has constant flexural rigidity EI.

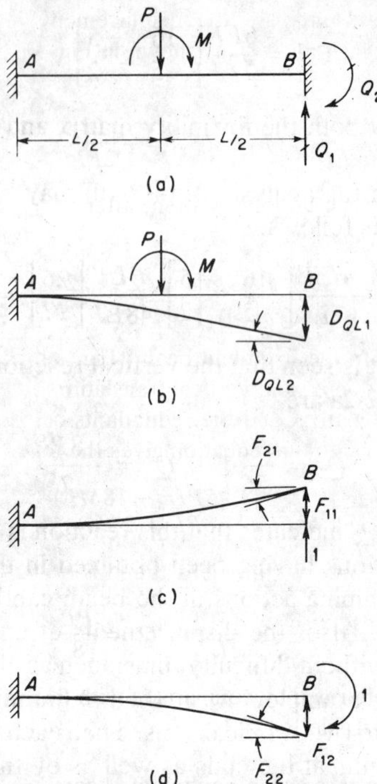

Fig. 2-3. Example 1: Fixed-end beam.

In order to begin the analysis by the flexibility method, two redundant actions must be selected. In this example the vertical reaction and the reactive moment at end B of the beam are selected as redundants and are denoted in Fig. 2-3a as Q_1 and Q_2, respectively. The redundant Q_1 is assumed to be positive in the upward direction, and Q_2 is assumed positive in the clockwise direction. Other possible choices for the redundants include the reactive moments at both ends and the bending moment and shearing force at any section along the beam.

For the redundants shown in Fig. 2-3a, the released structure consists of a cantilever beam (see Fig. 2-3b). The displacements in this beam corresponding to Q_1 and Q_2 and caused by the loads P and M are denoted by D_{QL1} and D_{QL2} in the figure. These displacements can be found with the aid of Table A-3 in the Appendix, and are as follows:

$$D_{QL1} = -\frac{5PL^3}{48EI} - \frac{3ML^2}{8EI}$$

$$D_{QL2} = \frac{PL^2}{8EI} + \frac{ML}{2EI}$$

Since D_{QL1} is downward, it is negative (opposite to Q_1); but D_{QL2} is positive because it is clockwise (same as Q_2). Therefore, the vector $\mathbf{D}_{QL}$ can be written as

$$\mathbf{D}_{QL} = \frac{L}{48EI}\begin{bmatrix}(-5PL^2 - 18ML)\\(6PL + 24M)\end{bmatrix}$$

The flexibility coefficients are the displacements of the released structure caused by unit values of Q_1 and Q_2, as shown in Figs. 2-3c and 2-3d. These coefficients are as follows:

$$F_{11} = \frac{L^3}{3EI} \qquad F_{12} = F_{21} = -\frac{L^2}{2EI} \qquad F_{22} = \frac{L}{EI}$$

The flexibility matrix **F** can now be written, after which its inverse can be determined; these matrices are:

$$\mathbf{F} = \frac{L}{6EI}\begin{bmatrix} 2L^2 & -3L \\ -3L & 6 \end{bmatrix} \qquad \mathbf{F}^{-1} = \frac{2EI}{L^3}\begin{bmatrix} 6 & 3L \\ 3L & 2L^2 \end{bmatrix}$$

The displacements in the fixed-end beam (Fig. 2-3a) corresponding to Q_1 and Q_2 are both zero, since there is no vertical translation and no rotation at support B. Therefore, $\mathbf{D}_Q$ is a null matrix, and the redundants can be found from Eq. (2-10). Substituting $\mathbf{F}^{-1}$ and $\mathbf{D}_{QL}$ into that equation gives the following:

$$\mathbf{Q} = -\frac{2EI}{L^3}\begin{bmatrix} 6 & 3L \\ 3L & 2L^2 \end{bmatrix}\frac{L}{48EI}\begin{bmatrix} (-5PL^2 - 18ML) \\ (6PL + 24M) \end{bmatrix} = \frac{1}{8L}\begin{bmatrix} (4PL + 12M) \\ (PL^2 + 2ML) \end{bmatrix}$$

Therefore, the reactive force and moment at end B of the beam (see Fig. 2-3a) are

$$Q_1 = \frac{P}{2} + \frac{3M}{2L} \qquad Q_2 = \frac{PL}{8} + \frac{M}{4}$$

These results can be confirmed by comparison with the formulas given in Appendix B (see Cases 1 and 2 of Table B-1).

Example 2. The three-span continuous beam shown in Fig. 2-4a has constant flexural rigidity EI and is acted upon by a uniform load w in span AB and concentrated loads P at the midpoints of spans BC and CD. Since the structure is statically indeterminate to the second degree, two redundant actions must be selected. In this example, the bending moments at joints B and C are chosen. When these moments are removed from the beam by inserting hinges at B and C, the released structure is seen to consist of three simple beams (Fig. 2-4b). The redundant moments Q_1 and Q_2 are shown acting in their positive directions in Fig. 2-4b. Each redundant consists of two couples, one acting on each adjoining span of the structure. For example, the left-hand couple in Q_1 acts on beam AB in the counterclockwise direction, while the right-hand couple in Q_1 acts on span BC in the clockwise direction. The positive direction of each Q corresponds to that of a bending moment which produces compression at the top of the beam. Therefore, a positive sign in the final solution for either Q_1 or Q_2 means that the redundant moment produces compression at the top of the beam; if negative, the redundant moment produces tension at the top of the beam.

The displacement corresponding to one of the redundant moments consists of the sum of two rotations, one in each adjoining span. For example, the displacement corresponding to Q_1 consists of the counterclockwise rotation at joint B of the right-hand end of member AB plus the clockwise rotation at joint B of the left-hand end of member BC. In a similar manner, the displacement corresponding to Q_2 is the sum of the two rotations at joint C.

The displacements D_{QL1} and D_{QL2} corresponding to Q_1 and Q_2, respectively, and caused by the loads acting on the released structure, are shown in Fig. 2-4c.

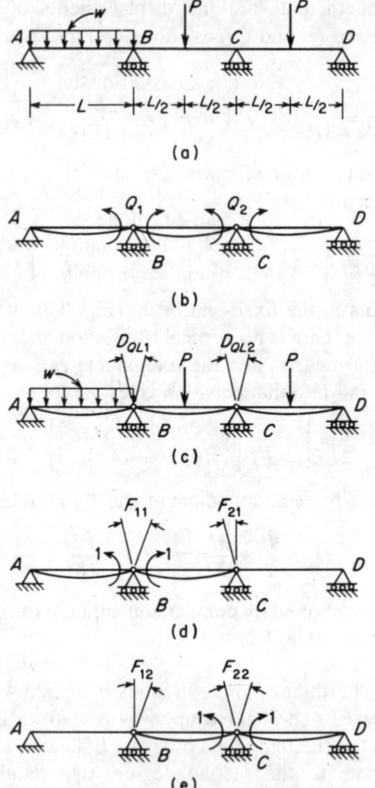

Fig. 2-4. Example 2: Continuous beam.

Since the counterclockwise rotation at end B of member AB due to the uniform load w is

$$\frac{wL^3}{24EI}$$

and since the clockwise rotation of end B of member BC due to the load P is

$$\frac{PL^2}{16EI}$$

it follows that the displacement D_{QL1} is

$$D_{QL1} = \frac{wL^3}{24EI} + \frac{PL^2}{16EI}$$

In a similar manner, the displacement D_{QL2} is seen to be

$$D_{QL2} = \frac{PL^2}{16EI} + \frac{PL^2}{16EI} = \frac{PL^2}{8EI}$$

Thus, the vector $\mathbf{D}_{QL}$ can be written as

$$\mathbf{D}_{QL} = \frac{L^2}{48EI}\begin{bmatrix}(2wL + 3P)\\ 6P\end{bmatrix}$$

The flexibility matrix must be found next. For this purpose, unit values of Q_1 and Q_2 are shown acting on the beams in Figs. 2-4d and 2-4e. The flexibility coefficient F_{11} (see Fig. 2-4d) is the sum of two rotations at joint B; one rotation is in span AB and the other is in span BC. Similarly, the coefficient F_{21} is the sum of the rotations at joint C. In Fig. 2-4d, however, the rotation in span CD is zero. Therefore, F_{21} is equal to the rotation in span BC alone. Similar comments apply to the flexibility coefficients shown in Fig. 2-4e. The rotations produced at the ends of a simply supported beam by a couple of unit value applied at one end of the beam are

$$\frac{L}{3EI} \quad \text{and} \quad \frac{L}{6EI}$$

at the near and far ends of the span, respectively, as given in Case 5, Table A-3. From these formulas the flexibility coefficients pictured in Figs. 2-4d and 2-4e can be obtained, as follows:

$$F_{11} = \frac{2L}{3EI} \qquad F_{12} = \frac{L}{6EI}$$

$$F_{21} = \frac{L}{6EI} \qquad F_{22} = \frac{2L}{3EI}$$

Therefore, the flexibility matrix is

$$\mathbf{F} = \frac{L}{6EI}\begin{bmatrix}4 & 1\\ 1 & 4\end{bmatrix}$$

and its inverse becomes

$$\mathbf{F}^{-1} = \frac{2EI}{5L}\begin{bmatrix}4 & -1\\ -1 & 4\end{bmatrix}$$

The displacements D_{Q1} and D_{Q2} in the original beam (Fig. 2-4a) corresponding to Q_1 and Q_2, respectively, are both zero because the beam is continuous across supports B and C. Therefore, the matrix $\mathbf{D}_Q$ is null and Eq. (2-10) can be used to determine the redundants, as shown:

$$\mathbf{Q} = -\frac{2EI}{5L}\begin{bmatrix}4 & -1\\ -1 & 4\end{bmatrix}\frac{L^2}{48EI}\begin{bmatrix}(2wL + 3P)\\ 6P\end{bmatrix} = -\frac{L}{120}\begin{bmatrix}(8wL + 6P)\\ (-2wL + 21P)\end{bmatrix}$$

Thus, the redundant bending moments Q_1 and Q_2 are given by the following formulas:

$$Q_1 = -\frac{wL^2}{15} - \frac{PL}{20} \qquad Q_2 = \frac{wL^2}{60} - \frac{7PL}{40}$$

Having obtained these moments at the supports, the remaining bending moments in the beam, as well as shearing forces and reactions, can be found by statics.

Example 3. The plane truss shown in Fig. 2-5a is statically indeterminate to the second degree. The horizontal reaction at the support B (positive to the right) and the axial force in bar AD (positive when tension) are selected as the redundants,

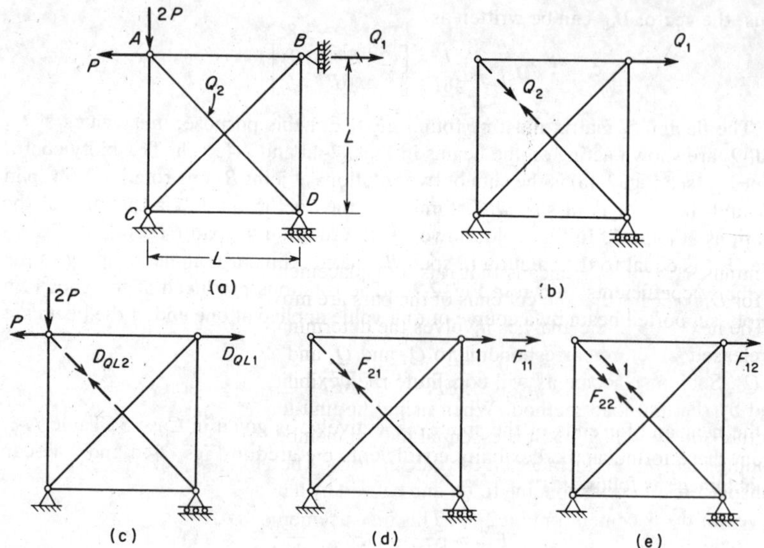

Fig. 2-5. Example 3: Plane truss.

resulting in the released structure shown in Fig. 2-5b. In the released structure the support at B has been removed, thereby releasing the reactive force Q_1, and the bar AD has been cut at some intermediate location in order to release the axial force Q_2 in the bar. Many other combinations of bar forces and reactions could be taken as redundants in this example, and each would give a different released structure. In all cases in which the force in a bar is selected as a redundant, the bar must be cut to obtain the released structure. The cut bar then remains as part of the released structure, since its deformations must be included in the calculation of displacements in the released structure. It should be observed that Q_2 consists of a pair of forces acting on the released structure. This type of redundant is analogous to those described in Example 2, where a redundant bending moment was considered as a pair of couples acting on the released structure (see Fig. 2-4b). From this discussion it follows that a displacement corresponding to Q_2 consists of a relative translation of the cut ends of bar AD. When the cut ends are displaced toward one another, the displacement is in the same direction as Q_2 and hence is positive. When the cut ends are displaced apart from one another, the displacement is negative.

The first step in the analysis consists of determining the displacements in the released structure corresponding to Q_1 and Q_2 and due to the loads P and $2P$ acting at joint A. These displacements are denoted as D_{QL1} and D_{QL2} and are represented by vectors in Fig. 2-5c. This manner of representing the displacements is used in lieu of drawing the structure in its displaced configuration, which may become difficult for complicated structures. The determination of these displacements may be carried out by the unit-load method, as illustrated in Example 1 of Art. A.2 of the Appendix. Assuming that all members of the truss have the same axial rigidity EA, it is found that the displacements due to the loads P and $2P$ are

$$D_{QL1} = -\frac{PL}{EA}(1 + 2\sqrt{2}) = -3.828\frac{PL}{EA}$$

$$D_{QL2} = -2\frac{PL}{EA}$$

Therefore, the vector $\mathbf{D}_{QL}$ is

$$\mathbf{D}_{QL} = -\frac{PL}{EA}\begin{bmatrix} 3.828 \\ 2 \end{bmatrix}$$

The minus sign for D_{QL1} means that it is a displacement to the left, and the minus sign for D_{QL2} means that the cut ends of the bars are moved apart from one another.

The next step in the analysis involves the determination of the displacements in the released structure corresponding to Q_1 and Q_2 and caused by unit values of Q_1 and Q_2. Such displacements will constitute the flexibility matrix $\mathbf{F}$ and also can be found by the unit-load method. When using the unit-load method for the released structure shown in Fig. 2-5, it is essential that all members of the released structure, including bar AD, be included in the calculations. The flexibility coefficient F_{11} is the displacement corresponding to Q_1 and caused by a unit value of Q_1 and is shown as a vector displacement in Fig. 2-5d. This displacement is

$$F_{11} = \frac{L}{EA}(1 + 2\sqrt{2}) = 3.828\frac{L}{EA}$$

The flexibility coefficient F_{21} is the displacement corresponding to Q_2 and due to a unit value of Q_1 (see Fig. 2-5d). Similarly, the coefficients F_{12} and F_{22} represent displacements in the released structure of Fig. 2-5e. When all of these displacements are obtained, the results are as follows:

$$F_{12} = F_{21} = \frac{L}{2EA}(4 + \sqrt{2}) = 2.707\frac{L}{EA}$$

$$F_{22} = \frac{2L}{EA}(1 + \sqrt{2}) = 4.828\frac{L}{EA}$$

Finally, the flexibility matrix can be formed and its inverse determined:

$$\mathbf{F} = \frac{L}{EA}\begin{bmatrix} 3.828 & 2.707 \\ 2.707 & 4.828 \end{bmatrix}$$

$$\mathbf{F}^{-1} = \frac{EA}{L}\begin{bmatrix} 0.4328 & -0.2426 \\ -0.2426 & 0.3431 \end{bmatrix}$$

Assuming that there are no support displacements in the truss, the redundants Q can be found by means of Eq. (2-10):

$$\mathbf{Q} = \frac{EA}{L}\begin{bmatrix} 0.4328 & -0.2426 \\ -0.2426 & 0.3431 \end{bmatrix}\frac{PL}{EA}\begin{bmatrix} 3.828 \\ 2 \end{bmatrix} = P\begin{bmatrix} 1.172 \\ -0.243 \end{bmatrix}$$

Therefore, the horizontal reactive force at B (Fig. 2-5a) is

$$Q_1 = 1.172P$$

and the axial force in bar AD is

$$Q_2 = -0.243P$$

The minus sign for Q_2 shows that the member is in compression. From the above results, one can calculate the remaining reactions and bar forces by statics.

Now assume that when the loads P and $2P$ act on the truss, support B moves a small distance s horizontally to the left. Therefore, the displacement D_{Q1} (see Eq. 2-7) is equal to minus s. The displacement D_{Q2} is zero, since it represents the relative displacement of the cut ends of bar AD in the original truss. The redundants $\mathbf{Q}$ can now be found from Eq. (2-9), as follows:

$$\mathbf{Q} = \frac{EA}{L}\begin{bmatrix} 0.4328 & -0.2426 \\ -0.2426 & 0.3431 \end{bmatrix}\left\{\begin{bmatrix} -s \\ 0 \end{bmatrix} + \frac{PL}{EA}\begin{bmatrix} 3.828 \\ 2 \end{bmatrix}\right\}$$

$$= \begin{bmatrix} -0.433\dfrac{sEA}{L} + 1.172P \\ 0.243\dfrac{sEA}{L} - 0.243P \end{bmatrix}$$

Therefore, the horizontal reaction at B is

$$Q_1 = -0.433\frac{sEA}{L} + 1.172P$$

and the force in bar AD is

$$Q_2 = 0.243\frac{sEA}{L} - 0.243P$$

In this example there was a redundant action (Q_1) corresponding to the support displacement at joint B. Of course, if other redundants had been selected, such as the forces in the two diagonal bars, the displacement at support B could not have been included in the vector $\mathbf{D_Q}$. Then it would be necessary to account for the support displacement by another means, such as that described later in Art. 2.4.

Example 4. The plane frame shown in Fig. 2-6a has fixed supports at A and C and is acted upon by the vertical load P at the midpoint of member AB. It is desired to analyze the frame, taking into account the effects of both flexural and axial deformations. The inclusion of axial effects is for illustrative purposes only; normally, in a frame of this type, only flexural effects need be considered, and the analysis would be simplified slightly. The members of the frame have constant flexural rigidity EI and constant axial rigidity EA. The structure is statically indeterminate to the third degree, and a suitable released structure (see Fig. 2-6b) is obtained by cutting the frame at joint B, thereby releasing two forces and a bending moment. These released actions are the redundants Q_1, Q_2, and Q_3, as shown in Fig. 2-6b. A displacement in the released structure corresponding to Q_1 consists of the horizontal translation of end B of member AB (taken positive to the right) plus the horizontal translation of end B of member BC (taken positive to the left). In other words, a displacement corresponding to Q_1 consists of the sum of two translations and represents a relative displacement between the two points labeled B in Fig. 2-6b. In a similar manner the displacements corresponding to Q_2 and Q_3 can be defined as the sum of two vertical translations and two rotations, respectively, at joint B.

The displacements in the released structure caused by the load P and corresponding to Q_1, Q_2, and Q_3 are shown in Fig. 2-6c. For example, the displacement D_{QL1} is shown in the figure to consist of two horizontal translations, as described in

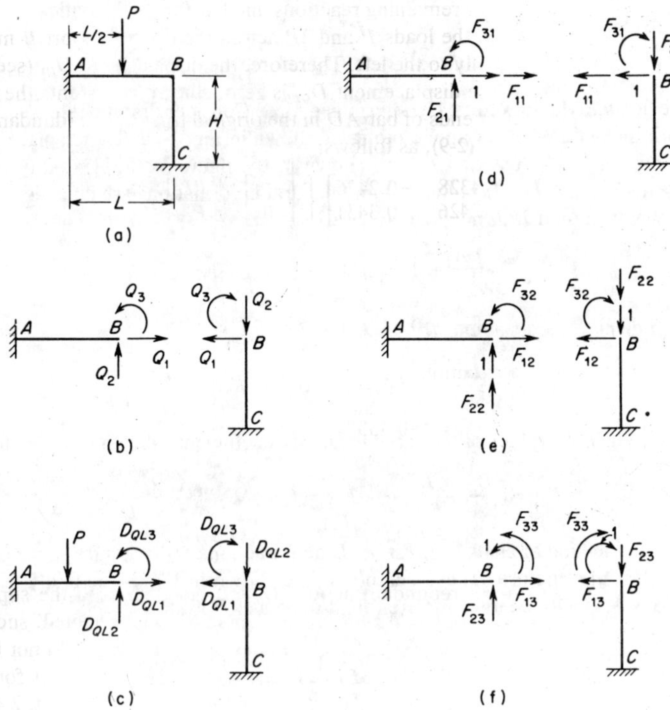

Fig. 2-6. Example 4: Plane frame.

the preceding paragraph. Similarly, the displacements D_{QL2} and D_{QL3} are shown as two vertical translations and two rotations, respectively. It is not difficult to calculate these displacements due to the force P, inasmuch as the released structure consists of two cantilever members. First, taking the member AB in Fig. 2-6c, it is seen that the displacements at end B are as follows:

$$(D_{QL1})_{AB} = 0 \qquad (D_{QL2})_{AB} = -\frac{5PL^3}{48EI} \qquad (D_{QL3})_{AB} = -\frac{PL^2}{8EI}$$

These expressions are based upon only the flexural deformations of member AB because there are no axial deformations. Secondly, the member BC in Fig. 2-6c must be considered. In this example there is no load on member BC and hence no displacements at end B; therefore,

$$(D_{QL1})_{BC} = (D_{QL2})_{BC} = (D_{QL3})_{BC} = 0$$

The final displacements caused by the load P can now be obtained by combining the above results:

$$D_{QL1} = 0 \qquad D_{QL2} = -\frac{5PL^3}{48EI} \qquad D_{QL3} = -\frac{PL^2}{8EI}$$

Therefore, the matrix $\mathbf{D}_{QL}$ is

$$\mathbf{D}_{QL} = \frac{PL^2}{48EI}\begin{bmatrix} 0 \\ -5L \\ -6 \end{bmatrix}$$

The flexibility matrix must be determined next. Consider first the released structure with the action $Q_1 = 1$ applied to it, as shown in Fig. 2-6d. The displacements corresponding to Q_1, Q_2, and Q_3 are shown in the figure as the flexibility coefficients F_{11}, F_{21}, and F_{31}. If both axial and flexural deformations are considered, the displacements at end B of member AB are

$$(F_{11})_{AB} = \frac{L}{EA} \qquad (F_{21})_{AB} = 0 \qquad (F_{31})_{AB} = 0$$

Also, the displacements at end B of member BC are

$$(F_{11})_{BC} = \frac{H^3}{3EI} \qquad (F_{21})_{BC} = 0 \qquad (F_{31})_{BC} = -\frac{H^2}{2EI}$$

Therefore, the final values of the three flexibility coefficients shown in Fig. 2-6d are

$$F_{11} = \frac{L}{EA} + \frac{H^3}{3EI} \qquad F_{21} = 0 \qquad F_{31} = -\frac{H^2}{2EI}$$

Next, the same kind of analysis must be made for actions $Q_2 = 1$ and $Q_3 = 1$ acting on the released structure. These conditions are shown in Figs. 2-6e and 2-6f, and the various flexibility coefficients are found to be as follows:

$$F_{12} = 0 \qquad F_{22} = \frac{L^3}{3EI} + \frac{H}{EA} \qquad F_{32} = \frac{L^2}{2EI}$$

$$F_{13} = -\frac{H^2}{2EI} \qquad F_{23} = \frac{L^2}{2EI} \qquad F_{33} = \frac{L}{EI} + \frac{H}{EI}$$

Finally, the flexibility matrix can be assembled:

$$\mathbf{F} = \begin{bmatrix} \frac{L}{EA} + \frac{H^3}{3EI} & 0 & -\frac{H^2}{2EI} \\ 0 & \frac{L^3}{3EI} + \frac{H}{EA} & \frac{L^2}{2EI} \\ -\frac{H^2}{2EI} & \frac{L^2}{2EI} & \frac{L}{EI} + \frac{H}{EI} \end{bmatrix} \qquad \text{(a)}$$

If axial deformations were to be neglected in the analysis, the two fractions in **F** containing the axial rigidity EA in the denominators would be omitted.

The next step in the solution is to obtain the inverse of the flexibility matrix and then to substitute it, as well as the matrix $\mathbf{D}_{QL}$, into Eq. (2-10). It is possible to use Eq. (2-10) in this example because the displacements D_Q are all equal to zero, inasmuch as joint B in the original frame (Fig. 2-6a) is a rigid connection. However, the inverse of the flexibility matrix **F** given above cannot be obtained conveniently in literal form, as was done in the preceding examples where **F** was a 2×2 matrix. For this reason the flexibility matrix in Eq. (a) will be simplified by letting $H = L$ and by introducing a nondimensional parameter γ, defined as

$$\gamma = \frac{I}{AL^2}$$

With this notation the flexibility matrix can be rewritten as

$$\mathbf{F} = \frac{L}{6EI}\begin{bmatrix} 2L^2(1+3\gamma) & 0 & -3L \\ 0 & 2L^2(1+3\gamma) & 3L \\ -3L & 3L & 12 \end{bmatrix} \qquad \text{(b)}$$

Note that when the cross-sectional area A becomes large, the parameter γ tends to zero. In that case the factor $(1 + 3\gamma)$ in Eq. (b) becomes unity.

The inverse of the flexibility matrix **F**, as given in Eq. (b), is found to be

$$\mathbf{F}^{-1} = \frac{EI}{2L^3 a_1 a_2}\begin{bmatrix} 3a_3 & -9 & 6La_1 \\ -9 & 3a_3 & -6La_1 \\ 6La_1 & -6La_1 & 4L^2a_1^2 \end{bmatrix}$$

This matrix contains three additional dimensionless parameters, which are defined as follows:

$$a_1 = 1 + 3\gamma \qquad a_2 = 1 + 12\gamma \qquad a_3 = 5 + 24\gamma$$

Substitution of $\mathbf{F}^{-1}$ and $\mathbf{D}_{QL}$ into Eq. (2-10) produces

$$\mathbf{Q} = \frac{-P}{32a_1a_2}\begin{bmatrix} 3a_4 \\ -a_5 \\ 2La_1a_4 \end{bmatrix}$$

where

$$a_4 = 1 - 12\gamma \qquad a_5 = 13 + 84\gamma$$

In terms of γ, the redundants are

$$Q_1 = -\frac{3P}{32}\frac{1-12\gamma}{(1+3\gamma)(1+12\gamma)}$$

$$Q_2 = \frac{13P}{32}\frac{1+84\gamma/13}{(1+3\gamma)(1+12\gamma)}$$

$$Q_3 = -\frac{PL}{16}\frac{1-12\gamma}{1+12\gamma}$$

The minus signs for Q_1 and Q_3 show that these actions are opposite in direction to the positive senses shown in Fig. 2-6b.

For typical plane frames the magnitude of γ is of the order 10^{-3}; so the factors $1 - 12\gamma$, etc., are approximately 1.0. If axial deformations are neglected altogether (that is, if $\gamma = 0$), the redundants become

$$Q_1 = -\frac{3P}{32} \qquad Q_2 = \frac{13P}{32} \qquad Q_3 = -\frac{PL}{16}$$

These values are sufficiently accurate for most practical purposes.

Example 5. The grid structure shown in Fig. 2-7a is in a horizontal plane (x-z plane) and carries a load P acting in the vertical direction. The supports of the grid at A and C are fixed, and the members AB and BC have length L, flexural rigidity EI, and torsional rigidity GJ (see Appendix A for definition of torsional rigidity). The redundants that are chosen for this example are released by cutting the grid at joint B (Fig. 2-7b), thereby giving a released structure in the form of two cantilever beams. Each redundant consists of a pair of actions, and these actions

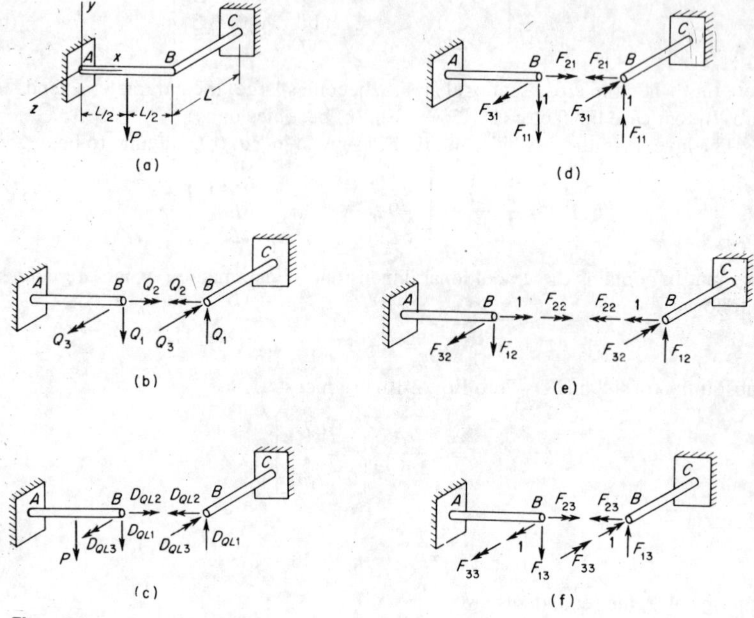

Fig. 2-7. Example 5: Grid.

are shown in their assumed positive senses in Fig. 2-7b. There are no other internal actions at joint B, because in a grid with only vertical loads there are no horizontal forces between members and no couples about a vertical axis.

When the load P is applied to the released structure (Fig. 2-7c), displacements D_{QL} are produced. These displacements are shown in their positive directions in the figure. Note that a translational displacement is denoted by a single-headed arrow, while a rotation is denoted by a double-headed arrow. This representation is analogous to the convention that is used when indicating forces and couples. The displacements shown in Fig. 2-7c can be obtained without difficulty, and then the matrix $\mathbf{D}_{QL}$ can be formed:

$$\mathbf{D}_{QL} = \frac{PL^2}{48EI}\begin{bmatrix} 5L \\ 0 \\ -6 \end{bmatrix}$$

The flexibility coefficients are the displacements shown in Figs. 2-7d, 2-7e, and 2-7f. For these figures, it is assumed that unit values of the redundants Q_1, Q_2, and Q_3 are applied one at a time to the released structure. The desired coefficients can be obtained from the figures, and the flexibility matrix is found to be

$$\mathbf{F} = \begin{bmatrix} \dfrac{2L^3}{3EI} & \dfrac{L^2}{2EI} & -\dfrac{L^2}{2EI} \\ \dfrac{L^2}{2EI} & \dfrac{L}{EI} + \dfrac{L}{GJ} & 0 \\ -\dfrac{L^2}{2EI} & 0 & \dfrac{L}{EI} + \dfrac{L}{GJ} \end{bmatrix}$$

This matrix can be written in a simpler form by introducing a nondimensional parameter ρ, defined as the ratio of the flexural and torsional rigidities:

$$\rho = \frac{EI}{GJ}$$

With this notation the flexibility matrix can be put into the following form:

$$\mathbf{F} = \frac{L}{6EI}\begin{bmatrix} 4L^2 & 3L & -3L \\ 3L & 6(1+\rho) & 0 \\ -3L & 0 & 6(1+\rho) \end{bmatrix}$$

The inverse of **F** is

$$\mathbf{F}^{-1} = \frac{EI}{2L^3 b_1 b_2}\begin{bmatrix} 12b_1^2 & -6Lb_1 & 6Lb_1 \\ -6Lb_1 & L^2 b_3 & -3L^2 \\ 6Lb_1 & -3L^2 & L^2 b_3 \end{bmatrix}$$

in which the following additional nondimensional parameters are used:

$$b_1 = 1 + \rho \qquad b_2 = 1 + 4\rho \qquad b_3 = 5 + 8\rho$$

Finally, substitution into Eq. (2-10) yields the vector of redundants:

$$\mathbf{Q} = \frac{-P}{16b_1 b_2}\begin{bmatrix} 2b_1 b_4 \\ -Lb_4 \\ -3L\rho \end{bmatrix}$$

where

$$b_4 = 2 + 5\rho$$

Thus, the redundants are as follows:

$$Q_1 = -\frac{P}{8}\frac{2+5\rho}{1+4\rho}$$

$$Q_2 = \frac{PL}{16}\frac{2+5\rho}{(1+\rho)(1+4\rho)}$$

$$Q_3 = \frac{3PL}{16}\frac{\rho}{(1+\rho)(1+4\rho)}$$

If the members AB and BC are torsionally very weak, then ρ can be considered as infinitely large. The above formulas then give the following values for the redundants (after dividing the numerators and denominators by ρ):

$$Q_1 = -\frac{5P}{32} \qquad Q_2 = Q_3 = 0$$

This result could also be obtained by assuming that a spherical hinge exists at B which transmits vertical force but not moment. Such a grid would be statically indeterminate to the first degree.

2.4 Temperature Changes, Prestrains, and Support Displacements. It is frequently necessary to include in the analysis of a structure not only the effects of loads but also the effects of temperature changes, prestrains of members, and displacements of one or more supports. In most cases these effects can be incorporated into the analysis by including them in the calculation of displacements in the released structure. As an example, consider the effects of temperature changes. If the changes are assumed to occur in the released structure, there will be displacements corresponding to the redundant actions Q. These displacements can be identified by the symbol D_{QT}, which is analogous to the symbol D_{QL} used previously to represent displacements in the released structure corresponding to the redundants and caused by the loads. The temperature displacements D_{QT} in the released structure may be due to either uniform changes in temperature or to differential changes in temperature, as discussed in Appendix A.

When the matrix $\mathbf{D}_{QT}$ of displacements due to temperature changes has been obtained, it can be added to the matrix $\mathbf{D}_{QL}$ of displacements due to loads to give the sum of all displacements in the released structure. Then these total displacements can be used in the equation of superposition (see Eq. 2-8) in place of $\mathbf{D}_{QL}$ alone. Therefore, the superposition equation becomes

$$\mathbf{D}_Q = \mathbf{D}_{QL} + \mathbf{D}_{QT} + \mathbf{FQ}$$

which can be solved for the vector $\mathbf{Q}$ of redundants as before. An illustration of the calculation of the vector $\mathbf{D}_{QT}$ to be used in this equation is given in a later example.

The effects of prestrains in any member of the structure can be handled in a manner analogous to that for temperature changes. By prestrain of a member is meant an initial deformation of the member due to any of various causes. For example, a truss member may be fabricated with a length greater or less than the theoretical length of the member, or a beam may be fabricated with an initial curvature. It can be seen intuitively that a truss member having prestrains consisting of an initial elongation will produce the same effects in the truss as if the member had been heated uniformly to a temperature that would produce the same increase in length. Thus, the method of analysis for prestrain effects is similar to that for temperature changes. The first step is to assume that the prestrains occur in the released structure. Then the displacements in the released structure corresponding to the redundants must be found. These displacements are denoted D_{QP}, signifying that they correspond to the redundants and are due to prestrain effects. The matrix $\mathbf{D}_{QP}$ for the prestrain displacements can then be added to the matrices $\mathbf{D}_{QL}$ and $\mathbf{D}_{QT}$ to give the sum of all displacements in the released structure. Then the sum of all displacements is included in the equation of superposition, as follows:

$$\mathbf{D}_Q = \mathbf{D}_{QL} + \mathbf{D}_{QT} + \mathbf{D}_{QP} + \mathbf{FQ}$$

As before, the superposition equation can be solved for the matrix of redundants $\mathbf{Q}$. An example involving the calculation of prestrain displacements will be given later.

Lastly, consider the possibility of known displacements occurring at the restraints (or supports) of a structure. There are two possibilities to be considered, depending upon whether or not the restraint displacement corresponds to one of the redundant actions Q. If the restraint displacement does correspond to a redundant, its effects can be taken into account by including the displacement in the vector $\mathbf{D_Q}$ of actual displacements in the structure. This procedure was discussed before in Art. 2.2 and was illustrated in Art. 2.3 in an example of a statically indeterminate truss (see Example 3). In a more general situation, however, there will be restraint displacements that do not correspond to any of the selected redundants. In that event, the effects of such restraint displacements must be incorporated in the analysis of the released structure in the same manner as in the case of temperature and prestrain effects. When the restraint displacements are assumed to occur in the released structure, there will be displacements D_{QR} corresponding to the redundants Q. When these displacements have been found, the matrix $\mathbf{D_{QR}}$ can be obtained. Then this matrix can be added to the other matrices representing displacements in the released structure.

The combination of all matrices representing displacements in the released structure will be denoted $\mathbf{D_{QC}}$ in future discussions. It can be expressed as follows:

$$\mathbf{D_{QC}} = \mathbf{D_{QL}} + \mathbf{D_{QT}} + \mathbf{D_{QP}} + \mathbf{D_{QR}} \tag{2-11}$$

Thus, the matrix $\mathbf{D_{QC}}$ contains displacements due to all causes, including loads, temperature changes, prestrain effects, and restraint displacements. With this notation, a generalized form of the superposition equation for the flexibility method becomes

$$\mathbf{D_Q} = \mathbf{D_{QC}} + \mathbf{FQ} \tag{2-12}$$

The superposition equation used previously (Eq. 2-8) can be considered as a special case of Eq. (2-12). When the latter equation is solved for $\mathbf{Q}$, the result is

$$\mathbf{Q} = \mathbf{F}^{-1}(\mathbf{D_Q} - \mathbf{D_{QC}}) \tag{2-13}$$

and this equation may be used in place of Eq. (2-9) when causes other than loads must be considered. Of course, in any particular analysis it is not likely that all of the matrices given in Eq. (2-11) will be of interest. Some examples of the use of the above equations will now be given.

Example 1. In order to illustrate the analysis of a structure when temperature changes are present, consider the two-span beam ABC in Fig. 2-8a. The beam is assumed to be subjected to a linear temperature gradient such that the top surface of the beam is at temperature T_2 while the lower surface is at temperature T_1. Previously, this same beam was analyzed for the effects of loads (see Fig. 2-2), and in that solution the reactions at supports B and C were taken as the redundants Q_1 and Q_2. These same redundants will be used again in solving for the thermal effects; therefore, the inverse of the flexibility matrix may be taken from the former solution (see Art. 2.2):

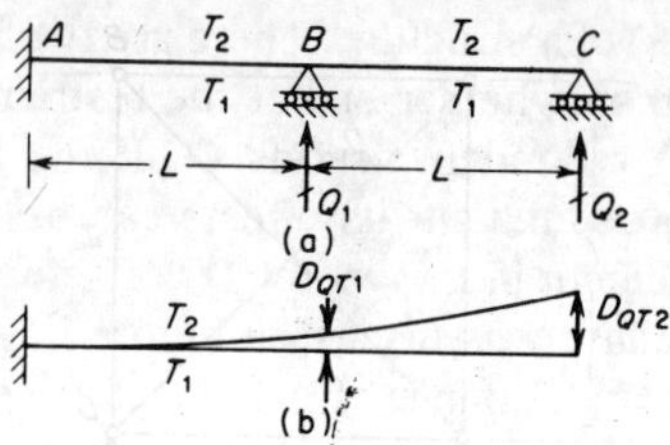

Fig. 2-8. Example 1: Continuous beam of Fig. 2-2.

$$\mathbf{F}^{-1} = \frac{6EI}{7L^3}\begin{bmatrix} 16 & -5 \\ -5 & 2 \end{bmatrix}$$

Also, the vector $\mathbf{D}_Q$ (see Eq. 2-13) giving the actual displacements in the beam is a null vector ($\mathbf{D}_Q = \mathbf{0}$).

The released structure for the beam *ABC* is the cantilever beam shown in Fig. 2-8b. If the temperature T_1 is greater than T_2, the released beam will deflect upward as indicated in the figure. The displacements corresponding to the redundants are denoted D_{QT1} and D_{QT2}. These displacements can be calculated by the unit-load method described in Appendix A (see Example 4 in Art. A.2). Thus, the displacements D_{QT1} and D_{QT2} are found to be

$$\frac{\alpha(T_1 - T_2)L^2}{2d} \quad \text{and} \quad \frac{2\alpha(T_1 - T_2)L^2}{d}$$

respectively. In these expressions α is the coefficient of thermal expansion for the material, and d is the depth of the beam. The vector $\mathbf{D}_{QT}$ is

$$\mathbf{D}_{QT} = \frac{\alpha(T_1 - T_2)L^2}{2d}\begin{bmatrix} 1 \\ 4 \end{bmatrix}$$

If only the effects of temperature are considered in the analysis of the beam, the vector $\mathbf{D}_{QT}$ becomes the vector $\mathbf{D}_{QC}$ in Eq. (2-13). Then the redundants as found from that equation are:

$$\mathbf{Q} = \frac{3EI\alpha(T_1 - T_2)}{7Ld}\begin{bmatrix} 4 \\ -3 \end{bmatrix}$$

The signs on the elements of $\mathbf{Q}$ show that the redundant reaction Q_1 is upward when T_1 is greater than T_2, while the redundant Q_2 is downward. If T_1 is less than T_2, the directions of the redundants will be reversed. Of course, if T_1 is equal to T_2, the redundants are zero.

If the combined effects of both loads and temperature are desired, the matrix $\mathbf{D}_{QC}$ in Eq. (2-13) is taken as the sum of $\mathbf{D}_{QT}$ (obtained above) and $\mathbf{D}_{QL}$ (obtained in Art. 2.2). Values of the redundants under the combined conditions will be the sum of the values obtained for temperature changes and loads taken separately.

Example 2. As a second example illustrating the effects of a temperature change, consider the plane truss shown previously in Fig. 2-5a and assume that bar *BD* has its temperature increased uniformly by an amount T. The resulting elongation will produce displacements D_{QT} in the released structure corresponding to the redundants. These displacements, denoted D_{QT1} and D_{QT2}, are shown in Fig. 2-9 in

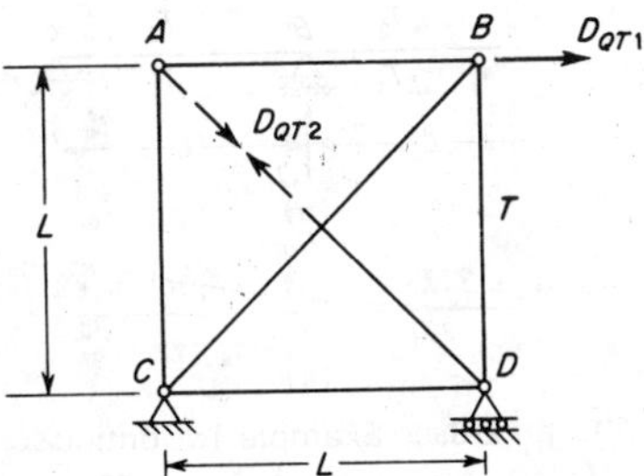

Fig. 2-9. Example 2: Plane truss of Fig. 2-5.

their positive directions (compare with Fig. 2-5c, which shows the displacements in the released structure due to loads). The displacements D_{QT} can be readily found by the unit-load method, as illustrated in Example 2 of Art. A.2. By this means the vector $\mathbf{D}_{QT}$ is found to be:

$$\mathbf{D}_{QT} = \alpha LT \begin{bmatrix} -1 \\ -\dfrac{1}{\sqrt{2}} \end{bmatrix}$$

The temperature effects can now be incorporated into the analysis of the truss by including the matrix $\mathbf{D}_{QT}$ in the calculation of $\mathbf{D}_{QC}$ (see Eq. 2-11).

If the member BD of the truss in Fig. 2-5a were fabricated with a length $L + e$ instead of L, the analysis could be handled in the same manner as shown above for a temperature change in the bar. The only difference is that the prestrain elongation e would replace the temperature elongation αLT for bar BD. Therefore, the vector of prestrain displacements in the released structure would be (see Example 2, Art. A.2):

$$\mathbf{D}_{QP} = e \begin{bmatrix} -1 \\ -\dfrac{1}{\sqrt{2}} \end{bmatrix}$$

A prestrain of any other bar could be handled in a similar fashion.

Example 3. This example illustrates how restraint displacements can be taken into account. Refer again to the two-span beam shown in Fig. 2-2a and assume that the following two support displacements occur. Joint A undergoes a known rotation in the clockwise direction of β radians, and joint B is displaced downward by a distance s. The displacement at B corresponds to one of the redundants and therefore is accounted for in the vector $\mathbf{D}_Q$, which now becomes

$$\mathbf{D}_Q = \begin{bmatrix} -s \\ 0 \end{bmatrix}$$

The minus sign is required in the first element inasmuch as Q_1 is positive when upward. The restraint displacement at support A is included in the analysis by means of the matrix $\mathbf{D}_{QR}$ (see Eq. 2-11). This matrix consists of the displacements in the released structure (see Fig. 2-10) when joint A is rotated through the clockwise angle β. Therefore, the displacements D_{QR1} and D_{QR2} corresponding to Q_1 and Q_2 are

$$-\beta L \quad \text{and} \quad -2\beta L$$

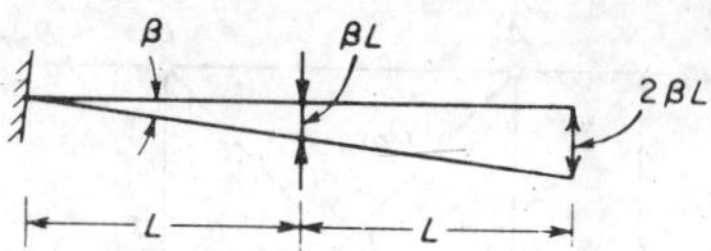

Fig. 2-10. Example 3: Support rotation in released structure.

respectively. These terms are negative because they are downward. Thus, the matrix $\mathbf{D}_{QR}$ is

$$\mathbf{D}_{QR} = \beta L \begin{bmatrix} -1 \\ -2 \end{bmatrix}$$

Finally, this matrix can be included in the calculation of $\mathbf{D}_{QC}$, after which Eq. (2-13) can be solved for the redundants. As in the other illustrations, the values of the redundants due to the combined effects will be equal to the sum of the values obtained separately.

2.5 Joint Displacements, Member End-Actions, and Support Reactions. In the preceding articles the emphasis was on finding redundant actions by the flexibility method. The redundant actions may be either internal stress resultants (such as axial forces and bending moments) or external reactions at points of support. In all cases it is possible to find other actions in the structure by using principles of static equilibrium after the redundants have been calculated. Such calculations would normally include the support reactions and the actions at the ends of each member (member end-actions). Furthermore, when all actions in the structure are available, it is possible to calculate all displacements. Usually the displacements of primary interest are the translations and rotations of the joints.

Instead of following the procedure described above, it is more systematic to incorporate the calculations for the joint displacements, member end-actions, and support reactions directly into the basic computations for the flexibility method. That is, the task of finding the various actions and displacements that are of interest can be performed in parallel with the computations for finding the redundants instead of postponing them as separate calculations to be performed after the redundants are determined.

In order to demonstrate the procedure for making a complete analysis of a structure, the two-span beam example given previously in Fig. 2-2 will now be extended. The beam to be analyzed is shown again in Fig. 2-11a. Assume that it is desired to calculate not only the redundants Q_1 and Q_2 for the beam, but also the joint displacements, member end-actions, and support reactions. The joint displacements in a structure will be denoted by the general symbol D_J, and numerical subscripts will be used to identify the individual joint displacements. For example, in the beam of Fig. 2-11a the two joint displacements to be found are the rotations at joints B and C. These displacements will be denoted D_{J1} and D_{J2}, respectively, and are assumed to be positive when counterclockwise, as shown in the figure.

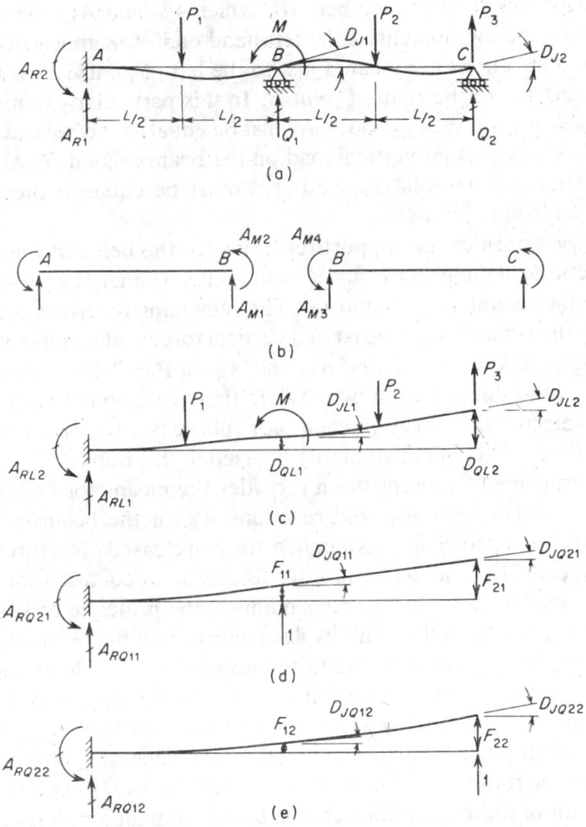

Fig. 2-11. Joint displacements, member end-actions, and support reactions.

The member end-actions are the couples and forces that act at the ends of a member when it is considered to be isolated from the remainder of the structure. For the beam under consideration, the end-actions are the bending moments and shearing forces at the ends of the members, as shown in Fig. 2-11b. These end-actions must be evaluated according to a specified sign convention, which may be either a deformation sign convention (related to how the member is deformed) or a statical sign convention (related to the direction of the action in space). The positive directions shown in Fig. 2-11b are based upon a statical sign convention that upward forces and counterclockwise moments are positive. In general, the end-actions are denoted by the symbol A_M and are distinguished from one another by numerical subscripts. In the example of Fig. 2-11 there are eight end-actions. However, it will be assumed arbitrarily that only the four actions labeled A_{M1}, A_{M2}, A_{M3}, and A_{M4} in Fig. 2-11b are to be calculated. The end-actions A_{M1} and A_{M2} are the shearing force and bending moment at the right-hand end of member AB, while A_{M3} and A_{M4} are the shearing force and bending moment at the left-hand end of member BC. Thus, the first two end-actions are located just to the left of joint B, and the last two are located just to the right of joint B. In this particular example, the sum of the shearing forces A_{M1} and A_{M3} must be equal to the redundant reaction Q_1 because there is no vertical load on the beam at joint B. Also, the sum of the bending moments A_{M2} and A_{M4} must be equal to the moment M acting as a load at joint B.

Finally, consider the support reactions for the beam in Fig. 2-11a. The two reactions at supports B and C will be determined automatically since they are the redundants Q_1 and Q_2. The remaining reactions, denoted generally by the symbol A_R, consist of a vertical force and a moment at support A. These reactions are labeled A_{R1} and A_{R2} in Fig. 2-11a and are assumed positive in the directions shown. While the sign convention for reactions may be selected arbitrarily in each particular case, the positive directions shown in the figure will customarily be used in this book.

The principle of superposition provides the means for finding displacements D_J, end-actions A_M, and reactions A_R for the beam in Fig. 2-11a. Previously, this principle was applied to the released structures shown in Figs. 2-11c, 2-11d, and 2-11e in order to obtain an equation for the redundants Q (see Eq. 2-8). In a similar manner, the principle of superposition may be used to obtain the joint displacements D_J in the beam of Fig. 2-11a. In order to accomplish this result, it is necessary to evaluate the displacements in the released structure (Figs. 2-11c, 2-11d, and 2-11e) corresponding to the displacements D_J. In the released structure subjected to the loads, these displacements are denoted by the general symbol D_{JL} and, in particular, the rotations at joints B and C are labeled D_{JL1} and D_{JL2}, respectively. Both of these quantities can be found from an analysis of the cantilever beam in Fig. 2-11c.

Next, the released structure subjected to unit values of the redundants must be considered (Figs. 2-11d and 2-11e). The displacements corresponding to D_J are denoted as D_{JQ}, in which the letter Q is used to indicate that these joint displacements are caused by unit values of the redundants. Consider, for example, the released structure subjected to a unit value of the redundant Q_1 (Fig. 2-11d). The joint displacements caused by this load are denoted D_{JQ11} and D_{JQ21}, in which the first numerical subscript identifies the particular displacement being considered and the second subscript denotes the redundant which is causing the displacement. In the same manner, joint displacements D_{JQ12} and D_{JQ22} caused by a unit value of Q_2 are shown in Fig. 2-11e.

The principle of superposition may now be used to obtain the displacements D_J in the actual beam. Superimposing the displacements from the beams in Figs. 2-11c, 2-11d, and 2-11e gives the displacements in the beam of Fig. 2-11a:

$$D_{J1} = D_{JL1} + D_{JQ11}Q_1 + D_{JQ12}Q_2$$
$$D_{J2} = D_{JL2} + D_{JQ21}Q_1 + D_{JQ22}Q_2$$

These equations can be expressed in simpler form by the following matrix equation:

$$\mathbf{D_J} = \mathbf{D_{JL}} + \mathbf{D_{JQ}Q} \tag{2-14}$$

in which the various matrices are

$$\mathbf{D_J} = \begin{bmatrix} D_{J1} \\ D_{J2} \end{bmatrix} \qquad \mathbf{D_{JL}} = \begin{bmatrix} D_{JL1} \\ D_{JL2} \end{bmatrix}$$
$$\mathbf{D_{JQ}} = \begin{bmatrix} D_{JQ11} & D_{JQ12} \\ D_{JQ21} & D_{JQ22} \end{bmatrix} \qquad \mathbf{Q} = \begin{bmatrix} Q_1 \\ Q_2 \end{bmatrix}$$

Of course, in a more general situation the matrices may be of greater order than in this example. If there are j joint displacements to be obtained, the order of the vectors $\mathbf{D_J}$ and $\mathbf{D_{JL}}$ will be $j \times 1$. If the number of redundants is q, so that the $\mathbf{Q}$ matrix is of order $q \times 1$, then the matrix $\mathbf{D_{JQ}}$ will be rectangular and of order $j \times q$. Equation (2-14) can be used to calculate the displacements $\mathbf{D_J}$ by matrix operations only, after the matrices $\mathbf{D_{JL}}$, $\mathbf{D_{JQ}}$, and $\mathbf{Q}$ have been obtained.

In a manner similar to that used in obtaining Eq. (2-14), the principle of superposition may be used to obtain the member end-actions A_M and the reactions A_R. In these cases the superposition equations are

$$\mathbf{A_M} = \mathbf{A_{ML}} + \mathbf{A_{MQ}Q} \tag{2-15}$$
$$\mathbf{A_R} = \mathbf{A_{RL}} + \mathbf{A_{RQ}Q} \tag{2-16}$$

in which $\mathbf{A_M}$ and $\mathbf{A_R}$ are vectors of member end-actions and reactions in the actual beam (Fig. 2-11a); $\mathbf{A_{ML}}$ and $\mathbf{A_{RL}}$ are vectors of member end-actions and reactions in the released structure due to loads; and $\mathbf{A_{MQ}}$ and $\mathbf{A_{RQ}}$ are matrices of end-actions and reactions in the released structure due to unit values of the redundants. In the example of Fig. 2-11 the matrices $\mathbf{A_M}$ and $\mathbf{A_{ML}}$ are of order 4×1 since there are four end-actions being considered; the matrices $\mathbf{A_R}$ and $\mathbf{A_{RL}}$ are of order 2×1 because there are two reactions being considered; and the matrices $\mathbf{A_{MQ}}$ and $\mathbf{A_{RQ}}$ are of order 4×2 and 2×2, respectively. In the general case in which there are m member end-actions, r reactions, and q redundants, the matrices $\mathbf{A_M}$ and $\mathbf{A_{ML}}$ are of order $m \times 1$, $\mathbf{A_{MQ}}$ is of order $m \times q$, $\mathbf{A_R}$ and $\mathbf{A_{RL}}$ are of order $r \times 1$, and $\mathbf{A_{RQ}}$ is of order $r \times q$.

From the above discussion it is seen that the steps to be followed in analyzing a structure by the flexibility method include rather extensive analyses of the released structure. With the loads on the released structure, it is necessary to find the actions and displacements that constitute the matrices $\mathbf{D_{QL}}$, $\mathbf{D_{JL}}$, $\mathbf{A_{ML}}$, and $\mathbf{A_{RL}}$. With the unit values of the redundants acting on the released structure it is necessary to determine the matrices $\mathbf{F}$, $\mathbf{D}_{JQ}$, $\mathbf{A}_{MQ}$, and $\mathbf{A}_{RQ}$. Then Eq. (2-8) is solved first, yielding the vector $\mathbf{Q}$ of redundants, after which Eqs. (2-14) through (2-16) can be evaluated for the vectors $\mathbf{D}_J$, $\mathbf{A}_M$, and $\mathbf{A}_R$. By this means all actions and displacements of interest in the actual structure can be found.

When the effects of temperature changes, prestrains, and support displacements must be taken into account, the only changes in the superposition equations are in the first terms on the right-hand sides of the equal signs. These terms represent the actions and displacements in the released structure and must include the effects of all influences. This situation was described in the preceding article for the displacements corresponding to the redundants, and Eq. (2-12) was derived as a generalized form of Eq. (2-8). Using the same approach, a more general form for Eq. (2-14) becomes the following:

$$\mathbf{D}_J = \mathbf{D}_{JC} + \mathbf{D}_{JQ}\mathbf{Q} \tag{2-17}$$

in which the vector $\mathbf{D}_{JC}$ represents the combined effects in the released structure and is given by the following expression:

$$\mathbf{D}_{JC} = \mathbf{D}_{JL} + \mathbf{D}_{JT} + \mathbf{D}_{JP} + \mathbf{D}_{JR} \tag{2-18}$$

In Eq. (2-18) the matrices $\mathbf{D}_{JT}$, $\mathbf{D}_{JP}$, and $\mathbf{D}_{JR}$ represent joint displacements due to temperature changes, prestrains, and restraint displacements, respectively. The restraint displacements that are considered in obtaining $\mathbf{D}_{JR}$ are those that do not correspond to redundants. Those that do correspond to redundants are represented in the matrix $\mathbf{D}_Q$.

There is no need to generalize Eqs. (2-15) and (2-16) to account for temperature changes, prestrains, and support displacements. None of these influences will produce any actions or reactions in a statically determinate released structure; instead, the structure will merely change its configuration to accommodate these effects. Such influences are propagated into the matrices $\mathbf{A}_M$ and $\mathbf{A}_R$ through the values of the redundants Q, which are obtained by solving Eq. (2-12).

In summary, the procedure in any particular example is to analyze the released structure for all causes and then to sum the appropriate matrices, as shown in Eqs. (2-11) and (2-18). Then the redundants are found from Eq. (2-13), after which the various actions and displacements may be obtained from Eqs. (2-15), (2-16), and (2-17).

Example. An extended solution for the two-span beam shown in Fig. 2-11 will now be given for the case of loads only. It is assumed that the object of the analysis is to calculate the various joint displacements D_J, member end-actions A_M, and reactions A_R that are shown in Figs. 2-11a and 2-11b. The beam has constant flexural rigidity EI and is acted upon by the loads P_1, M, P_2, and P_3, which are assumed to have the following values:

$$P_1 = 2P \qquad M = PL \qquad P_2 = P \qquad P_3 = P$$

When these loads act upon the released structure (Fig. 2-11c), the joint displacements D_{JL1} and D_{JL2} are found to be

$$D_{JL1} = \frac{5PL^2}{4EI} \qquad D_{JL2} = \frac{13PL^2}{8EI}$$

Therefore, the vector $\mathbf{D}_{JL}$ is

$$\mathbf{D}_{JL} = \frac{PL^2}{8EI}\begin{bmatrix} 10 \\ 13 \end{bmatrix}$$

The member end-actions in the beam of Fig. 2-11c can be found by static equilibrium. For instance, A_{ML1} and A_{ML2} are the shearing force and bending moment just to the left of the point where the couple M is applied. These quantities are equal to

$$-P_2 + P_3 \quad \text{and} \quad M - \frac{P_2L}{2} + P_3L$$

respectively, or

$$A_{ML1} = 0 \qquad A_{ML2} = \frac{3PL}{2}$$

Similarly, the shearing force and bending moment just to the right of point B in the beam of Fig. 2-11c are

$$A_{ML3} = P_2 - P_3 = 0 \qquad A_{ML4} = \frac{P_2L}{2} - P_3L = -\frac{PL}{2}$$

Thus, the matrix $\mathbf{A}_{ML}$ is

$$\mathbf{A}_{ML} = \begin{bmatrix} 0 \\ \frac{3PL}{2} \\ 0 \\ -\frac{PL}{2} \end{bmatrix}$$

Also, the reactions in the beam of Fig. 2-11c are

$$\mathbf{A}_{RL} = \begin{bmatrix} 2P \\ -\frac{PL}{2} \end{bmatrix}$$

The joint displacements due to unit values of the redundants are shown in Figs. 2-11d and 2-11e. These displacements can be readily calculated for the released structure and are given in the matrix $\mathbf{D}_{JQ}$:

$$\mathbf{D}_{JQ} = \frac{L^2}{2EI}\begin{bmatrix} 1 & 3 \\ 1 & 4 \end{bmatrix}$$

The member end-actions and reactions in the beams of Figs. 2-11d and 2-11e are found by statics and constitute the matrices $\mathbf{A}_{MQ}$ and $\mathbf{A}_{RQ}$:

$$\mathbf{A}_{MQ} = \begin{bmatrix} 1 & 1 \\ 0 & L \\ 0 & -1 \\ 0 & -L \end{bmatrix} \qquad \mathbf{A}_{RQ} = \begin{bmatrix} -1 & -1 \\ -L & -2L \end{bmatrix}$$

The matrices $\mathbf{D}_{QL}$ and $\mathbf{F}$ which appear in Eq. (2-8) were determined previously for the beam in Fig. 2-11 (see Art. 2.2) and the vector $\mathbf{Q}$ of redundants was found to be

$$\mathbf{Q} = \frac{P}{56}\begin{bmatrix} 69 \\ -64 \end{bmatrix}$$

Substitution of this matrix as well as the matrices $\mathbf{D}_{JL}$ and $\mathbf{D}_{JQ}$ into Eq. (2-14) gives the joint displacements in the actual beam:

$$\mathbf{D}_J = \frac{PL^2}{112EI}\begin{bmatrix} 17 \\ -5 \end{bmatrix}$$

This result shows that the rotation D_{J1} at joint B is in the counterclockwise sense and equal to

$$D_{J1} = \frac{17PL^2}{112EI}$$

while the rotation at joint C is

$$D_{J2} = -\frac{5PL^2}{112EI}$$

and is clockwise, as indicated by the negative sign.

The member end-actions $\mathbf{A}_M$ and reactions $\mathbf{A}_R$ are obtained by substituting the appropriate matrices given above into Eqs. (2-15) and (2-16). The results are

$$\mathbf{A}_M = \frac{P}{56}\begin{bmatrix} 5 \\ 20L \\ 64 \\ 36L \end{bmatrix} \qquad \mathbf{A}_R = \frac{P}{56}\begin{bmatrix} 107 \\ 31L \end{bmatrix}$$

which can be readily verified by static equilibrium.

2.6 Flexibilities of Prismatic Members. In the examples and problems of this chapter, certain flexibility coefficients (such as $L^3/3EI$) for prismatic members have been used frequently. They appear so often because the flexibility coefficients of a structure are calculated from the contributions of individual members. Therefore, it is worthwhile to construct *member flexibility matrices* for prismatic members in various types of framed structures. Such matrices are required in the formalized approach to the flexibility method described in the next article. They can also be inverted to obtain member stiffness matrices (see Art. 3.5).

It is necessary to define *member-oriented axes* for an individual member and to distinguish them from *structure-oriented axes* for the whole structure. (Such axes are also referred to as *local coordinates* and *global coordinates*, respectively.) Figures 2-12a through 2-12d show structure- and member-oriented axes for various types of framed structures. Each of the figures contains a typical member i that is connected to joints j and k of the structure. The axes labeled x, y, and z in each figure represent structure-oriented axes, that is, axes oriented in some convenient manner to the

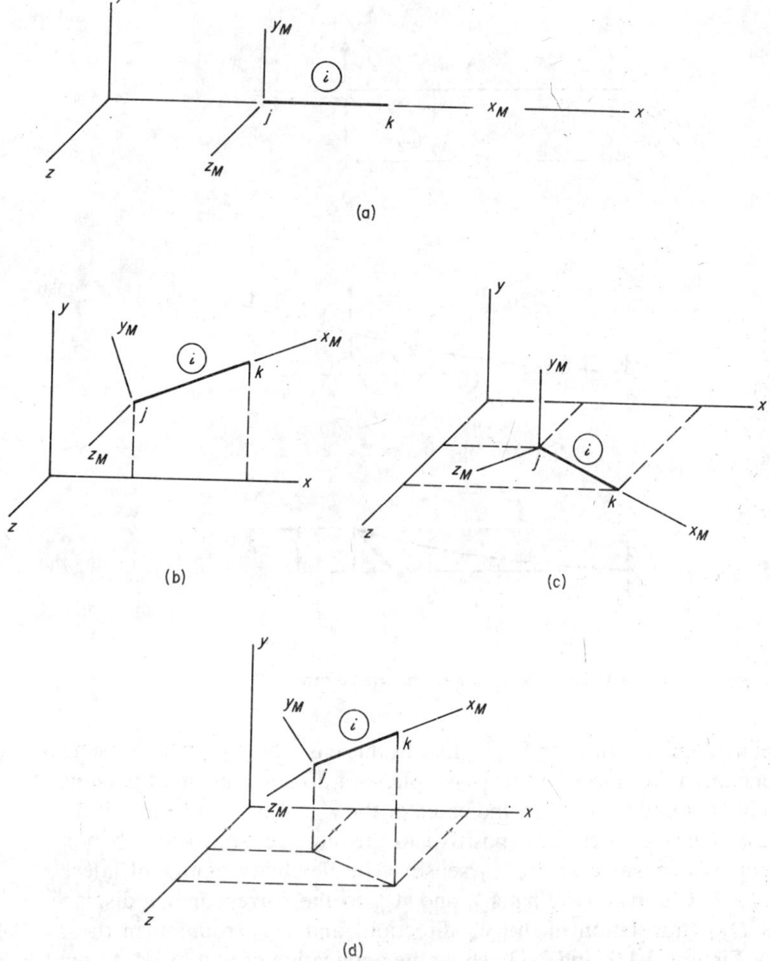

Fig. 2-12. Structure- and member-oriented axes for (a) beam member, (b) plane truss or plane frame member, (c) grid member, and (d) space truss or space frame member.

structure as a whole. On the other hand, the axes labeled x_M, y_M, and z_M are associated with the particular member under consideration and are called member-oriented axes. These axes are assumed to have their origin at the j end of the member. The x_M axis is aligned with the member axis and is taken positive from j to k. Axes y_M and z_M lie in the principal planes of bending for a member, but this arrangement is not essential for analyzing trusses.

Member flexibility matrices will now be developed for prismatic members that are restrained at their j ends and free at their k ends, starting with

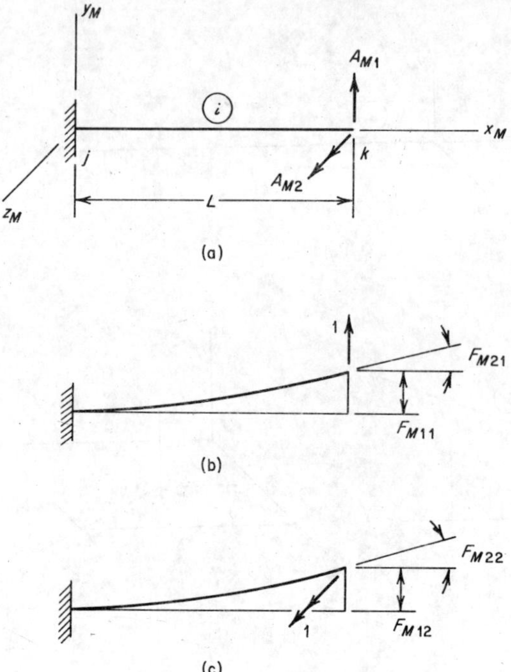

Fig. 2-13. Flexibilities for beam member.

the beam member shown in Fig. 2-13a. In this case the y_M axis is chosen so that bending takes place in the x_M-y_M plane (a principal plane of bending). Two kinds of end-actions are indicated at the k end of the member in Fig. 2-13a: a shearing force A_{M1} (positive in the y_M direction) and a bending moment A_{M2} (positive in the z_M sense). The flexibility matrix of interest here is a 2×2 array relating A_{M1} and A_{M2} to the corresponding displacements D_{M1} (translation in the y_M direction) and D_{M2} (rotation in the z_M sense). Figures 2-13b and 2-13c show the application of unit loads $A_{M1} = 1$ and $A_{M2} = 1$ to obtain the terms in the beam flexibility matrix $\mathbf{F}_{Mi}$, as follows:

$$\mathbf{F}_{Mi} = \begin{bmatrix} F_{M11} & F_{M12} \\ F_{M21} & F_{M22} \end{bmatrix} = \begin{bmatrix} \dfrac{L^3}{3EI} & \dfrac{L^2}{2EI} \\ \dfrac{L^2}{2EI} & \dfrac{L}{EI} \end{bmatrix} \tag{2-19}$$

These terms were found previously in the example at the end of Art. 1.11.

The truss member in Fig. 2-14a has only one end-action to be considered for the purpose of calculating member flexibilities, namely, the axial

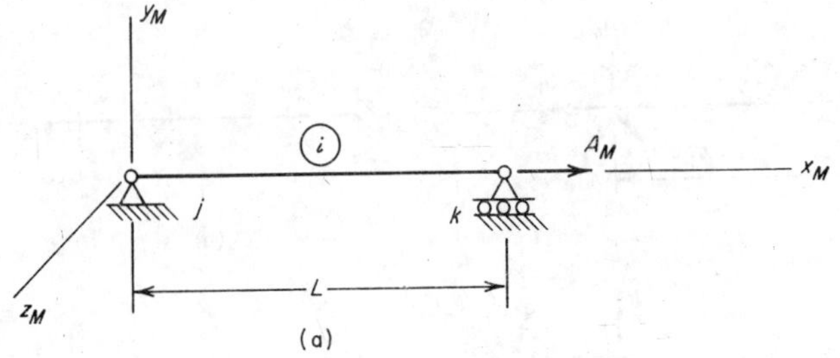

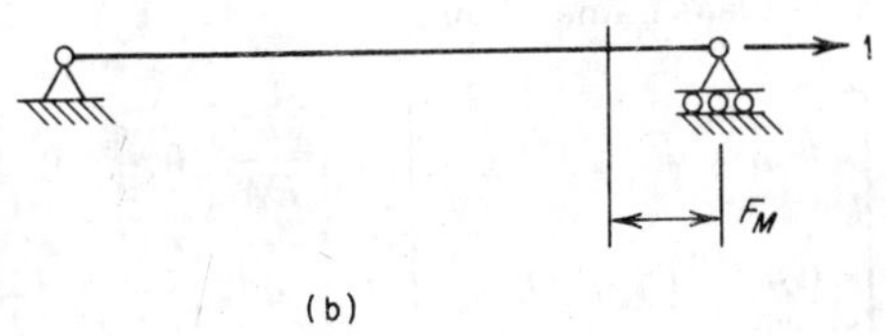

Fig. 2-14. Flexibility for truss member.

force A_M at the k end of the member (note that a positive value of A_M corresponds to a tensile force in the member). The corresponding displacement D_M consists of a translation in the x_M direction at the k end. Therefore, the single value of F_{Mi} for this case is

$$F_{Mi} = \frac{L}{EA} \tag{2-20}$$

which is the change in length of the member due to a unit axial load $A_M = 1$ (see Fig. 2-14b). This member flexibility pertains to either a plane truss or a space truss, and it is not influenced by the orientations of axes y_M and z_M.

Figure 2-15a shows a plane frame member that is fixed at the j end and free at the k end. As with the beam, the x_M-y_M plane is defined to be a principal plane of bending. If axial strains are to be neglected, the member flexibility matrix is the same as that for a beam (see Eq. 2-19). However, if axial strains are to be considered, the member flexibility matrix contains the term in Eq. (2-20) as well. Figure 2-15a shows the three member end-actions required for this purpose. They are the axial force A_{M1} in the x_M direction, the shearing force A_{M2}, in the y_M direction, and the bending moment A_{M3} in the z_M sense. Corresponding displacements D_{M1}, D_{M2}, and D_{M3} consist of translation in the x_M direction, translation in the y_M direction, and rotation in the z_M sense. Figures 2-15b, c, and d show the application of unit loads that produce the terms in the following 3×3 plane frame member flexibility matrix:

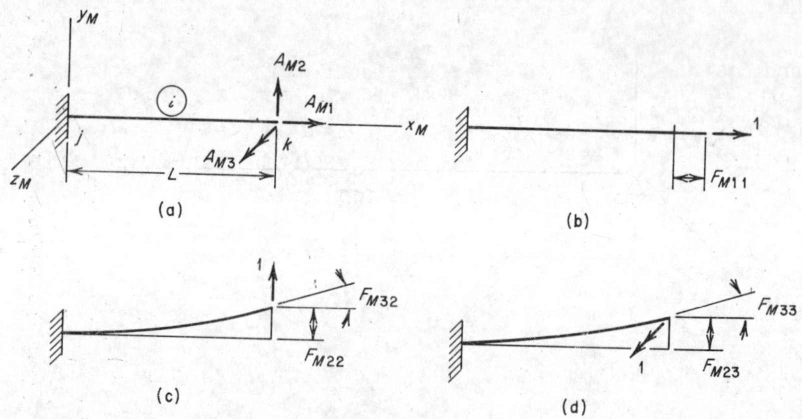

Fig. 2-15. Flexibilities for plane frame member.

$$\mathbf{F}_{Mi} = \begin{bmatrix} F_{M11} & F_{M12} & F_{M13} \\ F_{M21} & F_{M22} & F_{M23} \\ F_{M31} & F_{M32} & F_{M33} \end{bmatrix} = \begin{bmatrix} \dfrac{L}{EA} & 0 & 0 \\ 0 & \dfrac{L^3}{3EI} & \dfrac{L^2}{2EI} \\ 0 & \dfrac{L^2}{2EI} & \dfrac{L}{EI} \end{bmatrix} \tag{2-21}$$

It is seen that this equation is a combination of Eqs. (2-19) and (2-20).

The flexibility matrix for a grid member also can be obtained by augmenting that for a beam, because bending again occurs in the x_M-y_M plane. Figure 2-16a shows a grid member with a shearing force component A_{M1} (acting in the y_M direction), a torsional moment A_{M2} (acting in the x_M sense), and a bending moment A_{M3} (acting in the z_M sense). This sequence for numbering actions follows the rule of taking forces before moments

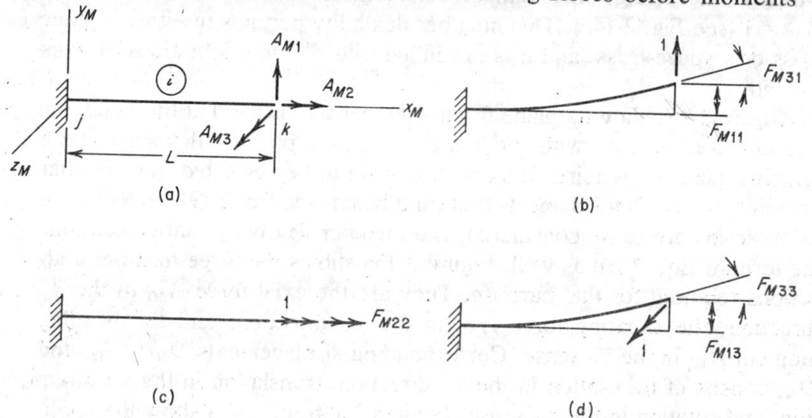

Fig. 2-16. Flexibilities for grid member.

(and corresponding translations before rotations). As illustrated in Figs. 2-16b, c, and d, the terms in the 3 × 3 flexibility matrix for a grid member are found to be

$$\mathbf{F}_{\mathrm{M}i} = \begin{bmatrix} F_{M11} & F_{M12} & F_{M13} \\ F_{M21} & F_{M22} & F_{M23} \\ F_{M31} & F_{M32} & F_{M33} \end{bmatrix} = \begin{bmatrix} \dfrac{L^3}{3EI} & 0 & \dfrac{L^2}{2EI} \\ 0 & \dfrac{L}{GJ} & 0 \\ \dfrac{L^2}{2EI} & 0 & \dfrac{L}{EI} \end{bmatrix} \tag{2-22}$$

In this case the first and third actions and displacements are coupled by flexural terms, whereas the second and third were coupled in Eq. (2-21).

To obtain the flexibility matrix for a space frame member, it is necessary to consider three force components and three moment components applied to the k end, as shown in Fig. 2-17a. For this type of member, both the x_M-y_M plane and the x_M-z_M plane are principal planes of bending. The

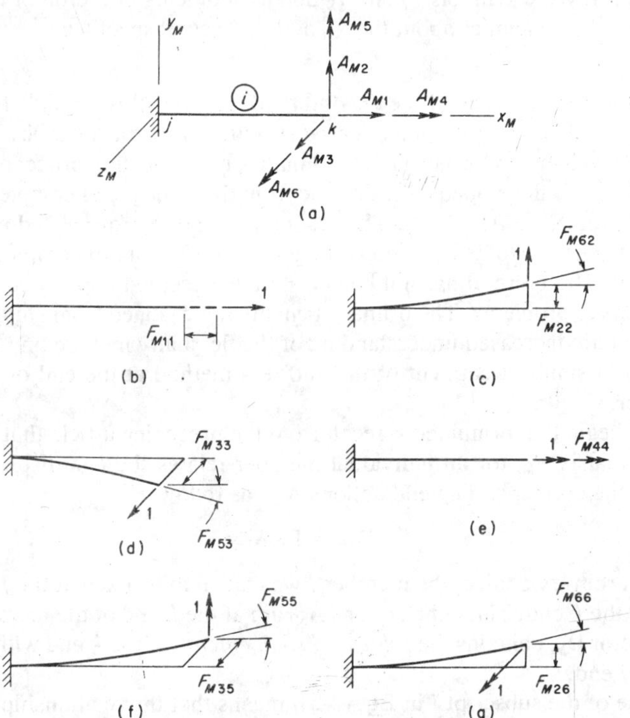

Fig. 2-17. Flexibilities for space frame member.

end-actions in Fig. 2-17a are numbered in the sequence x, y, and z, with forces taken before moments. Elements of the member flexibility matrix appear in Figs. 2-17b through 2-17g. These terms are readily evaluated (see Table A-3 in Appendix A) and placed in a 6×6 member flexibility matrix, as follows:

$$\mathbf{F}_{Mi} = \begin{bmatrix} \dfrac{L}{EA} & 0 & 0 & 0 & 0 & 0 \\ 0 & \dfrac{L^3}{3EI_Z} & 0 & 0 & 0 & \dfrac{L^2}{2EI_Z} \\ 0 & 0 & \dfrac{L^3}{3EI_Y} & 0 & \dfrac{-L^2}{2EI_Y} & 0 \\ 0 & 0 & 0 & \dfrac{L}{GJ} & 0 & 0 \\ 0 & 0 & \dfrac{-L^2}{2EI_Y} & 0 & \dfrac{L}{EI_Y} & 0 \\ 0 & \dfrac{L^2}{2EI_Z} & 0 & 0 & 0 & \dfrac{L}{EI_Z} \end{bmatrix} \tag{2-23}$$

In this array the symbols I_Y and I_Z denote moments of inertia of the cross section of the member about the y_M and z_M axes, respectively.

2.7 Formalization of the Flexibility Method. It is possible to assemble the flexibility matrix for a released structure from the flexibility matrices of individual members using a formal matrix multiplication procedure. Such a technique is developed in this article from the principle of complementary virtual work (see Art. 1.14). The reader will find the formalized approach described herein to be both interesting and orderly but not computationally efficient. Therefore, it has not been generally accepted in practice for solving realistic problems. The primary benefits to be gained from studying this material are increased understanding of the flexibility method and preparation for a similar treatment of the stiffness method at the end of the next chapter.

To begin the formulation, recall from the preceding article that the flexibility matrix $\mathbf{F}_{Mi}$ for an individual member relates the end-displacements $\mathbf{D}_{Mi}$ to the corresponding end-actions $\mathbf{A}_{Mi}$, as follows:

$$\mathbf{D}_{Mi} = \mathbf{F}_{Mi}\mathbf{A}_{Mi} \tag{2-24}$$

As an arbitrary choice, the member i was taken to be fixed at the j end and free at the k end. Thus, the actions $\mathbf{A}_{Mi}$ are at the k end of the member, and the vector $\mathbf{D}_{Mi}$ contains the *relative displacements* of the k end with respect to the j end.

Use of the subscript i in Eq. (2-24) means that the relationship pertains only to the i-th member in a given structure. If the equation is repeated for all of the members in a structure, the results can be expressed in the form

$$\begin{bmatrix} \mathbf{D}_{M1} \\ \mathbf{D}_{M2} \\ \mathbf{D}_{M3} \\ \cdots \\ \mathbf{D}_{Mi} \\ \cdots \\ \mathbf{D}_{Mm} \end{bmatrix} = \begin{bmatrix} \mathbf{F}_{M1} & \mathbf{0} & \mathbf{0} & \cdots & \mathbf{0} & \cdots & \mathbf{0} \\ \mathbf{0} & \mathbf{F}_{M2} & \mathbf{0} & \cdots & \mathbf{0} & \cdots & \mathbf{0} \\ \mathbf{0} & \mathbf{0} & \mathbf{F}_{M3} & \cdots & \mathbf{0} & \cdots & \mathbf{0} \\ \cdots & \cdots & \cdots & \cdots & \cdots & \cdots & \cdots \\ \mathbf{0} & \mathbf{0} & \mathbf{0} & \cdots & \mathbf{F}_{Mi} & \cdots & \mathbf{0} \\ \cdots & \cdots & \cdots & \cdots & \cdots & \cdots & \cdots \\ \mathbf{0} & \mathbf{0} & \mathbf{0} & \cdots & \mathbf{0} & \cdots & \mathbf{F}_{Mm} \end{bmatrix} \begin{bmatrix} \mathbf{A}_{M1} \\ \mathbf{A}_{M2} \\ \mathbf{A}_{M3} \\ \cdots \\ \mathbf{A}_{Mi} \\ \cdots \\ \mathbf{A}_{Mm} \end{bmatrix} \quad (2\text{-}25)$$

in which the subscript m equals the number of members. Equation (2-25) can be written more concisely as

$$\mathbf{D}_M = \mathbf{F}_M \mathbf{A}_M \quad (2\text{-}26)$$

where the symbol $\mathbf{F}_M$ represents a diagonal matrix of submatrices $\mathbf{F}_{Mi}$. One may think of the matrix $\mathbf{F}_M$ as the flexibility matrix for the *unassembled structure* because it merely consists of a collection of member flexibility matrices arranged in a diagonal pattern. For this reason it is given the name *unassembled flexibility matrix*. The size of $\mathbf{F}_M$ depends upon the number of members and the order of a typical submatrix $\mathbf{F}_{Mi}$. For example, if $\mathbf{F}_{Mi}$ is of order 3×3 (as in a grid), then the order of $\mathbf{F}_M$ is $3m \times 3m$. Also note that the vectors $\mathbf{A}_M$ and $\mathbf{D}_M$ in Eq. (2-26) contain end-actions $\mathbf{A}_{Mi}$ and relative end-displacements $\mathbf{D}_{Mi}$ for all of the members in the structure (see Eq. 2-25).

As the next step, all of the member end-actions in $\mathbf{A}_M$ will be related to certain structure actions $\mathbf{A}_S$ applied to the released structure. The vector of structure actions $\mathbf{A}_S$ consists of joint loads $\mathbf{A}_J$ (listed first) and redundant actions $\mathbf{A}_Q$ (listed second).* If loads are not applied at the joints, they must be converted to equivalent joint loads (see Art. 1.12) before this analysis can proceed. When the relationships between $\mathbf{A}_M$ and $\mathbf{A}_S$ are expressed in matrix form, the result is

$$\mathbf{A}_M = \mathbf{B}_{MS}\mathbf{A}_S = [\mathbf{B}_{MJ} \quad \mathbf{B}_{MQ}] \begin{bmatrix} \mathbf{A}_J \\ \mathbf{A}_Q \end{bmatrix} \quad (2\text{-}27)$$

The matrix $\mathbf{B}_{MS}$ in Eq. (2-27) is an action transformation matrix that relates $\mathbf{A}_M$ to $\mathbf{A}_S$ for the released structure. The terms in $\mathbf{B}_{MS}$ are found by elementary principles of static equilibrium. For this reason it is called the *equilibrium* (or *statics*) *matrix*. Note that the matrix $\mathbf{B}_{MS}$ is partitioned into submatrices $\mathbf{B}_{MJ}$ and $\mathbf{B}_{MQ}$, which relate $\mathbf{A}_M$ to $\mathbf{A}_J$ and $\mathbf{A}_Q$, respectively. Each column in the submatrix $\mathbf{B}_{MJ}$ consists of the member end-actions caused by a unit value of a joint load applied to the released structure. For completeness, terms in $\mathbf{B}_{MJ}$ are developed for all possible joint loads, regardless of whether such loads actually exist. Similarly, each column in the submatrix $\mathbf{B}_{MQ}$ contains member end-actions due to a unit value of a redundant applied to the released structure.

Now suppose that an arbitrary set of virtual loads $\delta\mathbf{A}_S$ is applied to the released structure. As with the vector $\mathbf{A}_S$, let $\delta\mathbf{A}_S$ consist of virtual joint

*The symbol $\mathbf{A}_Q$, used here to represent redundant actions, has the same meaning as the symbol $\mathbf{Q}$ used in previous articles.

loads $\delta\mathbf{A}_J$ and virtual redundants $\delta\mathbf{A}_Q$. Then an expression similar to Eq. (2-27) can be written for the resulting virtual member end-actions $\delta\mathbf{A}_M$, as follows:

$$\delta\mathbf{A}_M = \mathbf{B}_{MS}\,\delta\mathbf{A}_S = [\mathbf{B}_{MJ} \quad \mathbf{B}_{MQ}]\begin{bmatrix}\delta\mathbf{A}_J \\ \delta\mathbf{A}_Q\end{bmatrix} \tag{2-28}$$

The equilibrium matrix $\mathbf{B}_{MS}$ in this equation is the same as that in Eq. (2-27).

External complementary virtual work δW^* produced by the virtual loads $\delta\mathbf{A}_S$ and the actual displacements $\mathbf{D}_S$ may be expressed as

$$\delta W^* = \delta\mathbf{A}_S^T\mathbf{D}_S = [\delta\mathbf{A}_J^T \quad \delta\mathbf{A}_Q^T]\begin{bmatrix}\mathbf{D}_J \\ \mathbf{D}_Q\end{bmatrix} \tag{2-29}$$

where the symbols $\mathbf{D}_J$ and $\mathbf{D}_Q$ represent actual displacements corresponding to the joint loads and the redundants, respectively. In a similar manner the internal complementary virtual work δU^* generated by the virtual member end-actions $\delta\mathbf{A}_M$ and the actual relative end-displacements $\mathbf{D}_M$ takes the form

$$\delta U^* = \delta\mathbf{A}_M^T\mathbf{D}_M \tag{2-30}$$

By the principle of complementary virtual work, expressions (2-29) and (2-30) are equated to obtain

$$\delta\mathbf{A}_S^T\mathbf{D}_S = \delta\mathbf{A}_M^T\mathbf{D}_M \tag{2-31}$$

Substitution of Eqs. (2-26), (2-27), and (2-28) into the right-hand side of Eq. (2-31) yields

$$\delta\mathbf{A}_S^T\mathbf{D}_S = \delta\mathbf{A}_S^T\mathbf{B}_{MS}^T\mathbf{F}_M\mathbf{B}_{MS}\mathbf{A}_S \tag{2-32}$$

Because the virtual loads in $\mathbf{A}_S$ are arbitrary, this vector may be cancelled from both sides of Eq. (2-32) to produce

$$\mathbf{D}_S = \mathbf{F}_S\mathbf{A}_S \tag{2-33}$$

where

$$\mathbf{F}_S = \mathbf{B}_{MS}^T\mathbf{F}_M\mathbf{B}_{MS} \tag{2-34}$$

It is seen that the matrix $\mathbf{F}_S$ relates the displacements $\mathbf{D}_S$ to the actions $\mathbf{A}_S$ in Eq. (2-33). This flexibility matrix is formed by the congruence transformation in Eq. (2-34), using the $\mathbf{B}_{MS}$ matrix as the postmultiplier of $\mathbf{F}_M$ and the transposed matrix $\mathbf{B}_{MS}^T$ as the premultiplier of $\mathbf{F}_M$. By this operation the unassembled flexibility matrix $\mathbf{F}_M$ is transformed to the *assembled flexibility matrix* $\mathbf{F}_S$ for the whole structure.

In order to examine the assembled flexibility matrix $\mathbf{F}_S$ in more detail, it will be partitioned into submatrices pertaining to the joint loads $\mathbf{A}_J$, the redundant actions $\mathbf{A}_Q$, and their corresponding displacements, $\mathbf{D}_J$ and $\mathbf{D}_Q$.

This partitioning may be accomplished by writing the expanded form of Eq. (2-33) as

$$\begin{bmatrix} \mathbf{D}_J \\ \mathbf{D}_Q \end{bmatrix} = \begin{bmatrix} \mathbf{F}_{JJ} & \mathbf{F}_{JQ} \\ \mathbf{F}_{QJ} & \mathbf{F}_{QQ} \end{bmatrix} \begin{bmatrix} \mathbf{A}_J \\ \mathbf{A}_Q \end{bmatrix} \tag{2-35}$$

where

$$\mathbf{F}_{JJ} = \mathbf{B}_{MJ}^T \mathbf{F}_M \mathbf{B}_{MJ} \qquad \mathbf{F}_{JQ} = \mathbf{B}_{MJ}^T \mathbf{F}_M \mathbf{B}_{MQ}$$

$$\mathbf{F}_{QJ} = \mathbf{B}_{MQ}^T \mathbf{F}_M \mathbf{B}_{MJ} \qquad \mathbf{F}_{QQ} = \mathbf{B}_{MQ}^T \mathbf{F}_M \mathbf{B}_{MQ}$$

Note that each of the submatrices $\mathbf{F}_{JJ}$, $\mathbf{F}_{JQ}$, $\mathbf{F}_{QJ}$, and $\mathbf{F}_{QQ}$ is formed by a congruence transformation, using as operators $\mathbf{B}_{MJ}$ and $\mathbf{B}_{MQ}$ (the two parts of $\mathbf{B}_{MS}$ from Eq. 2-27). Because the submatrix $\mathbf{F}_{QQ}$ relates the displacements $\mathbf{D}_Q$ to the redundant actions $\mathbf{A}_Q$, it is recognized to be the same as the flexibility matrix $\mathbf{F}$ discussed in previous articles. Other tie-ins with earlier material will be made after the formalized solution is completely set up.

Unknowns in Eq. (2-35) consist of the joint displacements $\mathbf{D}_J$ and the redundant actions $\mathbf{A}_Q$. Therefore, it becomes advantageous to rewrite Eq. (2-35) as two separate matrix equations:

$$\mathbf{D}_J = \mathbf{F}_{JJ}\mathbf{A}_J + \mathbf{F}_{JQ}\mathbf{A}_Q \tag{2-36a}$$

$$\mathbf{D}_Q = \mathbf{F}_{QJ}\mathbf{A}_J + \mathbf{F}_{QQ}\mathbf{A}_Q \tag{2-36b}$$

Solving the second of these equations for the redundants gives

$$\mathbf{A}_Q = \mathbf{F}_{QQ}^{-1}(\mathbf{D}_Q - \mathbf{F}_{QJ}\mathbf{A}_J) \tag{2-37}$$

which symbolically represents the key that unlocks the whole problem. Once the redundants $\mathbf{A}_Q$ have been found, they can be substituted into Eq. (2-36a) to obtain the joint displacements $\mathbf{D}_J$.

In a complete analysis, the other items to be evaluated are the member end-actions $\mathbf{A}_M$ and the support reactions $\mathbf{A}_R$. Equation (2-27) provides the means for calculating the member end-actions caused by the joint loads and the redundants, using the $\mathbf{B}_{MS}$ matrix. It is important to realize, however, that equivalent joint loads may be contained in the vector $\mathbf{A}_J$ as well as actual joint loads. Therefore, the member end-actions computed from Eq. (2-27) must be added to initial fixed-end actions to obtain the final values of member end-actions (see Art. 1.12). This combination of effects can be stated in equation form as

$$\mathbf{A}_M = \mathbf{A}_{MF} + \mathbf{B}_{MJ}\mathbf{A}_J + \mathbf{B}_{MQ}\mathbf{A}_Q \tag{2-38}$$

where the symbol $\mathbf{A}_{MF}$ represents fixed-end actions.

In a similar manner, the support reactions caused by joint loads and redundants may be evaluated with an action transformation operator $\mathbf{B}_{RS}$ that relates $\mathbf{A}_R$ to $\mathbf{A}_S$, as follows:

$$\mathbf{A}_R = \mathbf{B}_{RS}\mathbf{A}_S = [\mathbf{B}_{RJ} \quad \mathbf{B}_{RQ}] \begin{bmatrix} \mathbf{A}_J \\ \mathbf{A}_Q \end{bmatrix} \tag{2-39}$$

In this expression the matrix $\mathbf{B}_{RS}$ is partitioned into submatrices $\mathbf{B}_{RJ}$ and $\mathbf{B}_{RQ}$, which relate $\mathbf{A}_R$ to $\mathbf{A}_J$ and $\mathbf{A}_Q$, respectively. Each column in the submatrix $\mathbf{B}_{RJ}$ consists of the support reactions due to a unit value of a joint load applied to the released structure. Similarly, each column in the submatrix $\mathbf{B}_{RQ}$ contains reactions caused by a unit value of a redundant applied to the released structure.

If actual or equivalent joint loads are applied directly to the supports, they must be taken into account by adding their negatives to the support reactions calculated from Eq. (2-39). These effects can be stated as

$$\mathbf{A}_R = -\mathbf{A}_{RC} + \mathbf{B}_{RJ}\mathbf{A}_J + \mathbf{B}_{RQ}\mathbf{A}_Q \tag{2-40}$$

where the symbol $\mathbf{A}_{RC}$ denotes combined loads (actual and equivalent) applied directly to the supports.

Equations of the flexibility method developed by superposition principles in previous articles will now be compared with their counterparts derived by complementary virtual work in this article. The comparable equations for calculating redundants are

$$\mathbf{Q} = \mathbf{F}^{-1}(\mathbf{D}_Q - \mathbf{D}_{QL}) \qquad \text{(2-9) repeated}$$

and

$$\mathbf{A}_Q = \mathbf{F}_{QQ}^{-1}(\mathbf{D}_Q - \mathbf{F}_{QJ}\mathbf{A}_J) \qquad \text{(2-37) repeated}$$

It is seen that $\mathbf{Q} = \mathbf{A}_Q$, $\mathbf{F} = \mathbf{F}_{QQ}$, and $\mathbf{D}_{QL} = \mathbf{F}_{QJ}\mathbf{A}_J$ (note the use of flexibility coefficients to obtain $\mathbf{D}_{QL}$).

Comparable equations for determining joint displacements are

$$\mathbf{D}_J = \mathbf{D}_{JL} + \mathbf{D}_{JQ}\mathbf{Q} \qquad \text{(2-14) repeated}$$

and

$$\mathbf{D}_J = \mathbf{F}_{JJ}\mathbf{A}_J + \mathbf{F}_{JQ}\mathbf{A}_Q \qquad \text{(2-36a) repeated}$$

Thus, $\mathbf{D}_{JL} = \mathbf{F}_{JJ}\mathbf{A}_J$ and $\mathbf{D}_{JQ} = \mathbf{F}_{JQ}$ (note the use of flexibility coefficients to obtain $\mathbf{D}_{JL}$).

For the purpose of evaluating member end-actions, the comparable equations are

$$\mathbf{A}_M = \mathbf{A}_{ML} + \mathbf{A}_{MQ}\mathbf{Q} \qquad \text{(2-15) repeated}$$

and

$$\mathbf{A}_M = \mathbf{A}_{MF} + \mathbf{B}_{MJ}\mathbf{A}_J + \mathbf{B}_{MQ}\mathbf{A}_Q \qquad \text{(2-38) repeated}$$

Therefore, $\mathbf{A}_{ML} = \mathbf{A}_{MF} + \mathbf{B}_{MJ}\mathbf{A}_J$ and $\mathbf{A}_{MQ} = \mathbf{B}_{MQ}$. In this instance the steps for obtaining $\mathbf{A}_{ML}$ as the sum of two parts are clearly evident.

Finally, the comparable equations for evaluating support reactions are

$$\mathbf{A}_R = \mathbf{A}_{RL} + \mathbf{A}_{RQ}\mathbf{Q} \qquad \text{(2-16) repeated}$$

and

$$\mathbf{A}_{R} = -\mathbf{A}_{RC} + \mathbf{B}_{RJ}\mathbf{A}_{J} + \mathbf{B}_{RQ}\mathbf{A}_{Q} \tag{2-40}$$
repeated

Hence, $\mathbf{A}_{RL} = -\mathbf{A}_{RC} + \mathbf{B}_{RJ}\mathbf{A}_{J}$ and $\mathbf{A}_{RQ} = \mathbf{B}_{RQ}$. Here the procedure for calculating $\mathbf{A}_{RL}$ as the sum of two terms is explicitly shown.

In summary, formalization of the flexibility method leads to the concept of the assembled flexibility matrix $\mathbf{F}_S$ for the whole structure. It is partitioned into submatrices associated with joint actions and redundant actions (see Eq. 2-35) and their corresponding displacements. The calculation of redundants and joint displacements by the formalized approach involves more extensive use of flexibility coefficients than before. In addition, the steps for evaluating member end-actions and support reactions become clearly evident. Therefore, one is less likely to make mistakes in analyzing a given structure.

It is implied that all joint displacements, member end-actions, and support reactions are to be calculated as a routine matter. This can be a disadvantage in large problems, where the selectivity inherent in the less formal approach of previous articles may be more appealing. In order to be selective in the formalized version, the analyst may omit columns of the $\mathbf{B}_{MS}$ and $\mathbf{B}_{RS}$ matrices that correspond to zero values of actual and equivalent joint loads. However, omitting columns of the $\mathbf{B}_{MS}$ matrix precludes the calculation of corresponding joint displacements by Eq. (2-36a) because the required rows of $\mathbf{F}_{JJ}$ and $\mathbf{F}_{JQ}$ will not be generated. It is also possible to omit rows of the $\mathbf{B}_{RS}$ matrix corresponding to support reactions that are not of interest. On the other hand, all rows of the $\mathbf{B}_{MS}$ matrix must be retained for the purpose of assembling the structure flexibility matrix from the member flexibility matrices.

Examples will now be given to show how the formalized version of the flexibility method can be applied to various types of framed structures. In each example it is assumed that all joint displacements, member end-actions, and support reactions are to be calculated. Only the effects of loads are considered, but the influences of temperature changes, prestrains, and support displacements can be readily included in the form of equivalent joint loads.

Example 1. The first example involves the analysis of the statically indeterminate truss shown in Fig. 2-18a. The truss has six members, as indicated by the circled numbers in the figure, and all members are assumed to have the same axial rigidity *EA*. There are two loads acting on the truss at joint *A*, and support restraints exist at joints *B*, *C*, and *D*. (The same truss was solved previously as Example 3 of Art. 2.3.)

In this problem all of the loads on the truss are in the form of joint loads; hence, it is not necessary to replace member loads by equivalent loads at the joints. However, if member loads are present, the determination of equivalent joint loads is quite simple. They can be found from the formulas for pinned-end actions for truss members, which are given in Table B-5 of Appendix B. The equivalent loads are in the form of forces only, since there are no bending moments produced at the ends of a truss member.

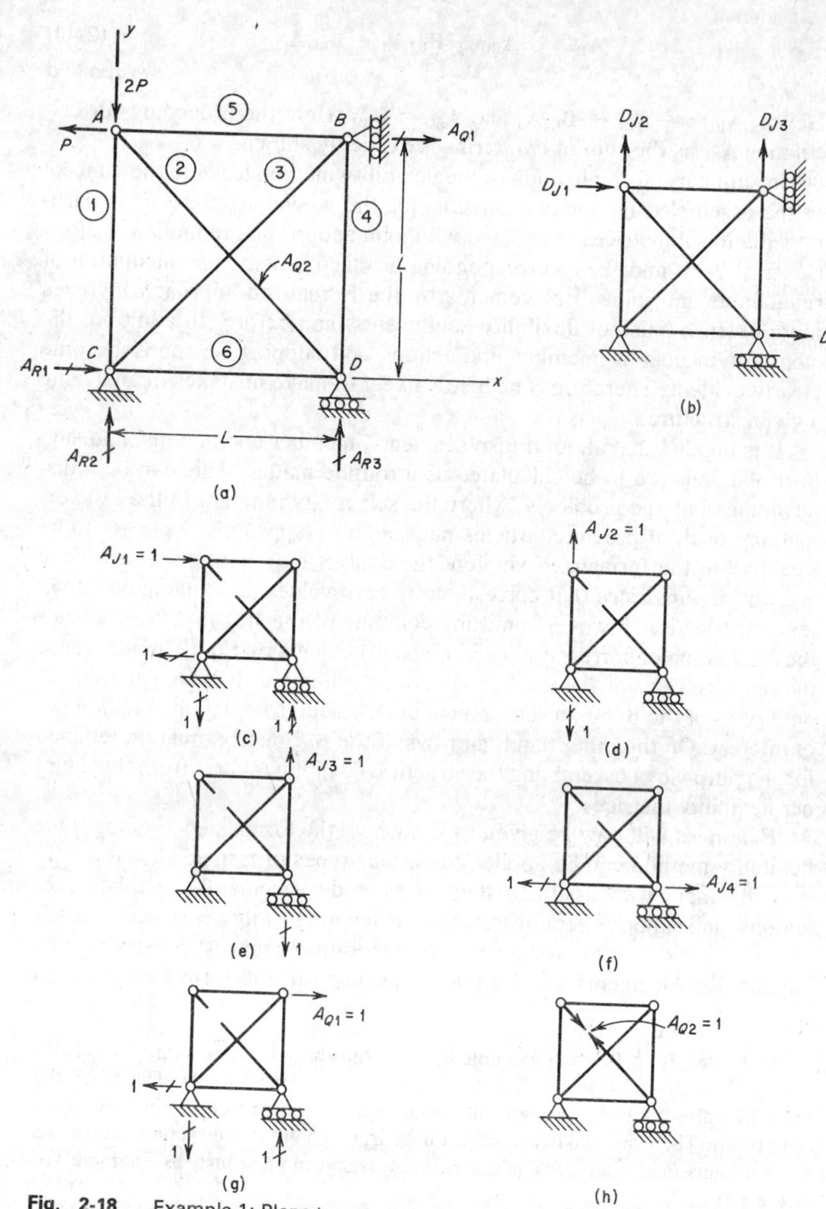

Fig. 2-18. Example 1: Plane truss.

The redundant actions A_{Q1} and A_{Q2} for the truss are selected as the horizontal reactive force at support B and the axial force in bar AD, respectively. The first of these redundants is assumed to be positive when in the positive x direction, and the second is positive when the member is in tension.

In addition to finding the redundants, it is assumed in this example that all member end-actions, reactions, and joint displacements are to be determined. The member end-actions are axial forces at the k ends of all the members, as shown in Fig. 2-14a for a typical member. While the member end-actions in general must be taken at a designated k end for each member, it can be seen that, in the case of a truss with joint loads only, the end-actions are the same at both ends. Therefore, in this example it is not necessary to designate specifically the end of each member at which the end-action is determined; instead, either end can be used.

The truss has four reactions to be determined; the reactions at joints C and D are denoted A_{R1}, A_{R2}, and A_{R3} (see Fig. 2-18a) while the remaining reaction is the redundant A_{Q1} itself. Also, the four unknown joint displacements (denoted D_{J1}, . . . , D_{J4}) are shown in Fig. 2-18b. The reactions and joint displacements are assumed positive when in the positive directions of the x and y axes.

The unassembled flexibility matrix $\mathbf{F}_M$ (see Eq. 2-25) for this example is

$$\mathbf{F}_M = \frac{L}{EA}\begin{bmatrix} 1 & 0 & 0 & 0 & 0 & 0 \\ 0 & 1.414 & 0 & 0 & 0 & 0 \\ 0 & 0 & 1.414 & 0 & 0 & 0 \\ 0 & 0 & 0 & 1 & 0 & 0 \\ 0 & 0 & 0 & 0 & 1 & 0 \\ 0 & 0 & 0 & 0 & 0 & 1 \end{bmatrix}$$

where the diagonal terms are flexibilities of the six members.

Figures 2-18c through 2-18h illustrate the procedure for finding the $\mathbf{B}_{MS}$ and $\mathbf{B}_{RS}$ matrices for the released structure. The first of these action transformation arrays consists of member forces due to unit values of A_{J1} through A_{Q2}. From statics it is found to be

$$\mathbf{B}_{MS} = \begin{bmatrix} \mathbf{B}_{MJ} & \mathbf{B}_{MQ} \end{bmatrix} = \left[\begin{array}{cccc|cc} 0 & 1 & 0 & 0 & 0 & -0.707 \\ 0 & 0 & 0 & 0 & 0 & 1 \\ 1.414 & 0 & 0 & 0 & 1.414 & 1 \\ -1 & 0 & 1 & 0 & -1 & -0.707 \\ -1 & 0 & 0 & 0 & 0 & -0.707 \\ 0 & 0 & 0 & 1 & 0 & -0.707 \end{array}\right]$$

In accordance with Eq. (2-27), this 6×6 matrix $\mathbf{B}_{MS}$ is partitioned column-wise into the 6×4 submatrix $\mathbf{B}_{MJ}$ and the 6×2 submatrix $\mathbf{B}_{MQ}$, pertaining to joint loads and redundants, respectively. The first column in $\mathbf{B}_{MJ}$ contains member forces caused by the unit load $A_{J1} = 1$ shown in Fig. 2-18c. In addition, the member forces listed in the second, third, and fourth columns of $\mathbf{B}_{MJ}$ are due to unit values of A_{J2}, A_{J3}, and A_{J4}, respectively (see Figs. 2-18d, e, and f). Similarly, the first and second columns of the submatrix $\mathbf{B}_{MQ}$ are member forces due to unit values of the redundants A_{Q1} and A_{Q2}, respectively (see Figs. 2-18g and h).

For later use in calculating reactions, the $\mathbf{B}_{RS}$ matrix can also be determined at this stage of the analysis:

$$\mathbf{B}_{RS} = \begin{bmatrix} \mathbf{B}_{RJ} & \mathbf{B}_{RQ} \end{bmatrix} = \left[\begin{array}{cccc|cc} -1 & 0 & 0 & -1 & -1 & 0 \\ -1 & -1 & 0 & 0 & -1 & 0 \\ 1 & 0 & -1 & 0 & 1 & 0 \end{array}\right]$$

This 3×6 array $\mathbf{B}_{RS}$ is partitioned into the 3×4 submatrix $\mathbf{B}_{RJ}$ and the 3×2

submatrix $\mathbf{B}_{RQ}$, as indicated by Eq. (2-39). Columns of the submatrix $\mathbf{B}_{RJ}$ contain support reactions caused by unit values of A_{J1} through A_{J4} (see Figs. 2-18c through f). Also, the columns of $\mathbf{B}_{RQ}$ consist of reactions due to unit values of A_{Q1} and A_{Q2} (see Figs. 2-18g and h).

After the above arrays have been found, the remainder of the solution is straightforward. From Eq. (2-34) the assembled flexibility matrix $\mathbf{F}_S$ is calculated by matrix multiplication as

$$\mathbf{F}_S = \mathbf{B}_{MS}^T \mathbf{F}_M \mathbf{B}_{MS} = \begin{bmatrix} \mathbf{F}_{JJ} & \mathbf{F}_{JQ} \\ \mathbf{F}_{QJ} & \mathbf{F}_{QQ} \end{bmatrix}$$

$$= \frac{L}{EA}\left[\begin{array}{cccc|cc} 4.828 & 0 & -1 & 0 & 3.828 & 3.414 \\ 0 & 1 & 0 & 0 & 0 & -0.707 \\ -1 & 0 & 1 & 0 & -1 & -0.707 \\ 0 & 0 & 0 & 1 & 0 & -0.707 \\ \hline 3.828 & 0 & -1 & 0 & 3.828 & 2.707 \\ 3.414 & -0.707 & -0.707 & -0.707 & 2.707 & 4.828 \end{array}\right]$$

which is partitioned in accordance with Eq. (2-35). Then the solution for the redundants $\mathbf{A}_Q$ may be obtained from Eq. (2-37), using a null matrix for $\mathbf{D}_Q$, as follows:

$$\mathbf{A}_Q = \mathbf{F}_{QQ}^{-1}(\mathbf{D}_Q - \mathbf{F}_{QJ}\mathbf{A}_J)$$

$$= -\frac{EA}{L}\begin{bmatrix} 0.4328 & -0.2426 \\ -0.2426 & 0.3431 \end{bmatrix}\begin{bmatrix} 3.828 & 0 & -1 & 0 \\ 3.414 & -0.707 & -0.707 & -0.707 \end{bmatrix}\begin{bmatrix} -1 \\ -2 \\ 0 \\ 0 \end{bmatrix}\frac{PL}{EA}$$

$$= \begin{bmatrix} 1.172 \\ -0.243 \end{bmatrix} P$$

These results for the redundants agree with those found earlier in Example 3 of Art. 2.3.

The next step in the solution consists of finding the joint displacements $\mathbf{D}_J$ from Eq. (2-36a):

$$\mathbf{D}_J = \mathbf{F}_{JJ}\mathbf{A}_J + \mathbf{F}_{JQ}\mathbf{A}_Q$$

$$= \frac{L}{EA}\begin{bmatrix} 4.828 & 0 & -1 & 0 \\ 0 & 1 & 0 & 0 \\ -1 & 0 & 1 & 0 \\ 0 & 0 & 0 & 1 \end{bmatrix}\begin{bmatrix} -1 \\ -2 \\ 0 \\ 0 \end{bmatrix} P + \frac{L}{EA}\begin{bmatrix} 3.828 & 3.414 \\ 0 & -0.707 \\ -1 & -0.707 \\ 0 & -0.707 \end{bmatrix}\begin{bmatrix} 1.172 \\ -0.243 \end{bmatrix} P$$

$$= \begin{bmatrix} -4.828 \\ -2 \\ 1 \\ 0 \end{bmatrix}\frac{PL}{EA} + \begin{bmatrix} 3.656 \\ 0.172 \\ -1 \\ 0.172 \end{bmatrix}\frac{PL}{EA} = \begin{bmatrix} -1.172 \\ -1.828 \\ 0 \\ 0.172 \end{bmatrix}\frac{PL}{EA}$$

Then member end-actions $\mathbf{A}_M$ (axial forces in this case) are obtained from Eq. (2-38), using a null matrix for $\mathbf{A}_{MF}$, as follows:

$$\mathbf{A}_M = \mathbf{A}_{MF} + \mathbf{B}_{MJ}\mathbf{A}_J + \mathbf{B}_{MQ}\mathbf{A}_Q$$

$$= \begin{bmatrix} 0 & 1 & 0 & 0 \\ 0 & 0 & 0 & 0 \\ 1.414 & 0 & 0 & 0 \\ -1 & 0 & 1 & 0 \\ -1 & 0 & 0 & 0 \\ 0 & 0 & 0 & 1 \end{bmatrix}\begin{bmatrix} -1 \\ -2 \\ 0 \\ 0 \end{bmatrix} P + \begin{bmatrix} 0 & -0.707 \\ 0 & 1 \\ 1.414 & 1 \\ -1 & -0.707 \\ 0 & -0.707 \\ 0 & -0.707 \end{bmatrix}\begin{bmatrix} 1.172 \\ -0.243 \end{bmatrix} P$$

$$
= \begin{bmatrix} -2 \\ 0 \\ -1.414 \\ 1 \\ 1 \\ 0 \end{bmatrix} P + \begin{bmatrix} 0.172 \\ -0.243 \\ 1.414 \\ -1 \\ 0.172 \\ 0.172 \end{bmatrix} P = \begin{bmatrix} -1.828 \\ -0.243 \\ 0 \\ 0 \\ 1.172 \\ 0.172 \end{bmatrix} P
$$

Finally, the support reactions $\mathbf{A}_{\mathrm{R}}$ may be calculated with Eq. (2-40). In this instance the matrix $\mathbf{A}_{\mathrm{RC}}$ is null, and the result is

$$
\mathbf{A}_{\mathrm{R}} = -\mathbf{A}_{\mathrm{RC}} + \mathbf{B}_{\mathrm{RJ}}\mathbf{A}_{\mathrm{J}} + \mathbf{B}_{\mathrm{RQ}}\mathbf{A}_{\mathrm{Q}}
$$

$$
= \begin{bmatrix} -1 & 0 & 0 & -1 \\ -1 & -1 & 0 & 0 \\ 1 & 0 & -1 & 0 \end{bmatrix} \begin{bmatrix} -1 \\ -2 \\ 0 \\ 0 \end{bmatrix} P + \begin{bmatrix} -1 & 0 \\ -1 & 0 \\ 1 & 0 \end{bmatrix} \begin{bmatrix} 1.172 \\ -0.243 \end{bmatrix} P
$$

$$
= \begin{bmatrix} 1 \\ 3 \\ -1 \end{bmatrix} P + \begin{bmatrix} -1.172 \\ -1.172 \\ 1.172 \end{bmatrix} P = \begin{bmatrix} -0.172 \\ 1.828 \\ 0.172 \end{bmatrix} P
$$

Thus, the complete analysis of the truss, including the determination of all joint displacements, all member end-actions, and all support reactions, has been accomplished by the formalized approach to the flexibility method.

Example 2. The two-span continuous beam shown in Fig. 2-19a is subjected to a uniform load w (equal to $4P/L$) in span AB and two concentrated forces P in span BC. Both members have the same length L, but the flexural rigidity of member AB is twice that of member BC. The objectives of the analysis are to determine end-actions, support reactions, and joint displacements for the beam.

Members AB and BC are denoted as members 1 and 2, respectively; and the k ends of these members are taken as ends B and C, respectively. Thus, the four end-actions used in the analysis are the shearing forces and bending moments at the right-hand ends of the members. These end-actions will be numbered sequentially from 1 to 4, and their positive directions are in accordance with the axes shown in the figure (see Fig. 2-13a for the sign convention for end-actions).

The redundant actions are selected as the reactive moment at A and the bending moment at B. Because there is a moment applied as a load at joint B when the actual loads are replaced by combined joint loads, ambiguity will result unless A_{Q2} is taken a small distance to one side of the joint. Therefore, in this example it is assumed that A_{Q2} is the bending moment to the right of joint B. The released structure associated with this selection of redundants is shown in Fig. 2-19b. Note that both redundants are assumed to be positive when they cause compression on the upper part of the beam.

The reactive forces in the y direction at the supports, as well as the two joint displacements (rotations at B and C), are identified in Fig. 2-19c. A fourth reactive action is the couple at support A, but since this reaction is chosen as one of the redundants it is not necessary to include it in the vector $\mathbf{A}_{\mathrm{R}}$.

The determination of the combined joint loads is illustrated in Fig. 2-19d, which shows the two fixed-end beams obtained by restraining the joints against displacements. The negatives of the fixed-end actions for these beams constitute the equivalent joint loads. Then, the equivalent loads are added to any actual joint loads

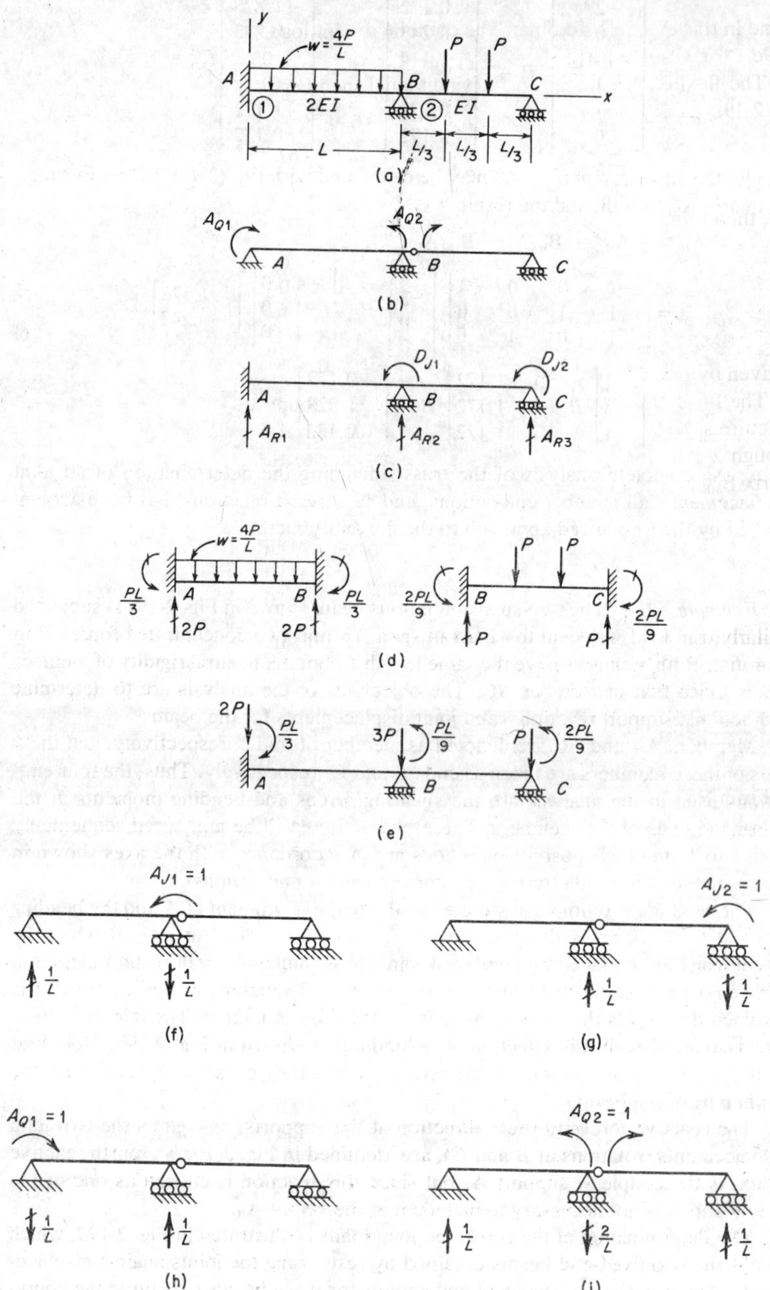

Fig. 2-19. Example 2: Continuous beam.

(none in this example) to obtain the combined joint loads, which are shown in Fig. 2-19e.

The flexibility matrices for the individual members can be found readily (see Eq. 2-19):

$$\mathbf{F}_{M2} = 2\mathbf{F}_{M1} = \frac{L}{12EI}\begin{bmatrix} 4L^2 & 6L \\ 6L & 12 \end{bmatrix}$$

and, therefore, the unassembled flexibility matrix $\mathbf{F}_M$ is

$$\mathbf{F}_M = \frac{L}{12EI}\begin{bmatrix} 2L^2 & 3L & 0 & 0 \\ 3L & 6 & 0 & 0 \\ 0 & 0 & 4L^2 & 6L \\ 0 & 0 & 6L & 12 \end{bmatrix}$$

as given by Eq. (2-25).

The $\mathbf{B}_{MS}$ and $\mathbf{B}_{RS}$ matrices for this example are found by analyzing the released structure subjected to the conditions $A_{J1} = 1$ through $A_{Q2} = 1$ shown in Figs. 2-19f through 2-19i. From the member end-actions due to these unit loads, the 4×4 matrix $\mathbf{B}_{MS}$ is found to be

$$\mathbf{B}_{MS} = \begin{bmatrix} \mathbf{B}_{MJ} & \mathbf{B}_{MQ} \end{bmatrix} = \frac{1}{L}\left[\begin{array}{cc|cc} -1 & 0 & 1 & -1 \\ L & 0 & 0 & L \\ 0 & -1 & 0 & 1 \\ 0 & L & 0 & 0 \end{array}\right]$$

Similarly, the support reactions shown in Figs. 2-19f through 2-19i are listed in the 3×4 matrix $\mathbf{B}_{RS}$, as follows:

$$\mathbf{B}_{RS} = \begin{bmatrix} \mathbf{B}_{RJ} & \mathbf{B}_{RQ} \end{bmatrix} = \frac{1}{L}\left[\begin{array}{cc|cc} 1 & 0 & 1 & 1 \\ -1 & 1 & -1 & -2 \\ 0 & -1 & 0 & 1 \end{array}\right]$$

From Eq. (2-34) the assembled flexibility matrix $\mathbf{F}_S$ is found to be

$$\mathbf{F}_S = \mathbf{B}_{MS}^T \mathbf{F}_M \mathbf{B}_{MS} = \begin{bmatrix} \mathbf{F}_{JJ} & \mathbf{F}_{JQ} \\ \mathbf{F}_{QJ} & \mathbf{F}_{QQ} \end{bmatrix} = \frac{L}{12EI}\left[\begin{array}{cc|cc} 2 & 0 & 1 & 2 \\ 0 & 4 & 0 & 2 \\ \hline 1 & 0 & 2 & 1 \\ 2 & 2 & 1 & 6 \end{array}\right]$$

With this matrix available, the solution for the redundants may be carried out, using a null matrix for $\mathbf{D}_Q$ in Eq. (2-37):

$$\mathbf{A}_Q = \mathbf{F}_{QQ}^{-1}(\mathbf{D}_Q - \mathbf{F}_{QJ}\mathbf{A}_J) = -\frac{12EI}{11L}\begin{bmatrix} 6 & -1 \\ -1 & 2 \end{bmatrix}\begin{bmatrix} 1 & 0 \\ 2 & 2 \end{bmatrix}\begin{bmatrix} 1 \\ 2 \end{bmatrix}\frac{PL^2}{(9)(12)EI} = \begin{bmatrix} 0 \\ -1 \end{bmatrix}\frac{PL}{9}$$

Then the joint displacements $\mathbf{D}_J$ can be determined from Eq. (2-36a), as follows:

$$\begin{aligned} \mathbf{D}_J &= \mathbf{F}_{JJ}\mathbf{A}_J + \mathbf{F}_{JQ}\mathbf{A}_Q \\ &= \frac{L}{12EI}\begin{bmatrix} 2 & 0 \\ 0 & 4 \end{bmatrix}\begin{bmatrix} 1 \\ 2 \end{bmatrix}\frac{PL}{9} + \frac{L}{12EI}\begin{bmatrix} 1 & 2 \\ 0 & 2 \end{bmatrix}\begin{bmatrix} 0 \\ -1 \end{bmatrix}\frac{PL}{9} \\ &= \begin{bmatrix} 1 \\ 4 \end{bmatrix}\frac{PL^2}{54EI} + \begin{bmatrix} -1 \\ -1 \end{bmatrix}\frac{PL^2}{54EI} = \begin{bmatrix} 0 \\ 1 \end{bmatrix}\frac{PL^2}{18EI} \end{aligned}$$

Next, the member end-actions $\mathbf{A}_M$ are calculated with Eq. (2-38). In this example the nonzero fixed-end actions in the vector $\mathbf{A}_{MF}$ are drawn from Fig. 2-19d. Thus,

$$\mathbf{A}_M = \mathbf{A}_{MF} + \mathbf{B}_{MJ}\mathbf{A}_J + \mathbf{B}_{MQ}\mathbf{A}_Q$$

$$= \begin{bmatrix} 18 \\ -3L \\ 9 \\ -2L \end{bmatrix} \frac{P}{9} + \frac{1}{L}\begin{bmatrix} -1 & 0 \\ L & 0 \\ 0 & -1 \\ 0 & L \end{bmatrix}\begin{bmatrix} 1 \\ 2 \end{bmatrix}\frac{PL}{9} + \frac{1}{L}\begin{bmatrix} 1 & -1 \\ 0 & L \\ 0 & 1 \\ 0 & 0 \end{bmatrix}\begin{bmatrix} 0 \\ -1 \end{bmatrix}\frac{PL}{9}$$

$$= \begin{bmatrix} 18 \\ -3L \\ 9 \\ -2L \end{bmatrix} \frac{P}{9} + \begin{bmatrix} -1 \\ L \\ -2 \\ 2L \end{bmatrix}\frac{P}{9} + \begin{bmatrix} 1 \\ -L \\ -1 \\ 0 \end{bmatrix}\frac{P}{9} = \begin{bmatrix} 6 \\ -L \\ 2 \\ 0 \end{bmatrix}\frac{P}{3}$$

Similarly, the support reactions $\mathbf{A}_R$ can be found with Eq. (2-40), using nonzero reactions in the vector $\mathbf{A}_{RC}$ drawn from Fig. 2-19e:

$$\mathbf{A}_R = -\mathbf{A}_{RC} + \mathbf{B}_{RJ}\mathbf{A}_J + \mathbf{B}_{RQ}\mathbf{A}_Q$$

$$= -\begin{bmatrix} -2 \\ -3 \\ -1 \end{bmatrix} P + \frac{1}{L}\begin{bmatrix} 1 & 0 \\ -1 & 1 \\ 0 & -1 \end{bmatrix}\begin{bmatrix} 1 \\ 2 \end{bmatrix}\frac{PL}{9} + \frac{1}{L}\begin{bmatrix} 1 & 1 \\ -1 & -2 \\ 0 & 1 \end{bmatrix}\begin{bmatrix} 0 \\ -1 \end{bmatrix}\frac{PL}{9}$$

$$= \begin{bmatrix} 18 \\ 27 \\ 9 \end{bmatrix}\frac{P}{9} + \begin{bmatrix} 1 \\ 1 \\ 2 \end{bmatrix}\frac{P}{9} + \begin{bmatrix} -1 \\ 2 \\ -1 \end{bmatrix}\frac{P}{9} = \begin{bmatrix} 3 \\ 5 \\ 1 \end{bmatrix}\frac{2P}{3}$$

The redundants $\mathbf{A}_Q$ calculated above pertain to the beam with combined joint loads (see Fig. 2-19e). However, these redundants are actually member end-actions; so their final values must account for initial fixed-end actions, as in Eq. (2-38). Thus,

$$(\mathbf{A}_Q)_{final} = \mathbf{A}_{QF} + \mathbf{A}_Q = \begin{bmatrix} -3 \\ -2 \end{bmatrix}\frac{PL}{9} + \begin{bmatrix} 0 \\ -1 \end{bmatrix}\frac{PL}{9} = \begin{bmatrix} -1 \\ -1 \end{bmatrix}\frac{PL}{3}$$

where the fixed-end actions in $\mathbf{A}_{QF}$ are obtained from Fig. 2-19d.

Example 3. The plane frame shown in Fig. 2-20a has constant EI for all members and is subjected to two concentrated forces. Only flexural effects are to be considered in the analysis, and the reactive force in the x direction at support D is chosen as the redundant action.

The j ends of members 1, 2, and 3 are selected arbitrarily as points A, B, and D, respectively. In accordance with this selection, the member end-actions at the k ends are the shearing forces and bending moments shown in Fig. 2-20b.

The joint displacements for the frame are also shown in Fig. 2-20b. They consist of the rotation at joint A, the translation and rotation at B, the rotation at C, and the rotation at D. Because axial deformations are neglected in this example, there is only one independent joint translation (D_{J2}). The reactions for the frame also are identified in Fig. 2-20b; only three reactions need to be considered, inasmuch as the fourth reaction is the redundant A_Q.

The equivalent joint loads are found by restraining all joints of the structure, calculating the fixed-end actions caused by the member loads, and then reversing their directions. In this example the only member load is the force P acting on member BC; thus, when joints B and C are restrained, the fixed-end actions are those shown in Fig. 2-20c. The negatives of these fixed-end actions are added to the actual joint loads to give the combined loads shown in Fig. 2-20d.

The member flexibility matrices are obtained by applying Eq. (2-19) to each member of the frame:

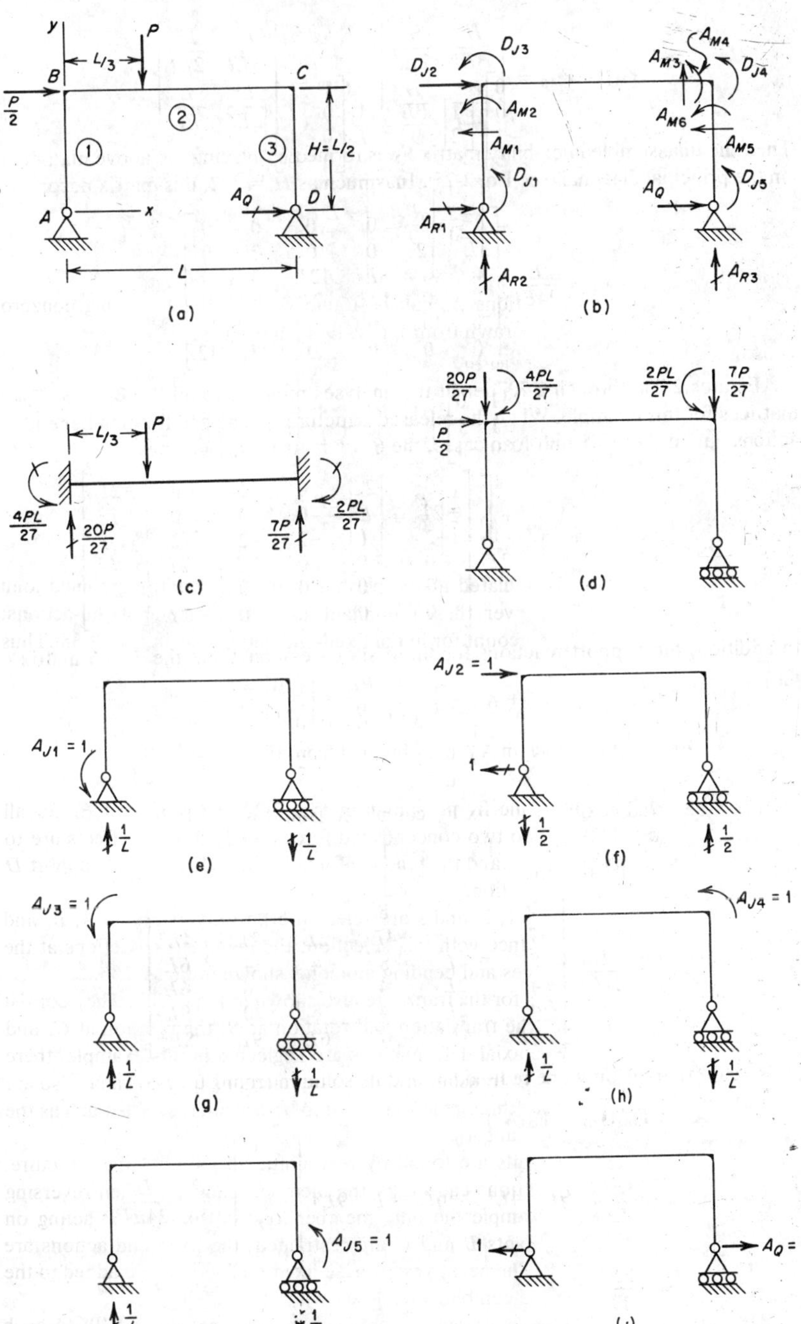

Fig. 2-20. Example 3: Plane frame (flexural effects only).

$$\mathbf{F}_{M1} = \mathbf{F}_{M3} = \begin{bmatrix} \dfrac{H^3}{3EI} & \dfrac{H^2}{2EI} \\ \dfrac{H^2}{2EI} & \dfrac{H}{EI} \end{bmatrix} \qquad \mathbf{F}_{M2} = \begin{bmatrix} \dfrac{L^3}{3EI} & \dfrac{L^2}{2EI} \\ \dfrac{L^2}{2EI} & \dfrac{L}{EI} \end{bmatrix}$$

Then the unassembled flexibility matrix $\mathbf{F}_M$ is formed by placing the above matrices on the principal diagonal (see Eq. 2-25). Inasmuch as $H = L/2$, this matrix becomes

$$\mathbf{F}_M = \frac{L}{24EI}\begin{bmatrix} L^2 & 3L & 0 & 0 & 0 & 0 \\ 3L & 12 & 0 & 0 & 0 & 0 \\ 0 & 0 & 8L^2 & 12L & 0 & 0 \\ 0 & 0 & 12L & 24 & 0 & 0 \\ 0 & 0 & 0 & 0 & L^2 & 3L \\ 0 & 0 & 0 & 0 & 3L & 12 \end{bmatrix}$$

Figures 2-20e through 2-20j depict the analyses required to fill the $\mathbf{B}_{MS}$ and $\mathbf{B}_{RS}$ matrices for this example. When the released structure is analyzed for member end-actions due to these six unit-load cases, the 6×6 matrix $\mathbf{B}_{MS}$ becomes

$$\mathbf{B}_{MS} = \begin{bmatrix}\mathbf{B}_{MJ} & \mathbf{B}_{MQ}\end{bmatrix} = \frac{1}{2L}\left[\begin{array}{ccccc|c} 0 & -2L & 0 & 0 & 0 & -2L \\ -2L & L^2 & 0 & 0 & 0 & L^2 \\ -2 & L & -2 & -2 & -2 & 0 \\ 0 & 0 & 0 & 2L & 2L & L^2 \\ 0 & 0 & 0 & 0 & 0 & 2L \\ 0 & 0 & 0 & 0 & -2L & -L^2 \end{array}\right]$$

In addition, the support reactions for these six cases constitute the 3×6 matrix $\mathbf{B}_{RS}$:

$$\mathbf{B}_{RS} = \begin{bmatrix}\mathbf{B}_{RJ} & \mathbf{B}_{RQ}\end{bmatrix} = \frac{1}{2L}\left[\begin{array}{ccccc|c} 0 & -2L & 0 & 0 & 0 & -1 \\ 2 & -L & 2 & 2 & 2 & 0 \\ -2 & L & -2 & -2 & -2 & 0 \end{array}\right]$$

The assembled flexibility matrix $\mathbf{F}_S$, calculated from Eq. (2-34), is

$$\mathbf{F}_S = \mathbf{B}_{MS}^T \mathbf{F}_M \mathbf{B}_{MS} = \begin{bmatrix}\mathbf{F}_{JJ} & \mathbf{F}_{JQ} \\ \mathbf{F}_{QJ} & \mathbf{F}_{QQ}\end{bmatrix}$$
$$= \frac{L}{24EI}\left[\begin{array}{ccccc|c} 20 & -7L & 8 & -4 & -4 & -9L \\ -7L & 3L^2 & -4L & 2L & 2L & 4L^2 \\ 8 & -4L & 8 & -4 & -4 & -6L \\ -4 & 2L & -4 & 8 & 8 & 6L \\ -4 & 2L & -4 & 8 & 20 & 9L \\ \hline -9L & 4L^2 & -6L & 6L & 9L & 8L^2 \end{array}\right]$$

In this case there is only one redundant, and its solution from Eq. (2-37) is

$$\mathbf{A}_Q = \mathbf{F}_{QQ}^{-1}(\mathbf{D}_Q - \mathbf{F}_{QJ}\mathbf{A}_J)$$
$$= -\frac{3EI}{L^3}\begin{bmatrix}-9L & 4L^2 & -6L & 6L & 9L\end{bmatrix}\begin{bmatrix}0 \\ 27 \\ -8L \\ 4L \\ 0\end{bmatrix}\left(\frac{P}{54}\right)\frac{L}{24EI}$$
$$= -\frac{5P}{12}$$

Next, the joint displacements $\mathbf{D}_J$ are determined using Eq. (2-36a), as follows:

$$\mathbf{D}_J = \mathbf{F}_{JJ}\mathbf{A}_J + \mathbf{F}_{JQ}\mathbf{A}_Q$$

$$= \frac{L}{24EI}\begin{bmatrix} 20 & -7L & 8 & -4 & -4 \\ -7L & 3L^2 & -4L & 2L & 2L \\ 8 & -4L & 8 & -4 & -4 \\ -4 & 2L & -4 & 8 & 8 \\ -4 & 2L & -4 & 8 & 20 \end{bmatrix}\begin{bmatrix} 0 \\ 27 \\ -8L \\ 4L \\ 0 \end{bmatrix}\frac{P}{54} + \frac{L}{24EI}\begin{bmatrix} -9L \\ 4L^2 \\ -6L \\ 6L \\ 9L \end{bmatrix}\left(-\frac{5P}{12}\right)$$

$$= \begin{bmatrix} -538 \\ 242L \\ -376 \\ 236 \\ 236 \end{bmatrix}\frac{PL^2}{2592EI} + \begin{bmatrix} 405 \\ -180L \\ 270 \\ -270 \\ -405 \end{bmatrix}\frac{PL^2}{2592EI} = \begin{bmatrix} -133 \\ 62L \\ -106 \\ -34 \\ -169 \end{bmatrix}\frac{PL^2}{2592EI}$$

Then the member end-actions are obtained from Eq. (2-38):

$$\mathbf{A}_M = \mathbf{A}_{MF} + \mathbf{B}_{MJ}\mathbf{A}_J + \mathbf{B}_{MQ}\mathbf{A}_Q$$

$$= \begin{bmatrix} 0 \\ 0 \\ 7 \\ -2L \\ 0 \\ 0 \end{bmatrix}\frac{P}{27} + \frac{1}{2L}\begin{bmatrix} 0 & -2L & 0 & 0 & 0 \\ -2L & L^2 & 0 & 0 & 0 \\ -2 & L & -2 & -2 & -2 \\ 0 & 0 & 0 & 2L & 2L \\ 0 & 0 & 0 & 0 & 0 \\ 0 & 0 & 0 & 0 & -2L \end{bmatrix}\begin{bmatrix} 0 \\ 27 \\ -8L \\ 4L \\ 0 \end{bmatrix}\frac{P}{54} + \frac{1}{2L}\begin{bmatrix} -2L \\ L^2 \\ 0 \\ L^2 \\ 2L \\ -L^2 \end{bmatrix}$$

$$= \begin{bmatrix} 0 \\ 0 \\ 56 \\ -16L \\ 0 \\ 0 \end{bmatrix}\frac{P}{216} + \begin{bmatrix} -108 \\ 54L \\ 70 \\ 16L \\ 0 \\ 0 \end{bmatrix}\frac{P}{216} + \begin{bmatrix} 90 \\ -45L \\ 0 \\ -45L \\ -90 \\ 45L \end{bmatrix}\frac{P}{216} = \begin{bmatrix} -2 \\ L \\ 14 \\ -5L \\ -10 \\ 5L \end{bmatrix}\frac{P}{24}$$

For this purpose, the fixed-end actions in $\mathbf{A}_{MF}$ are drawn from Fig. 2-20c. Finally, Eq. (2-40) gives the support reactions as

$$\mathbf{A}_R = -\mathbf{A}_{RC} + \mathbf{B}_{RJ}\mathbf{A}_J + \mathbf{B}_{RQ}\mathbf{A}_Q$$

$$= -\begin{bmatrix} 0 \\ -20 \\ -7 \end{bmatrix}\frac{P}{27} + \frac{1}{2L}\begin{bmatrix} 0 & -2L & 0 & 0 & 0 \\ 2 & -L & 2 & 2 & 2 \\ -2 & L & -2 & -2 & -2 \end{bmatrix}\begin{bmatrix} 0 \\ 27 \\ -8L \\ 4L \\ 0 \end{bmatrix}\frac{P}{24} + \begin{bmatrix} -1 \\ 0 \\ 0 \end{bmatrix}\left(-\frac{5P}{12}\right)$$

$$= \begin{bmatrix} 0 \\ 80 \\ 28 \end{bmatrix}\frac{P}{108} + \begin{bmatrix} -54 \\ -35 \\ 35 \end{bmatrix}\frac{P}{108} + \begin{bmatrix} 45 \\ 0 \\ 0 \end{bmatrix}\frac{P}{108} = \begin{bmatrix} -1 \\ 5 \\ 7 \end{bmatrix}\frac{P}{12}$$

where the terms in the vector $\mathbf{A}_{RC}$ are obtained from Fig. 2-20d. In this case the equivalent joint load $-20P/27$ at point B is transmitted directly to the support at point A. Similarly, the equivalent joint load $-7P/27$ at point C is transmitted directly to the support at point D.

Problems

The problems for Art. 2.3 are to be solved by the flexibility method using Eq. (2-9). In each problem the redundant actions are to be obtained, unless stated otherwise.

2.3-1. Determine the reactive moments at each end of the fixed-end beam shown in Fig. 2-3a, due to the force P and moment M acting at the middle of the span. The beam has constant flexural rigidity EI and length L. Select the reactive moments themselves as the redundant actions, and assume these moments are positive when they produce compression on the bottom of the beam. Take the first redundant at end A of the beam and the second at end B.

2.3-2. Analyze the two-span beam shown in Fig. 2-2a by taking the reactive moment at support A and the bending moment just to the left of support B as the redundants Q_1 and Q_2, respectively. Assume that these moments are positive when they produce compression on the top of the beam. Also, assume that the loads on the beam are $P_1 = 2P$, $M = PL$, $P_2 = P$, $P_3 = P$, and the flexural rigidity EI is constant.

2.3-3. Analyze the two-span beam of Fig. 2-2a if support B is displaced downward by a small distance s. Select the redundants to be the vertical reactions at supports B and C, as shown in Fig. 2-2a, and omit the effects of the loads in the analysis. Assume that EI is constant for both spans.

2.3-4. Find the redundant actions for the two-span beam of Fig. 2-2a using the released structure shown in Fig. 2-2b. Assume that EI is constant for the beam and that the loads are $P_1 = P$, $M = 0$, $P_2 = P$, $P_3 = P$. Number the redundants from left to right along the beam; also, assume that the redundant moment is positive when counterclockwise, and that the redundant force is positive when upward.

2.3-5. Determine the bending moments at supports B and C of the continuous beam shown in the figure, using these moments as the redundants Q_1 and Q_2, respectively. Assume that the redundants are positive when they produce compression on the top of the beam. The beam has constant flexural rigidity EI.

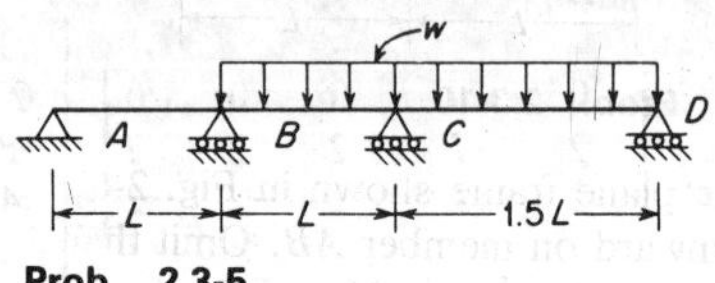

Prob. 2.3-5.

2.3-6. Find the bending moments at supports B and C of the continuous beam (see figure), using these moments as the redundants Q_1 and Q_2, respectively. Assume that Q_1 and Q_2 are positive when they produce compression on the top of the beam. The flexural rigidity of the beam is EI.

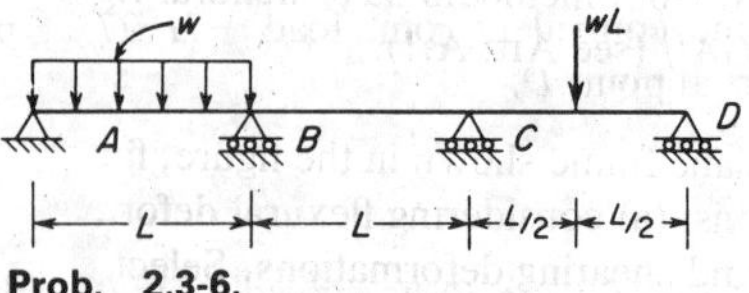

Prob. 2.3-6.

2.3-7. Analyze the plane truss shown in Fig. 2-5a by taking the forces in the two diagonal members AD and BC as the redundants Q_1 and Q_2, respectively. Assume that tension in a member is positive, and assume that there are no support displacements. All members have the same axial rigidity EA.

2.3-8. Solve the preceding problem using the force in member AD and the reaction at D as the redundants Q_1 and Q_2, respectively.

2.3-9. Find the forces in the members AE and CE of the truss shown in the figure by taking these member forces as the redundants Q_1 and Q_2, respectively. Consider tension in a member to be positive. The axial rigidity for the vertical and horizontal members is EA and for the diagonal members is $2EA$.

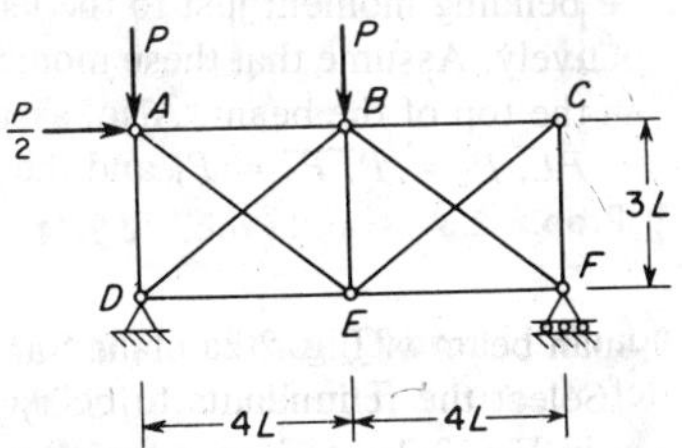

Prob. 2.3-9.

2.3-10. Find the forces in the members AB and BC of the truss in the figure, using these forces as the redundants Q_1 and Q_2, respectively. Assume that tensile force is positive. Let $H = 3L/4$ and $P_1 = P_2 = P$, and assume that all members have the same axial rigidity EA.

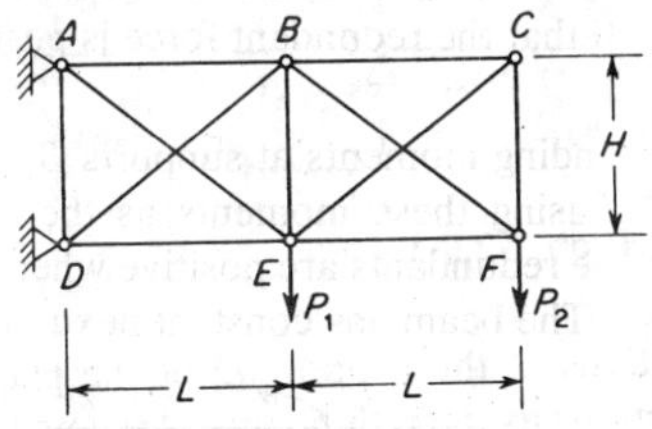

Prob. 2.3-10.

2.3-11. Analyze the plane frame shown in Fig. 2-6a due to a uniform load of intensity w acting downward on member AB. Omit the load P from the analysis, and let $H = L$. Use the redundants Q_1, Q_2, and Q_3 shown in Fig. 2-6b, and consider only the effects of flexural deformations.

2.3-12. Obtain the flexibility matrix $\mathbf{F}$ for the plane frame in Fig. 2-6a, corresponding to the redundants shown in Fig. 2-6b, by considering flexural, axial, and shearing deformations. Both members have flexural rigidity EI, axial rigidity EA, and shearing rigidity GA/f (see Art. A.1).

2.3-13. For the plane frame shown in the figure, find the flexibility matrix $\mathbf{F}$ for the following conditions: (a) considering flexural deformations only and (b) considering flexural, axial, and shearing deformations. Select the redundants Q_1, Q_2, and Q_3 as the axial force, shearing force, and bending moment, respectively, at the midpoint of member BC. Take these quantities as positive when in the same directions as the actions Q_1, Q_2, and Q_3 shown in Fig. 2-6b. Assume that all members of the frame have flexural rigidity EI, axial rigidity EA, and shearing rigidity GA/f.

2.3-14. Obtain the flexibility matrix **F** for the plane frame in the preceding problem if the redundants Q_1, Q_2, and Q_3 are the horizontal force (positive to the right), vertical force (positive upward), and moment (positive when counterclockwise), respectively, at support D. Consider only the effects of flexural deformations and assume that each member has constant EI.

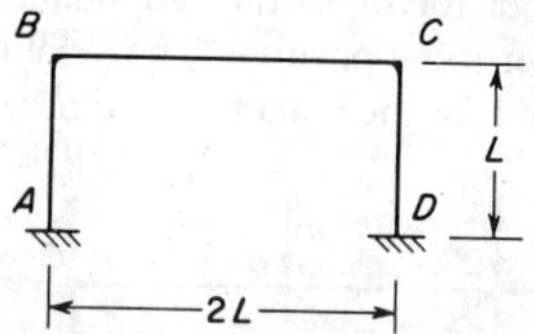

Prob. 2.3-13 and **Prob. 2.3-14.**

2.3-15. Find the redundants Q_1 and Q_2 for the plane frame shown in the figure, considering only flexural deformations. The flexural rigidity EI is the same for all members.

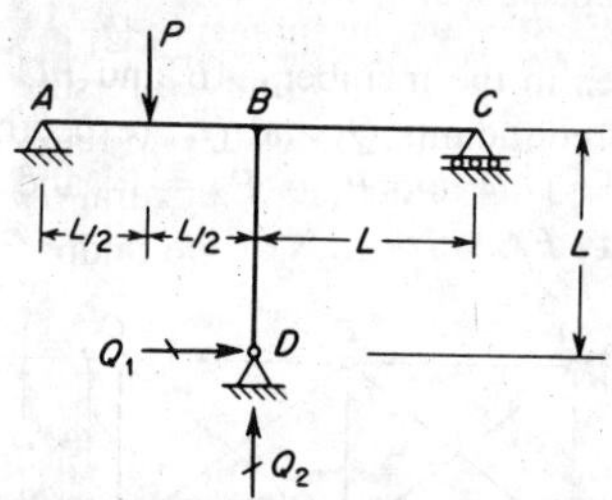

Prob. 2.3-15.

2.3-16. Find the reactions at the support D of the plane frame shown in the figure by taking those reactions as the redundants. Assume that Q_1 is the horizontal reaction (positive to the right) and Q_2 is the vertical reaction (positive upward). Consider only the effects of flexural deformations in the analysis. Let $H = L/3$; and assume that the flexural rigidity of member BC is $2EI$, whereas that for members AB and CD is EI.

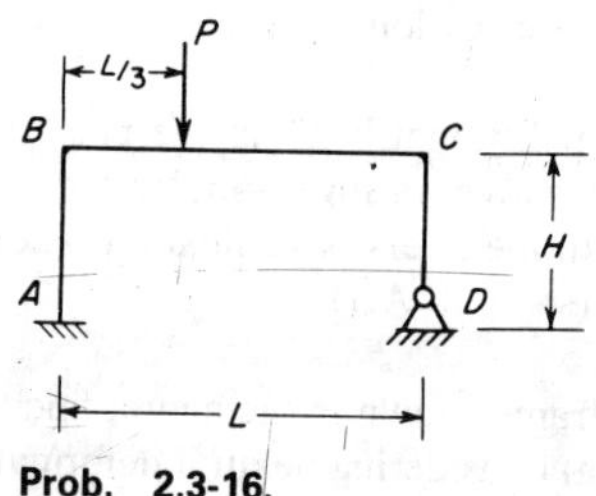

Prob. 2.3-16.

2.3-17. Obtain the flexibility matrix **F** for the grid shown in Fig. 2-7a, considering both flexural and torsional deformations, if the reactions at support C are taken as the redundants. Assume that Q_1 is the force in the positive y direction, Q_2 is the positive moment about the x axis, and Q_3 is the positive moment about the z axis.) The flexural and torsional rigidities of the members are EI and GJ, respectively.

2.3-18. Calculate the redundant reaction Q at support D for the horizontal grid shown in the figure. The grid is constructed of three members (AB, BC, and CD) that are rigidly joined at right angles and supported by simple supports at A, B, C, and D. Each member has flexural rigidity EI, torsional rigidity GJ, and length L. Assume that the loads P act at the midpoints of members AB and BC.

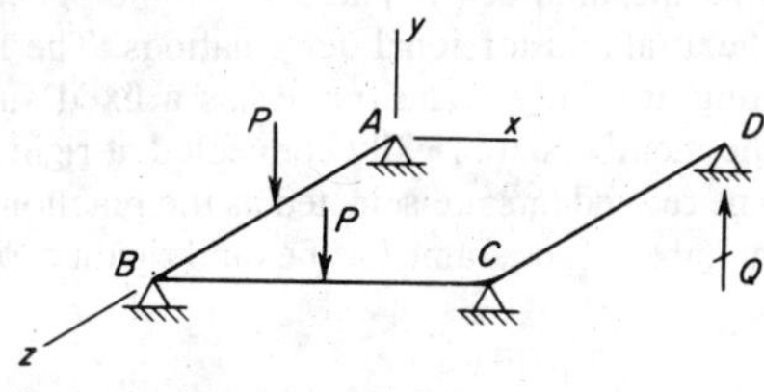

Prob. 2.3-18.

2.3-19. Determine the redundant moments Q_1 and Q_2 at support D of the grid shown in the figure. The supports at A and C are simple supports, and the support at D is a fixed support. The members of the grid are rigidly connected at joint B; also, members AB, BC, and BD each have flexural rigidity EI and torsional rigidity GJ (where $GJ = EI$). The load P acts at the midpoint of member BC, and $L_1 = L/2$.

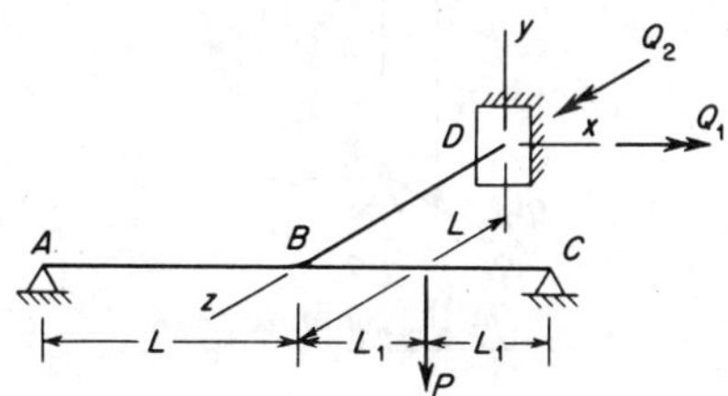

Prob. 2.3-19.

2.3-20. The space frame $ABCD$ has pin supports at A, C, and D; thus, each support is capable of resisting a force in any direction, but is not capable of trans-

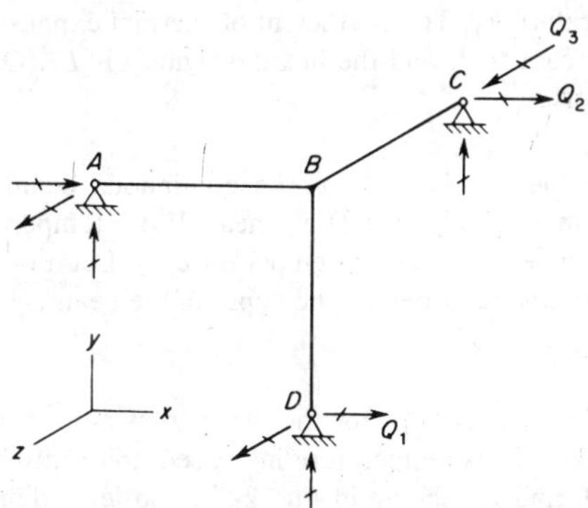

Prob. 2.3-20.

mitting a moment. The members are rigidly connected at joint B. Each member of the frame is of tubular cross section with length L, flexural rigidity EI, and axial rigidity EA. The effects of both flexural and axial deformations are to be considered. Obtain in literal form the flexibility matrix $\mathbf{F}$, assuming that the redundants Q_1, Q_2, and Q_3 are the reactions at D in the x direction, at C in the x direction, and at C in the z direction, respectively, as shown in the figure.

2.3-21. Determine the matrices $\mathbf{D}_{QL}$ and $\mathbf{F}$ for the space frame shown in the figure, considering flexural and torsional deformations. The load on the frame is a vertical force P acting at joint B. The frame has a fixed support at A and a pin support at D, and the members are rigidly connected at right angles to one another at joints B and C. The redundants are selected as the reactions at joint D, as shown in the figure. Each member of the frame has flexural rigidity EI and torsional rigidity GJ.

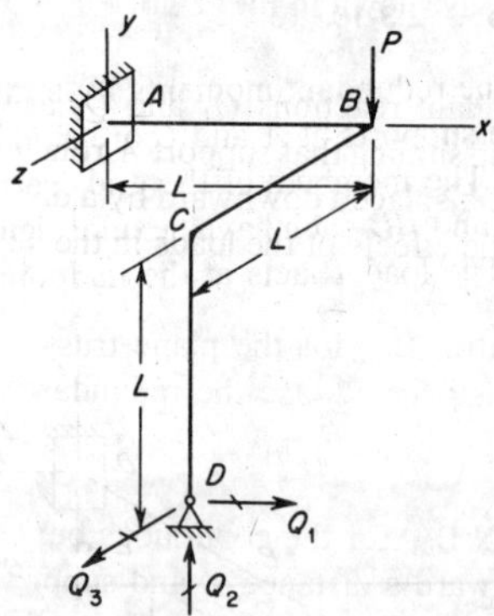

Prob. 2.3-21.

2.4-1. Find the redundant reactions Q_1 and Q_2 at support B of the fixed-end beam shown in Fig. 2-3a, assuming that the beam is subjected to a temperature differential such that the top of the beam is at temperature T_2 and the bottom of the beam is at temperature T_1. The coefficient of thermal expansion for the beam is α, the depth of the beam is d, and the flexural rigidity is EI. Omit the effects of the loads in the analysis.

2.4-2. Obtain the matrix $\mathbf{D}_{QT}$ for the continuous beam shown in Fig. 2-4a, assuming that members BC and CD are heated to a temperature T_1 on the lower surface while the upper surface is at temperature T_2. Let α denote the coefficient of thermal expansion, and let d denote the depth of the beam. (Use the redundants Q_1 and Q_2 shown in Fig. 2-4b.)

2.4-3. Obtain the matrix $\mathbf{D}_{QT}$ for the plane truss shown in Fig. 2-5a, assuming that the entire truss has its temperature increased uniformly by an amount T. Use the redundants Q_1 and Q_2 shown in Fig. 2-5b, and let α denote the coefficient of thermal expansion.

2.4-4. Find the redundant reactions Q_1 and Q_2 at support B of the fixed-end beam shown in Fig. 2-3a, assuming that the beam is constructed initially of two straight segments rigidly joined together but slightly out of alignment (see figure). The angle between the two halves of the beam is β, and the flexural rigidity of the beam is EI. Do not include the effects of the loads in the analysis.

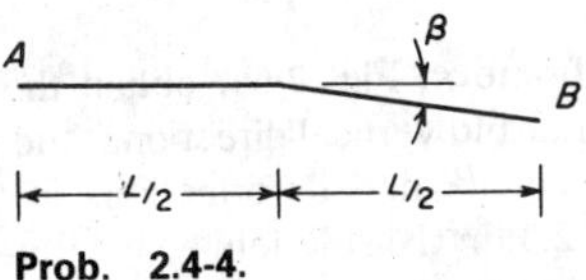

Prob. 2.4-4.

2.4-5. Obtain the matrix $\mathbf{D}_{QP}$ for the plane truss shown in Fig. 2-5a assuming that members AB and CD are constructed with lengths $L + e$ instead of L. Take the redundants Q_1 and Q_2 as shown in Fig. 2-5b.

2.4-6. Find the redundant reactions Q_1 and Q_2 at support B of the fixed-end beam shown in Fig. 2-3a, assuming that support A rotates β radians in the clockwise direction and support B is displaced downward by a distance s. The flexural rigidity of the beam is EI. Omit the effects of the loads in the analysis.

2.4-7. Obtain the matrix $\mathbf{D}_{QR}$ for the plane truss of Fig. 2-5a if support C is displaced downward a distance s. Use the redundants Q_1 and Q_2 shown in Fig. 2-5b.

2.4-8. Find the matrix $\mathbf{D}_{QR}$ for the continuous beam shown in Fig. 2-4a if support B is displaced downward a distance s_1 and support C is displaced downward a distance s_2. The redundants Q_1 and Q_2 are to be taken as shown in Fig. 2-4b.

2.4-9. Obtain the matrix $\mathbf{D}_{QC}$, representing the combined effects in the released structure, for the plane frame shown in Fig. 2-6a if, in addition to the load P, the frame has its temperature increased uniformly by an amount T, support A is displaced downward by an amount s, and support C rotates clockwise by an amount β. The members of the frame have flexural rigidity EI and coefficient of thermal expansion α. Take the redundants Q_1, Q_2, and Q_3 as shown in Fig. 2-6b.

Problems 2.5-1 to 2.5-3 are to be solved using Eqs. (2-14), (2-15), and (2-16). Assume that reactions and joint displacements are positive to the right, upward, and counterclockwise.

2.5-1. Find the joint displacements and support reactions for the continuous beam shown in Fig. 2-4a, assuming that the intensity w of the distributed load is such that $wL = P$. The beam has constant flexural rigidity EI. The four joint displacements and four reactions are to be numbered consecutively from left to right in the figure. Use the solution given in Example 2, Art. 2.3, for the redundants.

2.5-2. For the plane truss shown in Fig. 2-5a, obtain the horizontal and vertical displacements of joint A, the forces in members AB, AC, and BD, and the horizontal and vertical reactions at support C. Assume that there are no support displacements, and consider only the effects of the loads in the analysis. Assume that all members have axial rigidity EA and that tension in a member is positive. Number the displacements, member forces, and reactions in the order stated above. Use the solution given in Example 3, Art. 2.3, for the redundants.

2.5-3. For the plane frame in Fig. 2-6a, obtain the displacements of joint B (translations in the horizontal and vertical directions, and rotation) and the reactions at support C. Consider only flexural deformations in the analysis, and use the results of Example 4, Art. 2.3, in the solution.

Fundamentals of the Stiffness Method

3.1 Introduction. The *stiffness method* (also known as the *displacement method*) is the primary method used in matrix analysis of structures. One of its advantages over the flexibility method is that it is conducive to computer programming. Once the analytical model of a structure has been defined, no further engineering decisions are required in the stiffness method in order to carry out the analysis. In this respect it differs from the flexibility method, although the two approaches have similar mathematical forms. In the flexibility method the unknown quantities are redundant actions that must be arbitrarily chosen; but in the stiffness method the unknowns are the joint displacements in the structure, which are automatically specified. Thus, in the stiffness method the number of unknowns to be calculated is the same as the degree of kinematic indeterminacy of the structure.

In this chapter the stiffness method is developed on the basis of writing joint equilibrium equations in terms of stiffness coefficients and unknown joint displacements. A kinematically determinate version of the original structure (with the joint displacements restrained) proves to be useful for the purpose of characterizing such equilibrium conditions as action superposition equations. Because the stiffness method requires extensive use of restraint actions due to various causes, the material in Appendix B will be quite helpful. As in the preceding chapter, the effects of temperature changes, prestrains, and support displacements are considered as well as the effects of loads; and procedures for calculating member end-actions and support reactions are also developed. After the basic concepts of the stiffness method have been thoroughly discussed, the method is formalized into a procedure for assembling the overall stiffness matrix of the structure from individual member stiffness matrices. This approach will be further systematized (and simplified) in Chapter 4 as preparation for programming on a digital computer.

3.2 Stiffness Method. In order to illustrate the concepts of the stiffness method in their simplest form, consider the analysis of the beam in

Fig. 3-1a. This beam has a fixed support at A and a roller support at B; and it is subjected to a uniform load of intensity w. The beam is kinematically indeterminate to the first degree (if axial deformations are neglected) because the only unknown joint displacement is the rotation θ_B at joint B. The first phase of the analysis is to determine this rotation. Then the various actions and displacements throughout the beam can be determined, as will be shown later.

In the flexibility method a statically determinate released structure is obtained by altering the actual structure in such a manner that the selected redundant actions are zero. The analogous operation in the stiffness method is to obtain a kinematically determinate structure by altering the actual structure in such a manner that all unknown displacements are zero. Since the unknown displacements are the translations and rotations of the joints, they can be made equal to zero by restraining the joints of the structure against displacements of any kind. The structure obtained by restraining all joints of the actual structure is called the *restrained structure*. For the beam in Fig. 3-1a the restrained structure is obtained by restraining joint B against rotation. Thus, the restrained structure is the fixed-end beam shown in Fig. 3-1b.

When the loads act on the restrained beam (see Fig. 3-1b), there will be a couple M_B developed at support B. This reactive couple is in the clockwise direction and is given by the expression

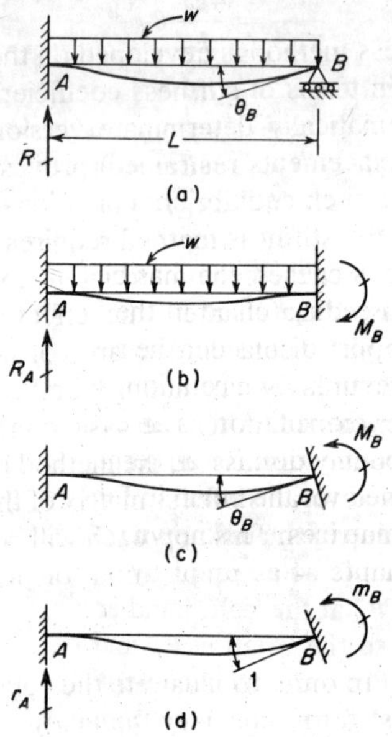

Fig. 3-1. Illustration of stiffness method.

$$M_B = \frac{wL^2}{12} \tag{3-1}$$

which can be found from the table of fixed-end moments given in Appendix B (see Table B-1). Note that the couple M_B is an action corresponding to the rotation θ_B, which is the unknown quantity in the analysis. Because there is no couple at joint B in the actual beam of Fig. 3-1a, it is necessary to consider next that the restrained beam is subjected to a couple equal and opposite to the couple M_B. Such a couple is shown acting on the beam in Fig. 3-1c. When the actions acting on the two beams in (b) and (c) are superimposed, they produce the actions on the actual beam. Thus, the analysis of the beam in Fig. 3-1a can be considered as the superposition of the analyses shown in Figs. 3-1b and 3-1c. It follows, therefore, that the rotation produced by the couple M_B in Fig. 3-1c is equal to θ_B, the unknown rotation in the actual beam.

The relation between the moment M_B and the rotation θ_B in the beam of Fig. 3-1c is

$$M_B = \frac{4EI}{L}\theta_B \tag{3-2}$$

in which EI is the flexural rigidity of the beam. Equation (3-2) is obtained from Case 3 of Table B-4. Equating the two expressions for the moment M_B from Eqs. (3-1) and (3-2) gives the equation

$$\frac{wL^2}{12} = \frac{4EI}{L}\theta_B$$

from which

$$\theta_B = \frac{wL^3}{48EI}$$

Thus, the rotation at joint B of the beam has been determined.

In a manner analogous to that used in the flexibility method, it is convenient in the above example to consider the restrained structure under the effect of a unit value of the unknown rotation. It is also more systematic to formulate the equation for the rotation as an equation of superposition and to use a consistent sign convention for all terms in the equation. This procedure will now be followed for the beam in Fig. 3-1.

The effect of a unit value of the unknown rotation is shown in Fig. 3-1d, where the restrained beam is acted upon by a couple m_B that produces a unit value of the rotation θ_B at the right-hand end. Since the moment m_B is an action corresponding to the rotation θ_B and caused by a unit value of that rotation (while all other joint displacements are zero), it is recognized that m_B is a *stiffness coefficient* for the restrained structure (see Art. 1.10). The value of the couple m_B (see Eq. 3-2) is

$$m_B = \frac{4EI}{L}$$

In formulating the equation of superposition the couples at joint B will be superimposed as follows. The couple in the restrained beam subjected to the load (Fig. 3-1b) will be added to the couple m_B (corresponding to a unit value of θ_B) multiplied by θ_B itself. The sum of these two terms must give the couple at joint B in the actual beam, which is zero in this example. All terms in the superposition equation will be expressed in the same sign convention, namely, that couples and rotations at joint B are positive when counterclockwise. According to this convention, the couple M_B in the beam of Fig. 3-1b is negative:

$$M_B = -\frac{wL^2}{12}$$

The equation for the superposition of moments at support B now becomes

$$M_B + m_B\theta_B = 0 \tag{3-3}$$

or

$$-\frac{wL^2}{12} + \frac{4EI}{L}\theta_B = 0$$

Solving this equation yields

$$\theta_B = \frac{wL^3}{48EI}$$

which is the same result as before. The positive sign for the result means that the rotation is counterclockwise.

The most essential part of the preceding solution consists of writing the action superposition equation (3-3), which expresses the fact that the moment at B in the actual beam is zero. Included in this equation are the moment caused by the loads on the restrained structure and the moment caused by rotating the end B of the restrained structure. The latter term in the equation was expressed conveniently as the product of the moment caused by a unit value of the unknown displacement (stiffness coefficient) times the unknown displacement itself. The two effects are summed algebraically, using the same sign convention for all terms in the equation. When the equation is solved for the unknown displacement, the sign of the result will give the true direction of the displacement. The equation may be referred to either as an *equation of action superposition* or as an *equation of joint equilibrium*. The latter name is used because the equation may be considered to express the equilibrium of moments at joint B.

Having obtained the unknown rotation θ_B for the beam, it is now possible to calculate other quantities, such as member end-actions and reactions. As an example, assume that the reactive force R acting at support A of the beam (Fig. 3-1a) is to be found. This force is the sum of the corresponding reactive force R_A at support A in Fig. 3-1b and θ_B times the force r_A in Fig. 3-1d, as shown in the following superposition equation:

$$R = R_A + \theta_B r_A$$

The forces R_A and r_A can be readily calculated for the restrained beam (see Case 6, Table B-1, and Case 3, Table B-4):

$$R_A = \frac{wL}{2} \qquad r_A = \frac{6EI}{L^2}$$

When these values, as well as the previously found value for θ_B, are substituted into the equation above, the result is

$$R = \frac{5wL}{8}$$

The same concepts can be used to calculate any other actions or displacements for the beam. However, in all cases the unknown joint displacements must be found first.

If a structure is kinematically indeterminate to more than one degree, a more organized approach to the solution, as well as a generalized notation, must be introduced. For this purpose, the same two-span beam used previously as an example in the flexibility method will be analyzed now by the stiffness method (see Fig. 3-2a). The beam has constant flexural rigidity EI and is subjected to the loads P_1, M, P_2, and P_3. Since rotations can occur at joints B and C, the structure is kinematically indeterminate to the second degree when axial deformations are neglected. Let the unknown rotations at these joints be D_1 and D_2, respectively, and assume that counterclockwise rotations are positive. These unknown displacements may be determined by solving equations of superposition for the actions at joints B and C, as described in the following discussion.

The first step in the analysis consists of applying imaginary restraints at the joints to prevent all joint displacements. The restrained structure which is obtained by this means is shown in Fig. 3-2b and consists of two fixed-end beams. The restrained structure is assumed to be acted upon by all of the loads except those that correspond to the unknown displacements. Thus, only the loads P_1, P_2, and P_3 are shown in Fig. 3-2b. All loads that correspond to the unknown joint displacements, such as the couple M in this example, are taken into account later. The moments A_{DL1} and A_{DL2} (Fig. 3-2b) are the actions of the restraints (against the restrained structure) corresponding to D_1 and D_2, respectively, and caused by loads acting on the structure. For example, the restraint action A_{DL1} is the sum of the reactive moment at B due to the load P_1 acting on member AB and the reactive moment at B due to the load P_2 acting on member BC. These actions can

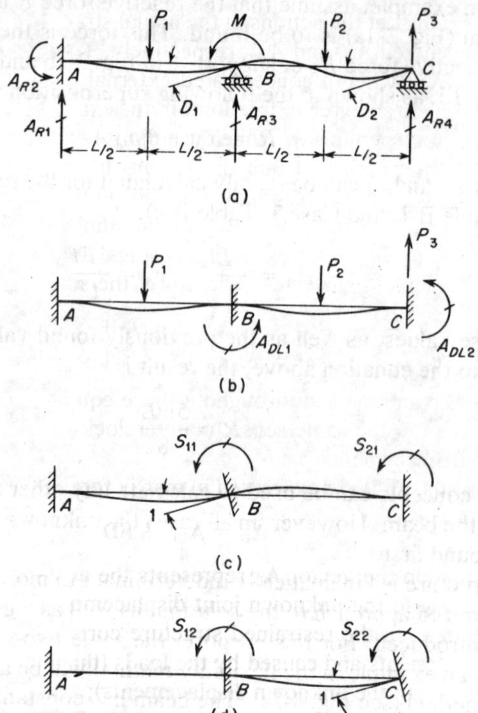

Fig. 3-2. Illustration of stiffness method.

be found with the aid of formulas for fixed-end moments in beams (see Appendix B), as illustrated later.

In order to generate the stiffness coefficients at joints B and C, unit values of the unknown displacements D_1 and D_2 are induced separately in the restrained structure. A unit displacement corresponding to D_1 consists of a unit rotation of joint B, as shown in Fig. 3-2c. The displacement D_2 remains equal to zero in this beam. Thus, the actions corresponding to D_1 and D_2 are the stiffness coefficients S_{11} and S_{21}, respectively. These stiffnesses consist of the couples exerted by the restraints on the beam at joints B and C, respectively. The calculation of these actions is not difficult when formulas for fixed-end moments in beams are available. Their determination in this example will be described later. The condition that D_2 is equal to unity while D_1 is equal to zero is shown in Fig. 3-2d. In the figure the stiffness S_{12} is the action corresponding to D_1 while the stiffness S_{22} is the action corresponding to D_2. Note that in each case the stiffness coefficient is the action that the artificial restraint exerts upon the structure.

Two superposition equations expressing the conditions pertaining to the moments acting on the original structure (Fig. 3-2a) at joints B and C may now be written. Let the actions in the actual structure corresponding to D_1 and D_2 be denoted A_{D1} and A_{D2}, respectively. These actions will be zero in all cases except when a concentrated external action is applied at a joint corresponding to a degree of freedom. In the example of Fig. 3-2, the action A_{D1} is equal to the couple M while the action A_{D2} is zero. The superposition equations express the fact that the actions in the original structure (Fig. 3-2a) are equal to the corresponding actions in the restrained structure due to the loads (Fig. 3-2b) plus the corresponding actions in the restrained structure due to the unit displacements (Figs. 3-2c and 3-2d) multiplied by the displacements themselves. Therefore, the superposition equations are

$$\begin{aligned} A_{D1} &= A_{DL1} + S_{11}D_1 + S_{12}D_2 \\ A_{D2} &= A_{DL2} + S_{21}D_1 + S_{22}D_2 \end{aligned} \qquad (3\text{-}4)$$

The sign convention used throughout these equations is that moments are positive when in the same sense (counterclockwise) as the corresponding unknown displacements.

When Eqs. (3-4) are expressed in matrix form, they become

$$\mathbf{A_D} = \mathbf{A_{DL}} + \mathbf{SD} \qquad (3\text{-}5)$$

In this equation the vector $\mathbf{A_D}$ represents the actions in the original beam corresponding to the unknown joint displacements $\mathbf{D}$; the vector $\mathbf{A_{DL}}$ represents actions in the restrained structure corresponding to the unknown joint displacements and caused by the loads (that is, all loads except those corresponding to the unknown displacements); and $\mathbf{S}$ is the stiffness matrix corresponding to the unknown displacements. The stiffness matrix $\mathbf{S}$ could also be denoted $\mathbf{A_{DD}}$, since it represents actions corresponding to the unknown joint displacements and caused by unit values of those displacements. For the example of Fig. 3-2 the matrices are as follows:

$$\mathbf{A_D} = \begin{bmatrix} A_{D1} \\ A_{D2} \end{bmatrix} \qquad \mathbf{A_{DL}} = \begin{bmatrix} A_{DL1} \\ A_{DL2} \end{bmatrix} \qquad \mathbf{S} = \begin{bmatrix} S_{11} & S_{12} \\ S_{21} & S_{22} \end{bmatrix} \qquad \mathbf{D} = \begin{bmatrix} D_1 \\ D_2 \end{bmatrix}$$

In general, these matrices will have as many rows as there are unknown joint displacements. Thus, if d represents the number of unknown displacements, the order of the stiffness matrix $\mathbf{S}$ is $d \times d$, while $\mathbf{A_D}$, $\mathbf{A_{DL}}$, and $\mathbf{D}$ are vectors of order $d \times 1$.

Subtracting $\mathbf{A_{DL}}$ from both sides of Eq. (3-5) and then premultiplying by $\mathbf{S}^{-1}$ gives the following equation for the unknown displacements:

$$\mathbf{D} = \mathbf{S}^{-1}(\mathbf{A_D} - \mathbf{A_{DL}}) \qquad (3\text{-}6)$$

This equation represents the solution for the displacements in matrix terms because the elements of $\mathbf{A_D}$, $\mathbf{A_{DL}}$, and $\mathbf{S}$ are either known or may be obtained from the restrained structure. Moreover, the member end-actions and reactions for the structure may be found after the joint displacements are known. The procedure for performing such calculations will be illustrated later.

At this point it can be observed that the term $-\mathbf{A}_{DL}$ in Eq. (3-6) represents a vector of *equivalent joint loads,* as described previously in Art. 1.12. Such loads are defined as the negatives of restraint actions corresponding to the unknown joint displacements, and they are caused by loads applied to members of the restrained structure. When these equivalent joint loads are added to the actual joint loads $\mathbf{A}_D$, the results are called *combined joint loads* (see Art. 1.12). Thus, the parentheses in Eq. (3-6) contain combined joint loads for the stiffness method of analysis.

In order to demonstrate the use of Eq. (3-6), the beam in Fig. 3-2a will be analyzed for the values of the loads previously given as

$$P_1 = 2P \qquad M = PL \qquad P_2 = P \qquad P_3 = P$$

When the loads P_1, P_2, and P_3 act upon the restrained structure (Fig. 3-2b), the actions A_{DL1} and A_{DL2}, corresponding to D_1 and D_2, respectively, are developed by the restraints at B and C. Since the couple M corresponds to one of the unknown displacements, it is taken into account later by means of the matrix $\mathbf{A}_D$. The actions A_{DL1} and A_{DL2} are found from the formulas for fixed-end moments (see Case 1, Table B-1):

$$A_{DL1} = -\frac{P_1L}{8} + \frac{P_2L}{8} = -\frac{PL}{8}$$

$$A_{DL2} = -\frac{P_2L}{8} = -\frac{PL}{8}$$

Therefore, the matrix $\mathbf{A}_{DL}$ is

$$\mathbf{A}_{DL} = \frac{PL}{8}\begin{bmatrix} -1 \\ -1 \end{bmatrix}$$

It may be observed from these calculations that the load P_3 does not enter into the matrix $\mathbf{A}_{DL}$, hence, it does not affect the calculations for the joint displacements. However, this load does affect the calculations for the reactions of the actual beam, which are given later.

The stiffness matrix $\mathbf{S}$ consists of the stiffness coefficients shown in Figs. 3-2c and 3-2d. Each coefficient is a couple corresponding to one of the unknown displacements and due to a unit value of one of the displacements. Elements of the first column of the stiffness matrix are shown in Fig. 3-2c, and elements of the second column in Fig. 3-2d. In order to find these coefficients, consider first the fixed-end beam shown in Fig. 3-3. This

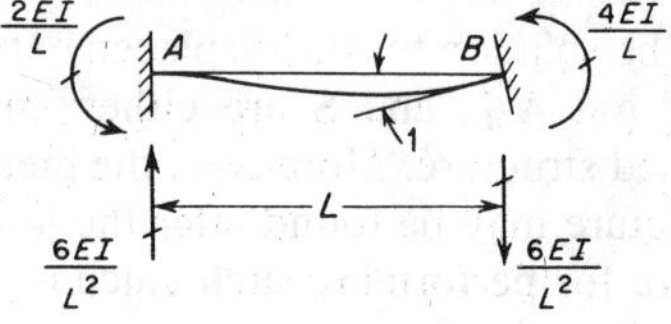

Fig. 3-3. Member stiffnesses for a beam member.

beam is subjected to a unit rotation at end B, and, as a result, the moment developed at end B is $4EI/L$ while the moment at the opposite end is $2EI/L$ (see Case 3, Table B-4). The reactive forces at the ends of the beam are each equal to $6EI/L^2$, and are also indicated in the figure. All of the actions shown in Fig. 3-3 are called *member stiffnesses,* because they are actions at the ends of the member due to a unit displacement of one end. The end of the beam which undergoes the unit displacement is sometimes called the *near end* of the beam, and the opposite end is called the *far end*. Thus, the member stiffnesses at ends A and B are sometimes referred to as the stiffnesses at the far and near ends of the beam. The subject of member stiffnesses is dealt with more extensively in Art. 3.5 and in Chapter 4.

The task of calculating the joint stiffnesses S_{11} and S_{21} in Fig. 3-2c may now be performed through the use of member stiffnesses. When the beam is rotated through a unit angle at joint B, a moment equal to $4EI/L$ is developed at B because of the rotation of the end of member AB. Also, a moment equal to $4EI/L$ is developed at B because of the rotation of the end of member BC. Thus, the total moment at B, equal to S_{11}, is

$$S_{11} = \frac{4EI}{L} + \frac{4EI}{L} = \frac{8EI}{L} \tag{a}$$

The stiffness S_{21} is the moment developed at joint C when joint B is rotated through a unit angle. Since joint C is at the far end of the member, the stiffness coefficient is

$$S_{21} = \frac{2EI}{L}$$

Both S_{11} and S_{21} are positive because they act in the counterclockwise sense. The stiffness coefficients S_{12} and S_{22} are shown in Fig. 3-2d. The first of these is equal to $2EI/L$, since it is an action at the far end of the member BC, while the latter is equal to $4EI/L$ since it is at the near end of the member.

The stiffness matrix $\mathbf{S}$ can be formed using the stiffness coefficients described above:

$$\mathbf{S} = \frac{EI}{L}\begin{bmatrix} 8 & 2 \\ 2 & 4 \end{bmatrix}$$

Each of the elements in $\mathbf{S}$ is a *joint stiffness,* inasmuch as it represents the action at one of the joints of the structure due to a unit value of a displacement at one of the joints. In this example, the joint stiffness S_{11} (see Fig. 3-2c) is the sum of the near-end member stiffnesses (see Eq. a) for the two members meeting at the joint. Similarly, the stiffness S_{22} is a near-end member stiffness. On the other hand, the stiffnesses S_{12} and S_{21} consist of far-end member stiffnesses for members which connect to a joint that is rotated. In a more general example, it will be found that stiffness elements on the principal diagonal are always composed of near-end stiffnesses while those off the diagonal may consist of either far-end or near-end stiffnesses, as will be seen in later examples. After the stiffness matrix **S** is determined, its inverse can be found:

$$\mathbf{S}^{-1} = \frac{L}{14EI}\begin{bmatrix} 2 & -1 \\ -1 & 4 \end{bmatrix}$$

The next matrix to be determined is the matrix $\mathbf{A}_D$, representing the actions in the actual structure corresponding to the unknown displacements. In this example the external load which corresponds to the rotation D_1 is the couple M (equal to PL) at joint B. There is no moment at joint C corresponding to D_2, and therefore the matrix $\mathbf{A}_D$ is

$$\mathbf{A}_D = \begin{bmatrix} PL \\ 0 \end{bmatrix}$$

Now that the matrices $\mathbf{A}_D$, $\mathbf{S}^{-1}$, and $\mathbf{A}_{DL}$ have been obtained, the matrix of displacements **D** in the actual structure can be found by substituting them into Eq. (3-6), as follows:

$$\mathbf{D} = \frac{L}{14EI}\begin{bmatrix} 2 & -1 \\ -1 & 4 \end{bmatrix}\left\{\begin{bmatrix} PL \\ 0 \end{bmatrix} - \frac{PL}{8}\begin{bmatrix} -1 \\ -1 \end{bmatrix}\right\} = \frac{PL^2}{112EI}\begin{bmatrix} 17 \\ -5 \end{bmatrix}$$

Thus, the rotations D_1 and D_2 at joints B and C are

$$D_1 = \frac{17PL^2}{112EI} \qquad D_2 = -\frac{5PL^2}{112EI} \tag{b}$$

These results agree with the joint displacements found by the flexibility method in the example of Art. 2.5.

The next step after finding the joint displacements is to determine the member end-actions and the reactions for the structure. As in the flexibility method, there are two approaches that can be followed when performing the calculations by hand. One approach is to obtain the end-actions and reactions by making separate calculations after the joint displacements have been found. The other approach is to set up the calculations in a systematic manner in parallel with the calculations for finding the displacements. In this chapter only the second approach will be utilized.

In order to show how the calculations are performed, consider again the two-span beam shown in Fig. 3-2. As in the flexibility method, the matrices of member end-actions and reactions in the actual structure (Fig. 3-2a) will be denoted $\mathbf{A}_M$ and $\mathbf{A}_R$, respectively. In the restrained structure subjected to the loads (Fig. 3-2b), the matrices of end-actions and reactions corresponding to $\mathbf{A}_M$ and $\mathbf{A}_R$ will be denoted $\mathbf{A}_{ML}$ and $\mathbf{A}_{RL}$, respectively. It should be noted again that when any reference is made to the loads on the restrained structure, it is assumed that all of the actual loads are taken into account except those that correspond to an unknown displacement. Thus, the joint load M shown in Fig. 3-2a does not appear on the restrained structure in Fig. 3-2b. However, all other loads are considered to act on the restrained beam in Fig. 3-2b, including the load P_3. This load does not affect the end-actions A_{ML} in the restrained structure, but the reactions A_{RL} are affected. Each of the matrices $\mathbf{A_M}$ and $\mathbf{A_{ML}}$ is of order $m \times 1$, assuming that m represents the number of member end-actions to be found. Similarly, the matrices $\mathbf{A_R}$ and $\mathbf{A_{RL}}$ are of order $r \times 1$, in which r denotes the number of reactions to be found.

In the restrained structure subjected to unit displacements (Figs. 3-2c and 3-2d), the matrices of end-actions and reactions will be denoted $\mathbf{A_{MD}}$ and $\mathbf{A_{RD}}$, respectively. The first column of each of the matrices will contain the actions obtained from the restrained beam in Fig. 3-2c while the second column is made up of actions obtained from the beam in Fig. 3-2d. In the general case the matrices $\mathbf{A_{MD}}$ and $\mathbf{A_{RD}}$ are of order $m \times d$ and $r \times d$, respectively, in which d represents the number of unknown displacements.

The superposition equations for the end-actions and reactions in the actual structure may now be expressed in matrix form:

$$\mathbf{A_M} = \mathbf{A_{ML}} + \mathbf{A_{MD}}\mathbf{D} \tag{3-7}$$

$$\mathbf{A_R} = \mathbf{A_{RL}} + \mathbf{A_{RD}}\mathbf{D} \tag{3-8}$$

The above two equations and Eq. (3-5) together constitute the three action superposition equations of the stiffness method. The complete solution of a structure consists of solving for the matrix $\mathbf{D}$ of displacements from Eq. (3-6) and then substituting into Eqs. (3-7) and (3-8) to determine $\mathbf{A_M}$ and $\mathbf{A_R}$. When this is done, all joint displacements, member end-actions, and reactions for the structure will be known.

Consider now the use of Eqs. (3-7) and (3-8) in the solution of the two-span beam shown in Fig. 3-2. The unknown displacements $\mathbf{D}$ have already been found (see Eqs. b), and all that remains is the determination of the matrices $\mathbf{A_{ML}}$, $\mathbf{A_{RL}}$, $\mathbf{A_{MD}}$, and $\mathbf{A_{RD}}$. Assume that the member end-actions to be calculated are the shearing force A_{M1} and moment A_{M2} at end B of member AB, and the shearing force A_{M3} and moment A_{M4} at end B of member BC. These are the same end-actions considered previously in the solution by the flexibility method (see Fig. 2-11b), and are selected solely for illustrative purposes. Also, assume that the reactions to be calculated are the force A_{R1} and couple A_{R2} at support A, and the forces A_{R3} and A_{R4} at supports B and C (see Fig. 3-2a). The first two of these reactions are the same as in the earlier solution, and the last two are the redundants from the earlier solution. All of these actions are assumed positive when either upward or counterclockwise.

In the restrained structure subjected to the loads (Fig. 3-2b), the end-actions and reactions in terms of the loads P_1, P_2, and P_3 are seen to be as follows:

$$A_{ML1} = \frac{P_1}{2} \quad A_{ML2} = -\frac{P_1 L}{8} \quad A_{ML3} = \frac{P_2}{2} \quad A_{ML4} = \frac{P_2 L}{8}$$

$$A_{RL1} = \frac{P_1}{2} \quad A_{RL2} = \frac{P_1 L}{8} \quad A_{RL3} = \frac{P_1}{2} + \frac{P_2}{2} \quad A_{RL4} = \frac{P_2}{2} - P_3$$

The values of the loads ($P_1 = 2P$, $P_2 = P$, $P_3 = P$) can now be substituted into these expressions, after which the matrices $\mathbf{A}_{ML}$ and $\mathbf{A}_{RL}$ can be formed:

$$\mathbf{A}_{ML} = \frac{P}{8}\begin{bmatrix} 8 \\ -2L \\ 4 \\ L \end{bmatrix} \qquad \mathbf{A}_{RL} = \frac{P}{4}\begin{bmatrix} 4 \\ L \\ 6 \\ -2 \end{bmatrix}$$

The matrices $\mathbf{A}_{MD}$ and $\mathbf{A}_{RD}$ are obtained from an analysis of the beams shown in Figs. 3-2c and 3-2d. For example, the member end-action A_{MD11} is the shearing force at end B of member AB due to a unit displacement corresponding to D_1 (Fig. 3-2c). Thus, this end-action is

$$A_{MD11} = -\frac{6EI}{L^2}$$

as can be seen from Fig. 3-3. The reaction A_{RD11} is the vertical force at support A in the beam of Fig. 3-2c, and is

$$A_{RD11} = \frac{6EI}{L^2}$$

In a similar manner, the other member end-actions and reactions can be found for the beam shown in Fig. 3-2c. These quantities constitute the first columns of the matrices $\mathbf{A}_{MD}$ and $\mathbf{A}_{RD}$. The terms in the second columns are found by similar analyses that are made for the beam shown in Fig. 3-2d. The results are as follows:

$$\mathbf{A}_{MD} = \frac{EI}{L^2}\begin{bmatrix} -6 & 0 \\ 4L & 0 \\ 6 & 6 \\ 4L & 2L \end{bmatrix} \qquad \mathbf{A}_{RD} = \frac{EI}{L^2}\begin{bmatrix} 6 & 0 \\ 2L & 0 \\ 0 & 6 \\ -6 & -6 \end{bmatrix}$$

Substituting the matrices $\mathbf{A}_{ML}$ and $\mathbf{A}_{MD}$ given above, as well as the matrix $\mathbf{D}$ obtained earlier, into Eq. (3-7) gives the following:

$$\mathbf{A}_M = \frac{P}{8}\begin{bmatrix} 8 \\ -2L \\ 4 \\ L \end{bmatrix} + \frac{EI}{L^2}\begin{bmatrix} -6 & 0 \\ 4L & 0 \\ 6 & 6 \\ 4L & 2L \end{bmatrix} \frac{PL^2}{112EI}\begin{bmatrix} 17 \\ -5 \end{bmatrix} = \frac{P}{56}\begin{bmatrix} 5 \\ 20L \\ 64 \\ 36L \end{bmatrix}$$

These results agree with those found previously by the flexibility method (see Art. 2.5). By substituting the matrices $\mathbf{A}_{RL}$, $\mathbf{A}_{RD}$, and $\mathbf{D}$ into Eq. (3-8) the reactions are found to be:

$$\mathbf{A}_R = \frac{P}{56}\begin{bmatrix} 107 \\ 31L \\ 69 \\ -64 \end{bmatrix}$$

These results also agree with those obtained previously by the flexibility method.

The method of solution described above for the two-span beam in Fig. 3-2 is quite general in its basic concepts, and the matrix equations (3-5) to (3-8) may be used in the solution of any type of framed structure. Also, the equations apply to structures having any number of degrees of kinematic indeterminacy n, in which case the order of the stiffness matrix $\mathbf{S}$ will be $n \times n$. Several examples illustrating the stiffness method are given in the following article.

3.3 Examples. The examples presented in this article illustrate the application of the stiffness method to several types of structures. In each example the object of the calculations is to determine the unknown joint displacements and certain selected member end-actions and reactions. Since the number of degrees of kinematic indeterminacy is small, the problems are suitable for solution by hand. All of the examples are solved in literal form in order to show clearly how the various terms in the matrices are obtained.

Example 1. The three-span continuous beam shown in Fig. 3-4a has fixed supports at A and D and roller supports at B and C; the length of the middle span is equal to 1.5 times the length of each end span. The loads on the beam are assumed to be two concentrated forces acting downward at the positions shown, a uniform load of intensity w acting on spans BC and CD, and a clockwise couple M applied at joint C. All members of the beam are assumed to have the same flexural rigidity EI.

The unknown joint displacements for the beam are the rotations at supports B and C, denoted D_1 and D_2, respectively, as shown in Fig. 3-4b. For illustrative purposes in this example, it will be assumed that the only member end-actions to be determined are the shearing force A_{M1} and the moment A_{M2} at the left-hand end of member AB, and the shearing force A_{M3} and moment A_{M4} at the left-hand end of member BC, as shown in Fig. 3-4b. The reactions to be found in this example are the vertical forces A_{R1} and A_{R2} at supports B and C, respectively. Other reactions could also be obtained if desired. All of the end-actions, reactions, and joint displacements are assumed to be positive when upward or counterclockwise.

The only load on the structure that corresponds to one of the unknown joint displacements is the couple M, which corresponds to the rotation D_2 (except that it is in the opposite sense). Therefore, the vector $\mathbf{A}_D$ of actions corresponding to the unknown displacements is

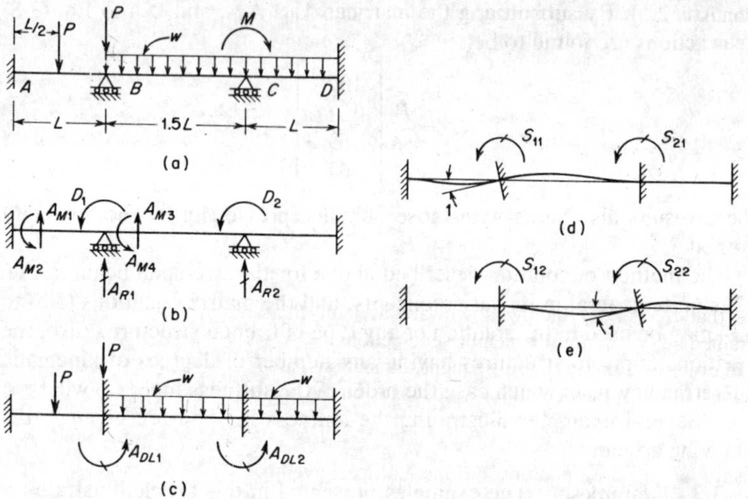

Fig. 3-4. Example 1: Continuous beam.

$$\mathbf{A}_{\mathrm{D}} = \begin{bmatrix} 0 \\ -M \end{bmatrix}$$

The remaining loads on the beam are taken into account by considering them to act on the restrained structure shown in Fig. 3-4c. This structure consists of fixed-end beams and is obtained by preventing joints B and C from rotating. The actions A_{DL1} and A_{DL2} exerted on the beam by the artificial restraints are the couples at supports B and C. Each of the couples is an action corresponding to a displacement D and caused by loads on the beam. The couples can be evaluated without difficulty by referring to the formulas for fixed-end moments given in Table B-1 of Appendix B (see Cases 1 and 6):

$$A_{DL1} = -\frac{PL}{8} + \frac{w(1.5L)^2}{12} = -\frac{PL}{8} + \frac{3wL^2}{16}$$

$$A_{DL2} = -\frac{w(1.5L)^2}{12} + \frac{wL^2}{12} = -\frac{5wL^2}{48}$$

Thus, the vector $\mathbf{A}_{\mathrm{DL}}$ becomes

$$\mathbf{A}_{\mathrm{DL}} = \frac{L}{48}\begin{bmatrix} -6P + 9wL \\ -5wL \end{bmatrix}$$

The end-actions A_{ML} and the reactions A_{RL} for the restrained beam of Fig. 3-4c can be determined also by referring to the table of fixed-end actions. For example, the end-action A_{ML1} is the shearing force at the left-hand end of member AB and is equal to $P/2$; similarly, the reaction A_{RL1} is the vertical reaction at support B, obtained as follows:

$$A_{RL1} = \frac{P}{2} + P + \frac{w(1.5L)}{2} = \frac{3P}{2} + \frac{3wL}{4}$$

By continuing in the same manner, all of the required actions in the restrained structure can be found. The resulting vectors $\mathbf{A}_{ML}$ and $\mathbf{A}_{RL}$ are:

$$\mathbf{A}_{ML} = \begin{bmatrix} \dfrac{P}{2} \\ \dfrac{PL}{8} \\ \dfrac{3wL}{4} \\ \dfrac{3wL^2}{16} \end{bmatrix} \qquad \mathbf{A}_{RL} = \begin{bmatrix} \dfrac{3P}{2} + \dfrac{3wL}{4} \\ \dfrac{5wL}{4} \end{bmatrix}$$

Note that the load P acting downward at joint B affects the reactions in the restrained beam but not the member end-actions.

In order to simplify the subsequent calculations, assume now that the following relationships exist between the various loads on the beam:

$$wL = P \qquad M = PL$$

Substitution of these relations into the matrices given in the preceding paragraphs yields

$$\mathbf{A}_D = PL \begin{bmatrix} 0 \\ -1 \end{bmatrix} \qquad \mathbf{A}_{DL} = \frac{PL}{48} \begin{bmatrix} 3 \\ -5 \end{bmatrix} \qquad \mathbf{A}_{ML} = \frac{P}{16} \begin{bmatrix} 8 \\ 2L \\ 12 \\ 3L \end{bmatrix} \qquad \mathbf{A}_{RL} = \frac{P}{4} \begin{bmatrix} 9 \\ 5 \end{bmatrix}$$

The next step in the solution is the analysis of the restrained beam for the effects of unit displacements corresponding to the unknowns. The two conditions to be taken into account are unit rotations at joints B and C, as illustrated in Figs. 3-4d and 3-4e. The four couples acting at joints B and C in these figures represent the elements of the stiffness matrix $\mathbf{S}$. With the formulas given in Fig. 3-3, each of these stiffnesses can be found without difficulty, as shown in the following calculations:

$$S_{11} = S_{22} = \frac{4EI}{L} + \frac{4EI}{1.5L} = \frac{20EI}{3L}$$

$$S_{12} = S_{21} = \frac{2EI}{1.5L} = \frac{4EI}{3L}$$

Therefore, the stiffness matrix $\mathbf{S}$ becomes

$$\mathbf{S} = \frac{4EI}{3L} \begin{bmatrix} 5 & 1 \\ 1 & 5 \end{bmatrix}$$

and the inverse matrix is

$$\mathbf{S}^{-1} = \frac{L}{32EI} \begin{bmatrix} 5 & -1 \\ -1 & 5 \end{bmatrix}$$

The joint displacements may now be found by substituting the matrices $\mathbf{S}^{-1}$, $\mathbf{A}_D$, and $\mathbf{A}_{DL}$ into Eq. (3-6) and evaluating $\mathbf{D}$, as follows:

$$\mathbf{D} = \frac{L}{32EI} \begin{bmatrix} 5 & -1 \\ -1 & 5 \end{bmatrix} \left\{ PL \begin{bmatrix} 0 \\ -1 \end{bmatrix} - \frac{PL}{48} \begin{bmatrix} 3 \\ -5 \end{bmatrix} \right\} = \frac{PL^2}{384EI} \begin{bmatrix} 7 \\ -53 \end{bmatrix}$$

This matrix gives the rotations at joints B and C of the continuous beam shown in Fig. 3-4a.

For the determination of the member end-actions and the reactions, it is necessary to consider again the restrained beams shown in Figs. 3-4d and 3-4e. In each of these beams there are end-actions and reactions that correspond to the end-actions and reactions selected previously and shown in Fig. 3-4b. These quantities for the beams with unit displacements are denoted A_{MD} and A_{RD}, respectively. For example, the end-action A_{MD11} is the shearing force at the left-hand end of member AB due to a unit value of D_1 (see Fig. 3-4d). The end-action A_{MD21} is the moment at the same location. In all cases the first subscript identifies the end-action itself and the second signifies the unit displacement that produces the action. The reactions in the beams of Figs. 3-4d and 3-4e follow a similar pattern, with A_{RD11} and A_{RD21} being the reactions at supports B and C, respectively, due to a unit value of the displacement D_1 (Fig. 3-4d). With this identification scheme in mind, and also using the formulas given in Fig. 3-3, it is not difficult to calculate the various end-actions and reactions. For example, the end-actions in the beam of Fig. 3-4d are the following:

$$A_{MD11} = \frac{6EI}{L^2} \qquad A_{MD21} = \frac{2EI}{L} \qquad A_{MD31} = \frac{6EI}{(1.5L)^2} = \frac{8EI}{3L^2} \qquad A_{MD41} = \frac{4EI}{1.5L} = \frac{8EI}{3L}$$

Also, the reactions in the same beam are

$$A_{RD11} = -\frac{6EI}{L^2} + \frac{6EI}{(1.5L)^2} = -\frac{10EI}{3L^2} \qquad A_{RD21} = -\frac{6EI}{(1.5L)^2} = -\frac{8EI}{3L^2}$$

Similarly, the end-actions and reactions for the beam in Fig. 3-4e can be found, after which the matrices $\mathbf{A}_{\mathrm{MD}}$ and $\mathbf{A}_{\mathrm{RD}}$ are constructed. These matrices are

$$\mathbf{A}_{\mathrm{MD}} = \frac{2EI}{3L^2}\begin{bmatrix} 9 & 0 \\ 3L & 0 \\ 4 & 4 \\ 4L & 2L \end{bmatrix} \qquad \mathbf{A}_{\mathrm{RD}} = \frac{2EI}{3L^2}\begin{bmatrix} -5 & 4 \\ -4 & 5 \end{bmatrix}$$

The final steps in the solution consist of calculating the matrices $\mathbf{A}_{\mathrm{M}}$ and $\mathbf{A}_{\mathrm{R}}$ for the end-actions and reactions in the original beam of Fig. 3-4a. These matrices are obtained by substituting into Eqs. (3-7) and (3-8) the matrices $\mathbf{D}$, $\mathbf{A}_{\mathrm{ML}}$, $\mathbf{A}_{\mathrm{MD}}$, $\mathbf{A}_{\mathrm{RL}}$, and $\mathbf{A}_{\mathrm{RD}}$, all of which have been determined above. The results are

$$\mathbf{A}_{\mathrm{M}} = \frac{P}{576}\begin{bmatrix} 351 \\ 93L \\ 248 \\ 30L \end{bmatrix} \qquad \mathbf{A}_{\mathrm{R}} = \frac{P}{576}\begin{bmatrix} 1049 \\ 427 \end{bmatrix}$$

Thus, all of the selected end-actions and reactions for the beam have been calculated.

Example 2. The continuous beam ABC shown in Fig. 3-5a has a fixed support at A, a roller support at B, and a guided support at C. Therefore, the only unknown joint displacements are the rotation at support B and the vertical translation at support C. These displacements are denoted D_1 and D_2, respectively, as identified in Fig. 3-5b. The beam has constant flexural rigidity EI and is subjected to the loads P_1 and P_2, acting at the positions shown in the figure. It will be assumed that the loads are given as follows:

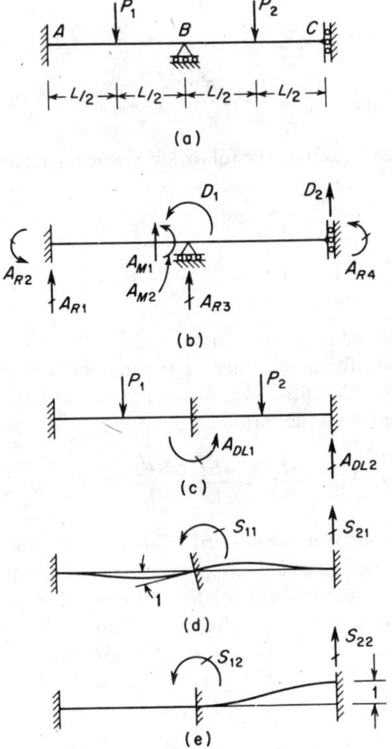

Fig. 3-5. Example 2: Continuous beam.

$$P_1 = 2P \qquad P_2 = P$$

For illustrative purposes in this example, certain member end-actions and all of the reactions for the beam are to be calculated. The selected end-actions are the shearing force A_{M1} and the moment A_{M2} at the right-hand end of member AB (see Fig. 3-5b), and the reactions are the vertical force A_{R1} and couple A_{R2} at support A, the vertical force A_{R3} at support B, and the couple A_{R4} at support C (see Fig. 3-5b). All actions and displacements are shown in their positive directions in the figure.

The restrained structure is obtained by preventing rotation at joint B and translation at joint C, thereby giving the two fixed-end beams shown in Fig. 3-5c. Due to the loads on this restrained structure, the actions corresponding to the unknown displacements D are

$$A_{DL1} = -\frac{P_1 L}{8} + \frac{P_2 L}{8} = -\frac{PL}{8} \qquad A_{DL2} = \frac{P_2}{2} = \frac{P}{2}$$

Also, the member end-actions in the same beam are

$$A_{ML1} = \frac{P_1}{2} = P \qquad A_{ML2} = -\frac{P_1 L}{8} = -\frac{PL}{4}$$

and the reactions are

$$A_{RL1} = \frac{P_1}{2} = P \qquad A_{RL2} = \frac{P_1 L}{8} = \frac{PL}{4}$$

$$A_{RL3} = \frac{P_1}{2} + \frac{P_2}{2} = \frac{3P}{2} \qquad A_{RL4} = -\frac{P_2 L}{8} = -\frac{PL}{8}$$

From the values given above, the following matrices that are required in the solution can be formed:

$$\mathbf{A}_{DL} = \frac{P}{8}\begin{bmatrix} -L \\ 4 \end{bmatrix} \qquad \mathbf{A}_{ML} = \frac{P}{4}\begin{bmatrix} 4 \\ -L \end{bmatrix} \qquad \mathbf{A}_{RL} = \frac{P}{8}\begin{bmatrix} 8 \\ 2L \\ 12 \\ -L \end{bmatrix}$$

After obtaining the matrices of actions due to loads, the next step is to analyze the restrained beam for unit values of the unknown displacements, as shown in Figs. 3-5d and 3-5e. The stiffnesses S_{11} and S_{21} caused by a unit rotation at joint B are readily obtained from the formulas given in Fig. 3-3, as follows:

$$S_{11} = \frac{4EI}{L} + \frac{4EI}{L} = \frac{8EI}{L} \qquad S_{21} = -\frac{6EI}{L^2}$$

In the case of a unit displacement corresponding to D_2 (Fig. 3-5e), it is necessary to have formulas for the forces and couples at the ends of a fixed-end beam subjected to a translation of one end relative to the other. The required formulas can be obtained from Table B-4 of Appendix B (see Case 2). When the translation is equal to unity, the couples at the ends are equal to $6EI/L^2$, and the forces are equal to $12EI/L^3$, as shown in Fig. 3-6. From these values, the stiffnesses S_{12} and S_{22} for the beam in Fig. 3-5e are seen to be as follows:

$$S_{12} = -\frac{6EI}{L^2} \qquad S_{22} = \frac{12EI}{L^3}$$

Therefore, the stiffness matrix can be constructed, and its inverse obtained:

$$\mathbf{S} = \frac{2EI}{L^3}\begin{bmatrix} 4L^2 & -3L \\ -3L & 6 \end{bmatrix} \qquad \mathbf{S}^{-1} = \frac{L}{30EI}\begin{bmatrix} 6 & 3L \\ 3L & 4L^2 \end{bmatrix}$$

The inverse matrix, as well as the matrix $\mathbf{A}_{DL}$ determined previously, can now be substituted into Eq. (3-6) in order to obtain the matrix $\mathbf{D}$ of unknown displacements. The matrix $\mathbf{A}_D$ appearing in the equation is a null matrix since there are no loads on the original beam corresponding to either D_1 or D_2. The solution for $\mathbf{D}$ is found to be

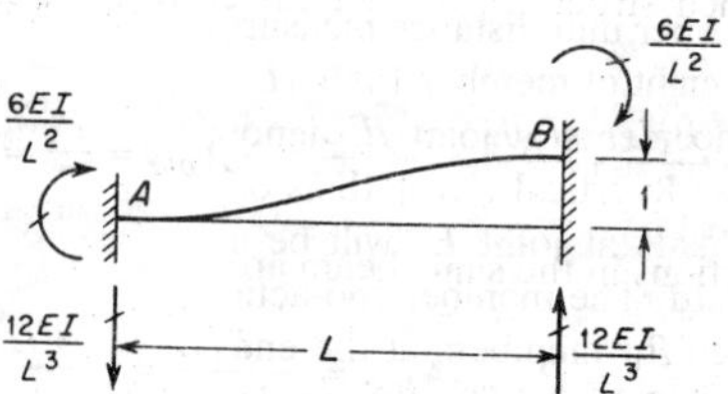

Fig. 3-6. Member stiffnesses for a beam member.

$$\mathbf{D} = \frac{PL^2}{240EI}\begin{bmatrix} -6 \\ -13L \end{bmatrix}$$

The matrices $\mathbf{A}_{MD}$ and $\mathbf{A}_{RD}$ which appear in Eqs. (3-7) and (3-8) represent the end-actions and reactions, respectively, in the restrained beams of Figs. 3-5d and 3-5e. The first column of each matrix is associated with a unit value of the displacement D_1 (Fig. 3-5d), and the second column with a unit value of D_2 (Fig. 3-5e). All of the elements in these matrices can be obtained with the aid of the formulas given in Figs. 3-3 and 3-6, and the results are as follows:

$$\mathbf{A}_{MD} = \frac{2EI}{L^2}\begin{bmatrix} -3 & 0 \\ 2L & 0 \end{bmatrix} \qquad \mathbf{A}_{RD} = \frac{2EI}{L^3}\begin{bmatrix} 3L & 0 \\ L^2 & 0 \\ 0 & -6 \\ L^2 & -3L \end{bmatrix}$$

Then the matrices $\mathbf{A}_M$ and $\mathbf{A}_R$ can be found by substituting the matrices $\mathbf{A}_{ML}$, $\mathbf{A}_{MD}$, $\mathbf{A}_{RL}$, $\mathbf{A}_{RD}$, and $\mathbf{D}$ into Eqs. (3-7) and (3-8), producing

$$\mathbf{A}_M = \frac{P}{20}\begin{bmatrix} 23 \\ -7L \end{bmatrix} \qquad \mathbf{A}_R = \frac{P}{20}\begin{bmatrix} 17 \\ 4L \\ 43 \\ 3L \end{bmatrix}$$

Thus, all of the desired member end-actions and support reactions, as well as the joint displacements, have been calculated.

Example 3. The purpose of this example is to illustrate the analysis of a plane truss by the stiffness method. The truss to be solved is shown in Fig. 3-7a and consists of four members meeting at a common joint E. This particular truss is selected because it has only two degrees of freedom for joint displacement, namely, the horizontal and vertical translations at joint E. However, most of the ensuing discussion pertaining to the solution of this truss is also applicable to more complicated trusses.

It is a convenience in the analysis to identify the bars of the truss numerically. Therefore, the members are numbered from 1 to 4 as shown by the numbers in circles in Fig. 3-7a. Also, for the purposes of general discussion it will be assumed that the four members have lengths L_1, L_2, L_3, and L_4, and axial rigidities EA_1, EA_2, EA_3, and EA_4, respectively. Later, all of these quantities will be given specific values in order that the solution may be carried to completion.

The loads on the truss consist of the two concentrated forces P_1 and P_2 acting at joint E, as well as the weights of the members. The weights act as uniformly distributed loads along the members and are assumed to be of intensity w_1, w_2, w_3, and w_4, respectively, for each of the four members. In all cases the intensity w is the weight of the member per unit distance measured along the axis of the member. For example, the total weight of member 1 is w_1L_1.

The unknown displacements at joint E, denoted D_1 and D_2 in Fig. 3-7b, are taken as the horizontal and vertical translations of the joint. These displacements, as well as the applied loads at joint E, will be assumed positive when directed toward the right or upward. The member end-actions to be calculated are selected as the axial forces in the four members at the ends A, B, C, and D, respectively. These actions are shown in Fig. 3-7b and are denoted A_{M1}, A_{M2}, A_{M3}, and A_{M4}. Because of the weights of the members, the axial forces at the other ends (that is, at

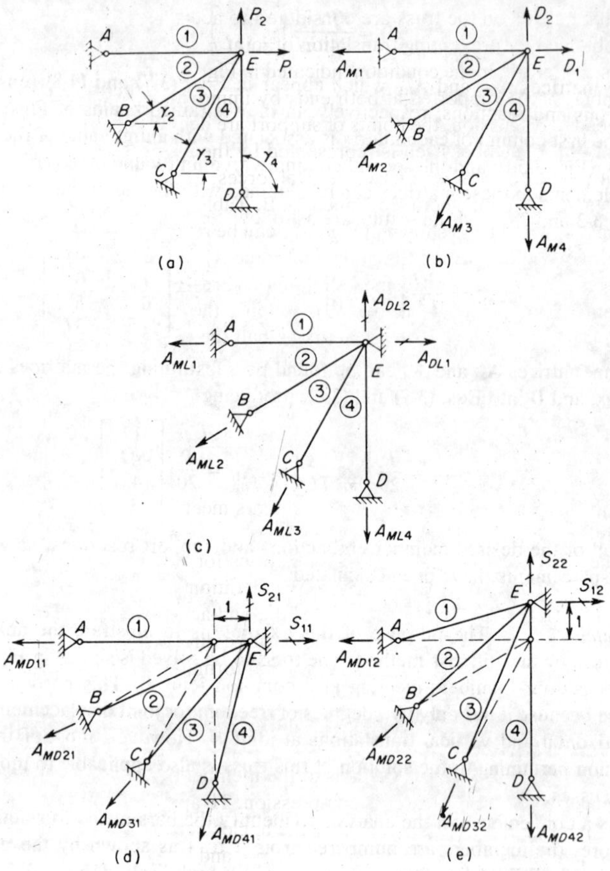

Fig. 3-7. Example 3: Plane truss.

joint E) will have different values from those at ends A, B, C, and D. The axial force at end E of each member could be included in the calculations if desired; also, the shearing forces at the ends could be included. However, the axial forces at E and the shearing forces are omitted in this example for simplicity. The end-actions are assumed positive when they produce tension in the members. It is superfluous to calculate reactions for the truss in Fig. 3-7 inasmuch as the reactions are the same as the end-actions. However, there are other situations, such as when several members of a truss are joined at one point of support, in which the reactions should be determined separately.

The loads P_1 and P_2 acting at joint E of the structure (Fig. 3-7a) are loads that correspond to the unknown displacements D_1 and D_2, respectively. Therefore, these loads appear in the vector $\mathbf{A}_D$ as follows:

$$\mathbf{A}_D = \begin{bmatrix} P_1 \\ P_2 \end{bmatrix}$$

The remaining loads on the truss are considered to act on the restrained structure, which is obtained by preventing translation of joint E (Fig. 3-7c). Each member of the truss is, therefore, in the condition indicated in Fig. 3-8, which shows an inclined truss member that is supported at both ends by immovable pin supports. For purposes of general discussion, the points of support are denoted A and B in the figure. The weight of the member itself is represented by the uniform load of intensity w. The reactions for the member are two vertical forces, each of which is equal to one-half the weight of the member (see Appendix B, Table B-5). Therefore, the actions A_{DL} for the restrained truss shown in Fig. 3-7c can be readily calculated. The actions to be determined at joint E are the horizontal force A_{DL1} and the vertical force A_{DL2}, which correspond to the displacements D_1 and D_2, respectively. Since the weights of the members produce no horizontal reactions, the action A_{DL1} must be zero. However, the action A_{DL2} will be equal to one-half the weight of all members meeting at joint E. Therefore, the vector $\mathbf{A}_{\mathrm{DL}}$ becomes

$$\mathbf{A}_{\mathrm{DL}} = \begin{bmatrix} 0 \\ \dfrac{w_1L_1}{2} + \dfrac{w_2L_2}{2} + \dfrac{w_3L_3}{2} + \dfrac{w_4L_4}{2} \end{bmatrix} = \begin{bmatrix} 0 \\ \dfrac{W}{2} \end{bmatrix}$$

The quantity W is the total weight of all members meeting at joint E, which in this example is also the total weight of the truss.

For the purpose of calculating the end-actions for the members, it is also necessary to obtain the vector $\mathbf{A}_{\mathrm{ML}}$ from a consideration of the restrained structure shown in Fig. 3-7c. This vector consists of the end-actions A_{ML1}, A_{ML2}, A_{ML3}, and A_{ML4}, which are shown in the figure in the positive directions (tension in the members). Each such quantity is given by the general formula

$$-\frac{wL}{2} \sin \gamma$$

in which γ is the angle between the axis of the member and the horizontal; and the minus sign indicates that the force is compression. To apply this formula to the truss in Fig. 3-7, it is necessary first to identify the angles between the members and the horizontal. Let these angles be denoted γ_1, γ_2, γ_3, and γ_4 for the four members 1, 2, 3, and 4, respectively. These angles are shown in Fig. 3-7a for all members except member 1, which is assumed to be horizontal in this particular example. Thus, the end-action A_{ML2}, for example, is given by the expression

$$A_{ML2} = -\frac{w_2L_2}{2} \sin \gamma_2$$

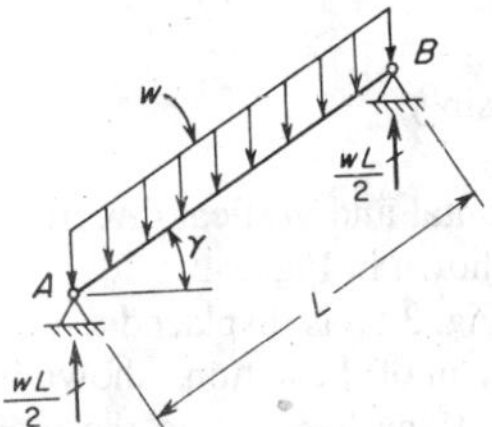

Fig. 3-8. End-actions for restrained truss member.

and it can be seen that the vector $\mathbf{A}_{ML}$ is the following:

$$\mathbf{A}_{ML} = -\frac{1}{2}\begin{bmatrix} w_1 L_1 \sin\gamma_1 \\ w_2 L_2 \sin\gamma_2 \\ w_3 L_3 \sin\gamma_3 \\ w_4 L_4 \sin\gamma_4 \end{bmatrix}$$

This vector can be evaluated in any particular case by substituting the appropriate values for each member of the truss. If it is assumed that member 1 is horizontal and member 4 is vertical, then the first and last elements of the vector can be simplified to zero and $-w_4L_4/2$, respectively.

To obtain the stiffness matrix **S**, it is necessary to impose unit displacements corresponding to D_1 and D_2 on the restrained structure, as shown in Figs. 3-7d and 3-7e, respectively. The actions corresponding to the joint displacements are the four joint stiffnesses S_{11}, S_{21}, S_{12}, and S_{22}. Each of these stiffnesses is shown in the figure acting in the positive direction. The joint stiffnesses are composed of contributions from each member of the truss; that is, from the stiffnesses of the members themselves. For example, S_{11} is the total force in the horizontal direction when a unit displacement in that direction is imposed upon the truss (see Fig. 3-7d) and consists of the sum of the horizontal components of the forces in all members of the truss. Thus, to obtain the joint stiffnesses shown in the figure it is first necessary to find the forces acting on the individual members when joint E is displaced.

The forces acting on a typical truss member due to a unit horizontal displacement of one end are shown in Fig. 3-9a. For purposes of general discussion, the lower and upper ends of the member are denoted as ends A and B, respectively. The upper end of the member is assumed to be moved a unit distance to the right, while all other end displacements are zero. As a result, the member becomes lengthened, and restraint forces are required at each end. The elongation of the member is determined from the displacements occurring at end B, as shown in Fig. 3-9b. From the triangle in the figure it is seen that the elongation of the member is $\cos\gamma$, in which γ is the angle of inclination of the member. Therefore, the tensile force in the member is (see Fig. 3-9a)

$$\frac{EA}{L}\cos\gamma$$

in which EA is the axial rigidity and L is the length of the member. The components of this axial force in the horizontal and vertical directions are also shown in the figure. These components, which are member stiffnesses, can be readily determined from the geometry of the figure.

In the case of a unit vertical displacement of end B of the member (Fig. 3-9c), the axial force in the member is

$$\frac{EA}{L}\sin\gamma$$

from which the components in the horizontal and vertical directions can be easily found. All of the forces for this case are shown in Fig. 3-9c.

If the lower end A of the member in Fig. 3-9a is displaced a unit distance to the right while the upper end B remains fixed, all of the actions shown in the figure will have their directions reversed. The same thing happens to the member shown in Fig. 3-9c if end A is displaced upward a unit distance while end B remains fixed.

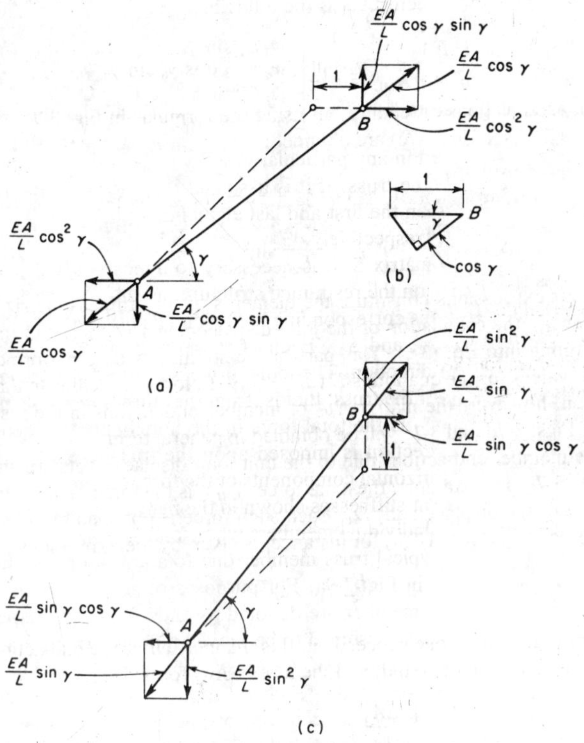

Fig. 3-9. Member stiffnesses for a plane truss member.

All of the formulas given in Fig. 3-9 are suitable for use in analyzing plane trusses by the stiffness method when the calculations are being performed by hand. Later, a more systematic approach to the use of member stiffnesses will be given, not only for trusses but also for other types of structures (see Art. 3.5 and Chapter 4).

Now consider again the calculation of the joint stiffnesses for the restrained truss shown in Figs. 3-7d and 3-7e. The stiffness S_{11} is composed of contributions from the various members of the truss. For example, the contribution to S_{11} from member 3 is (see Fig. 3-9a):

$$\frac{EA_3}{L_3}\cos^2\gamma_3$$

Similarly, the contribution of member 3 to the stiffness S_{21} is (see Fig. 3-9a):

$$\frac{EA_3}{L_3}\cos\gamma_3\sin\gamma_3 \ .$$

Expressions of the same form as the two expressions given above can be written for all four members of the truss. The sums of such expressions yield the stiffnesses S_{11} and S_{21}. Inasmuch as member 1 is horizontal ($\gamma_1 = 0$) and member 4 is vertical ($\gamma_4 = 90°$), the expressions for the stiffnesses are

$$S_{11} = \frac{EA_1}{L_1} + \frac{EA_2}{L_2}\cos^2\gamma_2 + \frac{EA_3}{L_3}\cos^2\gamma_3$$
$$S_{21} = \frac{EA_2}{L_2}\cos\gamma_2\sin\gamma_2 + \frac{EA_3}{L_3}\cos\gamma_3\sin\gamma_3$$

By following an analogous procedure but using the formulas in Fig. 3-9c, the stiffnesses S_{12} and S_{22} (see Fig. 3-7e) are obtained:

$$S_{12} = \frac{EA_2}{L_2}\sin\gamma_2\cos\gamma_2 + \frac{EA_3}{L_3}\sin\gamma_3\cos\gamma_3$$
$$S_{22} = \frac{EA_2}{L_2}\sin^2\gamma_2 + \frac{EA_3}{L_3}\sin^2\gamma_3 + \frac{EA_4}{L_4}$$

The above four expressions constitute the elements of the stiffness matrix **S**. The remaining steps in the calculation of the joint displacements D consist of inverting **S** and substituting into Eq. (3-6). This part of the solution will be performed later using specific values for the quantities EA, L, and γ. Before proceeding to that stage of the solution, however, the matrix $\mathbf{A}_{\mathbf{MD}}$ of member end-actions in the restrained structures of Figs. 3-7d and 3-7e will be obtained in general terms.

All of the member end-actions due to the unit joint displacements are shown in Figs. 3-7d and 3-7e. For example, the axial force A_{MD31} is the force in member 3 due to a unit value of the displacement D_1. Each such force is obtained by referring to Fig. 3-9a or 3-9c. The force A_{MD31}, for instance, is given by the expression

$$\frac{EA_3}{L_3}\cos\gamma_3$$

as found from Fig. 3-9a. If one proceeds in this manner, all of the end-actions due to unit displacements can be found, and the matrix $\mathbf{A}_{\mathbf{MD}}$ formed:

$$\mathbf{A}_{\mathbf{MD}} = \begin{bmatrix} \frac{EA_1}{L_1} & 0 \\ \frac{EA_2}{L_2}\cos\gamma_2 & \frac{EA_2}{L_2}\sin\gamma_2 \\ \frac{EA_3}{L_3}\cos\gamma_3 & \frac{EA_3}{L_3}\sin\gamma_3 \\ 0 & \frac{EA_4}{L_4} \end{bmatrix}$$

As mentioned previously, other end-actions, as well as reactions, may also be included in the analysis. Under such conditions, the appropriate formulas from Fig. 3-9 can be used in finding the values of the various actions due to the unit displacements. When all of the required matrices have been formed, Eqs. (3-6) through (3-8) are used in finding the resultant effects.

The preceding discussion has served to illustrate in general terms how the various matrices are obtained in a truss analysis. In order to complete the solution of the truss in Fig. 3-7, it is necessary to assume specific values for the quantities appearing in the matrices. Therefore, it will be assumed that all members have the same length L, the same axial rigidity EA, and the same weight w per unit length. Furthermore, it is assumed that the angles between adjoining members are 30 degrees, so that the angles between the members and the horizontal are

$$\gamma_1 = 0 \qquad \gamma_2 = 30° \qquad \gamma_3 = 60° \qquad \gamma_4 = 90°$$

Also, the total weight W of the truss is

$$W = 4wL$$

When these substitutions are made, the matrices described above simplify to the following:

$$\mathbf{A}_{DL} = wL\begin{bmatrix} 0 \\ 2 \end{bmatrix} \qquad \mathbf{S} = \frac{EA}{2L}\begin{bmatrix} 4 & \sqrt{3} \\ \sqrt{3} & 4 \end{bmatrix}$$

$$\mathbf{A}_{ML} = -\frac{wL}{4}\begin{bmatrix} 0 \\ 1 \\ \sqrt{3} \\ 2 \end{bmatrix} \qquad \mathbf{A}_{MD} = \frac{EA}{2L}\begin{bmatrix} 2 & 0 \\ \sqrt{3} & 1 \\ 1 & \sqrt{3} \\ 0 & 2 \end{bmatrix}$$

The inverse of $\mathbf{S}$ is

$$\mathbf{S}^{-1} = \frac{2L}{13EA}\begin{bmatrix} 4 & -\sqrt{3} \\ -\sqrt{3} & 4 \end{bmatrix}$$

and the vector $\mathbf{D}$, found from Eq. (3-6), is

$$\mathbf{D} = \frac{2L}{13EA}\begin{bmatrix} 4P_1 - \sqrt{3}P_2 + 2\sqrt{3}wL \\ -\sqrt{3}P_1 + 4P_2 - 8wL \end{bmatrix}$$

Into this vector can be substituted any particular values of the loads P_1 and P_2, as well as the weight w per unit length. Then the vector $\mathbf{A}_M$ of end-actions (or member forces) can be found by substituting the matrices $\mathbf{A}_{ML}$, $\mathbf{A}_{MD}$, and $\mathbf{D}$ into Eq. (3-7) and solving. If the weight of the truss is not included in the analysis, the matrices $\mathbf{A}_{DL}$ and $\mathbf{A}_{ML}$ become null.

As a particular case, assume that $P_1 = 0$, $P_2 = -P$, and $wL = P/10$. Then the vectors $\mathbf{D}$ and $\mathbf{A}_M$ are found to be

$$\mathbf{D} = \frac{12PL}{65EA}\begin{bmatrix} \sqrt{3} \\ -4 \end{bmatrix} \qquad \mathbf{A}_M = \frac{P}{520}\begin{bmatrix} 96\sqrt{3} \\ -61 \\ -157\sqrt{3} \\ -410 \end{bmatrix}$$

These results show that, under the assumed loading, joint E is displaced to the right and downward, and all members of the truss are subjected to compression except member 1.

Example 4. The analysis of the plane frame ABC shown in Fig. 3-10a is described in this example. This frame was analyzed previously by the flexibility method in Example 4 of Art. 2.3. When using the stiffness method, the first step is to determine the number of unknown joint displacements. In this example, the number of unknowns depends upon whether both flexural and axial deformations are considered in the analysis. If both types of deformation are included, joint B can

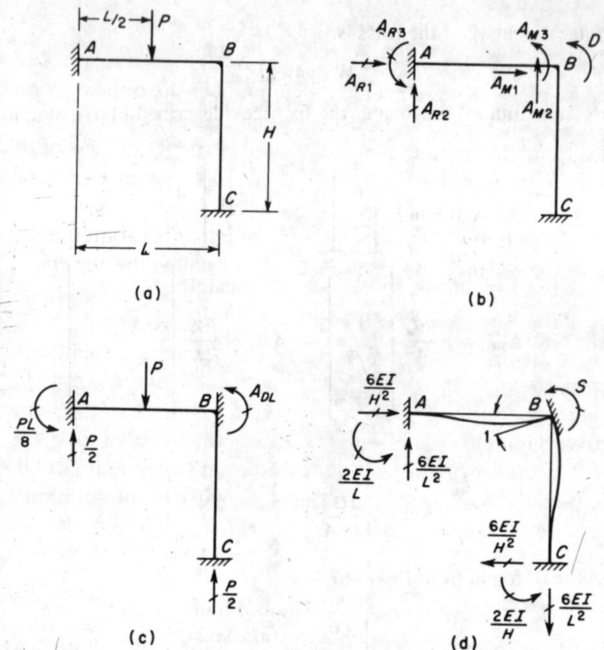

Fig. 3-10. Example 4: Plane frame (flexural effects only).

translate as well as rotate. An analysis of this kind is made later in the article. However, in a plane frame it is normally not necessary to include the axial deformations. If they are omitted, joint B will rotate but not translate, and there will be only one unknown displacement D (see Fig. 3-10b). In addition to finding this joint displacement, it will be assumed arbitrarily that the following end-actions and reactions are to be found as part of the solution: the axial force A_{M1}, shearing force A_{M2}, and moment A_{M3} at the right-hand end of member AB (see Fig. 3-10b) and the three reactions A_{R1}, A_{R2}, and A_{R3} at support A. Both members of the frame are assumed to have the same flexural rigidity EI.

The restrained structure (Fig. 3-10c) is obtained by preventing rotation of joint B. The action A_{DL} corresponding to the unknown displacement D and caused by the load P is

$$A_{DL} = -\frac{PL}{8}$$

The end-actions and reactions in the restrained structure due to the load P are

$$A_{ML1} = 0 \qquad A_{ML2} = \frac{P}{2} \qquad A_{ML3} = -\frac{PL}{8}$$

$$A_{RL1} = 0 \qquad A_{RL2} = \frac{P}{2} \qquad A_{RL3} = \frac{PL}{8}$$

All of the reactions in the restrained structure due to the load P are shown in Fig. 3-10c. It should be recalled at this stage of the calculations that the restraint at joint

B is a rotational restraint only, and hence it offers no restraint against translation. Therefore, horizontal and vertical forces can be transmitted through joint B from one member to the other. As a result, the vertical reactive force $P/2$ at the right-hand end of member AB is sustained at joint C rather than at joint B. The vectors $\mathbf{A}_{DL}$, $\mathbf{A}_{ML}$, and $\mathbf{A}_{RL}$ become

$$\mathbf{A}_{DL} = \frac{P}{8}\begin{bmatrix} -L \end{bmatrix} \qquad \mathbf{A}_{ML} = \frac{P}{8}\begin{bmatrix} 0 \\ 4 \\ -L \end{bmatrix} \qquad \mathbf{A}_{RL} = \frac{P}{8}\begin{bmatrix} 0 \\ 4 \\ L \end{bmatrix}$$

In this problem the vector $\mathbf{A}_D$ is a null vector containing one zero element because there is no couple applied as a load at joint B of the frame.

The stiffness matrix for the frame is found by imposing a unit displacement corresponding to D on the restrained structure (see Fig. 3-10d). From the figure it is seen that the stiffness S is

$$S = \frac{4EI}{L} + \frac{4EI}{H}$$

Therefore, the stiffness matrix $\mathbf{S}$ and its inverse are

$$\mathbf{S} = \left[4EI\left(\frac{1}{L} + \frac{1}{H}\right) \right] \qquad \mathbf{S}^{-1} = \left[\frac{LH}{4EI(L + H)} \right]$$

The end-actions and reactions for the restrained frame of Fig. 3-10d are evaluated next. Again recalling that there is only a rotational restraint at joint B, it is seen that the reactions for the restrained frame have the values shown in Fig. 3-10d. Therefore, the matrices of end-actions and reactions are as follows:

$$\mathbf{A}_{MD} = \begin{bmatrix} -\dfrac{6EI}{H^2} \\ -\dfrac{6EI}{L^2} \\ \dfrac{4EI}{L} \end{bmatrix} \qquad \mathbf{A}_{RD} = \begin{bmatrix} \dfrac{6EI}{H^2} \\ \dfrac{6EI}{L^2} \\ \dfrac{2EI}{L} \end{bmatrix}$$

All of the above matrices are now substituted into Eqs. (3-6), (3-7), and (3-8), producing the results

$$\mathbf{D} = \frac{PL^2}{32EI}\left[\frac{H}{L + H} \right]$$

$$\mathbf{A}_M = \frac{P}{16H(L + H)}\begin{bmatrix} -3L^2 \\ H(8L + 5H) \\ -2HL^2 \end{bmatrix} \qquad \mathbf{A}_R = \frac{P}{16H(L + H)}\begin{bmatrix} 3L^2 \\ H(8L + 11H) \\ HL(2L + 3H) \end{bmatrix}$$

In the special case when $H = L$, these results simplify to

$$D = \frac{PL^2}{64EI} \qquad \mathbf{A}_M = \frac{P}{32}\begin{bmatrix} -3 \\ 13 \\ -2L \end{bmatrix} \qquad \mathbf{A}_R = \frac{P}{32}\begin{bmatrix} 3 \\ 19 \\ 5L \end{bmatrix}$$

The three elements of the vector $\mathbf{A}_M$ have the same physical meanings as the redundant actions Q_1, Q_2, and Q_3 of Example 4, Art. 2.3 (compare Figs. 3-10b and 2-6b). Therefore, the values given above for the end-actions agree with those of Q_1, Q_2, and Q_3 obtained in the earlier example for the case when axial deformations are neglected.

When axial deformations are included in the analysis of the plane frame in Fig. 3-10a, three degrees of freedom must be recognized at joint B instead of only one. The displacements D_1, D_2, and D_3 for this case are shown in Fig. 3-11a. The restrained structure for this new analysis is obtained by preventing horizontal translation, vertical translation, and rotation at joint B (see Fig. 3-11b). The actions in this restrained structure corresponding to the unknown displacements and caused by the load P are shown in the figure as A_{DL1}, A_{DL2}, and A_{DL3}. These restraint actions are

$$A_{DL1} = 0 \qquad A_{DL2} = \frac{P}{2} \qquad A_{DL3} = -\frac{PL}{8}$$

Thus, the vector $\mathbf{A}_{DL}$ is

$$\mathbf{A}_{DL} = \frac{P}{8}\begin{bmatrix} 0 \\ 4 \\ -L \end{bmatrix}$$

As before, the vector $\mathbf{A}_D$ is a null vector because there are no forces or couples applied as loads at joint B.

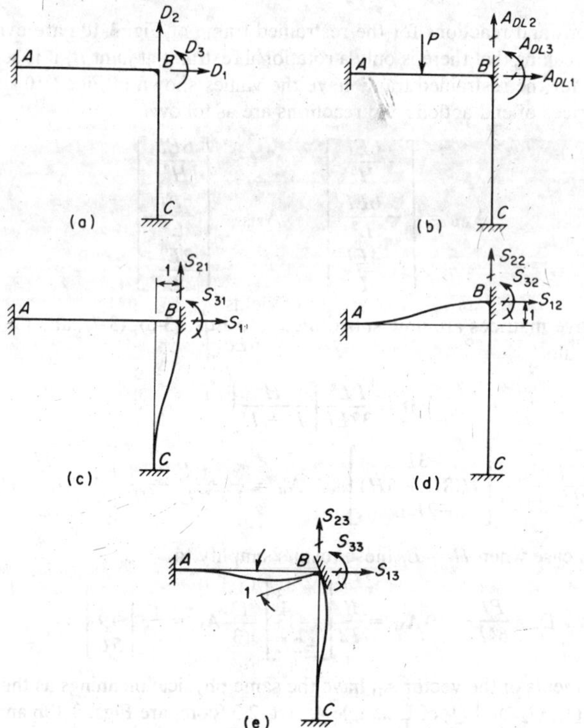

Fig. 3-11. Example 4: Plane frame (flexural and axial effects).

The elements of the stiffness matrix for the frame are pictured in Figs. 3-11c, d, and e. These figures show the actions developed at the restrained joint B when unit displacements corresponding to D_1, D_2, and D_3 are introduced. Each action can be found from the expressions for end-actions due to a unit displacement (see Figs. 3-3, 3-6, and 3-9). Thus, the stiffness matrix is

$$\mathbf{S} = \begin{bmatrix} \frac{EA}{L} + \frac{12EI}{H^3} & 0 & \frac{6EI}{H^2} \\ 0 & \frac{12EI}{L^3} + \frac{EA}{H} & -\frac{6EI}{L^2} \\ \frac{6EI}{H^2} & -\frac{6EI}{L^2} & \frac{4EI}{L} + \frac{4EI}{H} \end{bmatrix}$$

The inverse of this 3 × 3 matrix cannot be obtained easily in literal form. Therefore, the problem will be simplified by letting $H = L$ and by using the dimensionless parameter

$$\nu = \frac{AL^2}{I}$$

which is the reciprocal of the parameter γ used previously for analyzing this frame by the flexibility method. Then the stiffness matrix becomes

$$\mathbf{S} = \frac{EI}{L^3}\begin{bmatrix} 12+\nu & 0 & 6L \\ 0 & 12+\nu & -6L \\ 6L & -6L & 8L^2 \end{bmatrix}$$

and its inverse is

$$\mathbf{S}^{-1} = \frac{L}{8EIc_1c_2}\begin{bmatrix} 4L^2c_3 & -36L^2 & -6Lc_1 \\ -36L^2 & 4L^2c_3 & 6Lc_1 \\ -6Lc_1 & 6Lc_1 & c_1^2 \end{bmatrix}$$

where

$$c_1 = 12 + \nu \qquad c_2 = 3 + \nu \qquad c_3 = 5 + 2\nu$$

Substitution of $\mathbf{S}^{-1}$ and $\mathbf{A}_{DL}$ into Eq. (3-6) yields

$$\mathbf{D} = \frac{PL^2}{64EIc_1c_2}\begin{bmatrix} 6Lc_4 \\ -2Lc_5 \\ -c_1c_4 \end{bmatrix}$$

in which

$$c_4 = 12 - \nu \qquad c_5 = 4 + 13\nu$$

In terms of ν, the joint displacements are

$$D_1 = \frac{3PL^3}{32EI}\frac{12-\nu}{(12+\nu)(3+\nu)}$$

$$D_2 = \frac{-PL^3}{32EI}\frac{4+13\nu}{(12+\nu)(3+\nu)}$$

$$D_3 = \frac{-PL^2}{64EI}\frac{12-\nu}{3+\nu}$$

If the quantity ν is large, the values of D_1 and D_2 tend to zero; and the value of D_3 becomes $PL^2/64EI$, as in the first part of this example.

Example 5. The grid shown in Fig. 3-12a consists of two members (AB and BC) that are rigidly joined at B. The load on the grid is a concentrated force P acting at the midpoint of member AB. Each member is assumed to have constant flexural rigidity EI and torsional rigidity GJ. Because the supports at A and C are fixed, the only unknown joint displacements for the structure occur at joint B. These are a translation D_1 in the y direction, a rotation D_2 about the x axis, and a rotation D_3 about the z axis, as indicated in Fig. 3-12b. In the figure the positive direction of each displacement vector is assumed to be the same as the positive direction of one of the coordinate axes. As in earlier articles, a double-headed arrow is used to distinguish a rotation from a translation. The number of degrees of freedom of the grid happens to be the same as the degree of static indeterminacy (see Example 5, Art. 2.3).

To begin analyzing the grid by the stiffness method, an artificial restraint is supplied at joint B to prevent displacements corresponding to D_1, D_2, and D_3 (see Fig. 3-12c). When the loads act upon this restrained structure, the actions A_{DL} (corresponding to the displacements D) are developed at the restraint. These actions are evaluated readily by making use of the formulas for fixed-end actions in beams (see Appendix B). In the case of the load P acting on member AB, the restraint actions are

$$A_{DL1} = \frac{P}{2} \qquad A_{DL2} = 0 \qquad A_{DL3} = -\frac{PL}{8}$$

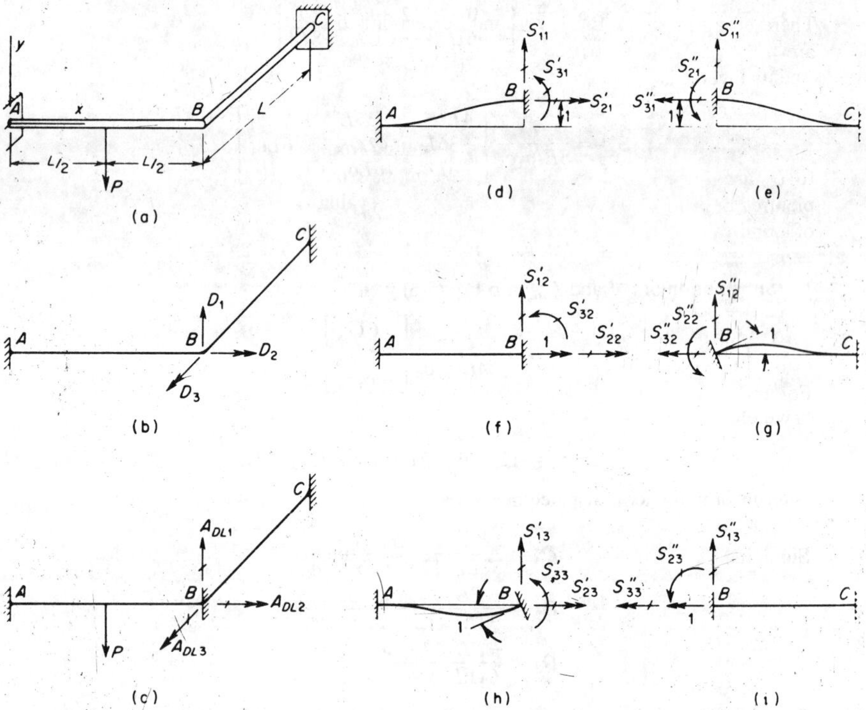

Fig. 3-12. Example 5: Grid.

and, hence, the vector $\mathbf{A}_{DL}$ is

$$\mathbf{A}_{DL} = \frac{P}{8}\begin{bmatrix} 4 \\ 0 \\ -L \end{bmatrix}$$

The vector $\mathbf{A}_D$, representing actions applied to the actual beam (Fig. 3-12a) corresponding to the unknown displacements, is a null vector because there are no concentrated forces or couples at joint B.

The stiffness matrix for the grid is found by analyzing the restrained structure for the effects of unit values of the unknown displacements. In the case of a unit value of D_1, joint B is displaced upward by a unit distance without rotating. Then the actions developed at the restraint corresponding to D_1, D_2, and D_3 are the joint stiffnesses S_{11}, S_{21}, and S_{31}, respectively. The effects of this unit translation on members AB and BC are shown separately in Figs. 3-12d and 3-12e. In Fig. 3-12d the contributions of member AB to the joint stiffnesses are shown with a single prime, while in Fig. 3-12e the contributions from member BC are shown with a double prime. From the figures it can be seen that the stiffness terms are as follows:

$$S'_{11} = \frac{12EI}{L^3} \qquad S'_{21} = 0 \qquad S'_{31} = -\frac{6EI}{L^2}$$

$$S''_{11} = \frac{12EI}{L^3} \qquad S''_{21} = \frac{6EI}{L^2} \qquad S''_{31} = 0$$

Therefore, the joint stiffnesses, found by summing the effects from both members, are

$$S_{11} = \frac{24EI}{L^3} \qquad S_{21} = \frac{6EI}{L^2} \qquad S_{31} = -\frac{6EI}{L^2}$$

In a similar manner the stiffnesses resulting from unit rotations corresponding to D_2 and D_3 can be found. The contributions from the individual members are pictured in Figs. 3-12f through 3-12i, and can be evaluated by inspection. In the case of a unit value of D_2 (see Figs. 3-12f and 3-12g) the contributions are

$$S'_{12} = 0 \qquad S'_{22} = \frac{GJ}{L} \qquad S'_{32} = 0$$

$$S''_{12} = \frac{6EI}{L^2} \qquad S''_{22} = \frac{4EI}{L} \qquad S''_{32} = 0$$

For a unit value of D_3 the contributions are (see Figs. 3-12h and 3-12i):

$$S'_{13} = -\frac{6EI}{L^2} \qquad S'_{23} = 0 \qquad S'_{33} = \frac{4EI}{L}$$

$$S''_{13} = 0 \qquad S''_{23} = 0 \qquad S''_{33} = \frac{GJ}{L}$$

Summing the individual terms given above yields the joint stiffnesses:

$$S_{12} = \frac{6EI}{L^2} \qquad S_{22} = \frac{4EI}{L} + \frac{GJ}{L} \qquad S_{32} = 0$$

$$S_{13} = -\frac{6EI}{L^2} \qquad S_{23} = 0 \qquad S_{33} = \frac{4EI}{L} + \frac{GJ}{L}$$

Finally, the stiffness matrix $\mathbf{S}$ can be expressed in the following form:

$$\mathbf{S} = \frac{EI}{L^3}\begin{bmatrix} 24 & 6L & -6L \\ 6L & (4+\eta)L^2 & 0 \\ -6L & 0 & (4+\eta)L^2 \end{bmatrix}$$

in which the dimensionless parameter η is

$$\eta = \frac{GJ}{EI}$$

The quantity η is the reciprocal of the parameter ρ used in solving this same grid by the flexibility method. The inverse matrix is

$$\mathbf{S}^{-1} = \frac{L}{24EId_1d_2}\begin{bmatrix} L^2d_1^2 & -6Ld_1 & 6Ld_1 \\ -6Ld_1 & 12d_3 & -36 \\ 6Ld_1 & -36 & 12d_3 \end{bmatrix}$$

in which the following additional parameters are used:

$$d_1 = 4 + \eta \qquad d_2 = 1 + \eta \qquad d_3 = 5 + 2\eta$$

The quantities η, d_1, d_2, and d_3 depend only on the ratio of the torsional rigidity GJ to the bending rigidity EI.

The displacements at joint B can be found by substituting the matrices $\mathbf{A}_{DL}$ and $\mathbf{S}^{-1}$ into Eq. (3-6) and solving for $\mathbf{D}$. This yields the result

$$\mathbf{D} = \frac{PL^2}{96EId_1d_2}\begin{bmatrix} -Ld_1d_3 \\ 6d_3 \\ -18 \end{bmatrix}$$

from which

$$D_1 = -\frac{PL^3}{96EI}\frac{5+2\eta}{1+\eta}$$

$$D_2 = \frac{PL^2}{16EI}\frac{5+2\eta}{(1+\eta)(4+\eta)}$$

$$D_3 = -\frac{3PL^2}{16EI}\frac{1}{(1+\eta)(4+\eta)}$$

If the members AB and BC are torsionally very weak, the grid can be considered to consist of two members joined at B by a hinge that is capable of transmitting a vertical force but not a couple. In the analysis of such a grid by the stiffness method, the only joint displacement to be treated as an unknown in the analysis is the translation in the y direction. Its value can be found from the result given above for D_1 by letting η become zero. Thus,

$$D_1 = -\frac{5PL^3}{96EI}$$

In general, the analysis of a grid consisting of beams that cross one another (thereby transmitting a vertical force but no moment at each crossing point) will have one unknown joint displacement at each such point.

End-actions and reactions for the grid can also be calculated, following the general techniques illustrated in the preceding examples. Such calculations are given as problems at the end of the chapter.

3.4 Temperature Changes, Prestrains, and Support Displacements. The effects of temperature changes, prestrain of members, and support displacements can be readily incorporated into the analysis of a structure by the stiffness method. A convenient procedure is to consider all such effects to occur in the restrained structure and to add the resulting actions to the actions produced by the loads. For example, in the restrained structure subjected to loads only, it is necessary to calculate the actions A_{DL} corresponding to the unknown displacements (see Eq. 3-5). When temperature changes are assumed to occur in the same restrained structure, additional actions corresponding to the unknown displacements may occur. Such actions will be denoted by the symbol A_{DT}, which is consistent with the symbol A_{DL} except that the cause is temperature rather than loads. The same idea can be applied to prestrains and restraint displacements, which will produce in the restrained structure the actions A_{DP} and A_{DR}, respectively. When all such actions have been determined for the restrained structure, the vectors $\mathbf{A}_{DL}$, $\mathbf{A}_{DT}$, $\mathbf{A}_{DP}$, and $\mathbf{A}_{DR}$ can be formed. These vectors are of order $n \times 1$, where n denotes the number of unknown displacements. The sum of these vectors represents the combination of all actions corresponding to the unknown displacements, and is denoted $\mathbf{A}_{DC}$. Hence, this vector is

$$\mathbf{A}_{DC} = \mathbf{A}_{DL} + \mathbf{A}_{DT} + \mathbf{A}_{DP} + \mathbf{A}_{DR} \tag{3-9}$$

The vector $\mathbf{A}_{DC}$ is included in the first of the three equations of superposition, in place of the matrix $\mathbf{A}_{DL}$ alone (see Eq. 3-5):

$$\mathbf{A}_{D} = \mathbf{A}_{DC} + \mathbf{SD} \tag{3-10}$$

This equation is a more general equation of the stiffness method and should be used instead of Eq. (3-5) whenever effects other than loads are to be included in the calculations. The equation can be solved for the displacements as follows:

$$\mathbf{D} = \mathbf{S}^{-1}(\mathbf{A}_{D} - \mathbf{A}_{DC}) \tag{3-11}$$

From this form it can be seen that the term $-\mathbf{A}_{DC}$ constitutes a vector of *equivalent joint loads* (see Art. 1.12).

Temperature changes, prestrains, and support displacements also affect the determination of member end-actions and reactions in a structure. The vectors of end-actions in the restrained structure due to these causes are denoted $\mathbf{A}_{MT}$, $\mathbf{A}_{MP}$, and $\mathbf{A}_{MR}$, respectively, and the combination of all such actions, including loads, is the vector

$$\mathbf{A}_{MC} = \mathbf{A}_{ML} + \mathbf{A}_{MT} + \mathbf{A}_{MP} + \mathbf{A}_{MR} \tag{3-12}$$

In an analogous manner the combination of all reactions due to these causes in the restrained structure gives the vector $\mathbf{A}_{RC}$:

$$\mathbf{A}_{RC} = \mathbf{A}_{RL} + \mathbf{A}_{RT} + \mathbf{A}_{RP} + \mathbf{A}_{RR} \tag{3-13}$$

in which the four matrices on the right-hand side of the equation represent reactions due to loads, temperature changes, prestrains, and restraint displacements, respectively. The equations of superposition for the member end-actions and the reactions now become

$$\mathbf{A_M} = \mathbf{A_{MC}} + \mathbf{A_{MD}}\,\mathbf{D} \tag{3-14}$$

$$\mathbf{A_R} = \mathbf{A_{RC}} + \mathbf{A_{RD}}\,\mathbf{D} \tag{3-15}$$

These equations may be considered as generalized forms of Eqs. (3-7) and (3-8), which were used previously when only loads acted on the structure. The use of the above equations will now be illustrated by extending the two-span beam example given in Art. 3.2 (see Fig. 3-2).

The beam to be analyzed is pictured again in Fig. 3-13a. The loads are not shown in the figure because their effects were considered in the earlier example. The unknown joint displacements D_1 and D_2 are indicated in Fig. 3-13a by the curved arrows. For illustrative purposes, consider first the effects of temperature changes, and assume that member BC is subjected to a linear temperature gradient such that the lower surface of the beam is at temperature T_1 while the upper surface is at temperature T_2. Member AB is assumed to remain at a constant temperature. When these temperature effects are assumed to occur in the restrained structure (see Fig. 3-13b), there will be moments developed at the ends of the member. Therefore, a restraint action A_{DT1} will be developed at joint B corresponding to the displacement D_1, and an action A_{DT2} will be developed at joint C corresponding to D_2. These actions are shown in their positive directions in the figure and can be evaluated from the expressions for fixed-end actions due to temperature changes given in Table B-2 of Appendix B. Thus,

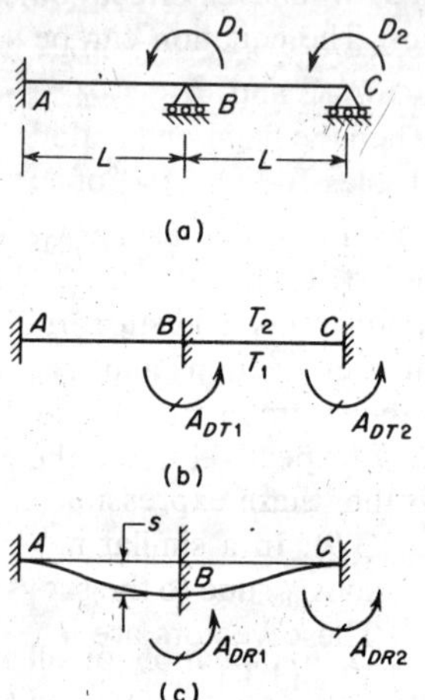

Fig. 3-13. Continuous beam with temperature differential and support displacement.

$$A_{DT1} = -A_{DT2} = \frac{\alpha EI(T_1 - T_2)}{d}$$

in which α is the coefficient of thermal expansion, EI is the flexural rigidity of the beam, and d is the depth of the beam. The vector $\mathbf{A}_{DT}$ can now be expressed as

$$\mathbf{A}_{DT} = \frac{\alpha EI(T_1 - T_2)}{d}\begin{bmatrix} 1 \\ -1 \end{bmatrix}$$

This vector can be added to the vector $\mathbf{A}_{DL}$ obtained in the example of Art. 3.2 to give the sum $\mathbf{A}_{DC}$ for use in Eq. (3-10), assuming that the beam is to be analyzed for the combined effects of the loads (see Fig. 3-2a) and the temperature differential in member BC. Of course, the displacements D calculated for both effects together will be the sum of the displacements obtained when the loads and the temperature effects are considered separately.

In a similar manner the member end-actions and the reactions due to the temperature change in member BC can be determined from the restrained beam in Fig. 3-13b. In this particular example the member end-actions are zero in span AB, and in member BC are the same fixed-end actions discussed above. Also, in this example all the reactions are zero inasmuch as the temperature differential does not produce any vertical force at joints B and C. However, in a more general situation there will be values for both the end-actions and the reactions; and these two sets of values are placed into the vectors $\mathbf{A}_{MT}$ and $\mathbf{A}_{RT}$, respectively. Then these quantities are added to the corresponding quantities caused by the loads, and the combined vectors are substituted into Eqs. (3-14) and (3-15).

The procedure is similar for prestrain effects and support displacements. In both cases three vectors of actions in the restrained structure are to be determined ($\mathbf{A}_{DP}$, $\mathbf{A}_{MP}$, $\mathbf{A}_{RP}$, and $\mathbf{A}_{DR}$, $\mathbf{A}_{MR}$, $\mathbf{A}_{RR}$). All of these vectors can be evaluated readily when the fixed-end actions are known. Such fixed-end actions are given in Tables B-3 and B-4 of Appendix B. Suppose, for example, that support B of the beam is known to undergo a downward displacement equal to s (see Fig. 3-13c). This results in actions A_{DR1} and A_{DR2}, which are evaluated as follows (see Table B-4):

$$A_{DR1} = \frac{6EIs}{L^2} - \frac{6EIs}{L^2} = 0 \qquad A_{DR2} = -\frac{6EIs}{L^2}$$

The vector $\mathbf{A}_{DR}$ is formed from these expressions and then included in the calculation of $\mathbf{A}_{DC}$ (see Eq. 3-9). In a similar manner the vectors of end-actions and reactions ($\mathbf{A}_{MR}$ and $\mathbf{A}_{RR}$) due to the support displacement shown in Fig. 3-13c can be found. These vectors are included in the calculations for $\mathbf{A}_{MC}$ and $\mathbf{A}_{RC}$ (see Eqs. 3-12 and 3-13).

As an example in which prestrain effects are to be determined, consider the analysis of the plane truss shown in Fig. 3-7. Previously, this truss was analyzed for the effects of the loads acting at joint E and for the weights of the members. Now assume that one of the members, such as member 3, is constructed with a length $L_3 + e$, instead of the theoretical length L_3. The effect of the additional increment of length is assumed to take place in the restrained truss shown in Fig. 3-7c. The result is that actions A_{DP1} and A_{DP2} are developed at joint E, corresponding to D_1 and D_2, respectively, and an axial force A_{MP3} is developed in member 3, corresponding to the action A_{M3}. All of these actions can be found without difficulty. For instance, the axial force is

$$A_{MP3} = -\frac{EA_3e}{L_3}$$

and the vector $\mathbf{A}_{\mathrm{MP}}$ becomes

$$\mathbf{A}_{\mathrm{MP}} = \begin{bmatrix} 0 \\ 0 \\ -\dfrac{EA_3e}{L_3} \\ 0 \end{bmatrix}$$

The horizontal and vertical components of the axial force in the member are used to obtain the actions at joint E in the restrained truss:

$$A_{DP1} = -\frac{EA_3e}{L_3}\cos\gamma_3 \qquad A_{DP2} = -\frac{EA_3e}{L_3}\sin\gamma_3$$

Therefore, the vector $\mathbf{A}_{\mathrm{DP}}$ is

$$\mathbf{A}_{\mathrm{DP}} = -\frac{EA_3e}{L_3}\begin{bmatrix} \cos\gamma_3 \\ \sin\gamma_3 \end{bmatrix}$$

The vectors $\mathbf{A}_{\mathrm{DP}}$ and $\mathbf{A}_{\mathrm{MP}}$, found above, are added to the vectors $\mathbf{A}_{\mathrm{DL}}$ and $\mathbf{A}_{\mathrm{ML}}$, found in the earlier example, to give the combined vectors $\mathbf{A}_{\mathrm{DC}}$ and $\mathbf{A}_{\mathrm{MC}}$. Then these vectors are used in Eqs. (3-11) and (3-14) when solving for the joint displacements and the member end-actions. The final results for the combined effects of loads and prestrains will be the sum of the results obtained for loads and prestrains occurring separately.

3.5 Stiffnesses of Prismatic Members. Commonly recurring stiffness coefficients (such as $4EI/L$) for prismatic members may be placed into *member stiffness matrices* for various types of framed structures. Terms in such a matrix are defined as holding actions at the ends of a member due to unit displacements of one end relative to the other. The type of member stiffness matrix $\mathbf{S}_{\mathrm{M}i}$ to be considered in this article is the inverse of the type of member flexibility matrix $\mathbf{F}_{\mathrm{M}i}$ discussed in Art. 2.6. Thus,

$$\mathbf{S}_{\mathrm{M}i} = \mathbf{F}_{\mathrm{M}i}^{-1} \tag{3-16}$$

Recall that $\mathbf{F}_{\mathrm{M}i}$ pertains to a member i that is fixed at the j end and free at the k end (see Figs. 2-13 through 2-17). Flexibilities in this matrix are defined as the relative displacements of the k end of the member with respect to the j end, due to unit actions at the k end. On the other hand, the inverse matrix $\mathbf{S}_{Mi}$ contains holding actions at the k end of the member, required for unit relative displacements of the k end with respect to the j end. Such member stiffnesses will be found directly by inducing unit displacements at the k ends of members for the same types of framed structures considered before. Formulas for the stiffness terms in $\mathbf{S}_{Mi}$ may be obtained from Table B-4 in Appendix B, and the results can be checked by the inversion process mentioned above. Member stiffness matrices developed in this manner will be used in the formalized approach to the stiffness method described in the next article.

Figure 3-14a shows a prismatic beam member i with both the j and k ends fixed. As before, the member-oriented axes are arranged so that the x_M-y_M plane is a principal plane of bending. Indicated by arrows at the k end of the member are the two kinds of displacements of interest: a translation D_{M1} in the y_M direction and a rotation D_{M2} in the z_M sense. The member stiffness matrix to be obtained is a 2 × 2 array relating D_{M1} and D_{M2} to the corresponding actions A_{M1} (force in the y_M direction) and A_{M2} (moment in the z_M sense). Figures 3-14b and 3-14c show the restraint actions at the k end of the member due to unit displacements $D_{M1} = 1$ and $D_{M2} = 1$. These terms form the beam stiffness matrix $\mathbf{S}_{Mi}$, as follows:

$$\mathbf{S}_{Mi} = \begin{bmatrix} S_{M11} & S_{M12} \\ S_{M21} & S_{M22} \end{bmatrix} = \begin{bmatrix} \dfrac{12EI}{L^3} & -\dfrac{6EI}{L^2} \\ -\dfrac{6EI}{L^2} & \dfrac{4EI}{L} \end{bmatrix} \tag{3-17}$$

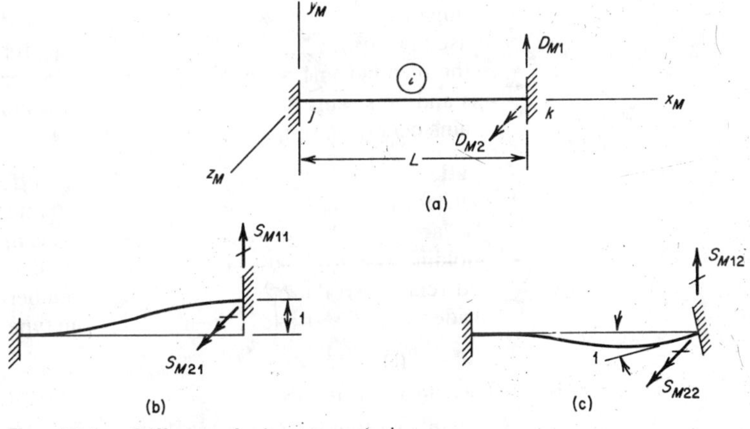

Fig. 3-14. Stiffnesses for beam member.

It is seen that this matrix is the inverse of the beam flexibility matrix $\mathbf{F}_{Mi}$ in Eq. (2-19).

For the truss member in Fig. 3-15a, only the axial translation D_M at the k end need be considered. The corresponding action A_M is the axial force at the k end, and the single member stiffness (see Fig. 3-15b) is found to be

$$S_{Mi} = \frac{EA}{L} \tag{3-18}$$

which is the reciprocal of the flexibility F_{Mi} in Eq. (2-20).

The stiffness matrix for a plane frame member is the same as that for a beam (see Eq. 3-17) if axial strains are to be neglected. Otherwise, if axial strains are to be considered, the term in Eq. (3-18) must also be included. In the latter case three displacement components at the k end of the member are required, as indicated in Fig. 3-16a. They are the translation D_{M1} in the x_M direction, the translation D_{M2} in the y_M direction, and the rotation D_{M3} in the z_M sense. Figures 3-16b, c, and d show unit values of these displacements and the corresponding restraint actions that constitute the following 3 × 3 array:

$$\mathbf{S}_{Mi} = \begin{bmatrix} S_{M11} & S_{M12} & S_{M13} \\ S_{M21} & S_{M22} & S_{M23} \\ S_{M31} & S_{M32} & S_{M33} \end{bmatrix} = \begin{bmatrix} \dfrac{EA}{L} & 0 & 0 \\ 0 & \dfrac{12EI}{L^3} & -\dfrac{6EI}{L^2} \\ 0 & -\dfrac{6EI}{L^2} & \dfrac{4EI}{L} \end{bmatrix} \tag{3-19}$$

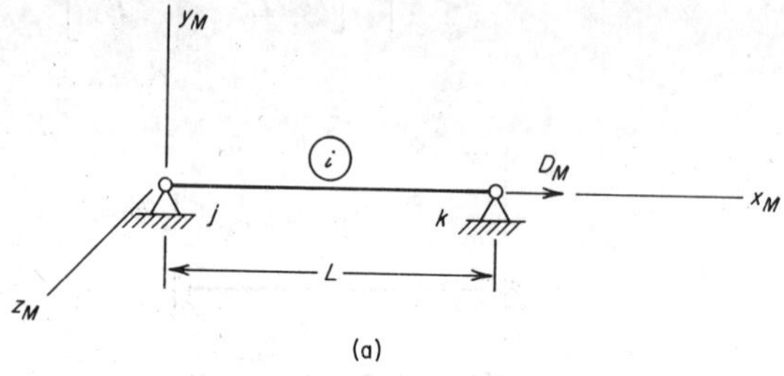

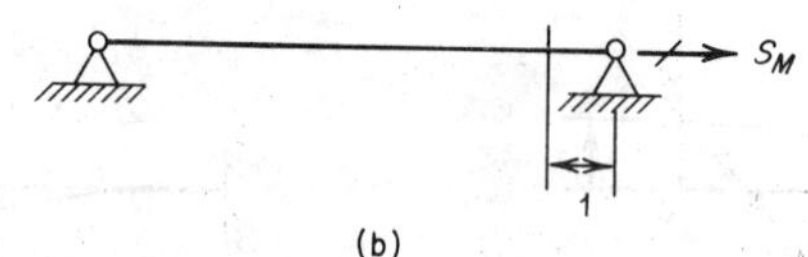

Fig. 3-15. Stiffness for truss member.

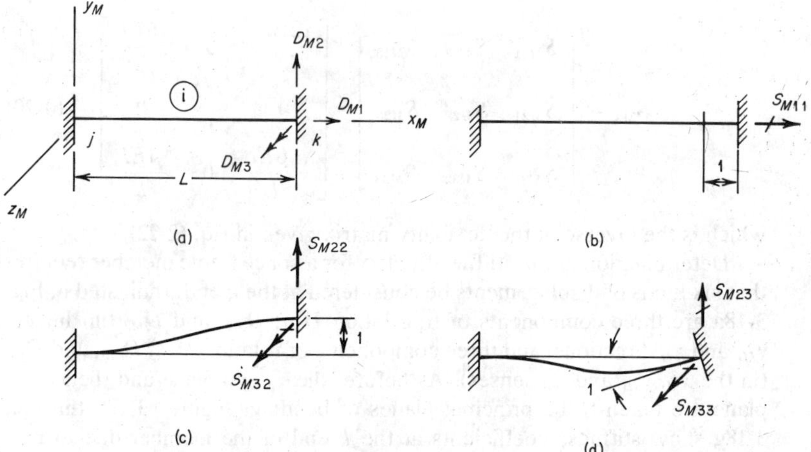

Fig. 3-16. Stiffnesses for plane frame member.

This plane frame member stiffness matrix is, of course, the inverse of the flexibility matrix in Eq. (2-21).

Figure 3-17a illustrates a grid member for which bending occurs in the x_M-y_M plane, as with a beam or plane frame member. Indicated at the k end of the member are a translation D_{M1} in the y_M direction, a rotation D_{M2} in the x_M sense, and a rotation D_{M3} in the z_M sense. Unit values of these displacements produce the stiffness terms shown in Figs. 3-17b, c, and d. The resulting 3×3 member stiffness matrix is

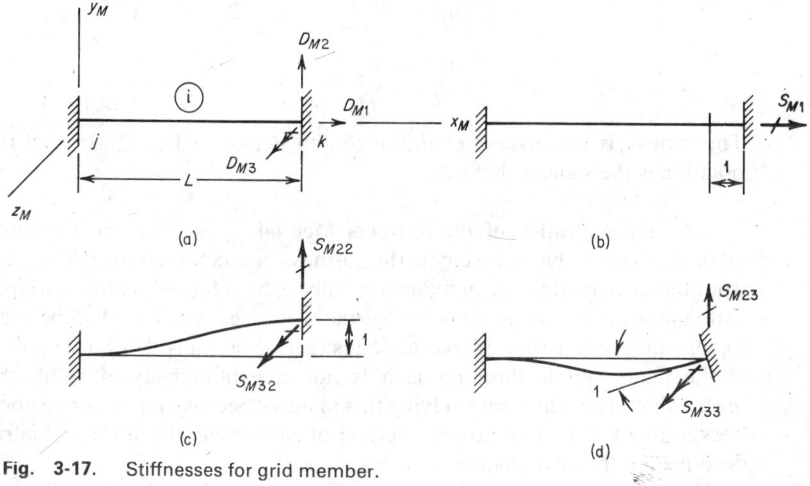

Fig. 3-17. Stiffnesses for grid member.

$$\mathbf{S}_{Mi} = \begin{bmatrix} S_{M11} & S_{M12} & S_{M13} \\ S_{M21} & S_{M22} & S_{M23} \\ S_{M31} & S_{M32} & S_{M33} \end{bmatrix} = \begin{bmatrix} \dfrac{12EI}{L^3} & 0 & -\dfrac{6EI}{L^2} \\ 0 & \dfrac{GJ}{L} & 0 \\ -\dfrac{6EI}{L^2} & 0 & \dfrac{4EI}{L} \end{bmatrix} \tag{3-20}$$

which is the inverse of the flexibility matrix given in Eq. (2-22).

Determination of the stiffness matrix for a space frame member requires that six kinds of displacements be considered at the k end. Indicated in Fig. 3-18a are three components of translation D_{M1}, D_{M2}, and D_{M3} (in the x_M, y_M, and z_M directions) and three components of rotation D_{M4}, D_{M5}, and D_{M6} (in the x_M, y_M, and z_M senses). As before, the x_M-y_M plane and the x_M-z_M plane are taken to be principal planes of bending. Figures 3-18b through 3-18g show stiffness coefficients at the k end of the member due to unit values of the displacements D_{M1} through D_{M6}. These terms become the nonzero elements in the following 6 × 6 member stiffness matrix:

$$\mathbf{S}_{Mi} = \begin{bmatrix} \dfrac{EA}{L} & 0 & 0 & 0 & 0 & 0 \\ 0 & \dfrac{12EI_Z}{L^3} & 0 & 0 & 0 & -\dfrac{6EI_Z}{L^2} \\ 0 & 0 & \dfrac{12EI_Y}{L^3} & 0 & \dfrac{6EI_Y}{L^2} & 0 \\ 0 & 0 & 0 & \dfrac{GJ}{L} & 0 & 0 \\ 0 & 0 & \dfrac{6EI_Y}{L^2} & 0 & \dfrac{4EI_Y}{L} & 0 \\ 0 & -\dfrac{6EI_Z}{L^2} & 0 & 0 & 0 & \dfrac{4EI_Z}{L} \end{bmatrix} \tag{3-21}$$

This matrix is the inverse of the flexibility matrix in Eq. (2-23), and the notation is the same as before.

3.6 Formalization of the Stiffness Method. As with the flexibility method, it is possible to assemble the stiffness matrix for a framed structure from the contributions of individual members by a formal matrix multiplication procedure. The principle of virtual work (see Art. 1.14) will be used for this purpose, assuming that all loads (actual or equivalent) are applied at the joints. While this approach is not computationally efficient, the reader should benefit from studying this material because it provides a good background for the more direct method of assembling the stiffness matrix described in the next chapter.

From the preceding article, the stiffness matrix $\mathbf{S}_{Mi}$ for an individual

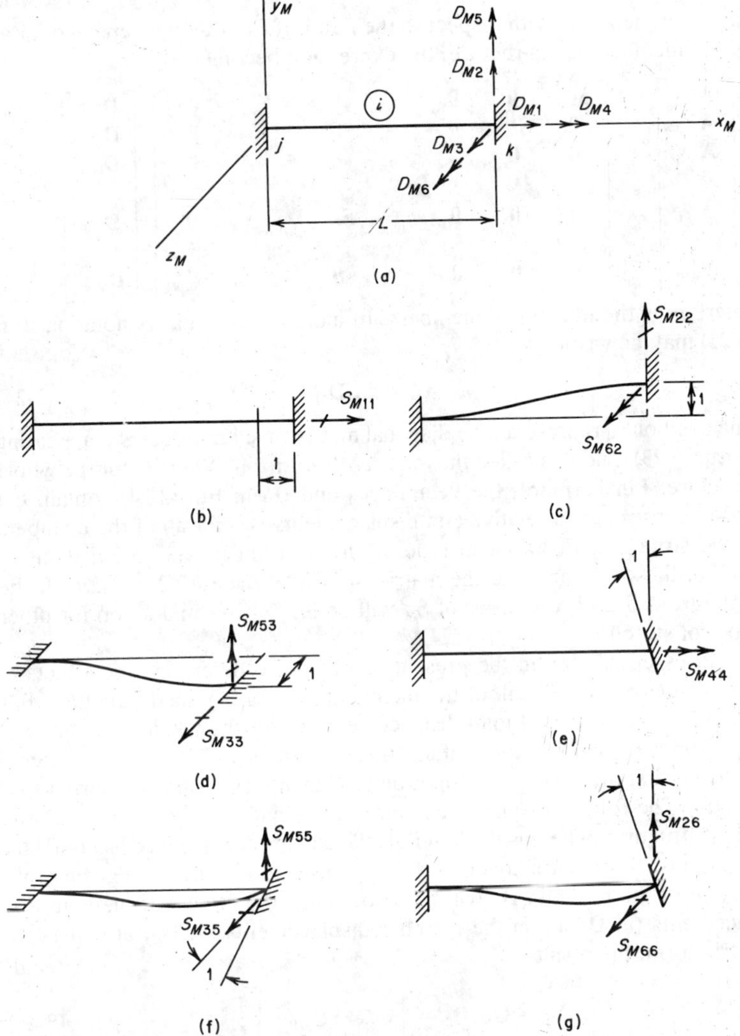

Fig. 3-18. Stiffnesses for space frame member.

member i relates the end-actions $\mathbf{A}_{Mi}$ to the corresponding end-displacements $\mathbf{D}_{Mi}$ by the expression

$$\mathbf{A}_{Mi} = \mathbf{S}_{Mi}\mathbf{D}_{Mi} \tag{3-22}$$

It should be recalled that the actions in $\mathbf{A}_{Mi}$ were chosen to be those at the k end of the member, and the vector $\mathbf{D}_{Mi}$ contains the relative displacements of the k end with respect to the j end. If Eq. (3-22) is repeated for all of the members in a structure, the expression becomes

$$\begin{bmatrix} \mathbf{A}_{M1} \\ \mathbf{A}_{M2} \\ \mathbf{A}_{M3} \\ \cdots \\ \mathbf{A}_{Mi} \\ \cdots \\ \mathbf{A}_{Mm} \end{bmatrix} = \begin{bmatrix} \mathbf{S}_{M1} & \mathbf{0} & \mathbf{0} & \cdots & \mathbf{0} & \cdots & \mathbf{0} \\ \mathbf{0} & \mathbf{S}_{M2} & \mathbf{0} & \cdots & \mathbf{0} & \cdots & \mathbf{0} \\ \mathbf{0} & \mathbf{0} & \mathbf{S}_{M3} & \cdots & \mathbf{0} & \cdots & \mathbf{0} \\ \cdots & \cdots & \cdots & \cdots & \cdots & \cdots & \cdots \\ \mathbf{0} & \mathbf{0} & \mathbf{0} & \cdots & \mathbf{S}_{Mi} & \cdots & \mathbf{0} \\ \cdots & \cdots & \cdots & \cdots & \cdots & \cdots & \cdots \\ \mathbf{0} & \mathbf{0} & \mathbf{0} & \cdots & \mathbf{0} & \cdots & \mathbf{S}_{Mm} \end{bmatrix} \begin{bmatrix} \mathbf{D}_{M1} \\ \mathbf{D}_{M2} \\ \mathbf{D}_{M3} \\ \cdots \\ \mathbf{D}_{Mi} \\ \cdots \\ \mathbf{D}_{Mm} \end{bmatrix} \tag{3-23}$$

where m is the number of members. In more concise matrix notation, Eq. (3-23) may be written as

$$\mathbf{A}_M = \mathbf{S}_M \mathbf{D}_M \tag{3-24}$$

The symbol $\mathbf{S}_M$ represents the diagonal matrix of submatrices $\mathbf{S}_{Mi}$ appearing in Eq. (3-23), and it is called the *unassembled stiffness matrix* for the whole structure. Furthermore, the vectors $\mathbf{A}_M$ and $\mathbf{D}_M$ in Eq. (3-24) contain the end-actions $\mathbf{A}_{Mi}$ and relative end-displacements $\mathbf{D}_{Mi}$ for all of the members in the structure. For a beam type of structure there will be $2m$ items in each of these vectors, and the matrix $\mathbf{S}_M$ will be of order $2m \times 2m$. If the structure is a grid, the order of $\mathbf{S}_M$ will be $3m \times 3m$, and so on for other types of structures.

An essential step in the present method is to determine the effects of joint displacements for all of the members in the restrained structure. For this purpose a vector of joint displacements $\mathbf{D}_J$ for the whole structure will be introduced. This vector contains free (but unknown) joint displacements $\mathbf{D}_F$ listed first and known restraint displacements $\mathbf{D}_R$ at the supports listed second. The former displacements are those that are free to occur in the actual structure, whereas the latter displacements are specified as boundary conditions. If there are no support displacements specified in a given problem, all of the terms in $\mathbf{D}_R$ will be zero. When the effects of the joint displacements (in $\mathbf{D}_J$) upon the member displacements (in $\mathbf{D}_M$) are cast into matrix form, the result is

$$\mathbf{D}_M = \mathbf{C}_{MJ} \mathbf{D}_J = \begin{bmatrix} \mathbf{C}_{MF} & \mathbf{C}_{MR} \end{bmatrix} \begin{bmatrix} \mathbf{D}_F \\ \mathbf{D}_R \end{bmatrix} \tag{3-25}$$

The symbol $\mathbf{C}_{MJ}$ in this equation represents a displacement transformation matrix relating $\mathbf{D}_M$ to $\mathbf{D}_J$ for the restrained structure. The elements of $\mathbf{C}_{MJ}$ are determined from compatibility considerations; so it is referred to as the *compatibility* (or *kinematics*) *matrix*. This array is partitioned column-wise into submatrices $\mathbf{C}_{MF}$ and $\mathbf{C}_{MR}$, which relate $\mathbf{D}_M$ to $\mathbf{D}_F$ and $\mathbf{D}_R$, respectively. Each column in the submatrix $\mathbf{C}_{MF}$ contains the member displacements caused by a unit value of an unknown joint displacement in the restrained structure. Similarly, each column in the submatrix $\mathbf{C}_{MR}$ consists of member displacements due to a unit value of a support displacement in the restrained structure. For completeness, terms in $\mathbf{C}_{MR}$ are developed for all support restraints, regardless of whether support displacements are zero or nonzero.

Next, consider an arbitrary set of small virtual displacements to be induced at all of the joints of the structure, including the points where it is supported. As with the vector $\mathbf{D}_J$, let the first part of $\delta\mathbf{D}_J$ contain virtual free displacements $\delta\mathbf{D}_F$; and let the second part consist of virtual support displacements $\delta\mathbf{D}_R$. The resulting virtual member end-displacements $\delta\mathbf{D}_M$ can now be written in terms of $\delta\mathbf{D}_J$ and its parts, as follows:

$$\delta\mathbf{D}_M = \mathbf{C}_{MJ}\delta\mathbf{D}_J = \begin{bmatrix}\mathbf{C}_{MF} & \mathbf{C}_{MR}\end{bmatrix}\begin{bmatrix}\delta\mathbf{D}_F \\ \delta\mathbf{D}_R\end{bmatrix} \tag{3-26}$$

The compatibility matrix $\mathbf{C}_{MJ}$ in this equation is the same as that in Eq. (3-25).

The external virtual work δW generated by the real joint actions $\mathbf{A}_J$ and the virtual joint displacements $\delta\mathbf{D}_J$ takes the form

$$\delta W = \mathbf{A}_J^T\delta\mathbf{D}_J = \begin{bmatrix}\mathbf{A}_F^T & \mathbf{A}_R^T\end{bmatrix}\begin{bmatrix}\delta\mathbf{D}_F \\ \delta\mathbf{D}_R\end{bmatrix} \tag{3-27}$$

in which the symbols $\mathbf{A}_F$ and $\mathbf{A}_R$ denote the parts of $\mathbf{A}_J$ corresponding to the free and restrained displacements, respectively. Similarly, the internal virtual work δU of the real member end-actions $\mathbf{A}_M$ and the virtual relative end-displacements $\delta\mathbf{D}_M$ may be stated as

$$\delta U = \mathbf{A}_M^T\delta\mathbf{D}_M \tag{3-28}$$

According to the principle of virtual work, expressions (3-27) and (3-28) must be equal. Thus,

$$\mathbf{A}_J^T\delta\mathbf{D}_J = \mathbf{A}_M^T\delta\mathbf{D}_M \tag{3-29}$$

When Eqs. (3-24), (3-25), and (3-26) are substituted into the right-hand side of Eq. (3-29), the result is

$$\mathbf{A}_J^T\delta\mathbf{D}_J = \mathbf{D}_J^T\mathbf{C}_{MJ}^T\mathbf{S}_M\mathbf{C}_{MJ}\delta\mathbf{D}_J \tag{3-30}$$

Cancelling the arbitrary vector $\delta\mathbf{D}_J$ and transposing Eq. (3-30) yields

$$\mathbf{A}_J = \mathbf{S}_J\mathbf{D}_J \tag{3-31}$$

where

$$\mathbf{S}_J = \mathbf{C}_{MJ}^T\mathbf{S}_M\mathbf{C}_{MJ} \tag{3-32}$$

Thus, the matrix $\mathbf{S}_J$ can be identified as the joint stiffness matrix relating the actions $\mathbf{A}_J$ to the displacements $\mathbf{D}_J$ in Eq. (3-31). It is formed by the congruence transformation in Eq. (3-32), using the $\mathbf{C}_{MJ}$ matrix as the postmultiplier of $\mathbf{S}_M$ and the transposed matrix $\mathbf{C}_{MJ}^T$ as the premultiplier. In this manner the unassembled stiffness matrix $\mathbf{S}_M$ is transformed to the *assembled stiffness matrix* $\mathbf{S}_J$ for all of the joints in the structure.

It is useful to partition the joint stiffness matrix $\mathbf{S}_J$ into submatrices pertaining to the free joint displacements $\mathbf{D}_F$, the support displacements $\mathbf{D}_R$, and their corresponding actions, $\mathbf{A}_F$ and $\mathbf{A}_R$. Thus, the expanded form of Eq. (3-31) is

$$\begin{bmatrix} \mathbf{A}_F \\ \mathbf{A}_R \end{bmatrix} = \begin{bmatrix} \mathbf{S}_{FF} & \mathbf{S}_{FR} \\ \mathbf{S}_{RF} & \mathbf{S}_{RR} \end{bmatrix} \begin{bmatrix} \mathbf{D}_F \\ \mathbf{D}_R \end{bmatrix} \tag{3-33}$$

where

$$\mathbf{S}_{FF} = \mathbf{C}_{MF}^T \mathbf{S}_M \mathbf{C}_{MF} \qquad \mathbf{S}_{FR} = \mathbf{C}_{MF}^T \mathbf{S}_M \mathbf{C}_{MR}$$

$$\mathbf{S}_{RF} = \mathbf{C}_{MR}^T \mathbf{S}_M \mathbf{C}_{MF} \qquad \mathbf{S}_{RR} = \mathbf{C}_{MR}^T \mathbf{S}_M \mathbf{C}_{MR}$$

Each of the submatrices $\mathbf{S}_{FF}$, $\mathbf{S}_{FR}$, $\mathbf{S}_{RF}$, and $\mathbf{S}_{RR}$ is obtained by a congruence transformation, using as operators $\mathbf{C}_{MF}$ and $\mathbf{C}_{MR}$ (the two parts of $\mathbf{C}_{MJ}$ from Eq. 3-25). At this point it can be seen that the submatrix $\mathbf{S}_{FF}$ relates the joint actions $\mathbf{A}_F$ to the free joint displacements $\mathbf{D}_F$. Therefore, it must be the same as the stiffness matrix $\mathbf{S}$ (without subscripts) discussed in previous articles. Such correlations with earlier material will be taken up in detail after all steps in the formalized approach have been covered.

The free joint displacements $\mathbf{D}_F$ and the support reactions $\mathbf{A}_R$ constitute the unknowns in Eq. (3-33). For convenience in solution, this equation will be rewritten as two separate matrix equations:

$$\mathbf{A}_F = \mathbf{S}_{FF}\mathbf{D}_F + \mathbf{S}_{FR}\mathbf{D}_R \tag{3-34a}$$

$$\mathbf{A}_R = \mathbf{S}_{RF}\mathbf{D}_F + \mathbf{S}_{RR}\mathbf{D}_R \tag{3-34b}$$

The first of these equations may be solved (symbolically) for the unknown joint displacements, as follows:

$$\mathbf{D}_F = \mathbf{S}_{FF}^{-1}(\mathbf{A}_F - \mathbf{S}_{FR}\mathbf{D}_R) \tag{3-35}$$

This expression represents the key step in the analysis, because the other quantities of interest have already been formulated in terms of the free displacements $\mathbf{D}_F$.

Support reactions may now be calculated from Eq. (3-34b). However, if actual or equivalent joint loads are applied directly to the supports, they should also be taken into account. This may be accomplished by adding their negatives to the results from Eq. (3-34b), as follows:

$$\mathbf{A}_R = -\mathbf{A}_{RC} + \mathbf{S}_{RF}\mathbf{D}_F + \mathbf{S}_{RR}\mathbf{D}_R \tag{3-36}$$

In this expression the symbol $\mathbf{A}_{RC}$ denotes combined loads (actual and equivalent) applied directly to the supports.

A formula for member end-actions caused by joint displacements can be obtained by substituting Eq. (3-25) into Eq. (3-24), yielding

$$\mathbf{A}_M = \mathbf{S}_M\mathbf{D}_M = \mathbf{S}_M\mathbf{C}_{MJ}\mathbf{D}_J \tag{a}$$

The form of this expression shows that the calculations proceed member-by-member, using the compatibility conditions in matrix $\mathbf{C}_{MJ}$ and individual member stiffnesses in $\mathbf{S}_M$. The member end-actions computed from Eq. (a) must be added to any initial fixed-end actions that may exist. Therefore, the final expression for $\mathbf{A}_M$ (in expanded form) becomes

$$\mathbf{A}_M = \mathbf{A}_{ML} + \mathbf{S}_M(\mathbf{C}_{MF}\mathbf{D}_F + \mathbf{C}_{MR}\mathbf{D}_R) \tag{3-37}$$

in which $\mathbf{A}_{ML}$ is a vector of fixed-end actions due to loads applied to the members.

Equations of the stiffness method developed by superposition principles in previous articles will now be correlated with similar expressions derived here by virtual work. The comparable equations for calculating unknown joint displacements (including the effects of support displacements) are

$$\mathbf{D} = \mathbf{S}^{-1}(\mathbf{A}_D - \mathbf{A}_{DL} - \mathbf{A}_{DR}) \qquad \text{(3-11) modified}$$

and

$$\mathbf{D}_F = \mathbf{S}_{FF}^{-1}(\mathbf{A}_F - \mathbf{S}_{FR}\mathbf{D}_R) \qquad \text{(3-35) repeated}$$

It is evident that $\mathbf{D} = \mathbf{D}_F$, $\mathbf{S} = \mathbf{S}_{FF}$, $\mathbf{A}_D - \mathbf{A}_{DL} = \mathbf{A}_F$, and $\mathbf{A}_{DR} = \mathbf{S}_{JR}\mathbf{D}_R$ (note the use of stiffness coefficients to obtain $\mathbf{A}_{DR}$). In either case the last term within the parentheses represents equivalent joint loads due to support displacements.

Comparable equations for determining support reactions are

$$\mathbf{A}_R = \mathbf{A}_{RL} + \mathbf{A}_{RD}\mathbf{D} + \mathbf{A}_{RR} \qquad \text{(3-15) modified}$$

and

$$\mathbf{A}_R = -\mathbf{A}_{RC} + \mathbf{S}_{RF}\mathbf{D}_F + \mathbf{S}_{RR}\mathbf{D}_R \qquad \text{(3-36) repeated}$$

Thus, $\mathbf{A}_{RL} = -\mathbf{A}_{RC}$, $\mathbf{A}_{RD} = \mathbf{S}_{RF}$, and $\mathbf{A}_{RR} = \mathbf{S}_{RR}\mathbf{D}_R$ (note the use of stiffness coefficients to obtain $\mathbf{A}_{RR}$).

In addition, the comparable equations for obtaining member end-actions are

$$\mathbf{A}_M = \mathbf{A}_{ML} + \mathbf{A}_{MD}\mathbf{D} + \mathbf{A}_{MR} \qquad \text{(3-14) modified}$$

and

$$\mathbf{A}_M = \mathbf{A}_{ML} + \mathbf{S}_M(\mathbf{C}_{MF}\mathbf{D}_F + \mathbf{C}_{MR}\mathbf{D}_R) \qquad \text{(3-37) repeated}$$

Hence, $\mathbf{A}_{MD} = \mathbf{S}_M\mathbf{C}_{MF}$ and $\mathbf{A}_{MR} = \mathbf{S}_M\mathbf{C}_{MR}\mathbf{D}_R$ (note the use of member stiffnesses and compatibility conditions to obtain $\mathbf{A}_{MD}$ and $\mathbf{A}_{MR}$).

In summary, formalization of the stiffness method leads to the concept of the assembled stiffness matrix $\mathbf{S}_J$ for all of the joints in the structure. It is partitioned into submatrices associated with unknown joint displacements $\mathbf{D}_F$ and specified support displacements $\mathbf{D}_R$ (see Eq. 3-33) as well as their corresponding actions. By this approach the calculation of support reactions and the effects of support displacements are automatically included within the basic formulation. Consequently, there is extensive use of stiffness coefficients associated with the supports.

To be selective in the formalized version, the analyst may omit columns of the $\mathbf{C}_{MR}$ matrix corresponding to zero values of support displacements. Of course, if all such displacements are zero, the $\mathbf{C}_{MR}$ part of the compatibility matrix can be omitted altogether. However, this would preclude the calculation of support reactions by Eq. (3-36) because the matrix $\mathbf{S}_{RF}$ (as well as $\mathbf{S}_{FR}$ and $\mathbf{S}_{RR}$) would not be generated.

Example 1. Figure 3-19a shows a continuous beam that was analyzed previously by the flexibility method (see Example 2 in Art. 2.7). Equivalent joint loads for this problem appear in Fig. 3-19b, which illustrates the restrained structure. In another figure of the same type (Fig. 3-19c), the six possible joint displacements are indicated. Two of these displacements (D_{F1} and D_{F2}) are free to occur, while four of them (D_{R1} through D_{R4}) are restrained. One of the latter displacements is specified to be nonzero ($D_{R3} = \delta$), but the others are all zero.

Stiffness matrices for the individual members of this continuous beam are found by Eq. (3-17) to be

$$\mathbf{S}_{M1} = 2\mathbf{S}_{M2} = \frac{2EI}{L^3}\begin{bmatrix} 12 & -6L \\ -6L & 4L^2 \end{bmatrix}$$

With these arrays as submatrices, the stiffness matrix $\mathbf{S}_M$ for the unassembled structure becomes

$$\mathbf{S}_M = \frac{2EI}{L^3}\begin{bmatrix} 12 & -6L & 0 & 0 \\ -6L & 4L^2 & 0 & 0 \\ 0 & 0 & 6 & -3L \\ 0 & 0 & -3L & 2L^2 \end{bmatrix}$$

as given by Eq. (3-23).

Figures 3-19d through 3-19i illustrate the six conditions of unit displacements required to generate the 4×6 compatibility matrix $\mathbf{C}_{MJ}$ for the restrained structure. In order to show clearly the meaning of each term in $\mathbf{C}_{MJ}$, it will be partitioned rowwise as well as column-wise. Thus,

$$\mathbf{C}_{MJ} = \begin{bmatrix} \mathbf{C}_{1F} & \mathbf{C}_{1R} \\ \mathbf{C}_{2F} & \mathbf{C}_{2R} \end{bmatrix} = \left[\begin{array}{cc|cccc} 0 & 0 & -1 & -L & 1 & 0 \\ 1 & 0 & 0 & -1 & 0 & 0 \\ \hline -L & 0 & 0 & 0 & -1 & 1 \\ -1 & 1 & 0 & 0 & 0 & 0 \end{array}\right]$$

in which the subscripts 1 and 2 refer to member numbers. Terms in the first column of $\mathbf{C}_{MJ}$ are the displacements at the k ends of members 1 and 2, relative to their j ends, due to the condition $D_{F1} = 1$. From Fig. 3-19d it is easy to see that for member 1 the translation and rotation of the k end, relative to the j end, are zero and unity, respectively. However, it is not obvious that for member 2 the translation and rotation of the k end, relative to the j end, are $-L$ and -1, respectively. As an aid to

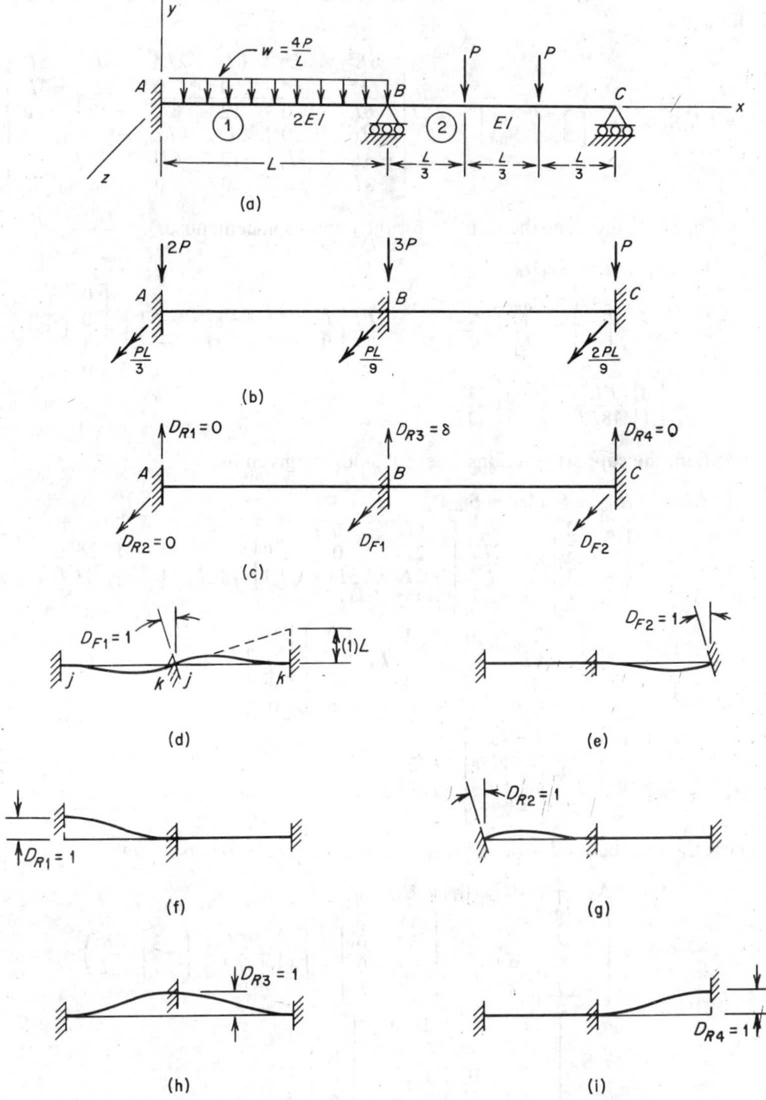

Fig. 3-19. Example 1: Continuous beam.

understanding this subtle point, a dashed line is drawn tangent to the j end of member 2 in Fig. 3-19d. Then it is seen that the translation of the k end, relative to the j end, is equal to $-(1)L$. In addition, the rotation of the k end of member 2, relative to its j end, is equal to -1. Terms in the other five columns of $\mathbf{C}_{MJ}$ can be obtained from Figs. 3-19e through 3-19i in a similar manner.

The joint stiffness matrix $\mathbf{S}_J$ for the assembled structure is found from Eq. (3-32) to be

$$\mathbf{S}_J = \mathbf{C}_{MJ}^T \mathbf{S}_M \mathbf{C}_{MJ} = \begin{bmatrix} \mathbf{S}_{FF} & \mathbf{S}_{FR} \\ \mathbf{S}_{RF} & \mathbf{S}_{RR} \end{bmatrix} = \frac{2EI}{L^3} \left[\begin{array}{cc|cccc} 6L^2 & L^2 & 6L & 2L^2 & -3L & -3L \\ L^2 & 2L^2 & 0 & 0 & 3L & -3L \\ \hline 6L & 0 & 12 & 6L & -12 & 0 \\ 2L^2 & 0 & 6L & 4L^2 & -6L & 0 \\ -3L & 3L & -12 & -6L & 18 & -6 \\ -3L & -3L & 0 & 0 & -6 & 6 \end{array}\right]$$

Then Eq. (3-35) provides the solution for the joint displacements as

$$\mathbf{D}_F = \mathbf{S}_{FF}^{-1}(\mathbf{A}_F - \mathbf{S}_{FR}\mathbf{D}_R)$$

$$= \frac{L}{22EI}\begin{bmatrix} 2 & -1 \\ -1 & 6 \end{bmatrix}\left(\frac{PL}{9}\begin{bmatrix} 1 \\ 2 \end{bmatrix} - \frac{2EI}{L^3}\begin{bmatrix} 6 & 2L^2 & -3L & -3L \\ 0 & 0 & 3L & -3L \end{bmatrix}\begin{bmatrix} 0 \\ 0 \\ \delta \\ 0 \end{bmatrix}\right)$$

$$= \begin{bmatrix} 0 \\ 1 \end{bmatrix}\frac{PL^2}{18EI} + \begin{bmatrix} -3 \\ 7 \end{bmatrix}\frac{3\delta}{11L}$$

In addition, the support reactions (see Eq. 3-36) are given by

$$\mathbf{A}_R = -\mathbf{A}_{RC} + \mathbf{S}_{RF}\mathbf{D}_F + \mathbf{S}_{RR}\mathbf{D}_R$$

$$= -\begin{bmatrix} -2 \\ -L/3 \\ -3 \\ -1 \end{bmatrix}P + \frac{2EI}{L^3}\begin{bmatrix} 6L & 0 \\ 2L^2 & 0 \\ -3L & 3L \\ -3L & -3L \end{bmatrix}\left(\begin{bmatrix} 0 \\ 1 \end{bmatrix}\frac{PL^2}{18EI} + \begin{bmatrix} -3 \\ 7 \end{bmatrix}\frac{3\delta}{11L}\right)$$

$$+ \frac{2EI}{L^3}\begin{bmatrix} 12 & 6L & -12 & 0 \\ 6L & 4L^2 & -6L & 0 \\ -12 & -6L & 18 & -6 \\ 0 & 0 & -6 & 6 \end{bmatrix}\begin{bmatrix} 0 \\ 0 \\ \delta \\ 0 \end{bmatrix}$$

$$= \begin{bmatrix} 6 \\ L \\ 10 \\ 2 \end{bmatrix}\frac{P}{3} + \begin{bmatrix} -40 \\ -17L \\ 63 \\ -22 \end{bmatrix}\frac{6EI\delta}{11L^3}$$

Finally, the member end-actions are obtained from Eq. (3-37), as follows:

$$\mathbf{A}_M = \mathbf{A}_{ML} + \mathbf{S}_M(\mathbf{C}_{MF}\mathbf{D}_F + \mathbf{C}_{MR}\mathbf{D}_R)$$

$$= \begin{bmatrix} 18 \\ -3L \\ 9 \\ -2L \end{bmatrix}\frac{P}{9} + \mathbf{S}_M\begin{bmatrix} 0 & 0 \\ 1 & 0 \\ -L & 0 \\ -1 & 1 \end{bmatrix}\left(\begin{bmatrix} 0 \\ 1 \end{bmatrix}\frac{PL^2}{18EI} + \begin{bmatrix} -3 \\ 7 \end{bmatrix}\frac{3\delta}{11L}\right)$$

$$+ \mathbf{S}_M\begin{bmatrix} -1 & -L & 1 & 0 \\ 0 & -1 & 0 & 0 \\ 0 & 0 & -1 & 1 \\ 0 & 0 & 0 & 0 \end{bmatrix}\begin{bmatrix} 0 \\ 0 \\ \delta \\ 0 \end{bmatrix}$$

$$= \begin{bmatrix} 6 \\ -L \\ 2 \\ 0 \end{bmatrix}\frac{P}{3} + \begin{bmatrix} 212 \\ -105L \\ -105 \\ 55L \end{bmatrix}\frac{12EI\delta}{11L^3}$$

The parts of this solution due to the applied loads in Fig. 3-19a are the same as those found previously by the flexibility method in Art. 2.7.

Example 2. A plane truss is to be analyzed for the vertical load and the horizontal support displacements shown in Fig. 3-20a. Of the eight possible joint displacements in this example, only the x and y translations at point A (denoted by D_{F1} and D_{F2}) are free to occur, while all others are restrained. To be selective in

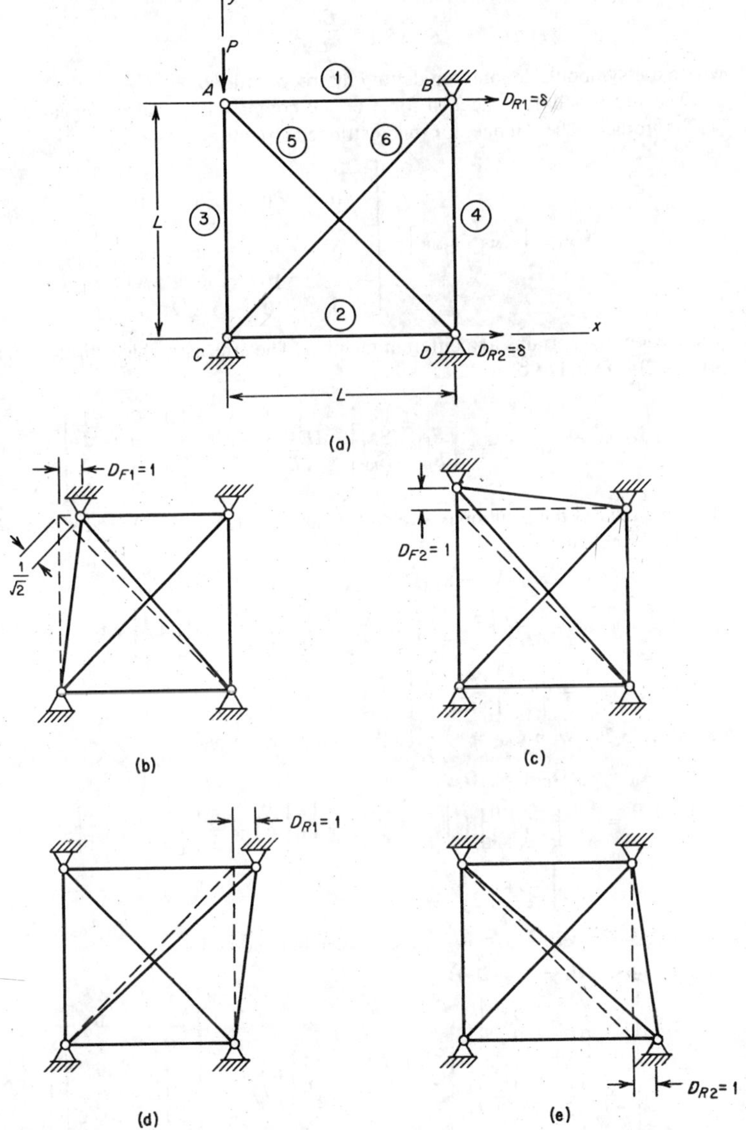

Fig. 3-20. Example 2: Plane truss.

the analysis, the only support reactions that will be determined are the horizontal forces at points B and D, corresponding to the specified support displacements $D_{R1} = D_{R2} = \delta$.

Cross-sectional areas of members 1 through 4 are equal to A, whereas those for members 5 and 6 are equal to $\sqrt{2}A$. Therefore, the axial stiffnesses of all members (Eq. 3-18) are EA/L, and the unassembled stiffness matrix (Eq. 3-23) is

$$\mathbf{S}_\mathrm{M} = \frac{EA}{L}\mathbf{I}_6$$

where the symbol $\mathbf{I}_6$ denotes an identity matrix of order 6.

The four conditions needed for a 6 × 4 compatibility matrix appear in Figs. 3-20b through 3-20e. Hence, for this restrained structure,

$$\mathbf{C}_\mathrm{MJ} = \begin{bmatrix}\mathbf{C}_\mathrm{MF} & \mathbf{C}_\mathrm{MR}\end{bmatrix} = \left[\begin{array}{cc|cc} -1 & 0 & 1 & 0 \\ 0 & 0 & 0 & 1 \\ 0 & 1 & 0 & 0 \\ 0 & 0 & 0 & 0 \\ -1/\sqrt{2} & 1/\sqrt{2} & 0 & 1/\sqrt{2} \\ 0 & 0 & 1/\sqrt{2} & 0 \end{array}\right]$$

where each row pertains to a different member. Then the assembled stiffness matrix (Eq. 3-32) is found to be

$$\mathbf{S}_\mathrm{J} = \mathbf{C}_\mathrm{MJ}^\mathrm{T}\mathbf{S}_\mathrm{M}\mathbf{C}_\mathrm{MJ} = \begin{bmatrix}\mathbf{S}_\mathrm{FF} & \mathbf{S}_\mathrm{FR} \\ \mathbf{S}_\mathrm{RF} & \mathbf{S}_\mathrm{RR}\end{bmatrix} = \frac{EA}{2L}\left[\begin{array}{cc|cc} 3 & -1 & -2 & -1 \\ -1 & 3 & 0 & 1 \\ \hline -2 & 0 & 3 & 0 \\ -1 & 1 & 0 & 3 \end{array}\right]$$

The remainder of the solution is straightforward. First, Eq. (3-35) produces the free joint displacements:

$$\begin{aligned}\mathbf{D}_\mathrm{F} &= \mathbf{S}_\mathrm{FF}^{-1}(\mathbf{A}_\mathrm{F} - \mathbf{S}_\mathrm{FR}\mathbf{D}_\mathrm{R}) \\ &= \frac{L}{4EA}\begin{bmatrix}3 & 1 \\ 1 & 3\end{bmatrix}\left(\begin{bmatrix}0 \\ -P\end{bmatrix} - \frac{EA}{2L}\begin{bmatrix}-2 & -1 \\ 0 & 1\end{bmatrix}\begin{bmatrix}1 \\ 1\end{bmatrix}\delta\right) \\ &= \begin{bmatrix}-1 \\ -3\end{bmatrix}\frac{PL}{4EA} + \begin{bmatrix}1 \\ 0\end{bmatrix}\delta\end{aligned}$$

Next, Eq. (3-36), with $\mathbf{A}_\mathrm{RC} = \mathbf{0}$, gives the support reactions:

$$\begin{aligned}\mathbf{A}_\mathrm{R} &= \mathbf{S}_\mathrm{RF}\mathbf{D}_\mathrm{F} + \mathbf{S}_\mathrm{RR}\mathbf{D}_\mathrm{R} \\ &= \frac{EA}{2L}\begin{bmatrix}-2 & 0 \\ -1 & 1\end{bmatrix}\left(\begin{bmatrix}-1 \\ -3\end{bmatrix}\frac{PL}{4EA} + \begin{bmatrix}1 \\ 0\end{bmatrix}\delta\right) + \frac{EA}{2L}\begin{bmatrix}3 & 0 \\ 0 & 3\end{bmatrix}\begin{bmatrix}1 \\ 1\end{bmatrix}\delta \\ &= \begin{bmatrix}1 \\ -1\end{bmatrix}\frac{P}{4} + \begin{bmatrix}1 \\ 2\end{bmatrix}\frac{EA\delta}{2L}\end{aligned}$$

Last, Eq. (3-37), with $\mathbf{A}_\mathrm{ML} = \mathbf{0}$, yields the member end-actions:

$$\begin{aligned}\mathbf{A}_\mathrm{M} &= \mathbf{S}_\mathrm{M}(\mathbf{C}_\mathrm{MF}\mathbf{D}_\mathrm{F} + \mathbf{C}_\mathrm{MR}\mathbf{D}_\mathrm{R}) \\ &= \mathbf{S}_\mathrm{M}\begin{bmatrix}-1 & 0 \\ 0 & 0 \\ 0 & 1 \\ 0 & 0 \\ -1/\sqrt{2} & 1/\sqrt{2} \\ 0 & 0\end{bmatrix}\left(\begin{bmatrix}-1 \\ -3\end{bmatrix}\frac{PL}{4EA} + \begin{bmatrix}1 \\ 0\end{bmatrix}\delta\right) + \mathbf{S}_\mathrm{M}\begin{bmatrix}1 & 0 \\ 0 & 1 \\ 0 & 0 \\ 0 & 0 \\ 0 & 1/\sqrt{2} \\ 1/\sqrt{2} & 0\end{bmatrix}\begin{bmatrix}1 \\ 1\end{bmatrix}\delta \\ &= \{1, 0, -3, 0, -\sqrt{2}, 0\}P/4 + \{0, 1, 0, 0, 0, 1/\sqrt{2}\}EA\delta/L\end{aligned}$$

Thus, only members 1, 3, and 5 are stressed by the load; and only members 2 and 6 are stressed by the support displacements.

Example 3. The third example consists of a rectangular plane frame with prismatic members, subjected to a lateral force P, as shown in Fig. 3-21a. There are no support displacements, and axial strains in the members are to be neglected. For members 1, 2, and 3 the j ends are taken to be at points A, C, and D, respectively. To be selective in this example, only the joint displacements and the member end-actions (at the k ends) will be determined. Support reactions at points C and D will be the same as the end-actions at the j ends of members 2 and 3.

When axial strains are neglected in a plane frame member, its stiffness matrix is the same as that for a beam member (see Eq. 3-17). Therefore, the member stiffness matrices for this problem are

$$\mathbf{S}_{M1} = \frac{EI}{L^3}\begin{bmatrix} 3 & -3L \\ -3L & 4L^2 \end{bmatrix} \qquad \mathbf{S}_{M2} = \mathbf{S}_{M3} = \frac{EI}{L^3}\begin{bmatrix} 12 & -6L \\ -6L & 4L^2 \end{bmatrix}$$

(Note that the flexural rigidity and the length of member 1 are double those of members 2 and 3.) Then the unassembled stiffness matrix (Eq. 3-23) becomes

$$\mathbf{S}_{M} = \frac{EI}{L^3}\begin{bmatrix} 3 & -3L & 0 & 0 & 0 & 0 \\ -3L & 4L^2 & 0 & 0 & 0 & 0 \\ 0 & 0 & 12 & -6L & 0 & 0 \\ 0 & 0 & -6L & 4L^2 & 0 & 0 \\ 0 & 0 & 0 & 0 & 12 & -6L \\ 0 & 0 & 0 & 0 & -6L & 4L^2 \end{bmatrix}$$

Figures 3-21b, c, and d show the three displacement patterns required to generate the compatibility submatrix $\mathbf{C}_{MF}$ for the restrained structure. Thus, for the displacements indicated,

$$\mathbf{C}_{MF} = \begin{bmatrix} \mathbf{C}_{1F} \\ \mathbf{C}_{2F} \\ \mathbf{C}_{3F} \end{bmatrix} = \begin{bmatrix} -2L & 0 & 0 \\ -1 & 1 & 0 \\ \hline 0 & 0 & -1 \\ 1 & 0 & 0 \\ \hline 0 & 0 & -1 \\ 0 & 1 & 0 \end{bmatrix}$$

which is partitioned row-wise according to members. The assembled stiffness submatrix $\mathbf{S}_{FF}$ (see Eq. 3-33) can now be determined as

$$\mathbf{S}_{FF} = \mathbf{C}_{MF}^{T}\mathbf{S}_{M}\mathbf{C}_{MF} = \frac{2EI}{L^3}\begin{bmatrix} 4L^2 & L^2 & 3L \\ L^2 & 4L^2 & 3L \\ 3L & 3L & 12 \end{bmatrix}$$

The solution for free joint displacements (Eq. 3-35) involves no support movements. Therefore, the results are simply

$$\mathbf{D}_{F} = \mathbf{S}_{FF}^{-1}\mathbf{A}_{F} = \frac{L}{84EI}\begin{bmatrix} 13 & -3 & -3L \\ -3 & 13 & -3L \\ -3L & -3L & 5L^2 \end{bmatrix}\begin{bmatrix} 0 \\ 0 \\ P \end{bmatrix} = \begin{bmatrix} -3 \\ -3 \\ 5L \end{bmatrix}\frac{PL^2}{84EI}$$

Similarly, member end-actions (Eq. 3-37) depend only upon the values of $\mathbf{D}_F$, as follows:

$$\begin{aligned} \mathbf{A}_{M} &= \mathbf{S}_{M}\mathbf{C}_{MF}\mathbf{D}_{F} = \mathbf{S}_{M}\{6L, 0, -5L, -3, -5L, -3\}PL^2/84EI \\ &= \{3, -3L, -7, 3L, -7, 3L\}P/14 \end{aligned}$$

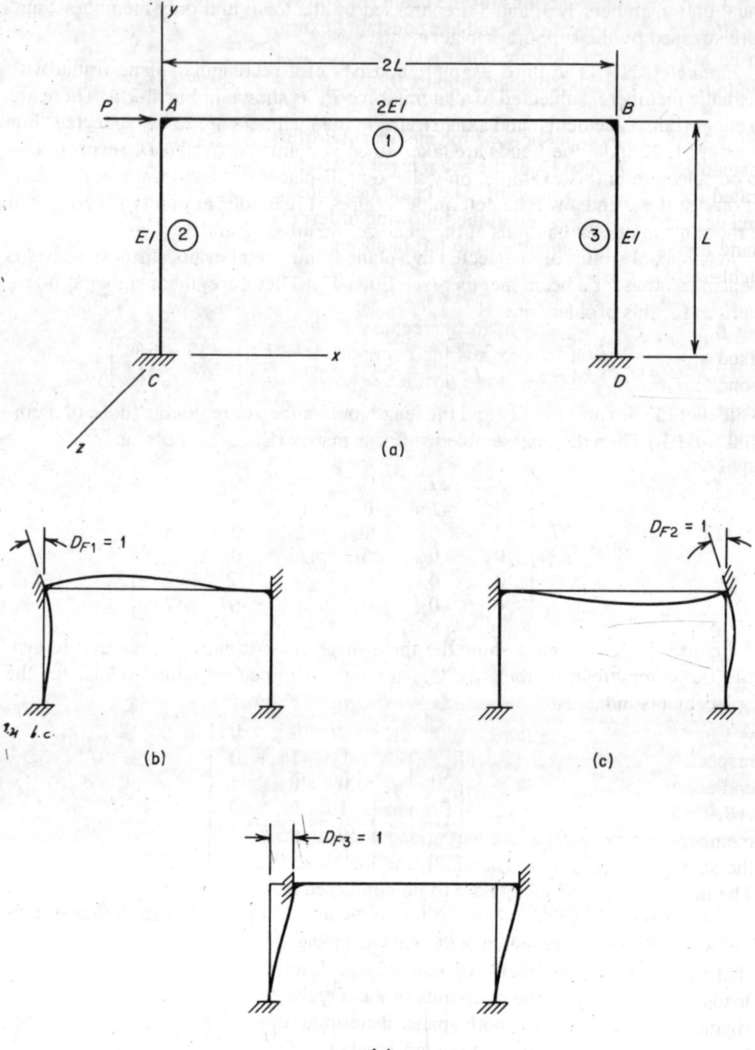

Fig. 3-21. Example 3: Plane frame (flexural effects only).

These values, together with principles of static equilibrium applied to members 2 and 3, are sufficient to determine the reactions at points C and D.

Problems. *The problems for Art. 3.3 are to be solved by the stiffness method using Eqs. (3-6) to (3-8). Assume all actions and displacements are positive either to the right, upward, or counterclockwise, unless stated otherwise.*

3.3-1. Find the reactions for the beam AB shown in Fig. 3-1a. The beam carries a uniform load of intensity w and has constant flexural rigidity EI. The reactions are to be taken in the following order: (1) the force at support A, (2) the moment at support A, and (3) the force at support B.

3.3-2. Find the end-actions for member AB of the beam shown in Fig. 3-1a if, instead of the uniform load, the beam is subjected to a vertical, downward, concentrated force P at the midpoint. The beam has constant flexural rigidity EI, and the end-actions are to be taken in the following order: (1) the shearing force at the left-hand end, (2) the moment at the left-hand end, and (3) the shearing force at the right-hand end.

3.3-3. Find the reactions for the beam AB shown in the figure. The beam has a fixed support at end A and a guided support at end B, and is subjected to two concentrated loads acting at the positions shown. Assume that the beam has constant flexural rigidity EI. The reactions are to be taken in the following order: (1) and (2), the vertical force and moment, respectively, at support A; and (3), the moment at support B.

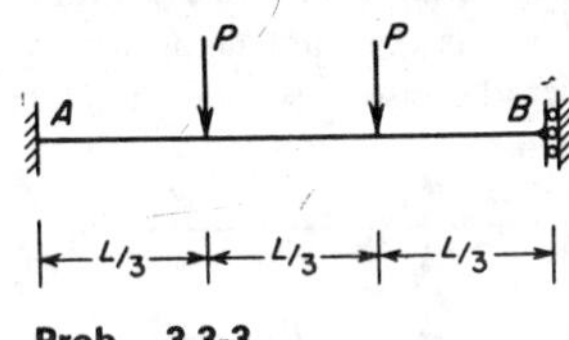

Prob. 3.3-3.

3.3-4. Analyze the two-span beam shown in Fig. 3-2a if $P_1 = P$, $M = PL$, $P_2 = 0$, and $P_3 = 0$. Assume that members AB and BC have lengths L and $1.5L$, respectively, and that the flexural rigidity EI is constant for both spans. Determine end-actions for the members as follows: (1) the shearing force at end A of member AB, (2) the moment at end A of member AB, (3) the shearing force at end B of member BC, and (4) the moment at end B of member BC. Determine reactions for the structure as follows: (1) and (2), the forces at supports B and C, respectively. The unknown displacements are to be numbered from left to right along the beam.

3.3-5. Analyze the beam ABC shown in Fig. 3-5a if $P_1 = 3P_2 = P$. Assume that the lengths of members AB and BC are L and $2L$, respectively, and that the loads P_1 and P_2 act at the midpoints of the members. Also assume that the flexural rigidity EI is constant for both spans. Determine the following end-actions for the beam: (1) the shearing force at the left-hand end of member AB and (2) the moment at the left-hand end of member AB. Determine the following reactions: (1) the force at support B and (2) the moment at support C. The unknown displacements are to be numbered from left to right along the beam.

3.3-6. Analyze the three-span beam shown in the figure if $L_1 = L_2 = L_3 = L$, $P_1 = P$, $P_2 = P_3 = 0$, $M = 0$, and $wL = P$. The flexural rigidity EI is constant for all members. Determine the following end-actions: (1) and (2), the shearing force and moment, respectively, at the left-hand end of member AB; (3) and (4), the shearing force and moment, respectively, at the right-hand end of member AB. Determine the following reactions: (1) and (2), the forces at supports B and C, respectively. The unknown displacements are to be numbered from left to right along the beam.

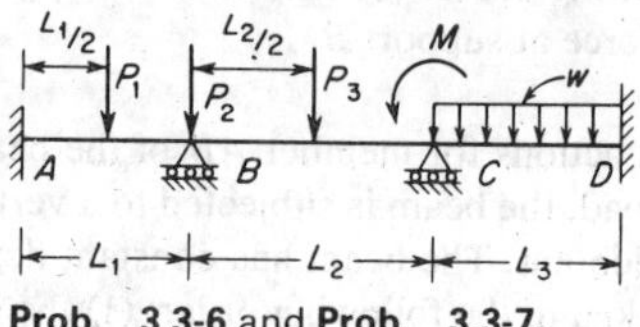

Prob. 3.3-6 and Prob. 3.3-7.

3.3-7. Analyze the three-span beam in the preceding problem if $L_1 = L_3 = L$, $L_2 = 2L$, $P_1 = P_2 = P_3 = P$, $M = PL$, and $wL = P$. The flexural rigidity for members *AB* and *CD* is *EI* and for member *BC* is 2*EI*. Determine the end-moments for all members, numbering the six actions consecutively from left to right. Also, determine reactions as follows: (1) and (2), the forces at supports *B* and *C*, respectively. The unknown displacements are to be numbered from left to right along the beam.

3.3-8. Obtain the stiffness matrix **S** for the continuous beam shown in the figure, assuming that the beam has constant flexural rigidity *EI*. The unknown displacements are to be numbered consecutively from left to right along the beam.

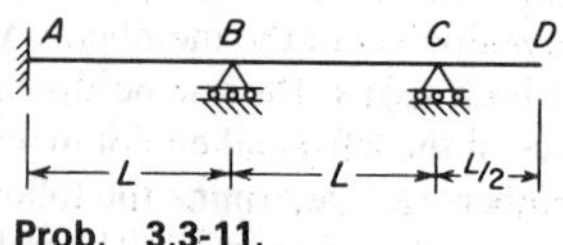

Prob. 3.3-8.

3.3-9. Solve the preceding problem for the beam of Prob. 2.3-5.

3.3-10. Solve Prob. 3.3-8 for a continuous beam on simple supports having five equal spans.

3.3-11. Obtain the stiffness matrix **S** for the beam shown in the figure, assuming that it has constant flexural rigidity *EI*. The unknown displacements are to be numbered from left to right along the beam with translation preceding rotation when both occur at the same joint.

Prob. 3.3-11.

3.3-12. Obtain the stiffness matrix **S** for the beam shown in the figure, assuming that the flexural rigidity of the middle span is twice that of the end spans. The unknown displacements are to be numbered from left to right in the figure.

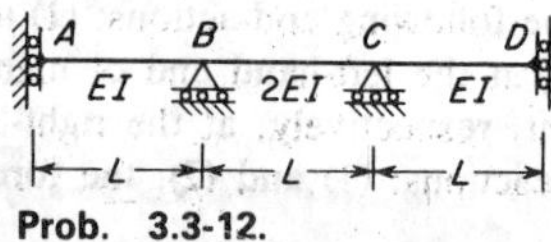

Prob. 3.3-12.

3.3-13. Obtain the stiffness matrix **S** for the four-span continuous beam shown in the figure if $L_1 = L_4 = L$, $L_2 = L_3 = 1.5L$, $EI_1 = EI_4 = EI$, and $EI_2 = EI_3 = 2EI$. The unknown displacements are to be numbered from left to right along the beam.

Prob. 3.3-13.

3.3-14. Find the reactions at the supports *A* and *C* for the beam with fixed ends shown in the figure. Assume that $EI_1 = 2EI$ and $EI_2 = EI$. Number the reactions in the following order: vertical force at *A*, moment at *A*, vertical force at *C*, and moment at *C*. (Hint: Perform the calculations by considering *AB* and *BC* as separate members of the structure.)

Prob. 3.3-14.

3.3-15. Find the axial forces in all members of the truss shown in the figure. The truss is subjected to a horizontal force *P* at joint *A*. Omit the weights of tl members from the analysis. Each member has length *L* and axial rigidity *EA* Assume that tensile force in a member is positive, and use the numbering systen shown in the figure.

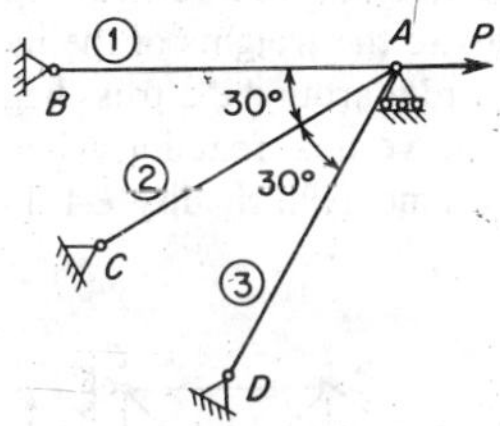

Prob. 3.3-15.

3.3-16. Calculate the axial forces in all members of the truss due to the force *P* only (see figure), assuming that *EA* is the same for all members. Assume that tensile

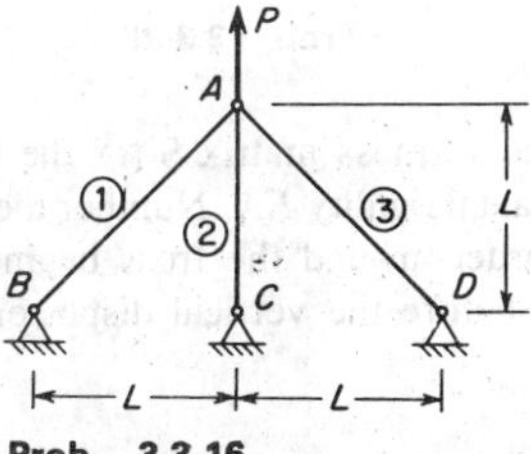

Prob. 3.3-16.

forces are positive, and use the numbering system shown in the figure. (Hint: Because of the symmetry of the structure and loading, there is only one unknown joint displacement.)

3.3-17. Solve the preceding problem if, in addition to the force P, the weights of all members are included in the analysis. Assume that each member has weight w per unit length. Calculate the axial forces at the upper ends of the members.

3.3-18. Find the axial forces in all members of the truss for Prob. 3.3-16 if there is a horizontal force P acting to the right at joint A (in addition to the upward force P). Assume that EA is the same for all members, and consider tension in a member to be positive. Omit the effects of the weights of the members.

3.3-19. Find the axial forces in the members of the truss shown in the figure if all members have the same length L and the same axial rigidity EA. The angles between the members are 45 degrees, and the load P makes an angle of 45 degrees with member 5, which is horizontal. Assume that tension in a member is positive. Omit the effects of the weights of the members.

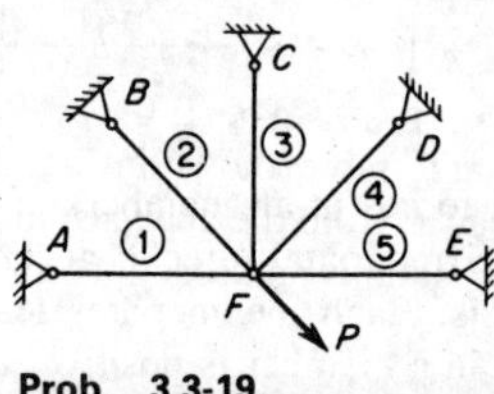

Prob. 3.3-19.

3.3-20. Calculate the horizontal and vertical reactions at the supports of the truss shown in the figure due to the weights of the members. Number the six reactions in a counterclockwise order around the truss beginning at joint A, and take the horizontal reaction before the vertical reaction when both occur at the same support. Each member has the same axial rigidity EA and the same weight w per unit length.

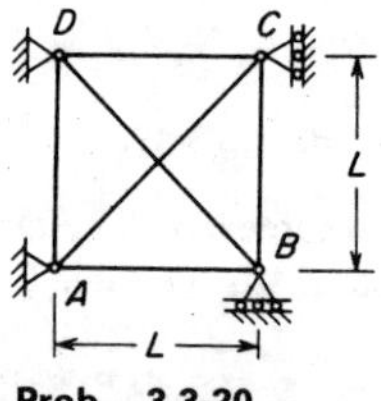

Prob. 3.3-20.

3.3-21. Construct the stiffness matrix **S** for the truss shown in the figure. All members have the same axial rigidity EA. Number the unknown joint displacements in a counterclockwise order around the truss beginning at joint B, and take the horizontal displacement before the vertical displacement when both occur at the same joint.

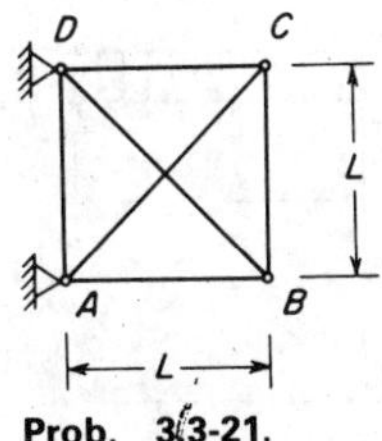

Prob. 3.3-21.

3.3-22. Obtain the stiffness matrix **S** for the truss shown in Fig. 2-5a, assuming that all members have the same axial rigidity EA. Number the joint displacements in a counterclockwise order around the truss beginning at joint D, and take the horizontal displacement before the vertical displacement when both occur at the same joint.

3.3-23. Obtain the stiffness matrix **S** for the truss of Prob. 2.3-9 if the axial rigidity for the vertical and horizontal members is EA and for the diagonal members is $2EA$. Number the joint displacements as described in the preceding problem, but begin with joint E.

3.3-24. Analyze the plane frame shown in Fig. 3-10a if a clockwise couple M acts at joint B. Omit the load P from the analysis, and consider only flexural deformations. Determine the end-actions and reactions shown in Fig. 3-10b.

3.3-25. Find the reactions at joints A and D for the plane frame shown in the figure, considering only flexural deformations. Assume that all members have flexural rigidity EI and that $L = 1.5H$. Number the reactions in the following order, first for joint A and then for joint D: horizontal force, vertical force, and moment.

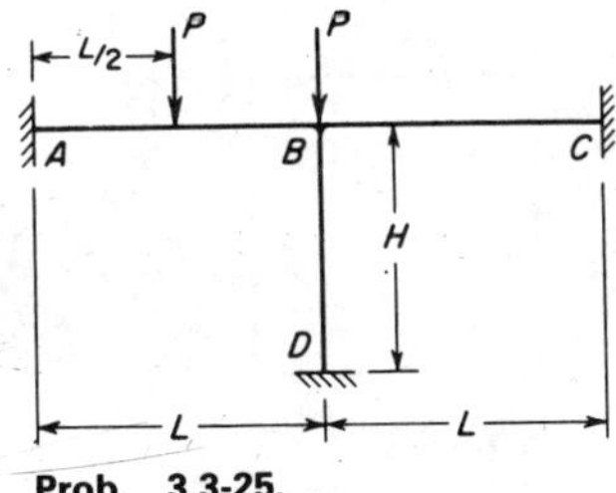

Prob. 3.3-25.

3.3-26. Analyze the plane frame shown in the figure considering only the effects of flexural deformations. Assume that $M = 2wL^2$, $H = L$, and that both members have flexural rigidity EI. Determine the following end-actions: (1) the axial force, (2) the shearing force, and (3) the moment at end B of member BC. Also determine the reactions at support C, taking the horizontal force before the vertical force.

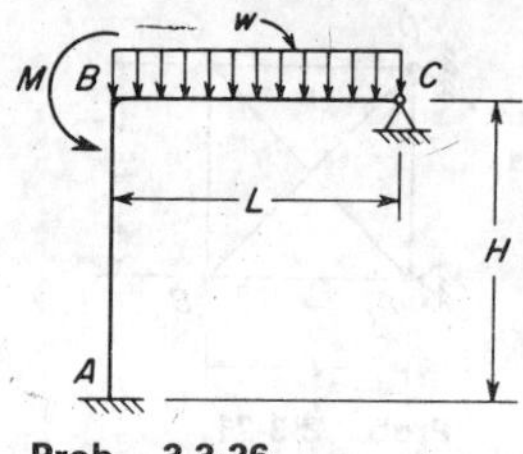

Prob. 3.3-26.

3.3-27. Obtain the stiffness matrix **S** for the plane frame of Prob. 2.3-13, considering (a) flexural deformations only and (b) flexural and axial deformations. Number the unknown displacements for part (a) as follows: horizontal displacement of B, rotation of B, and rotation of C. For part (b), number the displacements by taking joint B before joint C, and by taking the displacements at a joint in the following order: horizontal translation, vertical translation, and rotation. Assume that all members have flexural rigidity EI and axial rigidity EA.

3.3-28. Obtain the stiffness matrix **S** for the plane frame of Prob. 2.3-15 considering only flexural deformations. Number the unknown displacements in the same sequence that the joints are labeled.

3.3-29. Obtain the stiffness matrix **S** for the plane frame shown in the figure if only flexural deformations are taken into account. The flexural rigidities for the columns are EI_1 and for the beams are EI_2. Number the unknown joint displacements in the following order: (1) horizontal translation of beam AB, (2) horizontal translation of beam CD, (3) rotation of joint A, (4) rotation of B, (5) rotation of C, and (6) rotation of D.

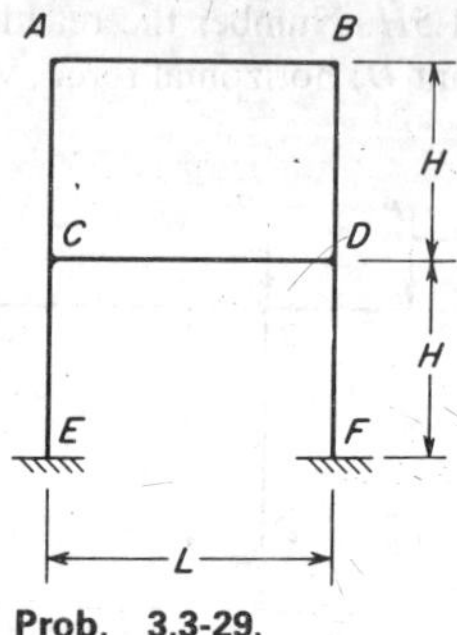

Prob. 3.3-29.

3.3-30. Find the reactions at support A for the grid in Fig. 3-12a. The load on the grid consists of the concentrated force P shown in the figure, and both members have the same flexural rigidity EI and torsional rigidity GJ. Number the reactions in the following order: (1) force in the y direction, (2) moment about the x axis, and (3) moment about the z axis. Assume all actions and displacements are positive when their vectors are in the positive directions of the coordinate axes.

3.3-31. Find the member end-actions at end C of member BC for the grid in Fig. 3-12a. The only load on the grid is the concentrated force P shown in the figure, and both members have flexural rigidity EI and torsional rigidity GJ. Number the end-actions as follows: (1) shearing force in the y direction, (2) bending moment about the x axis, and (3) twisting moment about the z axis. Use the sign convention described in the preceding problem.

3.3-32. Determine the displacements D_1, D_2, and D_3 at joint B of the grid shown in Fig. 3-12a due only to the weight of the members. Assume that each member has weight w per unit length, and that the rigidities EI and GJ are the same for both members.

3.3-33. Analyze the grid in Fig. 3-12a if there is a simple support at joint B. The support prevents translation in the y direction but does not offer any restraint against rotation of the joint. The load on the grid consists of the force P shown in the figure, and both members have the same flexural rigidity EI and torsional rigidity GJ. Number the unknown displacements in the following order: (1) rotation about the x axis and (2) rotation about the z axis. Determine the reactions at support A, using the numbering system and sign convention described in Prob. 3.3-30

3.3-34. Obtain the stiffness matrix **S** for the grid of Prob. 2.3-18. All members of the grid have the same flexural rigidity EI and torsional rigidity GJ. Number the unknown joint displacements in the same sequence that the joints are labeled, taking at each joint the rotation about the x axis before the rotation about the z axis. Assume all displacements are positive when their vectors are in the positive directions of the axes.

3.3-35. Determine the stiffness matrix **S** for the grid pictured in Prob. 2.3-19. Assume that the members AB, BC, and BD each have length L, flexural rigidity EI, and torsional rigidity GJ. Number the unknown joint displacements in the same sequence that the joints are labeled, taking the displacements at each joint (when they exist) in the following order: translation in the y direction, rotation about the x axis, and rotation about the z axis. Assume all displacements are positive when their vectors are in the positive directions of the axes.

In solving the problems for Art. 3.4, assume that all actions and displacements are positive either to the right, upward, or counterclockwise.

3.4-1. Find the reactions for the beam of Fig. 3-2a, due to a linear temperature gradient such that the lower surface of member AB is at temperature T_1 while the upper surface is at temperature T_2. Omit the loads from the analysis, and assume that both members have the same flexural rigidity EI. The depth of member AB is d, and the coefficient of thermal expansion is α. Take the reactions in the following order: vertical force at A, moment at A, force at B, and force at C.

3.4-2. Find the reactions for the beam of Fig. 3-2a if support B is displaced downward a distance s_1 and support C is displaced downward a distance s_2. Omit the loads from the analysis, and assume that both members have flexural rigidity EI. Take the reactions in the order described in the preceding problem.

3.4-3. Determine member end-actions and reactions for the beam of Fig. 3-4a if member BC initially has a sharp bend at the position shown in the accompanying figure (the angle β is a small angle). Omit the loads from the analysis, and assume all members have the same flexural rigidity EI. Determine the end-actions and reactions shown in Fig. 3-4b.

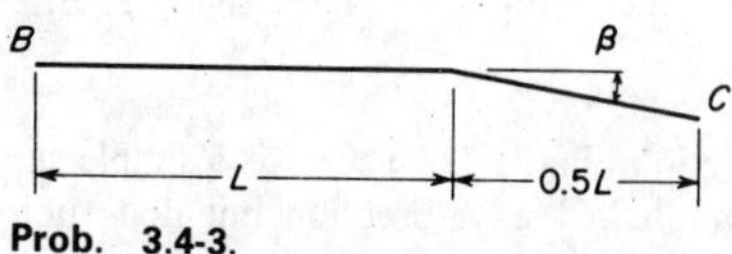

Prob. 3.4-3.

3.4-4. Find the end-actions and reactions shown for the beam in Fig. 3-5b if the entire beam has a linear temperature gradient such that the lower surface is at temperature T_1 while the upper surface is at temperature T_2. Omit the loads from the analysis. Assume that both members have flexural rigidity EI, depth d, and thermal coefficient α.

3.4-5. Solve Prob. 3.3-3 if support A rotates clockwise by an amount β, in which β is a small angle. Omit the loads from the analysis.

3.4-6. Determine the vectors $\mathbf{A}_{DC}$ and $\mathbf{A}_{RC}$ for the beam of Prob. 3.3-6 if, in addition to the loads, the following effects occur: support A is displaced downward a distance s, support D rotates counterclockwise through an angle β, and the entire beam has a linear temperature gradient (T_1 is the temperature on the bottom and T_2 is the temperature on the top).

3.4-7. Assume that the plane truss of Fig. 3-7a has its temperature increased uniformly by an amount T. Omitting the loads from the analysis, find the end-actions shown in Fig. 3-7b. The coefficient of thermal expansion is α, all members have the same axial rigidity EA and length L, and the angles between the members and the horizontal are $\gamma_1 = 0$, $\gamma_2 = 30°$, $\gamma_3 = 60°$, and $\gamma_4 = 90°$.

3.4-8. Solve Prob. 3.3-15 if member 2 initially has length $L + e$ instead of L. Omit the loads from the analysis.

3.4-9. Determine the vectors $\mathbf{A}_{DC}$ and $\mathbf{A}_{MC}$ for the truss of Prob. 3.3-16 if the following effects occur: support C is displaced upward a distance s, and members 1 and 3 have their temperature raised by an amount T. The coefficient of thermal expansion is α.

3.4-10. Determine the vector $\mathbf{A}_{DC}$ for the truss of Prob. 3.3-21, considering the following effects: support A is displaced to the left a distance s, member BC has an initial length $L + e$ instead of L, and the entire truss has its temperature increased by T. The coefficient of thermal expansion is α.

3.4-11. Find the vectors $\mathbf{A}_{DC}$ and $\mathbf{A}_{MC}$ for the plane frame in Fig. 3-11a if the following effects occur: member AB has its temperature increased uniformly by an amount T, member BC has a linear temperature gradient (T_1 is the temperature on the left and T_2 the temperature on the right), support C is displaced downward a distance s_1, and support A is displaced downward a distance s_2. Assume that both members have the same flexural rigidity EI and axial rigidity EA. Also, assume $H = L$, the coefficient of thermal expansion is α, and the depth of the members is d. The unknown displacements D and member end-actions A_M are to be taken as shown in Figs. 3-11a and 3-10b, respectively. Omit the load P from the analysis.

Computer-Oriented Direct Stiffness Method

4.1 Introduction. In the preceding chapter the stiffness method was developed initially by superposition of actions for the free displacement coordinates (Art. 3.2). Then the method was formalized and extended in Art. 3.6 using the compatibility matrix $\mathbf{C}_{MJ}$ and the virtual work concept. With this second approach the complete joint stiffness matrix $\mathbf{S}_J$ (for both free and restrained displacements) was assembled by the triple matrix multiplication given as Eq. (3-32). While the formal version is enlightening and well organized, it involves a large and sparse compatibility matrix containing terms that are not easy to evaluate correctly. Neither the generation of this matrix nor the multiplication process for assembly would be suitable for computer programming. A better methodology consists of drawing ideas from both approaches and adding a few computer-oriented techniques to evolve what is known as the *direct stiffness method*.

The primary objective of this chapter is to further develop the stiffness method into a form that may be readily programmed on a digital computer. In fact, the procedures established in this chapter have their counterparts in the form of flow charts in Chapter 5. In Art. 4.2 the direct stiffness method is described, and an outline is presented for the purpose of organizing the calculations into a sequence that is conducive to programming. Since member stiffnesses play an essential role in the analyses of all types of framed structures, this topic is treated next in Art. 4.3. Then other aspects of the direct stiffness method common to all types of framed structures are discussed in Arts. 4.4 through 4.7. These matters are demonstrated with a familiar example that has been solved previously. The remaining articles of the chapter deal with applications to various types of framed structures. For simplicity, only prismatic members and the effects of applied loads are considered in this chapter. Methods for handling nonprismatic members, temperature changes, prestrains, support displacements, and other effects are described in Chapter 6.

4.2 Direct Stiffness Method. The key to simplifying the assembly process for the joint stiffness matrix $\mathbf{S}_J$ is to use member stiffness matrices for actions and displacements at both ends of each member. If the member displacements are referenced to structural (global) coordinates, they coincide with joint displacements. In that case all of the geometric complications must be handled locally, and the transfer of member information to structural arrays is straightforward. That is, the stiffness matrix and the equivalent load vector can be assembled by direct addition instead of by matrix multiplication.

Thus, the assembly of the joint stiffness matrix, assuming m members, may be stated as

$$\mathbf{S}_{\mathrm{J}} = \sum_{i=1}^{m} \mathbf{S}_{\mathrm{MS}i} \tag{4-1}$$

In this expression the symbol $\mathbf{S}_{\mathrm{MS}i}$ represents the i-th member stiffness matrix with end-actions and displacements (for both ends) taken in the directions of structural axes. To be conformable for matrix addition, all such member stiffness matrices should be expanded to the same size as $\mathbf{S}_{\mathrm{J}}$ by augmenting them with rows and columns of zeros. However, this operation can be avoided in computer programming by merely placing terms from $\mathbf{S}_{\mathrm{MS}i}$ into $\mathbf{S}_{\mathrm{J}}$ where they belong.

Similarly, an equivalent joint load vector $\mathbf{A}_{\mathrm{E}}$ can be constructed from member contributions, as follows

$$\mathbf{A}_{\mathrm{E}} = -\sum_{i=1}^{m} \mathbf{A}_{\mathrm{MS}i} \tag{4-2}$$

In this instance $\mathbf{A}_{\mathrm{MS}i}$ is a vector of fixed-end actions in the directions of the structural axes at both ends of member i. As before, it is theoretically necessary to augment $\mathbf{A}_{\mathrm{MS}i}$ with zeros to make it conformable for matrix addition, but this step can be avoided in programming. The equivalent joint loads in Eq. (4-2) will be added to actual loads applied at the joints to form the total (or combined) load vector.

In order to separate terms pertaining to the free displacements of the structure from those for support restraints, it is necessary to rearrange and partition the stiffness and load matrices. This rearrangement of terms may be accomplished after the assembly is completed. However, in a computer program the rearrangement is more easily done when the information is transferred from small member arrays to large structural arrays.*

After the stiffness and load matrices have been rearranged, the solution proceeds in the manner already established in Art. 3.6. Thus, the basic solution for free joint displacements (see Eq. 3-35), due to loads only, becomes

*While it is possible to solve the equations of equilibrium "in-place" without rearrangement (see Appendix E), it is computationally more efficient to separate the independent equations from the dependent equations.

$$\mathbf{D}_{\mathrm{F}} = \mathbf{S}_{\mathrm{FF}}^{-1}\mathbf{A}_{\mathrm{FC}} \tag{4-3}$$

in which $\mathbf{A}_{\mathrm{FC}}$ is a vector of combined joint loads (actual and equivalent) corresponding to $\mathbf{D}_{\mathrm{F}}$. While it is symbolically convenient to imply inversion of the stiffness matrix $\mathbf{S}_{\mathrm{FF}}$ in Eq. (4-3), it is not efficient to actually calculate $\mathbf{S}_{\mathrm{FF}}^{-1}$ in a computer program. Instead, the coefficient matrix $\mathbf{S}_{\mathrm{FF}}$ should be factored and the solution for $\mathbf{D}_{\mathrm{F}}$ obtained in forward and backward sweeps. (The topics of factorization and solution are covered in Appendix D.)

Other items of interest consist of support reactions and member end-actions. Due to the effects of loads only, Eq. (3-36) gives the reactions

$$\mathbf{A}_{\mathrm{R}} = -\mathbf{A}_{\mathrm{RC}} + \mathbf{S}_{\mathrm{RF}}\mathbf{D}_{\mathrm{F}} \tag{4-4}$$

In addition, Eq. (3-37) for the member end-actions may be rewritten so that it pertains to only one member at a time, as follows:

$$\mathbf{A}_{\mathrm{M}i} = \mathbf{A}_{\mathrm{ML}i} + \mathbf{S}_{\mathrm{M}i}\mathbf{D}_{\mathrm{M}i} \tag{4-5}$$

All of the matrices in this expression are referenced to member axes. The action vector $\mathbf{A}_{\mathrm{ML}i}$ contains fixed-end actions (in the directions of member axes) due to loads applied on the member itself. In addition, the member stiffness matrix $\mathbf{S}_{\mathrm{M}i}$ has terms (in member directions) for both ends of the member. Consequently, the vector $\mathbf{D}_{\mathrm{M}i}$ must consist of displacements (in member directions) at both ends of the member. These displacements can be obtained by expressing the joint displacements at the j and k ends as components in the member directions.

A computer program for the analysis of a structure by the direct stiffness method divides conveniently into several phases. These phases are not the same as those in Chapter 3 for hand calculations. One difference is that when using a computer it is desirable to work in the early phases of the program with only the data pertaining to the structure. This part of the program includes the formation of the stiffness matrix, which is a property of the structure. Subsequently, the load data is manipulated, after which the final results of the analysis are computed. This sequence is particularly efficient if more than one load system is being considered, since the initial phases of the calculations need not be repeated. The steps to be considered in the subsequent discussions are the following:

(1) Identification of Structural Data. Information pertaining to the structure itself must be assembled and recorded. This information includes the number of members, the number of joints, the number of degrees of freedom, and the elastic properties of the material. The locations of the joints of the structure are specified by means of geometric coordinates. In addition, the section properties of each member in the structure must be given. Finally, the conditions of restraint at the supports of the structure must be identified. In computer programming, all such information is coded in some convenient way, as will be shown subsequently in this chapter and also in Chapter 5.

(2) Construction of Stiffness Matrix. The stiffness matrix is an inherent property of the structure and is based upon the structural data only. In computer programming it is convenient to obtain the joint stiffness matrix by summing contributions from individual member stiffness matrices (a discussion of complete member stiffnesses is given in Art. 4.3). The joint stiffness matrix to be considered is related to all possible joint displacements, including support displacements, as was discussed in Art. 3.6. This array shall be called the *over-all joint stiffness matrix*. Its formation and rearrangement are described in Arts. 4.4 and 4.6, respectively.

(3) Identification of Load Data. All loads acting on the structure must be specified in a manner suitable for computer programming. Both joint loads and member loads must be given. The former may be handled directly, but the latter are handled indirectly by supplying as data the fixed-end actions caused by the loads on the members.

(4) Construction of Load Vector. The fixed-end actions due to loads on members may be converted to *equivalent joint loads*, as described previously in Art. 1.12. These equivalent joint loads may then be added to the actual joint loads to produce a problem in which the structure is imagined to be loaded at the joints only. Formation and rearrangement of the load vector are described in Arts. 4.5 and 4.6, respectively.

(5) Calculation of Results. In the final phase of the analysis all of the joint displacements, reactions, and member end-actions are computed. The calculation of member end-actions proceeds member by member (see Eq. 4-5) instead of for the structure as a whole. Such calculations require the use of complete member stiffness matrices for member directions, a subject that is covered in the next article. A simple calculation of this type is demonstrated for illustrative purposes in Art. 4.7.

There are many possible variations in organizing the stiffness method for computer programming. The phases of analysis listed above constitute an orderly approach having certain essential features that are advantageous when dealing with large, complicated frameworks. These steps will be discussed and illustrated by means of a familiar example in the following articles.

4.3 Complete Member Stiffness Matrices. In order to construct the joint stiffness matrix with the summation process indicated by Eq. (4-1), it is necessary to generate complete member stiffness matrices $\mathbf{S}_{MSi}$ for structure-oriented axes. In addition, member stiffness matrices $\mathbf{S}_{Mi}$ for member-oriented axes are required for the purpose of calculating member end-actions with Eq. (4-5). It is always possible to obtain the member stiffnesses with respect to the member axes, as is done in this article, and then to transform these stiffnesses to the structure axes. The procedures for accomplishing this transformation are described later in the chapter for each type of structure.

In the special case of a continuous beam, the member axes are inherently parallel to those for the structure as a whole; so no transformation is necessary. Thus,

$$(\mathbf{S}_{Mi})_{\text{beam}} = (\mathbf{S}_{MSi})_{\text{beam}} \tag{a}$$

Figure 4-1 shows a prismatic space frame member i that is fully restrained at both ends, which are denoted as ends j and k. Orthogonal member-oriented axes also appear in the figure, with the origin located at point j. The x_M axis coincides with the centroidal axis of the member and is positive in the sense from j to k. The y_M and z_M axes are principal axes for the member; that is, the x_M-y_M and x_M-z_M planes are principal planes of bending. It is assumed that the shear center and the centroid of the member coincide so that twisting and bending of the member may occur independently of one another. This restriction is normally satisfied in framed structures; however, it is possible to make a more general analysis if necessary (see Art. 6.17).

Properties of the member shown in Fig. 4-1 will be defined in a systematic fashion for the purpose of computer programming for complex structures. Let L denote the length of the member and A_X the area of the cross section. (The subscript X is needed because the symbol A without subscript is used as an identifier for actions.) Principal moments of inertia of the cross section of the member with respect to the y_M and z_M axes are denoted I_Y and I_Z, respectively. Also, let the torsion constant for the member be I_X, which is the same as the constant J that appears in Chapters 2 and 3 and the Appendices. Since the symbol J has use as an index for joints of the structure in computer programming, the use of the symbol I_X for the torsion constant is more desirable. The torsion constant I_X is not to be

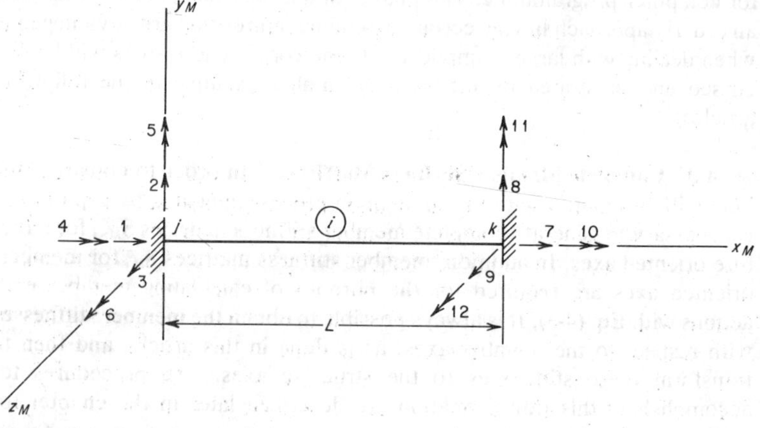

Fig. 4-1. Space frame member.

interpreted as the polar moment of inertia of the cross section except in the special case of a circular cylindrical member.

Member stiffnesses for the restrained member shown in Fig. 4-1 consist of actions exerted on the member by the restraints when unit displacements (translations and rotations) are imposed at each end of the member. Values of these restraint actions may be obtained from Table B-4 in Appendix B. The unit displacements are considered to be induced one at a time while all other end displacements are retained at zero; also, they are assumed to be positive in the x_M, y_M, and z_M directions. Thus, the positive senses of the three translations and the three rotations at each end of the member are indicated by arrows in Fig. 4-1. In the figure the single-headed arrows denote translations, and the double-headed arrows represent rotations. At joint j the translations are numbered 1, 2, and 3 and the rotations are numbered 4, 5, and 6. Similarly, at the k end of the member 7, 8, and 9 are translations and 10, 11, and 12 are rotations. In all cases the displacements are taken in the order x_M, y_M, and z_M, respectively.

The member stiffnesses for the twelve possible types of end displacements (shown in Fig. 4-1) are summarized pictorially in Fig. 4-2. In each case the various restraint actions (or member stiffnesses) are shown as vectors. An arrow with a single head represents a force vector, and an arrow with a double head represents a moment vector. All vectors are drawn in the positive senses, but in cases where the restraint actions are actually negative a minus sign precedes the expression for the stiffness coefficient.

In order to see how the member stiffnesses are determined, consider case (1) in Fig. 4-2. The restraint actions shown in the figure arise because of a unit translation of the j end of the member in the positive x_M direction. All other displacements are zero. This displacement causes a pure compressive force EA_X/L in the member. At the j end of the member this force is equilibrated by a restraint action of EA_X/L in the positive x_M direction, and at the k end of the member the restraint action has the same value but is in the negative x_M direction. All other restraint actions are zero in this case.

Case (2) in Fig. 4-2 involves a unit translation of the j end of the member in the positive y_M direction, while all other displacements are zero. This displacement causes both moment and shear in the member. At the j end, the restraint actions required to keep the member in equilibrium are a lateral force of $12EI_Z/L^3$ in the positive y_M direction and a moment $6EI_Z/L^2$ in the positive z_M sense (see Table B-4). At the k end of the member the restraint actions are the same except that the lateral force acts in the negative y_M direction.

All of the member stiffnesses shown in the figure are derived by determining the values of the restraint actions required to hold the distorted member in equilibrium. The reader should verify all of the expressions before proceeding further. These stiffnesses can be used to formulate stiffness matrices for the members of different types of structures. In the most

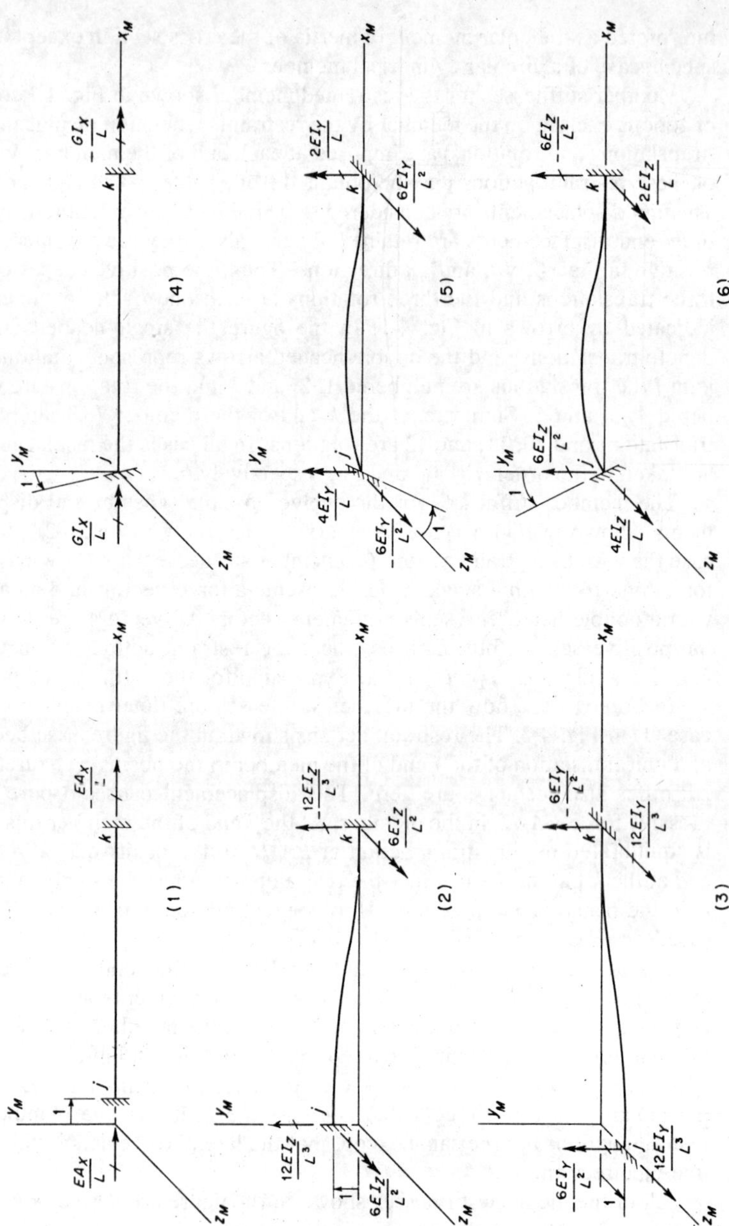
(1)
$\frac{EA_X}{L}$
$-\frac{EA_X}{L}$
(2)
$\frac{12EI_Z}{L^3}$
$\frac{6EI_Z}{L^2}$
$-\frac{12EI_Z}{L^3}$
$\frac{6EI_Z}{L^2}$
(3)
$-\frac{6EI_Y}{L^2}$
$\frac{12EI_Y}{L^3}$
$-\frac{6EI_Y}{L^2}$
$-\frac{12EI_Y}{L^3}$
(4)
$\frac{GI_X}{L}$
$-\frac{GI_X}{L}$
(5)
$\frac{4EI_Y}{L}$
$-\frac{6EI_Y}{L^2}$
$\frac{2EI_Y}{L}$
$\frac{6EI_Y}{L^2}$
(6)
$\frac{6EI_Z}{L^2}$
$\frac{4EI_Z}{L}$
$-\frac{6EI_Z}{L^2}$
$\frac{2EI_Z}{L}$
x_M
y_M
z_M
j
k
1

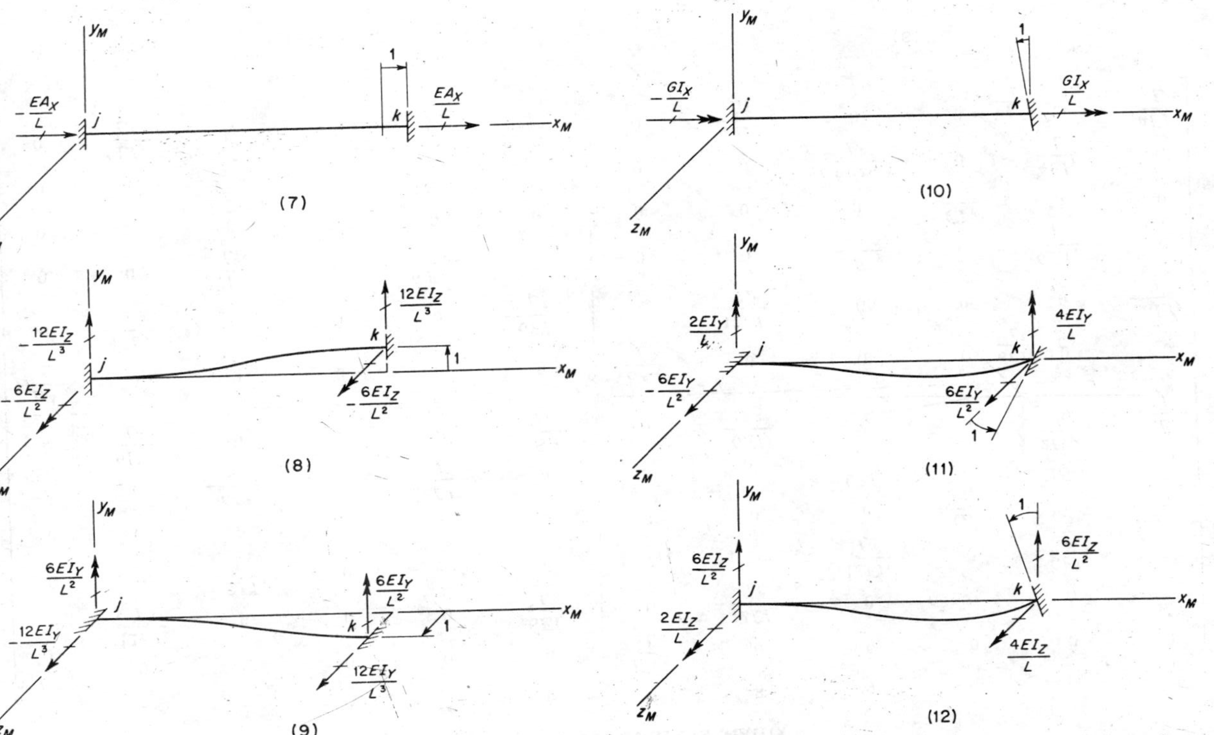

Fig. 4-2. Member stiffnesses: (1) unit x_M translation at *j*, (2) unit y_M translation at *j*, (3) unit z_M translation at *j*, (4) unit x_M rotation at *j*, (5) unit y_M rotation at *j*, (6) unit z_M rotation at *j*, (7) unit x_M translation at *k*, (8) unit y_M translation at *k*, (9) unit z_M translation at *k*, **(10) unit x_M rotation at *k*,** (11) unit y_M rotation at *k*, and (12) unit z_M rotation at *k*.

Table 4-1
Space Frame Member Stiffness Matrix

$$
\mathbf{S}_{Mi} = \begin{array}{c} 1 \\ 2 \\ 3 \\ 4 \\ 5 \\ 6 \\ 7 \\ 8 \\ 9 \\ 10 \\ 11 \\ 12 \end{array}
\left[\begin{array}{cccccc|cccccc}
1 & 2 & 3 & 4 & 5 & 6 & 7 & 8 & 9 & 10 & 11 & 12 \\
\frac{EA_X}{L} & 0 & 0 & 0 & 0 & 0 & -\frac{EA_X}{L} & 0 & 0 & 0 & 0 & 0 \\
0 & \frac{12EI_Z}{L^3} & 0 & 0 & 0 & \frac{6EI_Z}{L^2} & 0 & -\frac{12EI_Z}{L^3} & 0 & 0 & 0 & \frac{6EI_Z}{L^2} \\
0 & 0 & \frac{12EI_Y}{L^3} & 0 & -\frac{6EI_Y}{L^2} & 0 & 0 & 0 & -\frac{12EI_Y}{L^3} & 0 & -\frac{6EI_Y}{L^2} & 0 \\
0 & 0 & 0 & \frac{GI_X}{L} & 0 & 0 & 0 & 0 & 0 & -\frac{GI_X}{L} & 0 & 0 \\
0 & 0 & -\frac{6EI_Y}{L^2} & 0 & \frac{4EI_Y}{L} & 0 & 0 & 0 & \frac{6EI_Y}{L^2} & 0 & \frac{2EI_Y}{L} & 0 \\
0 & \frac{6EI_Z}{L^2} & 0 & 0 & 0 & \frac{4EI_Z}{L} & 0 & -\frac{6EI_Z}{L^2} & 0 & 0 & 0 & \frac{2EI_Z}{L} \\
\hline
-\frac{EA_X}{L} & 0 & 0 & 0 & 0 & 0 & \frac{EA_X}{L} & 0 & 0 & 0 & 0 & 0 \\
0 & -\frac{12EI_Z}{L^3} & 0 & 0 & 0 & -\frac{6EI_Z}{L^2} & 0 & \frac{12EI_Z}{L^3} & 0 & 0 & 0 & -\frac{6EI_Z}{L^2} \\
0 & 0 & -\frac{12EI_Y}{L^3} & 0 & \frac{6EI_Y}{L^2} & 0 & 0 & 0 & \frac{12EI_Y}{L^3} & 0 & \frac{6EI_Y}{L^2} & 0 \\
0 & 0 & 0 & -\frac{GI_X}{L} & 0 & 0 & 0 & 0 & 0 & \frac{GI_X}{L} & 0 & 0 \\
0 & 0 & -\frac{6EI_Y}{L^2} & 0 & \frac{2EI_Y}{L} & 0 & 0 & 0 & \frac{6EI_Y}{L^2} & 0 & \frac{4EI_Y}{L} & 0 \\
0 & \frac{6EI_Z}{L^2} & 0 & 0 & 0 & \frac{2EI_Z}{L} & 0 & -\frac{6EI_Z}{L^2} & 0 & 0 & 0 & \frac{4EI_Z}{L}
\end{array}\right]
$$

general case (a space frame), it is possible for the member to undergo any of the twelve displacements shown in Fig. 4-2. The stiffness matrix for such a member, denoted $\mathbf{S}_{Mi}$, is therefore of order 12 × 12, and each column in the matrix represents the actions caused by one of the unit displacements. The space frame member stiffness matrix is shown in Table 4-1; it is, of course, symmetrical. The rows and columns of the matrix are numbered down the side and across the top to assist the reader in identifying a partic-

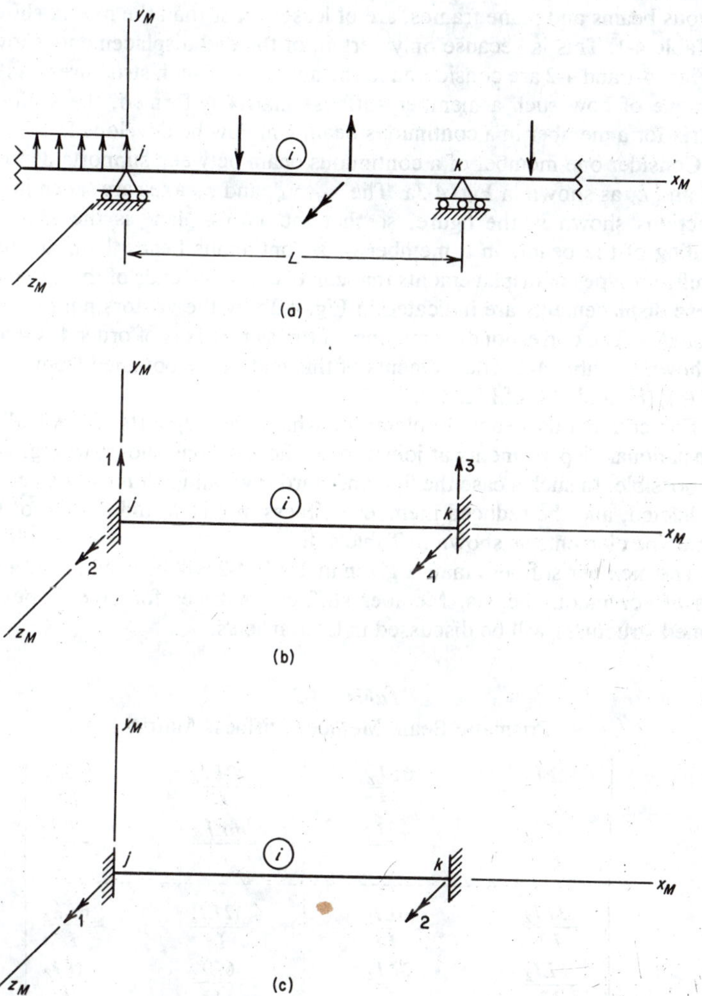

Fig. 4-3. Continuous beam member.

ular element. In addition, the matrix is partitioned in order to delineate the portions associated with the two ends of the member. Thus, the complete member stiffness matrix has the form

$$\mathbf{S}_{\mathrm{M}i} = \begin{bmatrix} \mathbf{S}_{\mathrm{M}jj} & \mathbf{S}_{\mathrm{M}jk} \\ \mathbf{S}_{\mathrm{M}kj} & \mathbf{S}_{\mathrm{M}kk} \end{bmatrix}_i \tag{b}$$

where the subscripts j and k attached to the submatrices refer to the ends of the member.

The member stiffness matrices needed for other structures, such as continuous beams and plane frames, are of lesser order than the matrix shown in Table 4-1. This is because only certain of the end displacements shown in Figs. 4-1 and 4-2 are considered in the analyses of such structures. As an example of how such a member stiffness matrix is formed, the stiffness matrix for a member in a continuous beam will now be developed.

Consider one member of a continuous beam between supports denoted as j and k, as shown in Fig. 4-3a. The x_M, y_M, and z_M axes are taken in the directions shown in the figure, so that the x_M-y_M plane is the plane of bending of the beam. In a member of a continuous beam there are four significant types of displacements that can occur at the ends of the member. These displacements are indicated in Fig. 4-3b by the vectors numbered 1 through 4. The corresponding member stiffness matrix is of order 4×4 and is shown in Table 4-2. The elements of this matrix are obtained from cases (2), (6), (8), and (12) of Fig. 4-2.

For continuous beam problems in which the supports do not allow translational displacements at joints, only the rotations shown in Fig. 4-3c are possible. In such a case the first and third rows and columns of $\mathbf{S}_\mathrm{M}$ could be deleted, and the reduced member stiffness matrix would consist of the remaining elements as shown in Table 4-3.

The member stiffness matrix given in Table 4-2 will be used in the analysis of continuous beams. Member stiffness matrices for other types of framed structures will be discussed in later articles.

Table 4-2

Prismatic Beam Member Stiffness Matrix

$$\mathbf{S}_{\mathrm{M}i} = \left[\begin{array}{cc|cc} \dfrac{12EI_Z}{L^3} & \dfrac{6EI_Z}{L^2} & -\dfrac{12EI_Z}{L^3} & \dfrac{6EI_Z}{L^2} \\ \dfrac{6EI_Z}{L^2} & \dfrac{4EI_Z}{L} & -\dfrac{6EI_Z}{L^2} & \dfrac{2EI_Z}{L} \\ \hline -\dfrac{12EI_Z}{L^3} & -\dfrac{6EI_Z}{L^2} & \dfrac{12EI_Z}{L^3} & -\dfrac{6EI_Z}{L^2} \\ \dfrac{6EI_Z}{L^2} & \dfrac{2EI_Z}{L} & -\dfrac{6EI_Z}{L^2} & \dfrac{4EI_Z}{L} \end{array}\right]$$

Table 4-3

Reduced Member Stiffness Matrix

$$\mathbf{S}_{Mi} = \left[\begin{array}{c|c} \frac{4EI_Z}{L} & \frac{2EI_Z}{L} \\ \hline \frac{2EI_Z}{L} & \frac{4EI_Z}{L} \end{array}\right]$$

4.4 Formation of Joint Stiffness Matrix. The process of forming the joint stiffness matrix will be explained in conjunction with the two-span beam example shown in Fig. 4-4a. This beam is the same one that was analyzed previously in Chapters 2 and 3, except that no particular loading system is indicated. As before, assume that the flexural rigidity EI_Z of the beam is constant throughout its length. In order to develop the necessary stiffness terms, consider the unloaded beam to be completely restrained at all joints, as shown in Fig. 4-4b. In this figure an *arbitrary numbering system* is indicated for the six possible joint displacements that could occur in the structure. The numbers shown will be called *joint displacement indexes*. At each joint, taken from left to right, the y translation is numbered before the z rotation. Displacements numbered 4 and 6 (rotations at

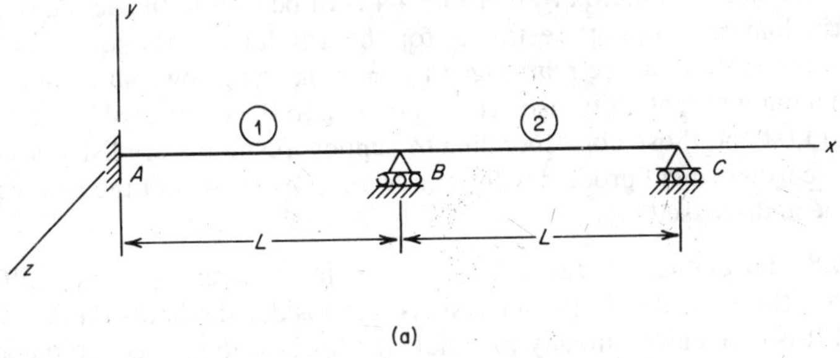

(a)

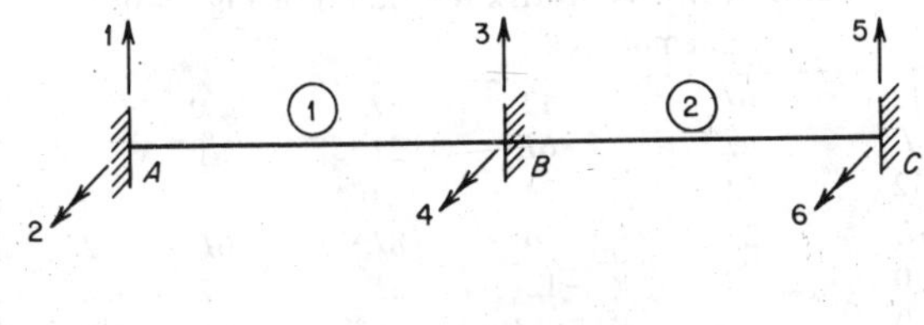

(b)

Fig. 4-4. Beam example.

B and C) are actually free to occur, but those numbered 1, 2, 3, and 5 are restrained by the supports.

Formation of the joint stiffness matrix $\mathbf{S}_J$ is accomplished by summing contributions from the complete member stiffness matrices for members 1 and 2 (see Eq. 4-1). The pattern of contributions is indicated in Table 4-4 by the two sections of $\mathbf{S}_J$ delineated with dashed lines and overlapping in the central portion. The 4 × 4 upper left-hand section receives contributions from the stiffness matrix of member AB (see Table 4-2, Art. 4.3), and the 4 × 4 lower right-hand section receives terms from the stiffness matrix of member BC. Together, these two member stiffness matrices (which are both the same in this example) constitute the joint stiffness matrix for the two-span beam. In the overlapping portion of $\mathbf{S}_J$, elements consist of the sums of contributions from members AB and BC. For example, the element S_{J33} (equal to $24EI_Z/L^3$) is the sum of S_{M33} (equal to $12EI_Z/L^3$) for member AB plus S_{M11} (equal to $12EI_Z/L^3$) for member BC. Similarly, S_{J34} (equal to 0) is the sum of S_{M34} (equal to $-6EI_Z/L^2$) for member AB plus S_{M12} (equal to $6EI_Z/L^2$) for member BC. Other elements of $\mathbf{S}_J$ outside of the overlapping portion but inside the dashed lines consist of single contributions from either member AB or member BC. Elements of $\mathbf{S}_J$ outside of the dashed lines are all zero due to the nature of the structure. Thus, the use of the arbitrary numbering system results in a joint stiffness matrix in which the contributions from member stiffnesses may be clearly seen.

If the stiffness matrix $\mathbf{S}_J$ in Table 4-4 is to be useful, the actual degrees of freedom and support restraints for the structure must be recognized. Then the matrix can be rearranged by interchanging rows and columns in such a manner that stiffnesses corresponding to the degrees of freedom are listed first and those corresponding to support restraints are listed second. This rearrangement process will be covered in Art. 4.6 after the load vector has been discussed.

4.5 Formation of Load Vector. After finding the joint stiffness matrix, the next step in the analysis is to consider the loads on the structure. It is convenient initially to handle the loads at the joints and the loads on the members separately. The reason for doing so is that the joint loads

Table 4-4

Joint Stiffness Matrix for Beam of Fig. 4-4b

Member AB

$$\mathbf{S}_J = \begin{bmatrix} 12 & 6L & -12 & 6L & 0 & 0 \\ 6L & 4L^2 & -6L & 2L^2 & 0 & 0 \\ -12 & -6L & 24 & 0 & -12 & 6L \\ 6L & 2L^2 & 0 & 8L^2 & -6L & 2L^2 \\ 0 & 0 & -12 & -6L & 12 & -6L \\ 0 & 0 & 6L & 2L^2 & -6L & 4L^2 \end{bmatrix} \frac{EI_Z}{L^3}$$

Member BC

and the member loads are treated in different ways. The joint loads are ready for immediate placement into a vector of actions to be used in the solution, but the loads on the members are taken into account by calculating the fixed-end actions that they produce. These fixed-end actions may then be transformed into equivalent joint loads and combined with the actual joint loads on the structure (see Art. 1.12 for a discussion of equivalent joint loads).

The loads applied at the joints may be listed in a vector $\mathbf{A}_\mathrm{J}$, which contains the applied loads corresponding to all possible joint displacements, including those at support restraints. The elements in $\mathbf{A}_\mathrm{J}$ are numbered in the same sequence as the joint displacements. For the two-span beam example, the arbitrary numbering system is shown in Fig. 4-4b. If this beam is subjected to the loads shown in Fig. 4-5a, the joint loads are the moment $M = PL$ at joint B and the force P at joint C. Therefore, the vector $\mathbf{A}_\mathrm{J}$ takes the form*

$$\mathbf{A}_\mathrm{J} = \{0, 0, 0, M, P, 0\}$$

The moment M and the force P in the vector $\mathbf{A}_\mathrm{J}$ are positive because M acts counterclockwise and P acts upward. In general, the joint loads on a continuous beam may be a force and a moment at every joint.

The remaining loads on the structure act on the members and are shown applied to the two members of the restrained structure in Fig. 4-5b. Also shown in this figure are the fixed-end actions at the ends of the members. For convenience, these fixed-end actions shall be compiled into a rectangular matrix $\mathbf{A}_\mathrm{ML}$ in which each column contains the end-actions for a given member. Thus, the format of $\mathbf{A}_\mathrm{ML}$ will be taken as follows:

$$\mathbf{A}_\mathrm{ML} = [\mathbf{A}_{\mathrm{ML}1}\ \mathbf{A}_{\mathrm{ML}2}\ \cdots\ \mathbf{A}_{\mathrm{ML}i}\ \cdots\ \mathbf{A}_{\mathrm{ML}m}] \tag{4-6}$$

In this arrangement each submatrix $\mathbf{A}_{\mathrm{ML}i}$ is a column vector pertaining to a given member i and is ready to be used in Eq. (4-5). The number of columns in $\mathbf{A}_\mathrm{ML}$ will always be equal to m, the number of members in the structure.

In a beam there are two significant types of fixed-end actions: shearing force and bending moment. Therefore, each column of the matrix $\mathbf{A}_\mathrm{ML}$ will contain for one member the shearing force and bending moment at the left end followed by the shear and moment at the right end. As shown in Fig. 4-5b, the fixed-end actions on member AB consist of the force P and the moment $PL/4$ at the left-hand end accompanied by the force P and the moment $-PL/4$ at the right-hand end. These actions constitute the first column of $\mathbf{A}_\mathrm{ML}$, as follows:

$$\mathbf{A}_{\mathrm{ML}1} = \{P,\ PL/4,\ P,\ -PL/4\}$$

*As mentioned previously, braces { } are used to denote a column vector that is written in a row to save space.

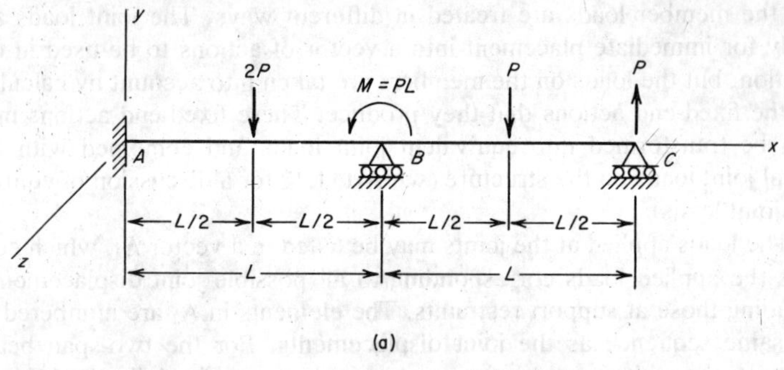

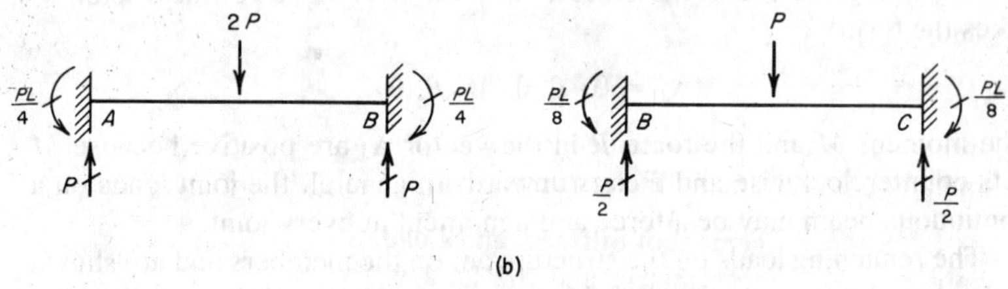

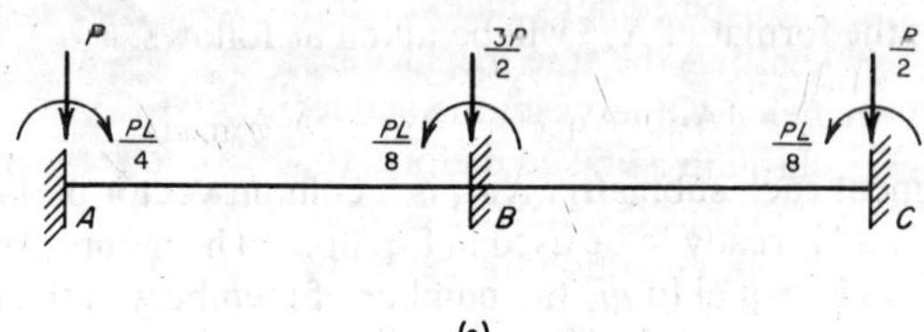

Fig. 4-5. Actual and equivalent joint loads.

The second column of $\mathbf{A}_{ML}$ is filled in a similar manner from the fixed-end actions for member *BC* (see Fig. 4-5b). This column is

$$\mathbf{A}_{ML2} = \{P/2, PL/8, P/2, -PL/8\}$$

When the fixed-end actions are reversed in direction, they constitute equivalent joint loads, which are assembled in accordance with Eq. (4-2). Since the member axes for a beam are always taken parallel to the structural axes, the following identity holds:

$$(\mathbf{A}_{MLi})_{beam} = (\mathbf{A}_{MSi})_{beam} \qquad \text{(a)}$$

That is, transformation of member end-actions to structural directions is not required for a beam. Thus, for the two-span example, Eq. (4-2) gives

$$\begin{aligned}\mathbf{A}_E &= -\{P, PL/4, P, -PL/4, 0, 0\} - \{0, 0, P/2, PL/8, P/2, -PL/8\}\\ &= \{-P, -PL/4, -3P/2, PL/8, -P/2, PL/8\}\end{aligned}$$

The equivalent joint loads are listed in this vector in the sequence denoted by the arbitrary numbering system in Fig. 4-4b. In addition, the terms in $\mathbf{A}_E$ are shown in Fig. 4-5c, where it is seen that only the actions at joint B consist of contributions from both members.

Actual joint loads (vector $\mathbf{A}_J$) may be added to equivalent joint loads (vector $\mathbf{A}_E$) to produce the vector of *combined joint loads* $\mathbf{A}_C$, as follows:

$$\mathbf{A}_C = \mathbf{A}_J + \mathbf{A}_E \tag{4-7}$$

Substituting into this expression the terms obtained above for $\mathbf{A}_J$ and $\mathbf{A}_E$ yields:

$$\mathbf{A}_C = \{-P, -PL/4, -3P/2, 9PL/8, P/2, PL/8\}$$

In the next article this combined load vector will be rearranged in a manner similar to that for the stiffness matrix.

4.6 Rearrangement of Stiffness and Load Arrays. As mentioned previously, the stiffness matrix and the load vector will be rearranged and partitioned so that terms pertaining to the degrees of freedom are separated from those for the support restraints. The beam example in Fig. 4-4a has two degrees of freedom, which are the rotations at points B and C. Figure 4-6 shows revised displacement indexes for the joints of the restrained structure. It is seen that the free displacements are numbered before the support displacements. Otherwise, the sequence of numbering is from left to right, with translations taken before rotations at each joint. These revised displacement indexes may be computed automatically by examining the actual restraint condition for each possible joint displacement. If it is not restrained, the displacement index must be reduced by the cumula-

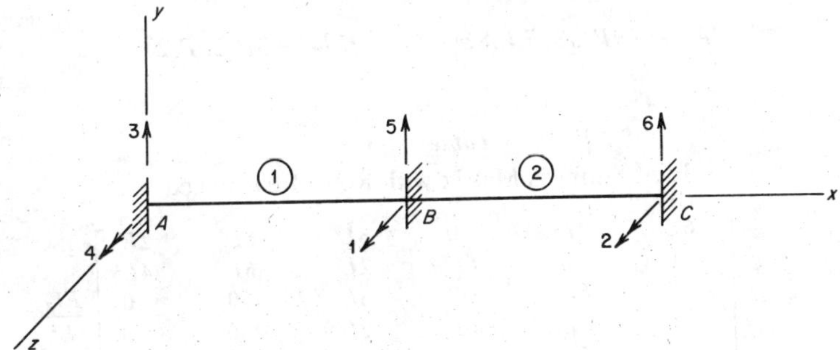

Fig. 4-6. Revised displacement indexes.

tive number of restraints encountered up to that point. In such a case, the new index is computed as

$$(j)_{new} = (j)_{old} - c_{jo} \tag{a}$$

in which c_{jo} represents the cumulative number of restraints for the old index. On the other hand, if the displacement under consideration is actually restrained, the new index is determined by the formula

$$(j)_{new} = n + c_{jo} \tag{b}$$

where n is the number of degrees of freedom. Expressions (a) and (b) will be implemented in a formal way later in this chapter (see Art. 4.9).

For the present example, a comparison of Fig. 4-6 and Fig. 4-4b shows that the actual degrees of freedom correspond to displacements 4 and 6 of the original numbering system. Therefore, the fourth and sixth rows and columns of the joint stiffness matrix $\mathbf{S_J}$ must be put into the first and second positions. As the first step, rows 4 and 6 of the matrix in Table 4-4 are switched to the first and second rows while all other rows are moved downward without changing their order. Then the array takes the form shown in Table 4-5. Next, the fourth and sixth columns in Table 4-5 are moved to the first two columns, and all other columns are moved to the right without changing their order. This manipulation produces the *rearranged joint stiffness matrix* shown in Table 4-6. Such a matrix is always symmetric and is partitioned in the following manner:

$$\mathbf{S_J} = \begin{bmatrix} \mathbf{S_{FF}} & \mathbf{S_{FR}} \\ \mathbf{S_{RF}} & \mathbf{S_{RR}} \end{bmatrix} \tag{4-8}$$

In this expression the subscripts F and R refer to free and restrained displacements, respectively.

Similarly, the combined load vector $\mathbf{A_C}$ for the example can be rearranged to conform to the numbering system of Fig. 4-6. In this instance the fourth and sixth terms in $\mathbf{A_C}$ are moved to the first and second positions while all others are moved toward the end without changing their order. This rearrangement gives

$$\mathbf{A_C} = \{9PL/8,\ PL/8,\ \vdots\ -P,\ -PL/4,\ -3P/2,\ P/2\}$$

Table 4-5

Joint Stiffness Matrix with Rows Rearranged

$$\mathbf{S_J} = \begin{bmatrix} 6L & 2L^2 & 0 & 8L^2 & -6L & 2L^2 \\ 0 & 0 & 6L & 2L^2 & -6L & 4L^2 \\ 12 & 6L & -12 & 6L & 0 & 0 \\ 6L & 4L^2 & -6L & 2L^2 & 0 & 0 \\ -12 & -6L & 24 & 0 & -12 & 6L \\ 0 & 0 & -12 & -6L & 12 & -6L \end{bmatrix} \frac{EI_Z}{L^3}$$

Table 4-6
Joint Stiffness Matrix with Columns Rearranged

$$\mathbf{S}_J = \left[\begin{array}{cc|cccc} 8L^2 & 2L^2 & 6L & 2L^2 & 0 & -6L \\ 2L^2 & 4L^2 & 0 & 0 & 6L & -6L \\ \hline 6L & 0 & 12 & 6L & -12 & 0 \\ 2L^2 & 0 & 6L & 4L^2 & -6L & 0 \\ 0 & 6L & -12 & -6L & 24 & -12 \\ -6L & -6L & 0 & 0 & -12 & 12 \end{array}\right] \frac{EI_Z}{L^3}$$

which is partitioned in the form

$$\mathbf{A}_C = \begin{bmatrix} \mathbf{A}_{FC} \\ \mathbf{A}_{RC} \end{bmatrix} \tag{4-9}$$

Thus, the combined loads corresponding to free joint displacements are separated from those corresponding to support restraints.

4.7 Calculation of Results. In the final phase of the analysis, matrices generated in previous steps are used to find unknown joint displacements $\mathbf{D}_F$, support reactions $\mathbf{A}_R$, and member end-actions $\mathbf{A}_{Mi}$ for each member. First, the unknown displacements are obtained from Eq. (4-3), which gives for the two-span beam

$$\mathbf{D}_F = \mathbf{S}_{FF}^{-1}\mathbf{A}_{FC} = \frac{L}{14EI_Z}\begin{bmatrix} 2 & -1 \\ -1 & 4 \end{bmatrix}\begin{bmatrix} 9 \\ 1 \end{bmatrix}\frac{PL}{8} = \frac{PL^2}{112EI_Z}\begin{bmatrix} 17 \\ -5 \end{bmatrix}$$

Next, the support reactions are found by substituting the matrices $\mathbf{A}_{RC}$, $\mathbf{S}_{RF}$, and $\mathbf{D}_F$ into Eq. (4-4). This substitution for the example yields the following results:

$$\mathbf{A}_R = \frac{P}{4}\begin{bmatrix} 4 \\ L \\ 6 \\ -2 \end{bmatrix} + \frac{2EI_Z}{L^2}\begin{bmatrix} 3 & 0 \\ L & 0 \\ 0 & 3 \\ -3 & -3 \end{bmatrix}\begin{bmatrix} 17 \\ -5 \end{bmatrix}\frac{PL^2}{112EI_Z} = \frac{P}{56}\begin{bmatrix} 107 \\ 31L \\ 69 \\ -64 \end{bmatrix}$$

The values of the displacements and the reactions given above are the same as those obtained previously in Art. 3.2.

In order to calculate member end-actions by Eq. (4-5), it is necessary to set up m displacement vectors $\mathbf{D}_{Mi}$ $(i = 1, 2, \ldots, m)$, each of which contains displacements of the ends of member i in the directions of member axes. At this stage of the analysis, all joint displacements are known and can be used to fill each of the vectors $\mathbf{D}_{Mi}$ in some systematic way. For two- and three-dimensional structures, a transformation of such displacements from structural directions to member directions is required. However, in a beam this difficulty is not encountered. That is,

$$(\mathbf{D}_{Mi})_{\text{beam}} = (\mathbf{D}_{MSi})_{\text{beam}} \tag{a}$$

where the identifier $\mathbf{D}_{MSi}$ represents the vector of displacements at the ends of member i in the structural directions. For the beam example the displacements at the ends of member AB are

$$\mathbf{D}_{M1} = \{0, 0, 0, 17\}PL^2/112EI_Z$$

and those for member BC are

$$\mathbf{D}_{M2} = \{0, 17, 0, -5\}PL^2/112EI_Z$$

The order in which the displacements are listed in these vectors follows the pattern given in Fig. 4-3b.

Having the above vectors on hand, one may then apply Eq. (4-5) twice (once for each member) in order to evaluate the final end-actions for members AB and BC. In both applications the member stiffness matrix $\mathbf{S}_{Mi}$ is the same and is given in Table 4-2 (see Art. 4.3). Thus, the end-actions become

$$\mathbf{A}_{M1} = \mathbf{A}_{ML1} + \mathbf{S}_{M1}\mathbf{D}_{M1} = \frac{P}{4}\begin{bmatrix} 4 \\ L \\ 4 \\ -L \end{bmatrix} + \mathbf{S}_{M1}\begin{bmatrix} 0 \\ 0 \\ 0 \\ 17 \end{bmatrix}\frac{PL^2}{112EI_Z} = \frac{P}{56}\begin{bmatrix} 107 \\ 31L \\ 5 \\ 20L \end{bmatrix}$$

and

$$\mathbf{A}_{M2} = \mathbf{A}_{ML2} + \mathbf{S}_{M2}\mathbf{D}_{M2} = \frac{P}{8}\begin{bmatrix} 4 \\ L \\ 4 \\ -L \end{bmatrix} + \mathbf{S}_{M2}\begin{bmatrix} 0 \\ 17 \\ 0 \\ -5 \end{bmatrix}\frac{PL^2}{112EI_Z} = \frac{P}{56}\begin{bmatrix} 64 \\ 36L \\ -8 \\ 0 \end{bmatrix}$$

It should be evident that these end-actions could also be used to compute support reactions as an alternative to using Eq. (4-4).

The above calculations may appear at first to be unnecessarily extensive and cumbersome, and indeed they are for the small structure used as an example. Their virtue, however, lies in the fact that they represent an orderly procedure that can be readily programmed for a digital computer. Other efficiencies not conducive to hand calculations will be seen in the computer programs of the next chapter.

4.8 Analysis of Continuous Beams. The continuous beams to be discussed in this article are assumed to consist of prismatic members rigidly connected to each other and supported at various points along their lengths. The joints of a continuous beam will be selected normally at points of support and at any free (or overhanging) ends. However, in those cases where there are changes in the cross section between points of support, so that the member consists of two or more prismatic segments, it is always possible to analyze the beam by considering a joint to exist at the change in section. This technique of solution will be illustrated in the example of the next article. Another method of analyzing a beam with changes in section, and which does not require the introduction of additional joints to the structure, is described later in Art. 6.12.

A continuous beam having m members and $m + 1$ joints is shown in Fig. 4-7a. The x-y plane is the plane of bending of the beam. The members are numbered above the beam in the figure, and the joints are numbered below the beam. In each case the numbering is from left to right along the beam. The central part of Fig. 4-7a shows the i-th member framing at the left end into a joint designated as joint j and at the right end into a joint designated as joint k. Note that the member and joint numbering systems are such that the joint number j must be equal numerically to the member number i, and the joint number k must be equal to $i + 1$. These particular relationships between the member numbers and the joint numbers are, of course, valid only for a continuous beam. Also, it can be seen that in this case the total number of joints, denoted henceforth as n_j, is always one more than the number of members; thus, $n_j = m + 1$.

Support restraints of two types may exist at any joint in a continuous beam. These are a restraint against translation in the y direction and a restraint against rotation about the z axis. For example, joint 1 of the beam in the figure is assumed to be fixed, and therefore is restrained against both translation and rotation; joints 2, j, and m are restrained against translation only; and joints k and $m + 1$ are not restrained at all.

In a continuous beam the displacements are due primarily to flexural deformations, and only such deformations will be considered in this article. The effects of shearing deformations can be included in the analysis if necessary, as described later in Art. 6.16. In either case, however, the omission of axial deformations means that a maximum of two possible displace-

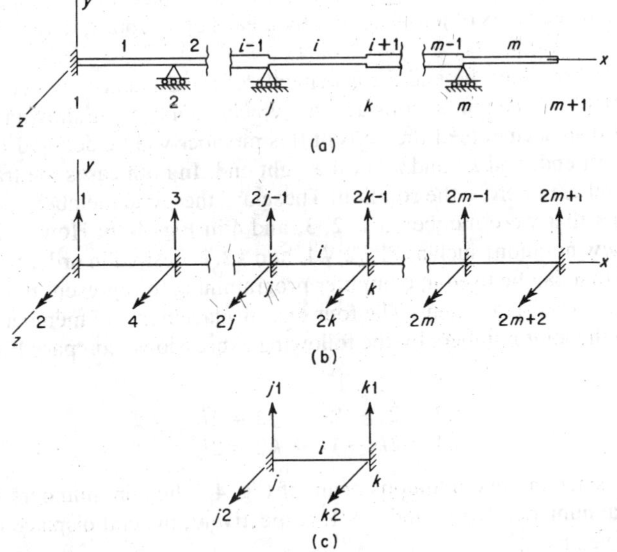

Fig. 4-7. Numbering system for a continuous beam.

ments may occur at any joint. These are the translation in the y direction and the rotation about the z axis. A numbering system for all of the possible joint displacements is indicated in Fig. 4-7b. Starting at the left-hand end of the structure and proceeding to the right, the translation and rotation at each joint are numbered in sequence. Since the translation at a particular joint is numbered before the rotation, it follows that in all cases the number representing the translation is equal to twice the joint number minus one, while the number for the rotation is twice the joint number. For instance, at joint j the translation and rotation are numbered $2j - 1$ and $2j$, respectively. Similar comments apply to the other joints, such as joints k, m, and $m + 1$. It is evident that the total number of possible joint displacements is twice the number of joints, or $2n_j$. Furthermore, if the total number of support restraints against translation and rotation is denoted n_r, the number of actual displacements, or degrees of freedom, is

$$n = 2n_j - n_r = 2m + 2 - n_r \tag{4-10}$$

in which n is the number of degrees of freedom.

The analysis of a continuous beam consists of setting up the necessary stiffness and load matrices and applying the equations of the direct stiffness method to the particular problem at hand. For this purpose, the most important matrix to be generated is the over-all joint stiffness matrix $\mathbf{S_J}$. The joint stiffness matrix consists of contributions from the beam stiffnesses $\mathbf{S}_{Mi}$ given previously in Art. 4.3 (see Table 4-2). For example, the i-th member in the continuous beam of Fig. 4-7 contributes to the stiffnesses of the joints j and k at the left and right ends of the member, respectively. Therefore, it is necessary to relate the end-displacements of member i to the displacements of joints j and k by means of an appropriate indexing system.

In order to relate the end-displacements of a particular member to the displacements of the joints, consider the typical member i as shown in Fig. 4-7c. The displacements of the ends of this member will be denoted $j1$ and $j2$ at the left end, and $k1$ and $k2$ at the right end. In both cases the translation is numbered before the rotation. These are the same member end-displacements that were numbered 1, 2, 3, and 4 in Fig. 4-3b. However, the use of a new notation, such as $j1$, $j2$, $k1$, and $k2$, is desired in order to have a symbol that can be used in computer programming to represent the number for the end-displacement. The four end-displacements of member i are related to the joint numbers by the following expressions (compare Figs. 4-7b and c):

$$\begin{aligned} j1 &= 2j - 1 \qquad & j2 &= 2j \\ k1 &= 2k - 1 \qquad & k2 &= 2k \end{aligned} \tag{4-11}$$

However, since in the continuous beam of Fig. 4-7 the joint numbers j and k are equal numerically to i and $i + 1$, respectively, the end-displacements are also given by

$$j1 = 2i - 1 \qquad = 2i$$
$$k1 = 2i + 1 \qquad = 2i + 2 \tag{4-12}$$

Thus, the above equations serve to index the possible joint displacements at the left and right ends of any member i in terms of either the joint numbers (Eqs. 4-11) or the member numbers (Eqs. 4-12). Such an indexing system is necessary for the purpose of constructing the joint stiffness matrix from the member stiffness matrices. The system also proves to be useful when calculating end-actions in the members due to displacements of the joints, as shown later.

As discussed previously, the over-all joint stiffness matrix $\mathbf{S_J}$ is made up of contributions from individual member stiffnesses. Hence, it is convenient to assess such contributions for one typical member i in the beam, and then to repeat the process for all members from 1 through m. A typical member i from a continuous beam is shown again in Fig. 4-8, with adjacent members $i - 1$ and $i + 1$ also indicated. In part (a) of the figure the beam is shown with a unit displacement corresponding to $j1$; that is, a y translation at the left end of the member. The four actions developed at joints j and k at the two ends of member i are stiffness coefficients S_J and are elements of the over-all joint stiffness matrix. Each such stiffness will have two subscripts as it appears in the joint stiffness matrix. The first subscript is the number (or index) that denotes the location of the action itself, and the second is the index for the unit displacement causing the action. Thus, the stiffness at joint j in the y direction has subscripts $j1$ and $j1$, meaning that the action corresponds to a displacement of type $j1$ and is caused by a unit displacement of type $j1$. This stiffness is denoted, therefore, by the symbol $(S_J)_{j1,j1}$, as shown in Fig. 4-8a. Of course, the actual value of the index $j1$ is obtained from Eqs. (4-12). Similarly, each of the remaining three stiffnesses in Fig. 4-8a has a first subscript that identifies the type of displacement to which the stiffness corresponds. However, each of them has the same second subscript, denoting the unit displacement of type $j1$.

The joint stiffnesses due to the remaining three possible joint displacements at the ends of member i are shown in Figs. 4-8b, 4-8c, and 4-8d. In each figure the four joint stiffnesses are shown with the appropriate subscripts that indicate (1) the type of action and (2) the unit displacement. The reader should verify the notation for each of the stiffnesses shown in Fig. 4-8.

The next step is to express the joint stiffness coefficients shown in Fig. 4-8 in terms of the various member stiffnesses that contribute to the joint stiffnesses. This step requires that the member stiffnesses be obtained from Table 4-2, which gives the stiffnesses for a continuous beam member (see Fig. 4-3b for the indexing system to Table 4-2). For example, the contribution to the joint stiffness $(S_J)_{j1,j1}$ from member $i - 1$ (see Fig. 4-8a) is the stiffness S_{M33} for that member. Similarly, the contribution to $(S_J)_{j1,j1}$ from member i is the stiffness S_{M11} for member i. These two contributions will

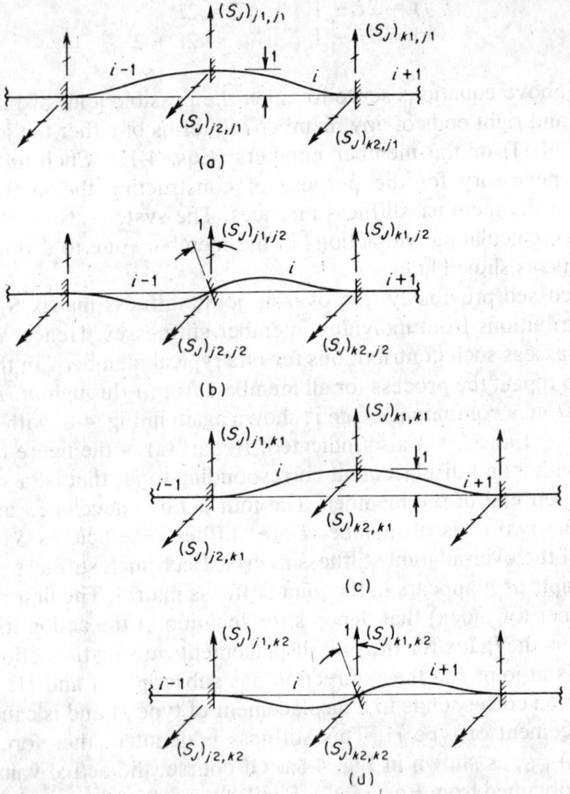

Fig. 4-8. Joint stiffnesses for a continuous beam

be denoted $(S_{M33})_{i-1}$ and $(S_{M11})_i$, respectively. In general, the contribution of one member to a particular joint stiffness will be denoted by appending the member subscript to the member stiffness itself. The latter quantity is obtained from the appropriate member stiffness matrix, which is Table 4-2 for a continuous beam having two possible displacements at each joint. From this discussion it can be seen that the joint stiffness coefficients shown in Fig. 4-8a are given in terms of the member stiffnesses by the following expressions:

$$\begin{aligned}(S_J)_{j1,j1} &= (S_{M33})_{i-1} + (S_{M11})_i \\ (S_J)_{j2,j1} &= (S_{M43})_{i-1} + (S_{M21})_i \\ (S_J)_{k1,j1} &= (S_{M31})_i \\ (S_J)_{k2,j1} &= (S_{M41})_i\end{aligned} \tag{4-13}$$

which represent the transfer of elements of the first column of the member stiffness matrix to the appropriate locations in $\mathbf{S_J}$. The first two joint stiffnesses consist of the sum of contributions from members $i-1$ and i. The last two stiffnesses involve contributions from member i only. This pattern of multiple terms when the stiffness is at the near end of the member, and single terms when the stiffness is at the far end of the member, is typical of all types of framed structures. Of course, there are occasional exceptions, such as at the end of a continuous beam, where there is only a single contribution even though the stiffness is at the near end.

Expressions that are analogous to Eqs. (4-13) are easily obtained for a unit rotation about the z axis at joint j. This rotation is a unit displacement of type $j2$ for member i, as shown in Fig. 4-8b. The expressions for the joint stiffnesses in terms of the member stiffnesses are (from the second column of the member stiffness matrix):

$$\begin{aligned}(S_J)_{j1,j2} &= (S_{M34})_{i-1} + (S_{M12})_i \\ (S_J)_{j2,j2} &= (S_{M44})_{i-1} + (S_{M22})_i \\ (S_J)_{k1,j2} &= (S_{M32})_i \\ (S_J)_{k2,j2} &= (S_{M42})_i\end{aligned} \tag{4-14}$$

Similarly, for a unit y translation at joint k (see Fig. 4-8c) the stiffnesses are (from the third column):

$$\begin{aligned}(S_J)_{j1,k1} &= (S_{M13})_i \\ (S_J)_{j2,k1} &= (S_{M23})_i \\ (S_J)_{k1,k1} &= (S_{M33})_i + (S_{M11})_{i+1} \\ (S_J)_{k2,k1} &= (S_{M43})_i + (S_{M21})_{i+1}\end{aligned} \tag{4-15}$$

Finally, the expressions for a unit z rotation at joint k (see Fig. 4-8d) are (from the fourth column):

$$\begin{aligned}(S_J)_{j1,k2} &= (S_{M14})_i \\ (S_J)_{j2,k2} &= (S_{M24})_i \\ (S_J)_{k1,k2} &= (S_{M34})_i + (S_{M12})_{i+1} \\ (S_J)_{k2,k2} &= (S_{M44})_i + (S_{M22})_{i+1}\end{aligned} \tag{4-16}$$

Equations (4-13) through (4-16) show that the sixteen elements of the 4 $\times$ 4 member stiffness matrix $[S_M]_i$ for member i contribute to sixteen of the joint stiffnesses in a very regular pattern. This pattern may be observed schematically in Fig. 4-9b, which indicates the formation of the joint stiffness matrix for a six-span continuous beam shown with restrained joints in Fig. 4-9a. For this structure the number of joints is seven, the number of possible joint displacements is fourteen, and therefore the joint stiffness matrix is of order 14 $\times$ 14. The indexing scheme is shown down the left-hand edge and across the upper edge of the matrix in Fig. 4-9. The contributions of individual members are indicated by the hatched blocks, each of which is of the order 4 $\times$ 4. The blocks are numbered in their upper right-hand corners to identify the members which are being considered. The overlapping blocks, which are of order 2 $\times$ 2 in this example, denote elements of $\mathbf{S_J}$ which receive contributions from two adjacent members. All elements outside the shaded blocks are zero.

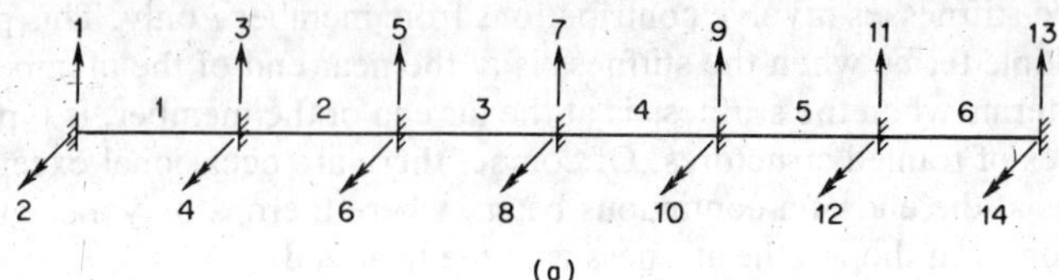

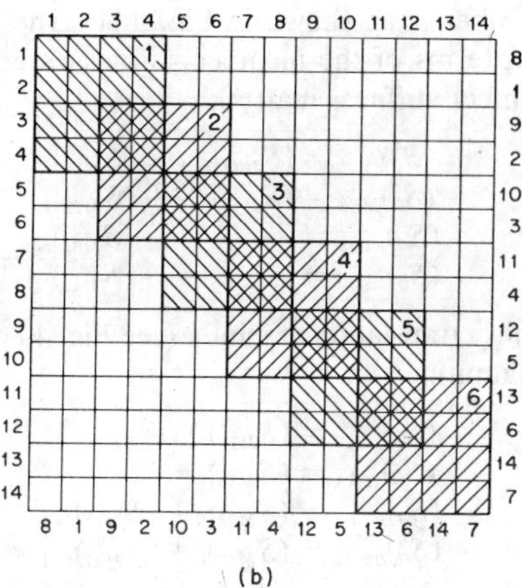

Fig. 4-9. Joint stiffness matrix for a continuous beam.

The stiffness matrix illustrated in Fig. 4-9b is the over-all matrix for all possible joint displacements. In order to analyze a given beam, however, the matrix must be rearranged into the partitioned form given by Eq. (4-8). Suppose, for example, that the actual beam has simple supports at all joints, as shown in Fig. 4-10a. The rearranged and partitioned stiffness matrix for this case is indicated in Fig. 4-10b. To obtain this rearranged matrix, rows and columns of the original matrix have been switched in proper sequence in order to place the stiffnesses pertaining to the actual degrees of freedom in the first seven rows and columns. At the same time, the stiffnesses pertaining to the support restraints have been placed in the last seven rows and columns. As an aid in the rearranging process, the new row and column designations are listed in Fig. 4-9b down the right side and across the bottom of the array. The rearrangement of the original stiffness matrix in this manner is equivalent to numbering the degrees of freedom and support restraints as shown in Fig. 4-10a. However, the general approach in this chapter, as explained earlier, is to generate the stiffness matrix using a numbering system like that shown in Fig. 4-9a, and then to rearrange rows and columns as was done in the example of Figs. 4-9 and 4-10.

In summary, the procedure to be followed in generating the joint stiff-

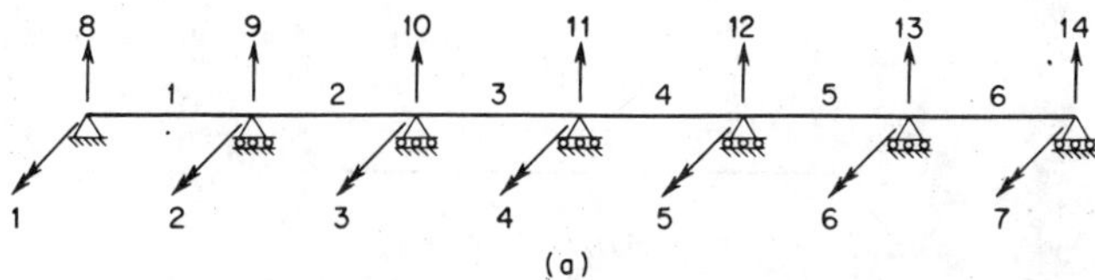

(a)

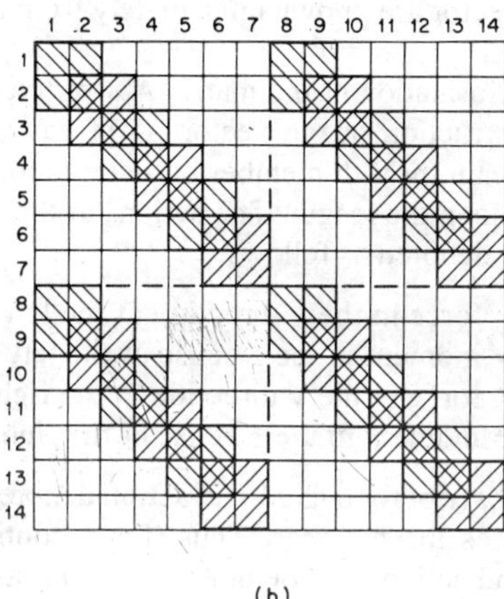

(b)

Fig. 4-10. Rearranged joint stiffness matrix for a continuous beam.

ness matrix $\mathbf{S}_J$ consists of taking the members in sequence and evaluating their contributions one at a time. For the typical member i, the procedure is as follows. First, the possible displacements at ends j and k of the member are related to the member number by calculating the indexes $j1, j2, k1$, and $k2$ (see Eqs. 4-12). Then the member stiffness matrix $\mathbf{S}_{Mi}$ is generated, and the elements of this matrix are transferred to $\mathbf{S}_J$ as indicated by the terms with subscripts i in Eqs. (4-13) through (4-16). After all members have been processed in this manner, the matrix $\mathbf{S}_J$ is complete. It may then be rearranged and partitioned in accordance with Eq. (4-8).

After obtaining the stiffness matrix, the next step is to obtain the load vectors. Consider first the vector $\mathbf{A}_J$ of actual joint loads. This vector contains $2n_j$ elements, each of which corresponds to one of the possible joint displacements shown in Fig. 4-7b. Thus, at each joint there are two possible applied loads, namely, a force in the y direction and a couple in the z sense. These loads are shown in Fig. 4-11 and are denoted $(A_J)_{2k-1}$ and $(A_J)_{2k}$, respectively. The subscripts used in identifying these actions are the same as the numbering system for the possible joint displacements (see Fig. 4-7b). Thus, the vector $\mathbf{A}_J$ takes the form

$$\mathbf{A}_J = \{(A_J)_1, (A_J)_2, \ldots, (A_J)_{2k-1}, (A_J)_{2k}, \ldots, (A_J)_{2m+1}, (A_J)_{2m+2}\} \quad (4\text{-}17)$$

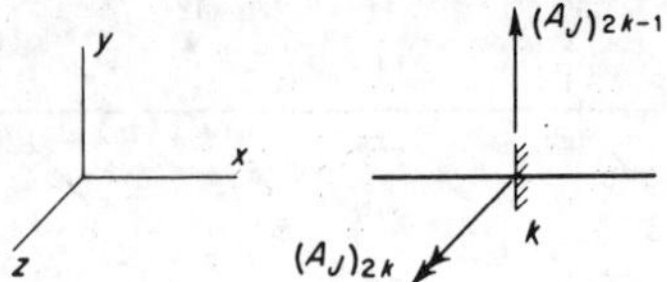

Fig. 4-11. Joint loads for a continuous beam.

The elements in this vector are known immediately from the given loads on the beam.

Consider next the formation of the matrix $\mathbf{A}_{\mathrm{ML}}$ of fixed-end actions due to loads and the construction of the vector $\mathbf{A}_{\mathrm{E}}$ of equivalent joint loads. Figure 4-12b shows again the i-th member of a continuous beam, but with lateral loads applied along the length. The actions at the ends of member i with its ends fixed are denoted as follows:

$$
\begin{aligned}
(A_{ML})_{1,i} &= \text{force in the } y \text{ direction at the left end}\\
(A_{ML})_{2,i} &= \text{moment in the } z \text{ sense at the left end}\\
(A_{ML})_{3,i} &= \text{force in the } y \text{ direction at the right end}\\
(A_{ML})_{4,i} &= \text{moment in the } z \text{ sense at the right end}
\end{aligned}
$$

In general, the first subscript for a fixed-end action denotes the action itself, and the second denotes the member. Thus, for a continuous beam, the matrix $\mathbf{A}_{\mathrm{ML}}$ of fixed-end actions will be of order $4 \times m$, as follows:

$$
\mathbf{A}_{\mathrm{ML}} = \begin{bmatrix} (A_{ML})_{1,1} & \cdots & (A_{ML})_{1,i} & \cdots & (A_{ML})_{1,m} \\ (A_{ML})_{2,1} & \cdots & (A_{ML})_{2,i} & \cdots & (A_{ML})_{2,m} \\ (A_{ML})_{3,1} & \cdots & (A_{ML})_{3,i} & \cdots & (A_{ML})_{3,m} \\ (A_{ML})_{4,1} & \cdots & (A_{ML})_{4,i} & \cdots & (A_{ML})_{1,m} \end{bmatrix} \tag{4-18}
$$

The elements in this matrix are fixed-end actions that can be found from the formulas given in Appendix B.

The vector $\mathbf{A}_{\mathrm{E}}$ of equivalent joint loads may be constructed from the elements of the matrix $\mathbf{A}_{\mathrm{ML}}$. The process for doing this can be visualized by referring again to Fig. 4-12. The member i shown in part (b) of the figure contributes to the equivalent loads at joints j and k, which are at the ends of the member. The equivalent loads at these joints are shown in Figs. 4-12a and 4-12c, and are denoted by the same subscripts used previously

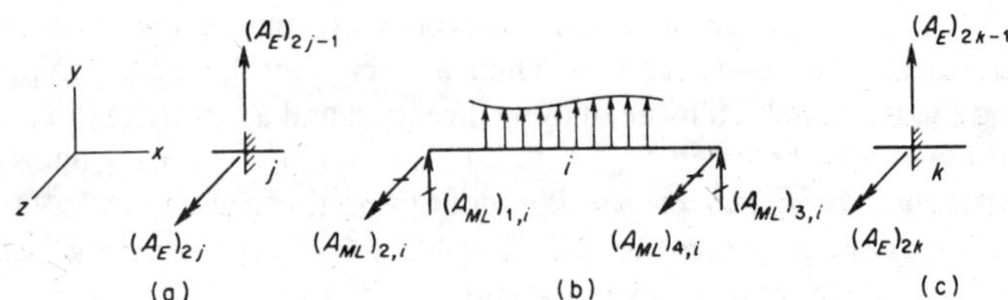

Fig. 4-12. Loads on a continuous beam member.

for the possible joint displacements (see Fig. 4-7b). Thus, the vector $\mathbf{A}_E$ has the general form

$$\mathbf{A}_E = \{(A_E)_1, (A_E)_2, \ldots, (A_E)_{2j-1}, (A_E)_{2j}, (A_E)_{2k-1}, (A_E)_{2k}, \ldots, (A_E)_{2m+2}\} \tag{4-19}$$

The individual elements of the vector $\mathbf{A}_E$ consist of contributions from the two adjoining members. Consider first the force $(A_E)_{2j-1}$ in the y direction at joint j (see Fig. 4-12a). This action, which also may be denoted $(A_E)_{2i-1}$, is composed of the negative of the end-action $(A_{ML})_{1,i}$ from member i (see Fig. 4-12b) and a similar contribution from the adjacent member $i-1$ on the left. The latter contribution is the negative of the force $(A_{ML})_{3i,-1}$ at the right-hand end of member $i-1$. By analogous reasoning the expressions for the other three equivalent loads can be obtained, thus giving the following results:

$$\begin{aligned}
(A_E)_{2j-1} &= (A_E)_{2i-1} = -(A_{ML})_{3,i-1} - (A_{ML})_{1,i} \\
(A_E)_{2j} &= (A_E)_{2i} = -(A_{ML})_{4,i-1} - (A_{ML})_{2,i} \\
(A_E)_{2k-1} &= (A_E)_{2i+1} = -(A_{ML})_{3,i} - (A_{ML})_{1,i+1} \\
(A_E)_{2k} &= (A_E)_{2i+2} = -(A_{ML})_{4,i} - (A_{ML})_{2,i+1}
\end{aligned} \tag{4-20}$$

The method for obtaining the vector $\mathbf{A}_E$ consists of taking the members in sequence and evaluating their contributions one at a time. Thus, successive columns of the matrix $\mathbf{A}_{ML}$ are considered, and the elements of each column are transferred to the appropriate elements in the vector $\mathbf{A}_E$ as indicated by the terms with subscripts i in Eqs. (4-20). After all members of the beam have been considered in this fashion, the vector $\mathbf{A}_E$ is complete. It may be added to the vector $\mathbf{A}_J$ according to Eq. (4-7) to form the vector $\mathbf{A}_C$ of combined joint loads. Lastly, the vector $\mathbf{A}_C$ can be rearranged so that the first part becomes $\mathbf{A}_{FC}$ and the second part becomes $\mathbf{A}_{RC}$, as explained in Art. 4.6 (see Eq. 4-9).

After the required matrices and vectors have been formulated by the methods described above, the solution for the free joint displacements $\mathbf{D}_F$ and the support reactions $\mathbf{A}_R$ consists simply of substituting into Eqs. (4-3) and (4-4), respectively, and performing the indicated matrix multiplications. After finding the joint displacements $\mathbf{D}_F$, which correspond only to the degrees of freedom of the structure, it is convenient to form the overall joint displacement vector $\mathbf{D}_J$ containing elements that correspond to all possible joint displacements (see Fig. 4-7b). For a continuous beam, the vector $\mathbf{D}_J$ will be order $2n_j \times 1$. It will contain values given by the vector $\mathbf{D}_F$ (corresponding to degrees of freedom), and the remaining elements (corresponding to support restraints) will be zero. In general, the form of the vector $\mathbf{D}_J$ is

$$\mathbf{D}_J = \{(D_J)_1, (D_J)_2, \ldots, (D_J)_{2j-1}, (D_J)_{2j}, (D_J)_{2k-1}, (D_J)_{2k}, \ldots, (D_J)_{2m+2}\} \tag{4-21}$$

Elements from this vector are used in calculating member end-actions, as described in the following.

To evaluate member end-actions, Eq. (4-5) must be applied once for each member of the structure. When written in detailed form, the equation is the following:

$$\begin{bmatrix} (A_M)_{1,i} \\ (A_M)_{2,i} \\ (A_M)_{3,i} \\ (A_M)_{4,i} \end{bmatrix} = \begin{bmatrix} (A_{ML})_{1,i} \\ (A_{ML})_{2,i} \\ (A_{ML})_{3,i} \\ (A_{ML})_{4,i} \end{bmatrix} + \begin{bmatrix} S_{M11} & S_{M12} & S_{M13} & S_{M14} \\ S_{M21} & S_{M22} & S_{M23} & S_{M24} \\ S_{M31} & S_{M32} & S_{M33} & S_{M34} \\ S_{M41} & S_{M42} & S_{M43} & S_{M44} \end{bmatrix} \begin{bmatrix} D_{M1} \\ D_{M2} \\ D_{M3} \\ D_{M4} \end{bmatrix} \tag{4-22}$$

The vector $\mathbf{A}_{\mathrm{ML}i}$ is obtained from the i-th column of the matrix $\mathbf{A}_{\mathrm{ML}}$ given previously, and the stiffness matrix $\mathbf{S}_{\mathrm{M}i}$ is obtained from Table 4-2. The vector $\mathbf{D}_{\mathrm{M}i}$ represents the end-displacements for member i. These displacements are obtained from the vector $\mathbf{D}_{\mathrm{J}}$ by taking the four consecutive displacements that are associated with member i. In general, the four displacements shown in Eq. (4-22), that is, D_{M1}, D_{M2}, D_{M3}, and D_{M4}, are equal to the displacements $(D_J)_{j1}$, $(D_J)_{j2}$, $(D_J)_{k1}$, and $(D_J)_{k2}$, respectively, from the vector $\mathbf{D}_{\mathrm{J}}$. Thus, the four end-displacements for any member can be extracted from the vector $\mathbf{D}_{\mathrm{J}}$ without difficulty.

Equation (4-22) can be further expanded by substituting into it the elements of $\mathbf{S}_{\mathrm{M}i}$ and $\mathbf{D}_{\mathrm{M}i}$ and performing the matrix multiplication. When this is done, the following four equations are obtained:

$$\begin{aligned}
(A_M)_{1,i} &= (A_{ML})_{1,i} + \frac{12EI_{Zi}}{L_i^3}[(D_J)_{j1} - (D_J)_{k1}] \\
&\qquad + \frac{6EI_{Zi}}{L_i^2}[(D_J)_{j2} + (D_J)_{k2}] \\
(A_M)_{2,i} &= (A_{ML})_{2,i} + \frac{6EI_{Zi}}{L_i^2}[(D_J)_{j1} - (D_J)_{k1}] \\
&\qquad + \frac{4EI_{Zi}}{L_i}\left[(D_J)_{j2} + \frac{1}{2}(D_J)_{k2}\right] \\
(A_M)_{3,i} &= (A_{ML})_{3,i} - \frac{12EI_{Zi}}{L_i^3}[(D_J)_{j1} - (D_J)_{k1}] \\
&\qquad - \frac{6EI_{Zi}}{L_i^2}[(D_J)_{j2} + (D_J)_{k2}] \\
(A_M)_{4,i} &= (A_{ML})_{4,i} + \frac{6EI_{Zi}}{L_i^2}[(D_J)_{j1} - (D_J)_{k1}] \\
&\qquad + \frac{4EI_{Zi}}{L_i}\left[\frac{1}{2}(D_J)_{j2} + (D_J)_{k2}\right]
\end{aligned} \tag{4-23}$$

Equations (4-22) and (4-23) are equivalent, of course, and either may be used for the purpose of computing the member end-actions.

The analysis of continuous beams using the highly organized method described above is demonstrated by an example in the next article. In the

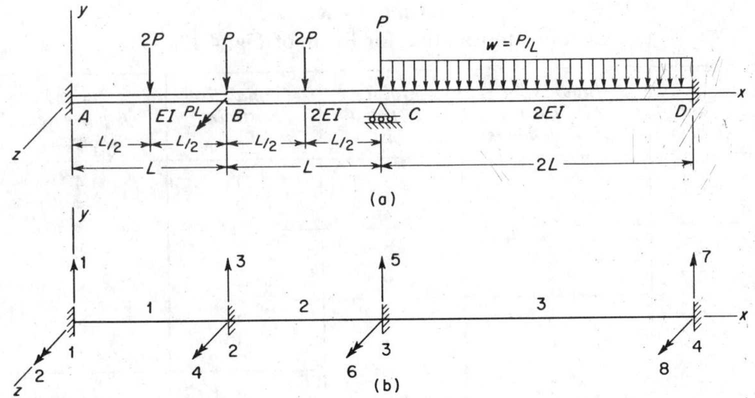

Fig. 4-13. Example (continuous beam).

solution, the steps described in this article are followed as much as possible. Naturally, such procedures are very cumbersome for a hand solution, but they are used here deliberately in order to illustrate the way in which the solution is carried out by means of a computer program. A computer program for the analysis of continuous beams is presented in Art. 5.6.

4.9 Example. The continuous beam shown in Fig. 4-13a is to be analyzed as described in the preceding article. The beam is restrained against translation at support C and against both translation and rotation at points A and D. At point B the flexural rigidity of the beam changes from EI to $2EI$. Therefore, point B is taken to be a joint in the structure.

The member and joint numbering systems are shown in Fig. 4-13b, and it is seen that the number of members m is three, the number of joints n_j is four, and the number of support restraints n_r is five. Therefore, the number of degrees of freedom n is equal to three (see Eq. 4-10).

The member properties and the joint displacement indexes computed from Eq. (4-12) are given in Table 4-7. The moment of inertia and length of each member can be considered as given data (or input) to the analysis. The remaining quantities in the table are computed from the member numbers. Joint information for the beam

Table 4-7

Member Information for Beam of Fig. 4-13a

Member Number	*Joint Numbers at Ends of Member*		*Indexes for End-Displacements of Member*				I_Z	*Length*
i	j	k	$j1$	$j2$	$k1$	$k2$		
1	1	2	1	2	3	4	I	L
2	2	3	3	4	5	6	$2I$	L
3	3	4	5	6	7	8	$2I$	$2L$

Table 4-8

Joint Information for Beam of Fig. 4-13a

Joint Number	Indexes for All Possible Displacements	Joint Restraint List	Cumulative Restraint List	Revised Displacement Indexes
1	1 2	1 1	1 2	4 5
2	3 4	0 0	2 2	1 2
3	5 6	1 0	3 3	6 3
4	7 8	1 1	4 5	7 8

is indicated in Table 4-8. In this table the indexes of all possible displacements are listed for each joint of the beam. This is followed by a *joint restraint list,* in which the numeral 1 is used to indicate the existence of a restraint, while a zero indicates no restraint (or a degree of freedom). Cumulative addition of the values in the restraint list produces the *cumulative restraint list* given in the next column. The last column of the table shows the joint displacement indexes when they are revised so that the degrees of freedom are numbered before the actual restraints. These revised indexes are obtained by applying expressions (a) and (b) in Art. 4.6. Thus, the two columns in Table 4-8 containing joint numbers and the restraint list can be considered as input, while the other three columns can be derived from the data in the input columns.

In preparation for generating the over-all joint stiffness matrix $\mathbf{S}_J$, the member stiffness matrices for all three members of the structure are given in Table 4-9. These are formed from Table 4-2 given in Art. 4.3. The elements of the first member stiffness matrix $\mathbf{S}_{M1}$ for member 1 are transferred to $\mathbf{S}_J$ according to Eqs. (4-13)

Table 4-9

Member Stiffness Matrices

$$\mathbf{S}_{M1} = \frac{EI}{L^3}\begin{bmatrix} 12 & 6L & -12 & 6L \\ 6L & 4L^2 & -6L & 2L^2 \\ -12 & -6L & 12 & -6L \\ 6L & 2L^2 & -6L & 4L^2 \end{bmatrix}$$

$$\mathbf{S}_{M2} = \frac{EI}{L^3}\begin{bmatrix} 24 & 12L & -24 & 12L \\ 12L & 8L^2 & -12L & 4L^2 \\ -24 & -12L & 24 & -12L \\ 12L & 4L^2 & -12L & 8L^2 \end{bmatrix}$$

$$\mathbf{S}_{M3} = \frac{EI}{L^3}\begin{bmatrix} 3 & 3L & -3 & 3L \\ 3L & 4L^2 & -3L & 2L^2 \\ -3 & -3L & 3 & -3L \\ 3L & 2L^2 & -3L & 4L^2 \end{bmatrix}$$

through (4-16). The contributions to $\mathbf{S_J}$ from member 1 appear in Table 4-10 within the upper left-hand portion enclosed by dashed lines. In a similar manner, elements from $\mathbf{S_{M2}}$ and $\mathbf{S_{M3}}$ are transferred to $\mathbf{S_J}$ as shown by the two remaining portions of Table 4-10 that are delineated by dashed lines. In the regions where the portions enclosed by dashed lines overlap, the elements shown in the table are the sums of two terms, one from each of two member stiffness matrices. Note that all elements in the table are to be multiplied by the factor EI/L^3. The numbering system for the stiffness matrix appears on the left side and across the top of the matrix and is in accordance with the numbering system shown in Fig. 4-13b.

Next, the matrix $\mathbf{S_J}$ is rearranged according to the revised displacement indexes given in the last column of Table 4-8. This list is indicated by the numbers on the right side and across the bottom of the matrix in Table 4-10. The third, fourth, and sixth rows and columns are to be shifted to the first three positions, and all other rows and columns are to be shifted downward and to the right without changing their order. The resulting rearranged joint stiffness matrix is presented in Table 4-11. This matrix is partitioned, as shown by the dashed lines, in the manner given by Eq. (4-8).

The 3×3 stiffness matrix $\mathbf{S_{FF}}$ is drawn from Table 4-11 as follows:

$$\mathbf{S_{FF}} = \frac{EI}{L^3}\begin{bmatrix} 36 & 6L & 12L \\ 6L & 12L^2 & 4L^2 \\ 12L & 4L^2 & 12L^2 \end{bmatrix}$$

Inversion of this matrix produces

$$\mathbf{S_{FF}^{-1}} = \frac{L}{756EI}\begin{bmatrix} 32L^2 & -6L & -30L \\ -6L & 72 & -18 \\ -30L & -18 & 99 \end{bmatrix}$$

Thus, all of the calculations involving the properties of the structure have been completed, and information concerning the loads on the structure may now be processed.

Inspection of Fig. 4-13a shows that loads are applied at joints B and C only (joints 2 and 3 in Fig. 4-13b). These joint loads are listed in Table 4-12 in a manner

Table 4-10

Joint Stiffness Matrix for Beam of Fig. 4-13

(All elements in the matrix are to be multiplied by the factor EI/L^3)

	1	2	3	4	5	6	7	8	
1	12	$6L$	-12	$6L$	0	0	0	0	4
2	$6L$	$4L^2$	$-6L$	$2L^2$	0	0	0	0	5
3	-12	$-6L$	36	$6L$	-24	$12L$	0	0	1
4	$6L$	$2L^2$	$6L$	$12L^2$	$-12L$	$4L^2$	0	0	2
5	0	0	-24	$-12L$	27	$-9L$	-3	$3L$	6
6	0	0	$12L$	$4L^2$	$-9L$	$12L^2$	$-3L$	$2L^2$	3
7	0	0	0	0	-3	$-3L$	3	$-3L$	7
8	0	0	0	0	$3L$	$2L^2$	$-3L$	$4L^2$	8
	4	5	1	2	6	3	7	8	

Table 4-11
Rearranged Joint Stiffness Matrix for Beam of Fig. 4-13
(All elements in the matrix are to be multiplied by the factor EI/L^3)

	1	2	3	4	5	6	7	8
1	36	$6L$	$12L$	-12	$-6L$	-24	0	0
2	$6L$	$12L^2$	$4L^2$	$6L$	$2L^2$	$-12L$	0	0
3	$12L$	$4L^2$	$12L^2$	0	0	$-9L$	$-3L$	$2L^2$
4	-12	$6L$	0	12	$6L$	0	0	0
5	$-6L$	$2L^2$	0	$6L$	$4L^2$	0	0	0
6	-24	$-12L$	$-9L$	0	0	27	-3	$3L$
7	0	0	$-3L$	0	0	-3	3	$-3L$
8	0	0	$2L^2$	0	0	$3L$	$-3L$	$4L^2$

that is suitable for input to a computer program. Next, these actions are placed in the vector $\mathbf{A}_J$ as follows (see Eq. 4-17):

$$\mathbf{A}_J = \{0, 0, -P, PL, -P, 0, 0, 0\}$$

Fixed-end actions $\mathbf{A}_{ML}$ are tabulated row-wise in Table 4-13, which is transposed with respect to the arrangement in Eq. (4-18). The elements of $\mathbf{A}_{ML}$ are transferred to the vector of equivalent joint loads $\mathbf{A}_E$ as indicated by Eqs. (4-20). First, the fixed-end actions for member 1 are transferred to the first four elements of $\mathbf{A}_E$; then the actions for member 2 are transferred to elements three to six of $\mathbf{A}_E$; and lastly, the actions for member 3 are transferred to the last four elements of $\mathbf{A}_E$. When the vector has been generated in this manner, the result is:

$$\mathbf{A}_E = \{-P, -PL/4, -2P, 0, -2P, -PL/12, -P, PL/3\}$$

The vectors $\mathbf{A}_J$ and $\mathbf{A}_E$ are now combined using Eq. (4-7) to obtain the vector $\mathbf{A}_C$:

$$\mathbf{A}_C = \{-P, -PL/4, -3P, PL, -3P, -PL/12, -P, PL/3\}$$

This vector must be rearranged by placing the third, fourth, and sixth elements into the first three positions and moving all other elements toward the end without changing their order. This rearrangement results in the following vector:

$$\mathbf{A}_C = \{-3P, PL, -PL/12, -P, -PL/4, -3P, -P, PL/3\}$$

in which the first three elements are the vector $\mathbf{A}_{FC}$:

$$\mathbf{A}_{FC} = \{-3P, PL, -PL/12\}$$

and the last five are the elements of $\mathbf{A}_{RC}$ (see Eq. 4-9):

$$\mathbf{A}_{RC} = \{-P, -PL/4, -3P, -P, PL/3\}$$

Table 4-12
Actions Applied at Joints

Joint	*Force in y Direction*	*Moment in z Sense*
1	0	0
2	$-P$	PL
3	$-P$	0
4	0	0

Table 4-13
Fixed-End Actions Due to Loads

Member	$(A_{ML})_{1,i}$	$(A_{ML})_{2,i}$	$(A_{ML})_{3,i}$	$(A_{ML})_{4,i}$
1	P	$PL/4$	P	$-PL/4$
2	P	$PL/4$	P	$-PL/4$
3	P	$PL/3$	P	$-PL/3$

Having all of the required matrices on hand, one may complete the solution by first calculating the free joint displacements $\mathbf{D}_F$ using Eq. (4-3):

$$\mathbf{D}_F = \mathbf{S}_{FF}^{-1}\mathbf{A}_{FC} = \frac{L}{756EI}\begin{bmatrix} 32L^2 & -6L & -30L \\ -6L & 72 & -18 \\ -30L & -18 & 99 \end{bmatrix}\begin{bmatrix} -36 \\ 12L \\ -L \end{bmatrix}\frac{P}{12}\cdot$$

$$= \frac{PL^2}{3024EI}\begin{bmatrix} -398L \\ 366 \\ 255 \end{bmatrix}$$

The vector $\mathbf{D}_F$ may now be used in obtaining the vector $\mathbf{D}_J$ of all possible joint displacements by referring to the original and revised displacement indexes shown in Table 4-8. These indexes show that $D_{J3} = D_1$, $D_{J4} = D_2$, and $D_{J6} = D_3$. All other elements of $\mathbf{D}_J$ are zero because they correspond to support restraints. Thus, the vector $\mathbf{D}_J$ becomes the following:

$$\mathbf{D}_J = \frac{PL^2}{3024EI}\{0, 0, -398L, 366, 0, 255, 0, 0\}$$

Next, the support reactions $\mathbf{A}_R$ may be determined from Eq. (4-4). For this purpose the matrix $\mathbf{S}_{RF}$ is obtained from the lower left-hand portion of Table 4-11. The calculation for $\mathbf{A}_R$ is as follows:

$$\mathbf{A}_R = -\mathbf{A}_{RC} + \mathbf{S}_{RF}\mathbf{D}_F$$

$$= \frac{P}{12}\begin{bmatrix} 12 \\ 3L \\ 36 \\ 12 \\ -4L \end{bmatrix} + \frac{EI}{L^3}\begin{bmatrix} -12 & 6L & 0 \\ -6L & 2L^2 & 0 \\ -24 & -12L & -9L \\ 0 & 0 & -3L \\ 0 & 0 & 2L^2 \end{bmatrix}\begin{bmatrix} -398L \\ 366 \\ 255 \end{bmatrix}\frac{PL^2}{3024EI}$$

$$= \frac{P}{1008}\begin{bmatrix} 3332 \\ 1292L \\ 3979 \\ 753 \\ -166L \end{bmatrix}$$

Finally, the member end-actions $\mathbf{A}_{Mi}$ are obtained by repeated applications of Eq. (4-5) (see also Eq. 4-22). Since there are three members in this example, three calculations of this type are required. For member 1 the calculations are as follows:

$$\mathbf{A}_{M1} = \mathbf{A}_{ML1} + \mathbf{S}_{M1}\mathbf{D}_{M1}$$

$$= \frac{P}{4}\begin{bmatrix} 4 \\ L \\ 4 \\ -L \end{bmatrix} + \mathbf{S}_{M1}\begin{bmatrix} 0 \\ 0 \\ -398L \\ 366 \end{bmatrix}\frac{PL^2}{3024EI} = \frac{P}{252}\begin{bmatrix} 833 \\ 323L \\ -329 \\ 258L \end{bmatrix}$$

In the above equation the vector $\mathbf{D}_{M1}$ consists of the first four elements of $\mathbf{D}_J$. Similarly, the equations for $\mathbf{A}_{M2}$ and $\mathbf{A}_{M3}$ are

$$\mathbf{A}_{M2} = \mathbf{A}_{ML2} + \mathbf{S}_{M2}\mathbf{D}_{M2}$$

$$= \frac{P}{4}\begin{bmatrix} 4 \\ L \\ 4 \\ -L \end{bmatrix} + \mathbf{S}_{M2}\begin{bmatrix} -398L \\ 366 \\ 0 \\ 255 \end{bmatrix}\frac{PL^2}{3024EI} = \frac{P}{252}\begin{bmatrix} 77 \\ -6L \\ 427 \\ -169L \end{bmatrix}$$

$$\mathbf{A}_{M3} = \mathbf{A}_{ML3} + \mathbf{S}_{M3}\mathbf{D}_{M3}$$

$$= \frac{P}{3}\begin{bmatrix} 3 \\ L \\ 3 \\ -L \end{bmatrix} + \mathbf{S}_{M3}\begin{bmatrix} 0 \\ 255 \\ 0 \\ 0 \end{bmatrix}\frac{PL^2}{3024EI} = \frac{P}{1008}\begin{bmatrix} 1263 \\ 676L \\ 753 \\ -166L \end{bmatrix}$$

in which the vector $\mathbf{D}_{M2}$ is made up of the third through the sixth elements of $\mathbf{D}_J$, and the vector $\mathbf{D}_{M3}$ consists of the last four elements of $\mathbf{D}_J$. Thus, all of the joint displacements, support reactions, and member end-actions are determined, and the problem is considered to be completed.

4.10 Plane Truss Member Stiffnesses. Determination of the complete member stiffness matrix for a typical truss member is a preliminary to

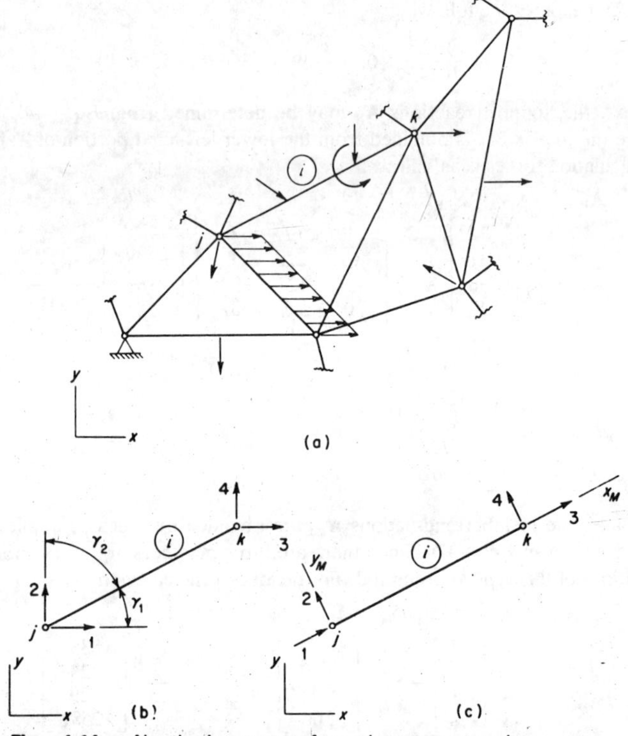

Fig. 4-14. Numbering system for a plane truss member.

the analysis of plane trusses. A typical member i within a plane truss is shown in Fig. 4-14a. The joints at the ends of this member are denoted as joints j and k. The plane truss itself is assumed to lie in the x-y plane, where x and y are reference axes for the structure. The joint translations are the unknown displacements in the analysis, and all of these translations may be expressed conveniently by their components in the x and y directions. For the typical member i, the positive directions of the four displacement components at its ends (with respect to the structure-oriented axes) are indicated in Fig. 4-14b.

It is convenient when dealing with inclined members in a framed structure to make use of direction cosines. The direction cosines for the member shown in Fig. 4-14b are the cosines of the angles γ_1 and γ_2 between the axis of the member and the x and y axes, respectively. These angles will always be taken at the j end of the member. The corresponding direction cosines are denoted as C_X and C_Y and are given as follows:

$$C_X = \cos \gamma_1 \qquad C_Y = \cos \gamma_2 = \sin \gamma_1$$

The direction cosines for the member can also be expressed in terms of the coordinates of the joints j and k. If the x and y coordinates of joints j and k are denoted by (x_j, y_j) and (x_k, y_k), respectively, the direction cosines become

$$C_X = \frac{x_k - x_j}{L} \qquad C_Y = \frac{y_k - y_j}{L} \tag{4-24}$$

in which L is the length of the member. The length L may be computed from the coordinates of the joints at the ends of the member as follows:

$$L = \sqrt{(x_k - x_j)^2 + (y_k - y_j)^2} \tag{4-25}$$

Both direction cosines are positive when the member is oriented as shown in Fig. 4-14b, that is, when the x and y coordinates of joint k are greater than those for joint j. If the angle γ_1 is greater than 90°, the formulas given above are still valid, and one or both of the direction cosines will be negative.

In the analysis of a plane truss, as in the case of any other type of framed structure, it is convenient to generate the joint stiffness matrix $\mathbf{S_J}$ by assessing the contributions from member stiffnesses. In the case of a continuous beam this operation is straightforward (see Art. 4.8) because the member-oriented axes may be taken parallel to, or coincident with, the structure-oriented axes. Thus, the beam member stiffness matrix for member-oriented axes, as derived in Art. 4.3, can be used directly when obtaining joint stiffnesses for the structure axes. For any other type of framed structure, however, the member axes will not necessarily be parallel to the structure axes. For example, the x-y axes in Fig. 4-14b are structure-oriented axes, whereas the member axes x_M and y_M are assumed to be along the axis of the member and perpendicular to the member, as shown in Fig. 4-14c. The stiffness matrix $\mathbf{S}_{\mathbf{M}i}$ for these member-oriented axes can be

Table 4-14

Plane Truss Member Stiffness Matrix for Member Axes (Fig. 4-14c)

$$\mathbf{S}_{\mathrm{M}i} = \begin{bmatrix} \mathbf{S}_{\mathrm{M}jj} & \mathbf{S}_{\mathrm{M}jk} \\ \mathbf{S}_{\mathrm{M}kj} & \mathbf{S}_{\mathrm{M}kk} \end{bmatrix} = \frac{EA_X}{L}\left[\begin{array}{cc|cc} 1 & 0 & -1 & 0 \\ 0 & 0 & 0 & 0 \\ \hline -1 & 0 & 1 & 0 \\ 0 & 0 & 0 & 0 \end{array}\right]$$

obtained readily by referring to cases (1) and (7) of Fig. 4-2. With the numbering system shown in Fig. 4-14c for the end-displacements with respect to the x_M-y_M axes, it becomes evident that the stiffness matrix $\mathbf{S}_{\mathrm{M}i}$ for those axes has the form shown in Table 4-14. In the table, the axial rigidity of the member is denoted by EA_X, and the matrix is partitioned as before.

However, since the matrix $\mathbf{S}_J$ is based upon axes oriented to the structure, it also becomes necessary to obtain member stiffnesses for the structure axes. The member stiffnesses for the structure axes may be found in either of two ways. The first method consists of a direct formulation of the stiffnesses. In this approach, unit displacements in the directions of the structure axes (see Fig. 4-14b) are induced at the ends of the member, and the corresponding restraint actions in the same directions are calculated. These actions become the elements of the member stiffness matrix for the structure axes. The second method for finding the member stiffness matrix consists of first obtaining the stiffness matrix for member-oriented axes (see Table 4-14) and then transforming this matrix to the structure-oriented axes by a process of rotation of axes. With an appropriate transformation matrix, the rotation of axes may be executed by matrix multiplications, as will be described in a later article.

The member stiffness matrix for a plane truss member will now be developed by the direct method. For this purpose it is necessary to consider unit displacements in the x and y directions at both ends of the member. The first such displacement is shown in Fig. 4-15a and consists of a unit translation in the x direction at the j end of the member. As a result of this displacement, an axial force is induced in the member. This force can be calculated from the axial shortening of the member, which is numerically equal to the x direction cosine C_X for the member (see Fig. 4-15a). The axial compressive force in the member due to this change of length is equal to

$$\frac{EA_X}{L} C_X$$

The restraint actions at the ends of the member in the x and y directions, which are equal to the components of the axial force, are the desired member stiffnesses for the structure axes. Such stiffnesses are identified by the symbol S_{MS} (see Fig. 4-15a) in order to distinguish them from the stiffnesses S_M for the member axes (see Table 4-14). The numbering system for the stiffnesses S_{MS} is shown in Fig. 4-14b. The restraint action at end j in the x

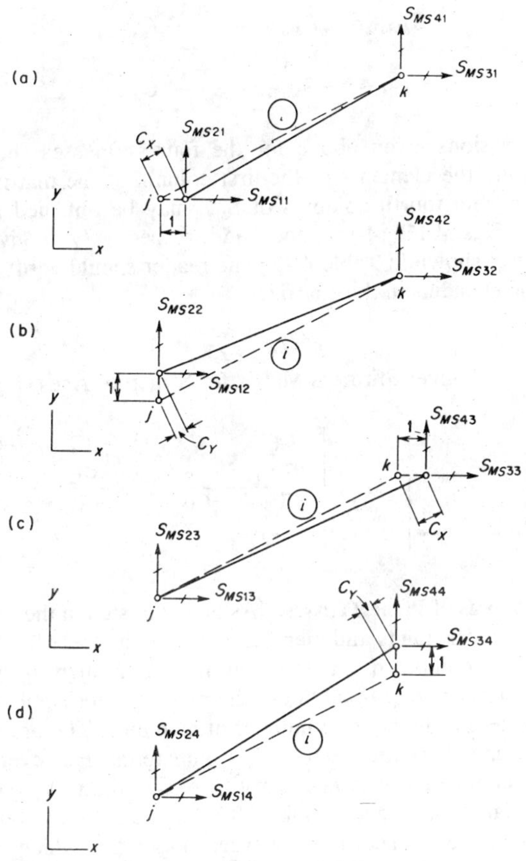

Fig. 4-15. Plane truss member stiffnesses for structure axes.

direction, denoted as S_{MS11}, must be equal to the x component of the force in the member. Therefore, this stiffness is equal to the axial force times the x direction cosine, as follows:

$$S_{MS11} = \frac{EA_X}{L} C_X^2$$

Also, the restraint action at j in the y direction is equal to the y component of the force in the member:

$$S_{MS21} = \frac{EA_X}{L} C_X C_Y$$

The restraint actions at the k end of the member in Fig. 4-15a are readily found by static equilibrium, as follows:

$$S_{MS31} = -S_{MS11} = -\frac{EA_X}{L} C_X^2$$

$$S_{MS41} = -S_{MS21} = -\frac{EA_X}{L} C_X C_Y$$

The expressions given above for the four stiffnesses shown in Fig. 4-15a constitute the elements of the first column of the matrix $\mathbf{S}_{MSi}$. The second, third, and fourth columns of $\mathbf{S}_{MSi}$ may be obtained in a similar manner from Figs. 4-15b, 4-15c, and 4-15d, respectively, to give the 4×4 stiffness matrix shown in Table 4-15. The reader should verify for himself the remaining elements in this matrix.

Table 4-15

Plane Truss Member Stiffness Matrix for Structure Axes (Fig. 4-14b)

$$\mathbf{S}_{MSi} = \begin{bmatrix} \mathbf{S}_{MSjj} & \mathbf{S}_{MSjk} \\ \mathbf{S}_{MSkj} & \mathbf{S}_{MSkk} \end{bmatrix} = \frac{EA_X}{L} \left[\begin{array}{cc|cc} C_X^2 & C_XC_Y & -C_X^2 & -C_XC_Y \\ C_XC_Y & C_Y^2 & -C_XC_Y & -C_Y^2 \\ \hline -C_X^2 & -C_XC_Y & C_X^2 & C_XC_Y \\ -C_XC_Y & -C_Y^2 & C_XC_Y & C_Y^2 \end{array} \right]$$

4.11 Analysis of Plane Trusses. As an initial step in the analysis of a plane truss, all of the joints and members must be numbered. The joints of the structure are numbered consecutively from 1 through n_j, where n_j is the total number of joints. In addition, the members are numbered from 1 through m, where m is the total number of members. The order in which the joints and members are numbered is immaterial. However, after the numbering is completed, it is necessary to record the two joint numbers that are associated with each member. This association of joint numbers with member numbers is necessary in order to ascertain which elements of the joint stiffness matrix $\mathbf{S}_J$ and the equivalent load vector $\mathbf{A}_E$ receive contributions from each member.

To illustrate the arbitrary numbering of the joints and members, a plane truss is shown in Fig. 4-16a. In the figure the joint numbers appear adjacent to the joints, while the member numbers are enclosed in circles adjacent to the members. The numbering system for this structure is shown in Table 4-16. It is necessary for purposes of analysis to identify a j end and a k end

Table 4-16

Listing of Members and Joints for Truss of Fig. 4-16a

Member	1	2	3	4	5	6	7	8	9
Joint j	2	6	4	3	4	1	6	1	2
Joint k	5	5	1	5	2	5	3	6	1

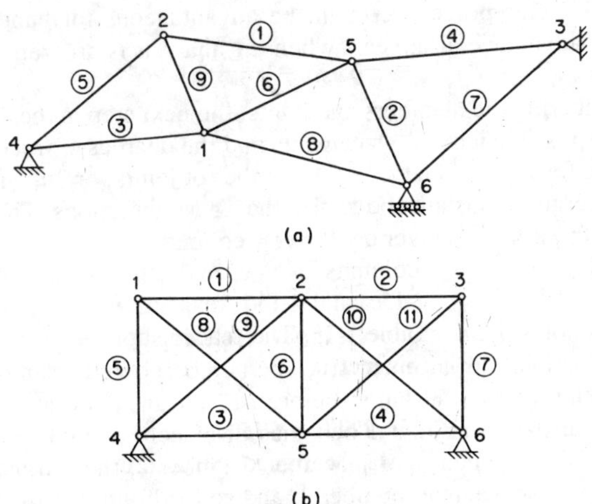

Fig. 4-16. Numbering systems for members and joints of plane trusses.

for each member, as shown in the table, although the selection itself is arbitrary.

While the methods of analysis in this chapter do not require any particular order of numbering joints and members, it is only natural to number them in a systematic pattern whenever feasible. For example, in some structures a natural order may be from left to right or from top to bottom, as shown by the plane truss in Fig. 4-16b. The geometry of this truss suggests that a systematic pattern for numbering the joints is to proceed from left to right across the top and then from left to right across the bottom. A systematic pattern for the members consists of numbering consecutively the horizontal members, the vertical members, and then the diagonal members, as shown in the figure. The numbering system for this structure is summarized in Table 4-17. Note again that it is necessary to identify one end of each member as being the j end and one end as the k end.

Many possible schemes for numbering the joints and members of a truss can be organized, depending mainly on personal preference. If it is desired to improve the efficiency of the solution, the numbering system may be selected to minimize the amount of rearrangement of the over-all joint stiff-

Table 4-17

Listing of Members and Joints for Truss of Fig. 4-16b

Member	1	2	3	4	5	6	7	8	9	10	11
Joint j	1	2	4	5	1	2	3	1	4	2	5
Joint k	2	3	5	6	4	5	6	5	2	6	3

ness matrix $\mathbf{S}_J$. This procedure might be advantageous for hand calculations, but the advantage disappears when the analysis is programmed for a computer.

After numbering the members and joints, the next step in the analysis is to identify all possible joint displacements and the degrees of freedom. The number of the former will be twice the number of joints, or $2n_j$, since each joint may undergo a translation in both the x and y directions. The number of degrees of freedom n is given by the expression

$$n = 2n_j - n_r \tag{4-26}$$

in which n_r denotes the number of support restraints.

The possible joint displacements will be numbered in the same order as the joints, taking the x translation before the y translation at each joint. Thus, the x translation at joint 1 becomes displacement number 1, the y translation at joint 1 becomes displacement number 2, the x translation at joint 2 becomes displacement number 3, and so forth, until at the last joint the x and y translations are numbered $2n_j - 1$ and $2n_j$, respectively. In general, at joint j of the truss the x and y translations have the indexes $2j - 1$ and $2j$, respectively.

In order to construct the stiffness matrix $\mathbf{S}_J$ from the member stiffnesses, it is useful to relate the indexes for the possible joint displacements to the end displacements for a particular member, as was done in the analysis of continuous beams. For this purpose consider the typical member i, shown in Fig. 4-17, which frames into joints j and k at the ends. The x and y axes in the figure are assumed to be structure-oriented axes. The end-displacements of this member may be identified by the indexes $j1$, $j2$, $k1$, and $k2$, as shown in the figure. These indexes for the member are related to the joint numbers j and k as follows:

$$\begin{aligned} j1 &= 2j - 1 \qquad & j2 &= 2j \\ k1 &= 2k - 1 \qquad & k2 &= 2k \end{aligned} \tag{4-27}$$

These relations follow directly from the numbering system for the possible joint displacements, which is described above.

The joint stiffness matrix consists of contributions from the individual

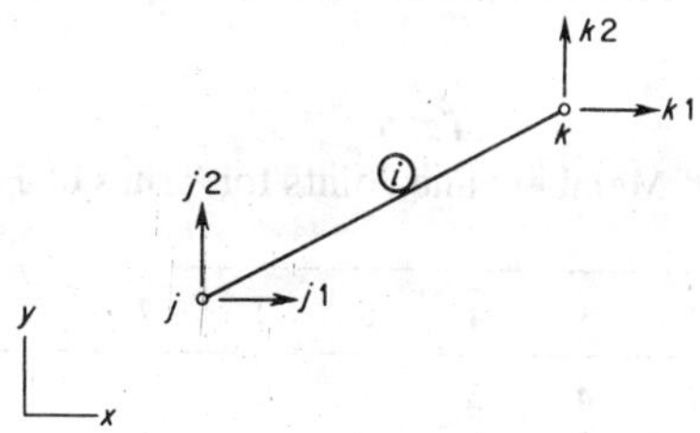

Fig. 4-17. End-displacements for plane truss member.

members. Therefore, the member stiffness matrix $\mathbf{S}_{MS\,i}$ (see Art. 4.10, Table 4-15) for each member must be determined and its elements placed in the proper positions in the matrix $\mathbf{S}_J$. In order to see how this is accomplished, consider again a typical member i (Fig. 4-18a). This member contributes to the stiffnesses of joints j and k at its ends. If a unit displacement in the x direction is induced in the restrained structure, as shown in the figure, there will be restraint actions in the x and y directions at both joints. At joint j the action in the x direction is the joint stiffness $(S_J)_{j1,j1}$, and the action in the y direction is the stiffness $(S_J)_{j2,j1}$. Similarly, the actions at joint k are denoted $(S_J)_{k1,j1}$ and $(S_J)_{k2,j1}$. The two actions at joint j consist of contributions from the member i, plus contributions from all other members that frame into joint j. The latter contributions will be denoted for simplicity by the indefinite symbol ΣS_{MS}. The contributions from member i, however, will be written in precise form since these are the terms of current interest.

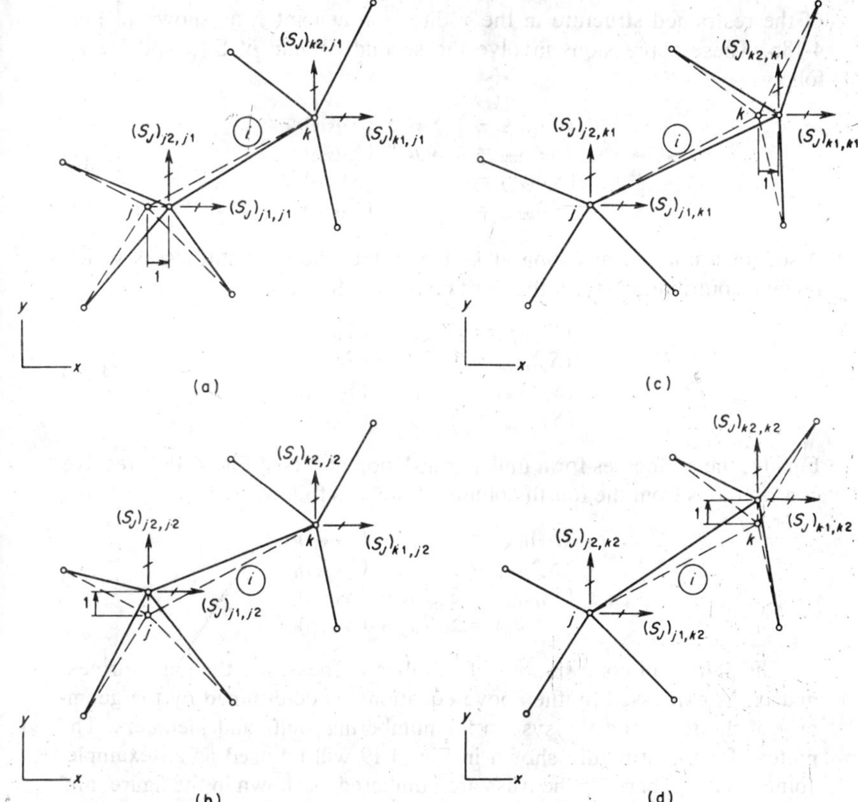

Fig. 4-18. Joint stiffnesses for a plane truss.

In the case of the stiffness $(S_J)_{j1,j1}$ the contribution from member i is the member stiffness S_{MS11}, and in the case of the stiffness $(S_J)_{j2,j1}$ the contribution is S_{MS21}. For the stiffnesses at joint k, only contributions from member i are involved. These contributions are S_{MS31} and S_{MS41} for the stiffnesses in the x and y directions, respectively, at joint k. Thus, the expressions for the joint stiffnesses shown in Fig. 4-18a are:

$$\begin{aligned}(S_J)_{j1,j1} &= \Sigma S_{MS} + (S_{MS11})_i \\ (S_J)_{j2,j1} &= \Sigma S_{MS} + (S_{MS21})_i \\ (S_J)_{k1,j1} &= (S_{MS31})_i \\ (S_J)_{k2,j1} &= (S_{MS41})_i\end{aligned} \tag{4-28}$$

Each of the above stiffnesses is due to a unit x translation at joint j, and each receives a contribution from the first column of the matrix $\mathbf{S}_{\mathbf{MS}i}$ for member i.

Expressions similar to Eqs. (4-28) may be written for a unit translation of the restrained structure in the y direction at joint j, as shown in Fig. 4-18b. These expressions involve the second column of $\mathbf{S}_{\mathbf{MS}i}$ and are as follows:

$$\begin{aligned}(S_J)_{j1,j2} &= \Sigma S_{MS} + (S_{MS12})_i \\ (S_J)_{j2,j2} &= \Sigma S_{MS} + (S_{MS22})_i \\ (S_J)_{k1,j2} &= (S_{MS32})_i \\ (S_J)_{k2,j2} &= (S_{MS42})_i\end{aligned} \tag{4-29}$$

Also, for a unit x translation at k (Fig. 4-18c) the joint stiffnesses which receive contributions from the third column of $\mathbf{S}_{\mathbf{MS}i}$ are:

$$\begin{aligned}(S_J)_{j1,k1} &= (S_{MS13})_i \\ (S_J)_{j2,k1} &= (S_{MS23})_i \\ (S_J)_{k1,k1} &= \Sigma S_{MS} + (S_{MS33})_i \\ (S_J)_{k2,k1} &= \Sigma S_{MS} + (S_{MS43})_i\end{aligned} \tag{4-30}$$

Finally, the stiffnesses for a unit y translation at k (see Fig. 4-18d) receive contributions from the fourth column of $\mathbf{S}_{\mathbf{MS}i}$, as follows:

$$\begin{aligned}(S_J)_{j1,k2} &= (S_{MS14})_i \\ (S_J)_{j2,k2} &= (S_{MS24})_i \\ (S_J)_{k1,k2} &= \Sigma S_{MS} + (S_{MS34})_i \\ (S_J)_{k2,k2} &= \Sigma S_{MS} + (S_{MS44})_i\end{aligned} \tag{4-31}$$

The pattern of contributions of member stiffnesses to the joint stiffness matrix, as expressed by the above equations, is determined by the geometry of the truss and the system for numbering joints and members. The pattern for the structure shown in Fig. 4-19 will be used as an example. Joints and members for the truss are numbered as shown in the figure, and the possible joint displacements (shown by arrows) are numbered according to Eqs. (4-27). For this structure there are six members and four joints.

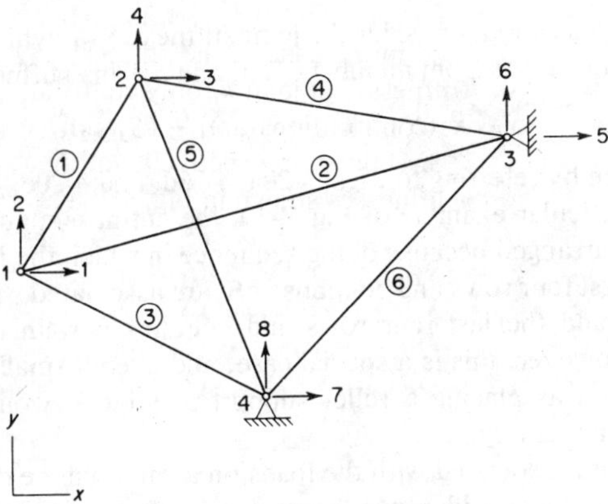

Fig. 4-19. Example (plane truss).

and the over-all joint stiffness matrix is of order 8 × 8. The positions of the member contributions to the joint stiffness matrix are indicated by crosses in Figs. 4-20a through 4-20f, and the final stiffness matrix is a composite of these contributions. As an example, the joint stiffness S_{J11} is made up of contributions from members 1, 2, and 3, as shown in Fig. 4-20a, 4-20b, and 4-20c. Therefore, the stiffness is:

$$S_{J11} = (S_{MS11})_1 + (S_{MS11})_2 + (S_{MS11})_3$$

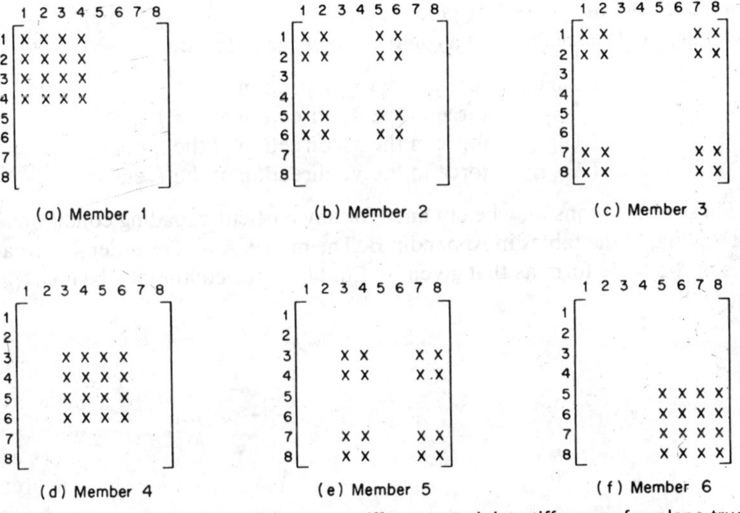

Fig. 4-20. Contributions of member stiffnesses to joint stiffnesses for plane truss example.

As another illustration, consider the joint stiffness S_{J34}, which is composed of member stiffnesses from members 1, 4, and 5. This stiffness is

$$S_{J34} = (S_{MS34})_1 + (S_{MS12})_4 + (S_{MS12})_5$$

as can be seen by referring to Figs. 4-20a, 4-20d, and 4-20e.

In the particular example of Fig. 4-19, the stiffness matrix $\mathbf{S}_J$ does not have to be rearranged because of the sequence in which the joints are numbered. The first four rows and columns of $\mathbf{S}_J$ are associated with the degrees of freedom, and the last four rows and columns pertain to the support restraints. However, this is a special case, and even a small change in the structure, such as placing a roller support at joint 3, would make a rearrangement necessary.

The vectors associated with the loads on a truss will be discussed next. The first one to be considered is the vector $\mathbf{A}_J$ of actions (or loads) applied at the joints. At a typical joint k, two orthogonal force components may exist, as shown in Fig. 4-21. The action $(A_J)_{2k-1}$ is the force in the positive x direction, and the action $(A_J)_{2k}$ is the force in the positive y direction. The vector $\mathbf{A}_J$, therefore, will have the following form:

$$\mathbf{A}_J = \{(A_J)_1, (A_J)_2, \ldots, (A_J)_{2k-1}, (A_J)_{2k}, \ldots, (A_J)_{2n_j}\} \quad (4\text{-}32)$$

in which there are $2n_j$ elements corresponding to the $2n_j$ possible joint displacements.

Consider next the matrix of actions $\mathbf{A}_{ML}$ due to loads acting on the members when the joints of the truss are restrained against translation. Loads acting on member i are illustrated in Fig. 4-22b, which also shows a set of member-oriented axes x_M and y_M. The actions $\mathbf{A}_{MLi}$ for member i are defined with respect to the x_M and y_M axes, as follows:

$$\begin{aligned}
(A_{ML})_{1,i} &= \text{force in the } x_M \text{ direction at the } j \text{ end} \\
(A_{ML})_{2,i} &= \text{force in the } y_M \text{ direction at the } j \text{ end} \\
(A_{ML})_{3,i} &= \text{force in the } x_M \text{ direction at the } k \text{ end} \\
(A_{ML})_{4,i} &= \text{force in the } y_M \text{ direction at the } k \text{ end}
\end{aligned}$$

These end-actions may be obtained for any particular loading conditions by referring to the tables in Appendix B. The matrix $\mathbf{A}_{ML}$ is of order $4 \times m$ and is of the same form as that given by Eq. (4-18) for continuous beams.

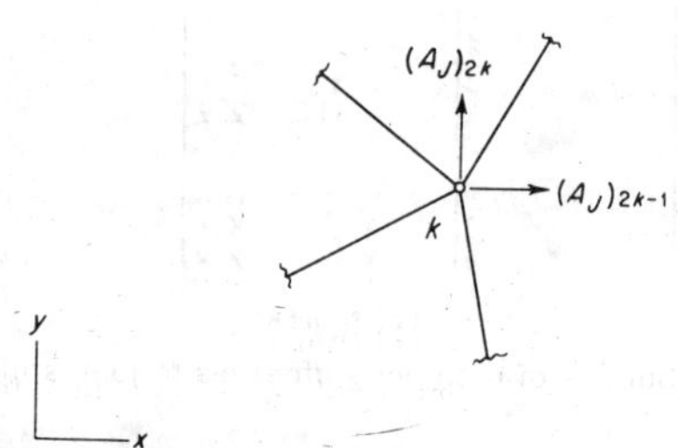

Fig. 4-21. Joint loads for a plane truss.

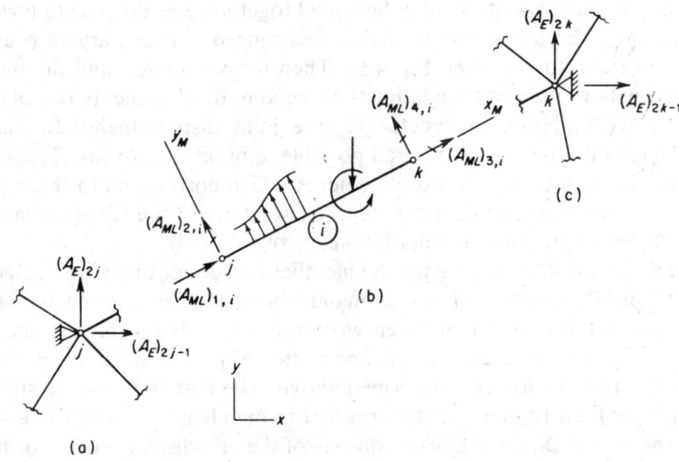

Fig. 4-22. Loads on a plane truss member.

The vector $\mathbf{A}_E$ of equivalent joint loads may be constructed from the elements of the matrix $\mathbf{A}_{ML}$. This vector is of the same form as that given by Eq. (4-32), except that A_E replaces A_J. To illustrate the calculation of the equivalent joint loads, consider the action $(A_E)_{2j-1}$ shown in Fig. 4-22a. This action is made up of contributions from member i, plus contributions from all other members meeting at the joint. The latter contributions will be denoted by the indefinite symbol ΣA_{MS}. The contributions from member i, however, can be readily expressed in terms of the end-actions in the i-th column of $\mathbf{A}_{ML}$. All that is necessary is to take the components in the x direction of the end-actions at joint j, and to reverse their signs. Thus, the contribution of member i to the equivalent load $(A_E)_{2j-1}$ is

$$-C_{Xi}(A_{ML})_{1,i} + C_{Yi}(A_{ML})_{2,i}$$

in which C_{Xi} and C_{Yi} are the direction cosines for member i. By proceeding in this manner, the equivalent loads at joints j and k (see Figs. 4-22a and 4-22c) can be found:

$$\begin{aligned}
(A_E)_{2j-1} &= -\Sigma A_{MS} - C_{Xi}(A_{ML})_{1,i} + C_{Yi}(A_{ML})_{2,i} \\
(A_E)_{2j} &= -\Sigma A_{MS} - C_{Yi}(A_{ML})_{1,i} - C_{Xi}(A_{ML})_{2,i} \\
(A_E)_{2k-1} &= -\Sigma A_{MS} - C_{Xi}(A_{ML})_{3,i} + C_{Yi}(A_{ML})_{4,i} \\
(A_E)_{2k} &= -\Sigma A_{MS} - C_{Yi}(A_{ML})_{3,i} - C_{Xi}(A_{ML})_{4,i}
\end{aligned} \tag{4-33}$$

These equations give the contributions of member i to the equivalent joint loads, as indicated earlier in Eq. (4-2). By applying the equations consecutively to all members of the truss, the equivalent joint loads can be obtained for all joints.

The contributions of member i to the vector $\mathbf{A}_E$ may also be ascertained by the method of rotation of axes. This procedure is demonstrated for plane truss structures in Art. 4.14.

The vectors $\mathbf{A}_J$ and $\mathbf{A}_E$ may be added together (see Eq. 4-7) to form the vector $\mathbf{A}_C$. The latter vector is then rearranged if necessary in order to isolate the vector $\mathbf{A}_{FC}$ (see Eq. 4-9). Then the vector $\mathbf{A}_{FC}$ and the inverse matrix $\mathbf{S}_{FF}^{-1}$ are used in calculating the free joint displacements $\mathbf{D}_F$, as given by Eq. (4-3). Next, the vector of free joint displacements $\mathbf{D}_F$ can be expanded into the vector $\mathbf{D}_J$ of all possible joint displacements. The vector $\mathbf{D}_J$ has $2n_j$ elements, and those elements which correspond to the support restraints are zero. As the next step, the solution for the support reactions is obtained in the usual manner by means of Eq. (4-4).

The final task of solving for the member end-actions may be carried out by means of Eq. (4-5). This equation must be applied once for each member of the structure. Note that when written in expanded form, the equation has the same form as Eq. (4-22). The vector $\mathbf{A}_{ML\,i}$ is obtained from the i-th column of the matrix $\mathbf{A}_{ML}$ described above. The matrix $\mathbf{S}_{M\,i}$ is the stiffness matrix for the i-th member with respect to member axes (see Table 4-14).

The vector $\mathbf{D}_{M\,i}$ in Eq. (4-5) consists of the end-displacements for member i in the directions of the member axes. These displacements can be calculated from the joint displacements D_J, which are in the directions of the structure axes. For instance, the first element D_{M1} in the vector $\mathbf{D}_{M\,i}$ represents the displacement of joint j in the x_M direction. This displacement is given by the following expression:

$$D_{M1} = (D_J)_{j1}C_{Xi} + (D_J)_{j2}C_{Yi}$$

in which $(D_J)_{j1}$ and $(D_J)_{j2}$ are the displacements of joint j in the x and y directions, respectively (see Fig. 4-17). Similarly, the displacement of joint j in the y_M direction can be expressed in terms of $(D_J)_{j1}$ and $(D_J)_{j2}$ by the following:

$$D_{M2} = -(D_J)_{j1}C_{Yi} + (D_J)_{j2}C_{Xi}$$

In addition, the end-displacements at joint k may be obtained as follows:

$$D_{M3} = (D_J)_{k1}C_{Xi} + (D_J)_{k2}C_{Yi}$$
$$D_{M4} = -(D_J)_{k1}C_{Yi} + (D_J)_{k2}C_{Xi}$$

The above expressions for the elements of $\mathbf{D}_{M\,i}$ can now be substituted into Eq. (4-5). When similar substitutions for the elements of $\mathbf{S}_{M\,i}$ are made, the equation can be expanded into the following four separate equations:

$$\begin{aligned}
(A_M)_{1,i} &= (A_{ML})_{1,i} + \frac{EA_{Xi}}{L_i}\{[(D_J)_{j1} - (D_J)_{k1}]C_{Xi} \\
&\qquad + [(D_J)_{j2} - (D_J)_{k2}]C_{Yi}\} \\
(A_M)_{2,i} &= (A_{ML})_{2,i} \\
(A_M)_{3,i} &= (A_{ML})_{3,i} - \frac{EA_{Xi}}{L_i}\{[(D_J)_{j1} - (D_J)_{k1}]C_{Xi} \\
&\qquad + [(D_J)_{j2} - (D_J)_{k2}]C_{Yi}\} \\
(A_M)_{4,i} &= (A_{ML})_{4,i}
\end{aligned} \tag{4-34}$$

These equations may now be applied repeatedly for all members of the truss. For each member the displacements D_J which appear in the equations must be extracted in the appropriate manner from the vector $\mathbf{D}_J$ of all joint displacements.

Another method for obtaining the vector $\mathbf{D}_{Mi}$ which appears in Eq. (4-5) is to use rotation of axes. This method is described later in Art. 4.14.

An example analysis of a plane truss structure appears in the next article, and a computer program for the analysis of plane trusses is presented in Art. 5.7.

4.12 Example. The plane truss shown in Fig. 4-23a is to be analyzed using the methods described in the previous article. The truss is restrained at points C and D by hinge supports which prevent translations in both the x and y directions. The loads on the truss consist of both joint loads and member loads.

An arbitrary system for numbering the members and joints is given in Fig. 4-23b, which shows the restrained structure. Member numbers are indicated in

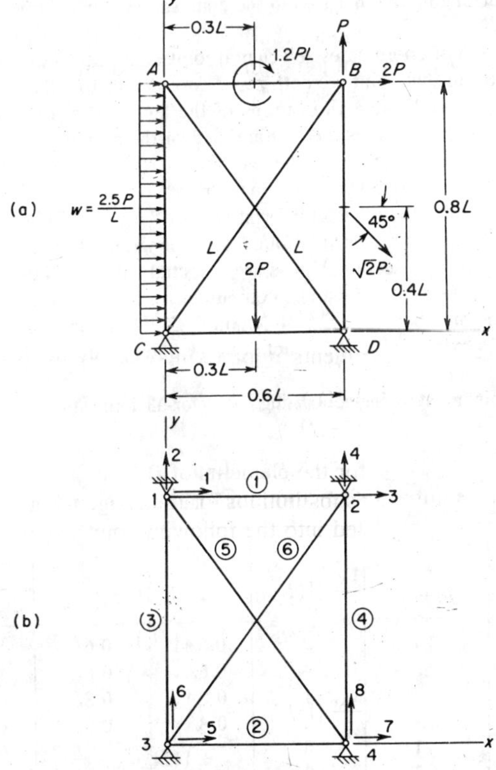

Fig. 4-23. Example (plane truss).

Table 4-18

Joint Information for Truss of Fig. 4-23a

Joint	Coordinates		Restraint List	
	x	y	x	y
1	0	$0.8L$	0	0
2	$0.6L$	$0.8L$	0	0
3	0	0	1	1
4	$0.6L$	0	1	1

circles adjacent to the members, and joint numbers are indicated by numbers adjacent to the joints. The numbering system for joint displacements is represented by numbers adjacent to the arrows which denote the positive directions of the possible displacements. Of course, the displacement numbering system derives from the joint numbering system according to Eqs. (4-27). Note that the system shown in Fig. 4-23b for numbering the joints (and hence the displacements) obviates the necessity of rearranging the matrices in the analysis. However, this will not be the case in general.

With the origin of coordinates selected at joint 3, as shown in Fig. 4-23a, the x and y coordinates of that point are both zero. Coordinates for all of the joints in the structure are given in Table 4-18 in terms of the length L of one of the diagonal members. Table 4-18 also gives the restraint list for the structure. As in the continuous beam example, the integer 1 in the restraint list indicates that a restraint exists, and the presence of a zero indicates that a restraint does not exist.

Table 4-19 contains the member information for the truss of Fig. 4-23a (note that the cross-sectional areas are proportional to the lengths). The member numbers, joint numbers, and cross-sectional areas are essential data, but the lengths of members and their direction cosines can be computed from the coordinates of the joints at the ends of the members (see Eqs. 4-24 and 4-25). Note that the arbitrary choice of which end of a member is to be denoted j or k determines the signs of the direction cosines.

As a preliminary step for generating the over-all joint stiffness matrix $\mathbf{S}_J$, the

Table 4-19

Member Information for Truss of Fig. 4-23a

Member	Joint j	Joint k	Area	Length	Direction Cosines	
					C_X	C_Y
1	1	2	$0.6A_X$	$0.6L$	1.0	0
2	3	4	$0.6A_X$	$0.6L$	1.0	0
3	3	1	$0.8A_X$	$0.8L$	0	1.0
4	4	2	$0.8A_X$	$0.8L$	0	1.0
5	1	4	A_X	L	0.6	−0.8
6	3	2	A_X	L	0.6	0.8

Table 4-20

Member Stiffness Matrices for Structure Axes

$$\mathbf{S}_{MS1} = \frac{EA_X}{L}\begin{bmatrix} 1 & 0 & -1 & 0 \\ 0 & 0 & 0 & 0 \\ -1 & 0 & 1 & 0 \\ 0 & 0 & 0 & 0 \end{bmatrix}\begin{matrix} 1 \\ 2 \\ 3 \\ 4 \end{matrix} \qquad \mathbf{S}_{MS2} = \frac{EA_X}{L}\begin{bmatrix} 1 & 0 & \boxed{-1} & 0 \\ 0 & 0 & 0 & 0 \\ -1 & 0 & 1 & 0 \\ 0 & 0 & 0 & 0 \end{bmatrix}\begin{matrix} \boxed{5} \\ 6 \\ 7 \\ 8 \end{matrix}$$
$$\begin{matrix} 1 & 2 & 3 & 4 \end{matrix} \qquad\qquad \begin{matrix} 5 & 6 & \boxed{7} & 8 \end{matrix}$$

$$\mathbf{S}_{MS3} = \frac{EA_X}{L}\begin{bmatrix} 0 & 0 & 0 & 0 \\ 0 & 1 & 0 & -1 \\ 0 & 0 & 0 & 0 \\ 0 & -1 & 0 & 1 \end{bmatrix}\begin{matrix} 5 \\ 6 \\ 1 \\ 2 \end{matrix} \qquad \mathbf{S}_{MS4} = \frac{EA_X}{L}\begin{bmatrix} 0 & 0 & 0 & 0 \\ 0 & 1 & 0 & -1 \\ 0 & 0 & 0 & 0 \\ 0 & -1 & 0 & 1 \end{bmatrix}\begin{matrix} 7 \\ 8 \\ 3 \\ 4 \end{matrix}$$
$$\begin{matrix} 5 & 6 & 1 & 2 \end{matrix} \qquad\qquad \begin{matrix} 7 & 8 & 3 & 4 \end{matrix}$$

$$\mathbf{S}_{MS5} = \frac{EA_X}{L}\begin{bmatrix} 0.36 & -0.48 & -0.36 & 0.48 \\ -0.48 & 0.64 & 0.48 & -0.64 \\ -0.36 & 0.48 & 0.36 & -0.48 \\ 0.48 & -0.64 & -0.48 & 0.64 \end{bmatrix}\begin{matrix} 1 \\ 2 \\ 7 \\ 8 \end{matrix}$$
$$\begin{matrix} 1 & 2 & 7 & 8 \end{matrix}$$

$$\mathbf{S}_{MS6} = \frac{EA_X}{L}\begin{bmatrix} 0.36 & 0.48 & -0.36 & -0.48 \\ 0.48 & 0.64 & -0.48 & -0.64 \\ -0.36 & -0.48 & 0.36 & 0.48 \\ -0.48 & -0.64 & 0.48 & 0.64 \end{bmatrix}\begin{matrix} 5 \\ 6 \\ 3 \\ 4 \end{matrix}$$
$$\begin{matrix} 5 & 6 & 3 & 4 \end{matrix}$$

individual member stiffness matrices $\mathbf{S}_{MSi}$ are summarized in Table 4-20. These matrices are formed from Table 4-15 given in Art. 4.10. The elements of each of these matrices may be transferred to the appropriate locations in $\mathbf{S}_J$ by calculating the indexes $j1$, $j2$, $k1$, and $k2$ given by Eqs. (4-27) and then making the transfers according to Eqs. (4-28) through (4-31). As an aid in this process, the numerical values of the indexes $j1$, $j2$, $k1$, and $k2$ are given in Table 4-20 down the right-hand side and across the bottom of each matrix $\mathbf{S}_{MSi}$. As an example of the transferring process, consider the element $(S_{MS13})_2$, which is the element in the first row and third column of the matrix $\mathbf{S}_{MSi}$ for member 2. This element is encircled in Table 4-20, and its value is $-EA_X/L$. The row and column indexes of $\mathbf{S}_J$ for the element are also encircled in Table 4-20. They indicate that the element is to be transferred to the position in the fifth row and the seventh column of the matrix $\mathbf{S}_J$. This element is encircled in Table 4-21, which shows the appearance of the matrix $\mathbf{S}_J$ after the

Table 4-21

Joint Stiffness Matrix for Truss of Fig. 4-23a

$$\mathbf{S}_J = \frac{EA_X}{L}\left[\begin{array}{cccc|cccc} 1.36 & -0.48 & -1.00 & 0 & 0 & 0 & -0.36 & 0.48 \\ -0.48 & 1.64 & 0 & 0 & 0 & -1.00 & 0.48 & -0.64 \\ -1.00 & 0 & 1.36 & 0.48 & -0.36 & -0.48 & 0 & 0 \\ 0 & 0 & 0.48 & 1.64 & -0.48 & -0.64 & 0 & -1.00 \\ \hline 0 & 0 & -0.36 & -0.48 & 1.36 & 0.48 & \boxed{-1.00} & 0 \\ 0 & -1.00 & -0.48 & -0.64 & 0.48 & 1.64 & 0 & 0 \\ -0.36 & 0.48 & 0 & 0 & -1.00 & 0 & 1.36 & -0.48 \\ 0.48 & -0.64 & 0 & -1.00 & 0 & 0 & -0.48 & 1.64 \end{array}\right]$$

Table 4-22
Inverse of Stiffness Matrix

$$\mathbf{S}_{FF}^{-1} = \frac{L}{EA_X}\begin{bmatrix} 2.503 & 0.733 & 2.053 & -0.601 \\ 0.733 & 0.824 & 0.601 & -0.176 \\ 2.053 & 0.601 & 2.503 & -0.733 \\ -0.601 & -0.176 & -0.733 & 0.824 \end{bmatrix}$$

transferring process is completed. The matrix is partitioned in accordance with Eq. (4-8), and the 4 × 4 stiffness matrix $\mathbf{S}_{FF}$ appears automatically in the upper left-hand portion with no necessity for rearrangement in this example. The inverse of the matrix $\mathbf{S}_{FF}$ is shown in Table 4-22 for use in subsequent calculations.

The loads applied to the structure appear in Fig. 4-23a. Those applied at joints are listed in Table 4-23, and the actions at the ends of the members in the restrained structure (caused by loads on members) are summarized in Table 4-24. The joint loads are placed in the vector $\mathbf{A}_J$ (see Eq. 4-32):

$$\mathbf{A}_J = P\{0, 0, 2, 1, 0, 0, 0, 0\}$$

Actions in the rows of Table 4-24 form columns of the matrix $\mathbf{A}_{ML}$. Next, the elements of $\mathbf{A}_{ML}$ are transferred to the vector of equivalent joint loads $\mathbf{A}_E$, as indicated by Eqs. (4-33). The resulting vector is as follows:

$$\mathbf{A}_E = P\{1.0, 2.0, 0.5, -2.5, 1.0, -1.0, 0.5, -1.5\}$$

To understand how the terms in vector $\mathbf{A}_E$ are obtained, consider the loaded members one at a time. As an example, member 4 has its j end at joint 4 and its k end at joint 2. Consequently, its direction cosines are $C_X = 0$ and $C_Y = 1.0$ (see Table 4-19). For this member Eqs. (4-33) yield

$$\begin{aligned} (A_E)_7 &= -\Sigma A_{MS} + (A_{ML})_{2,4} \\ (A_E)_8 &= -\Sigma A_{MS} - (A_{ML})_{1,4} \\ (A_E)_3 &= -\Sigma A_{MS} + (A_{ML})_{4,4} \\ (A_E)_4 &= -\Sigma A_{MS} - (A_{ML})_{3,4} \end{aligned}$$

Thus, the contributions of member 4 to $\mathbf{A}_E$ are

$$\mathbf{A}_{E4} = P\{0, 0, 0.5, -0.5, 0, 0, 0.5, -0.5\}$$

and so on for the other loaded members.

Then the vectors $\mathbf{A}_J$ and $\mathbf{A}_E$ are added (see Eq. 4-7) to form the combined load vector $\mathbf{A}_C$:

Table 4-23
Actions Applied at Joints

Joint	*Force in x Direction*	*Force in y Direction*
1	0	0
2	$2P$	P
3	0	0
4	0	0

Table 4-24

Actions at Ends of Restrained Members Due to Loads

Member	$(A_{ML})_{1,i}$	$(A_{ML})_{2,i}$	$(A_{ML})_{3,i}$	$(A_{ML})_{4,i}$
1	0	$-2P$	0	$2P$
2	0	P	0	P
3	0	P	0	P
4	$0.5P$	$0.5P$	$0.5P$	$0.5P$
5	0	0	0	0
6	0	0	0	0

$$\mathbf{A}_C = P\{1.0, 2.0, 2.5, -1.5, 1.0, -1.0, 0.5, -1.5\}$$

As mentioned earlier, there is no need to rearrange this vector. The first four elements constitute the vector

$$\mathbf{A}_{FC} = P\{1.0, 2.0, 2.5, -1.5\}$$

and the last four elements are

$$\mathbf{A}_{RC} = P\{1.0, -1.0, 0.5, -1.5\}$$

The solution may be completed by first calculating the free joint displacements $\mathbf{D}_F$, using Eq. (4-3), with the following result:

$$\mathbf{D}_F = \mathbf{S}_{FF}^{-1}\mathbf{A}_{FC} = \frac{PL}{EA_X}\{10.001, 4.147, 10.611, -4.020\}$$

The vector $\mathbf{D}_J$ for this structure contains the vector $\mathbf{D}_F$ in the first part and zeros in the latter part:

$$\mathbf{D}_J = \frac{PL}{EA_X}\{10.001, 4.147, 10.611, -4.020, 0, 0, 0, 0\}$$

In the next step the support reactions $\mathbf{A}_R$ are computed using Eq. (4-4) with the matrix $\mathbf{S}_{RF}$ obtained from the lower left-hand portion of Table 4-21:

$$\mathbf{A}_R = -\mathbf{A}_{RC} + \mathbf{S}_{RF}\mathbf{D}_F = P\{-2.89, -5.67, -2.11, 7.67\}$$

As the last step in the analysis, the member end-actions $\mathbf{A}_{Mi}$ are obtained for each member by applying either Eq. (4-5) or Eqs. (4-34). The results of these calculations are summarized in Table 4-25. This step completes the analysis of the plane truss structure for the given loads.

Table 4-25

Final Member End-Actions

Member	$(A_M)_{1,i}$	$(A_M)_{2,i}$	$(A_M)_{3,i}$	$(A_M)_{4,i}$
1	$-0.610P$	$-2.000P$	$0.610P$	$2.000P$
2	0.0	$1.000P$	0.0	$1.000P$
3	$-4.147P$	$1.000P$	$4.147P$	$1.000P$
4	$4.520P$	$0.500P$	$-3.520P$	$0.500P$
5	$2.683P$	0.0	$-2.683P$	0.0
6	$-3.150P$	0.0	$3.150P$	0.0

4.13 Rotation of Axes in Two Dimensions. As mentioned previously in Art. 4.10, the direct method of formulating member stiffnesses is satisfactory for continuous beams and trusses. On the other hand, a method employing rotation of axes is well suited to more complex structures. In this article rotation of axes for two-dimensional vectors is formulated on a geometric basis, and in the next article the method is applied in the analysis of plane trusses. Then in Art. 4.15 the subject of rotation of axes in three dimensions is discussed, and in subsequent articles other types of structures are treated using this technique.

In order to begin the discussion, consider an action A, which lies in the x-y plane (see Fig. 4-24). Two sets of orthogonal axes with origin at 0 are shown in the figure. The x_S, y_S axes will later be taken parallel to a set of structure reference axes, and the x_M, y_M set will be taken as a pair of member-oriented axes. The x_M, y_M axes are rotated from the x_S, y_S axes by the angle γ. Let the direction cosines of the x_M axis with respect to the axes x_S and y_S be λ_{11} and λ_{12}, respectively. It is evident from Fig. 4-24 that these direction cosines may be expressed in terms of the angle γ as follows:

$$\lambda_{11} = \cos\gamma \qquad \lambda_{12} = \cos(90° - \gamma) = \sin\gamma \tag{a}$$

Also, let the direction cosines of the y_M axis with respect to the axes x_S and y_S be λ_{21} and λ_{22}, respectively. These direction cosines may also be expressed in terms of the angle γ:

$$\lambda_{21} = \cos(90° + \gamma) = -\sin\gamma \qquad \lambda_{22} = \cos\gamma \tag{b}$$

For any one of the direction cosines above, the first subscript refers to the x_M, y_M axes, and the second subscript refers to the x_S, y_S axes. Moreover, the number 1 denotes the x direction (either x_M or x_S), and the number 2 denotes the y direction (either y_M or y_S). For example, λ_{12} is the direction cosine of the x_M axis with respect to the y_S axis.

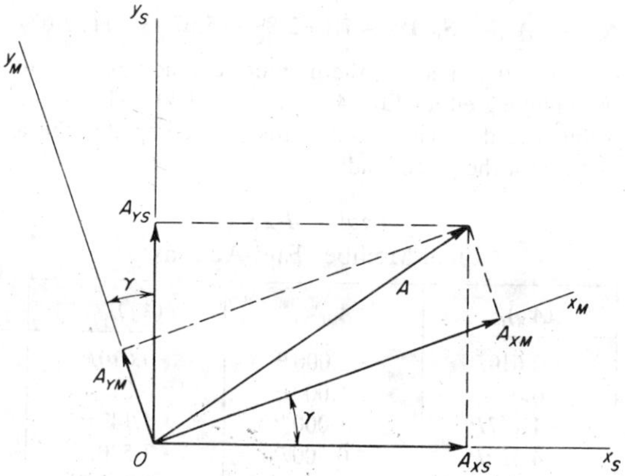

Fig. 4-24. Rotation of axes in two dimensions.

The action A may be resolved into two orthogonal components A_{XS} and A_{YS} in the x_S and y_S directions, respectively, as shown in Fig. 4-24. Alternatively, A may also be resolved into two orthogonal components A_{XM} and A_{YM} in the directions x_M and y_M, respectively, as shown also in the figure. The latter set of components may be expressed in terms of the former set by inspection of the geometry in Fig. 4-24. It may be observed that A_{XM} is equal to the sum of the projections of A_{XS} and A_{YS} on the x_M axis. In addition, A_{YM} is equal to the sum of the projections of A_{XS} and A_{YS} on the y_M axis. Therefore, the expressions for A_{XM} and A_{YM} are:

$$\begin{aligned} A_{XM} &= \lambda_{11}A_{XS} + \lambda_{12}A_{YS} \\ A_{YM} &= \lambda_{21}A_{XS} + \lambda_{22}A_{YS} \end{aligned} \tag{c}$$

In matrix form, these equations become:

$$\begin{bmatrix} A_{XM} \\ A_{YM} \end{bmatrix} = \begin{bmatrix} \lambda_{11} & \lambda_{12} \\ \lambda_{21} & \lambda_{22} \end{bmatrix} \begin{bmatrix} A_{XS} \\ A_{YS} \end{bmatrix} \tag{4-35}$$

Substitution of expressions (a) and (b) into Eq. (4-35) yields the following alternative form:

$$\begin{bmatrix} A_{XM} \\ A_{YM} \end{bmatrix} = \begin{bmatrix} \cos\gamma & \sin\gamma \\ -\sin\gamma & \cos\gamma \end{bmatrix} \begin{bmatrix} A_{XS} \\ A_{YS} \end{bmatrix} \tag{4-36}$$

Equations (4-35) and (4-36) may also be stated in concise form by the expression

$$\mathbf{A}_\mathrm{M} = \mathbf{R}\mathbf{A}_\mathrm{S} \tag{4-37}$$

In this equation, $\mathbf{A}_\mathrm{M}$ is a vector consisting of the components of the action A parallel to the x_M, y_M axes, $\mathbf{A}_S$ is a vector containing the components of the action A parallel to the x_S, y_S axes, and $\mathbf{R}$ is a matrix of direction cosines which will be referred to as the *rotation matrix*. As shown by Eqs. (4-35) and (4-36), the rotation matrix in a two-dimensional problem is as follows:

$$\mathbf{R} = \begin{bmatrix} \lambda_{11} & \lambda_{12} \\ \lambda_{21} & \lambda_{22} \end{bmatrix} = \begin{bmatrix} \cos\gamma & \sin\gamma \\ -\sin\gamma & \cos\gamma \end{bmatrix} \tag{4-38}$$

It is also possible to express the x_S, y_S set of components of the action A in terms of the x_M, y_M set of components. This transformation may be accomplished by observing that A_{XS} is equal to the sum of the projections of A_{XM} and A_{YM} on the x_S axis, and that A_{YS} is equal to the sum of the projections of A_{XM} and A_{YM} on the y_S axis. Thus,

$$\begin{aligned} A_{XS} &= \lambda_{11}A_{XM} + \lambda_{21}A_{YM} \\ A_{YS} &= \lambda_{12}A_{XM} + \lambda_{22}A_{YM} \end{aligned} \tag{d}$$

When expressed in matrix form, these equations become:

$$\begin{bmatrix} A_{XS} \\ A_{YS} \end{bmatrix} = \begin{bmatrix} \lambda_{11} & \lambda_{21} \\ \lambda_{12} & \lambda_{22} \end{bmatrix} \begin{bmatrix} A_{XM} \\ A_{YM} \end{bmatrix} \tag{4-39}$$

or

$$\begin{bmatrix} A_{XS} \\ A_{YS} \end{bmatrix} = \begin{bmatrix} \cos\gamma & -\sin\gamma \\ \sin\gamma & \cos\gamma \end{bmatrix} \begin{bmatrix} A_{XM} \\ A_{YM} \end{bmatrix} \tag{4-40}$$

Equations (4-39) and (4-40) may be represented concisely by the matrix equation

$$\mathbf{A}_S = \mathbf{R}^T \mathbf{A}_M \tag{4-41}$$

in which $\mathbf{R}^T$ is the transpose of the rotation matrix $\mathbf{R}$.

Finally, from Eqs. (4-37) and (4-41) it is apparent that the transpose of $\mathbf{R}$ is equal to its inverse:

$$\mathbf{R}^T = \mathbf{R}^{-1} \tag{4-42}$$

Therefore, the rotation matrix $\mathbf{R}$ is an orthogonal matrix.*

Since small displacements as well as actions may be treated as vectors, the relationships formulated above for the action A may be applied equally well to a displacement D. Thus, equations similar to Eqs. (4-37) and (4-41) can be written for displacements, as follows:

$$\mathbf{D}_M = \mathbf{R}\mathbf{D}_S \tag{4-43}$$

$$\mathbf{D}_S = \mathbf{R}^T \mathbf{D}_M \tag{4-44}$$

In these equations the vector $\mathbf{D}_M$ consists of the components of the displacement D parallel to the member axes, and the vector $\mathbf{D}_S$ contains the components of the displacement D parallel to the structure axes.

The concepts of rotation of axes discussed above are applied to plane truss structures in the next article.

4.14 Application to Plane Truss Members. A typical plane truss member i, framing into joints j and k at the ends, is shown in Fig. 4-25. For

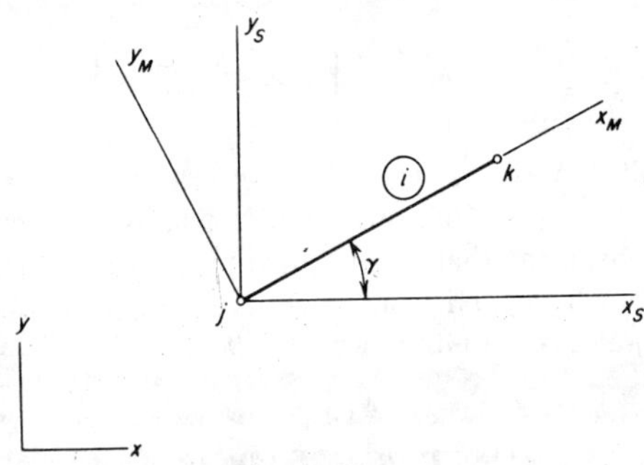

Fig. 4-25. Rotation of axes for plane truss member

*For further discussion of orthogonal matrices, see any text on matrix algebra; for example, *Matrix Algebra for Engineers*, by the authors, D. Van Nostrand Co., Inc., Princeton, N.J., 1965.

convenience, the member is oriented in such a manner that its direction cosines are positive. The member axes x_M, y_M are rotated through the angle γ from the axes x_S, y_S, which are parallel to the reference axes x, y for the structure as a whole. For the purpose of rotational transformations it is immaterial whether one refers to the x_S, y_S set of axes or the parallel x, y set.

As a preliminary step in the use of rotation of axes in a plane truss analysis, the rotation matrix **R** will be expressed in terms of the direction cosines of the member i. Inspection of Fig. 4-25 shows that the direction cosines of the member in terms of the angle γ are:

$$C_X = \cos\gamma \qquad C_Y = \sin\gamma$$

Then the rotation matrix (see Eq. 4-38) becomes:

$$\mathbf{R} = \begin{bmatrix} \cos\gamma & \sin\gamma \\ -\sin\gamma & \cos\gamma \end{bmatrix} = \begin{bmatrix} C_X & C_Y \\ -C_Y & C_X \end{bmatrix} \tag{4-45}$$

Next, the action-displacement relationships in the x_M and y_M directions at the ends of member i may be expressed by the following:

$$\mathbf{A}_{\mathrm{M}i} = \mathbf{S}_{\mathrm{M}i}\mathbf{D}_{\mathrm{M}i} \tag{4-46}$$

Equation (4-46) is the same as Eq. (4-5) when the vector $\mathbf{A}_{\mathrm{ML}i}$, which appears in the latter equation, is null. Thus, Eq. (4-46) gives the actions at the ends of the member due to the displacements of the ends. The member stiffness matrix $\mathbf{S}_{\mathrm{M}i}$ (see Eq. 4-46) was developed in Art. 4.10 and presented in Table 4-14. The objective now is to transform this matrix for member axes into the matrix $\mathbf{S}_{\mathrm{MS}i}$ for structure axes (see Table 4-15).

As a preliminary step in transforming to structure axes, Eq. (4-46) may be written in partitioned form (omitting the subscript i), as follows:

$$\begin{bmatrix} A_{M1} \\ A_{M2} \\ \hline A_{M3} \\ A_{M4} \end{bmatrix} = \left[\begin{array}{cc|cc} S_{M11} & S_{M12} & S_{M13} & S_{M14} \\ S_{M21} & S_{M22} & S_{M23} & S_{M24} \\ \hline S_{M31} & S_{M32} & S_{M33} & S_{M34} \\ S_{M41} & S_{M42} & S_{M43} & S_{M44} \end{array}\right] \begin{bmatrix} D_{M1} \\ D_{M2} \\ \hline D_{M3} \\ D_{M4} \end{bmatrix} \tag{a}$$

The subscripts 1, 2, 3, and 4 used in this equation refer to the member-oriented directions shown in Fig. 4-14c. Equation (a) may also be written in the following manner:

$$\begin{bmatrix} \mathbf{A}_{\mathrm{M}j} \\ \mathbf{A}_{\mathrm{M}k} \end{bmatrix} = \begin{bmatrix} \mathbf{S}_{\mathrm{M}jj} & \mathbf{S}_{\mathrm{M}jk} \\ \mathbf{S}_{\mathrm{M}kj} & \mathbf{S}_{\mathrm{M}kk} \end{bmatrix} \begin{bmatrix} \mathbf{D}_{\mathrm{M}j} \\ \mathbf{D}_{\mathrm{M}k} \end{bmatrix} \tag{b}$$

In this equation the subscripts j and k attached to the submatrices refer to the j and k ends of the member. The terms $\mathbf{A}_{\mathrm{M}j}$, $\mathbf{A}_{\mathrm{M}k}$, $\mathbf{D}_{\mathrm{M}j}$, and $\mathbf{D}_{\mathrm{M}k}$ in Eq. (b) represent two-dimensional vectors (either actions or displacements) at the ends of the member in the directions of member axes (see Fig. 4-14c). Therefore, they can be expressed with respect to the structure axes (see Fig. 4-14b) by using the appropriate rotation formulas from the preceding article. These formulas are Eqs. (4-37) and (4-43). When these relations are substituted into Eq. (b), it becomes

$$\begin{bmatrix} \mathbf{RA}_{Sj} \\ \mathbf{RA}_{Sk} \end{bmatrix} = \begin{bmatrix} \mathbf{S}_{Mjj} & \mathbf{S}_{Mjk} \\ \mathbf{S}_{Mkj} & \mathbf{S}_{Mkk} \end{bmatrix} \begin{bmatrix} \mathbf{RD}_{Sj} \\ \mathbf{RD}_{Sk} \end{bmatrix} \tag{c}$$

The submatrices $\mathbf{A}_{Sj}$, $\mathbf{A}_{Sk}$, $\mathbf{D}_{Sj}$, and $\mathbf{D}_{Sk}$ represent the two-dimensional action and displacement vectors at the ends of the member with respect to the structure axes.

An equivalent form of Eq. (c) is the following:

$$\begin{bmatrix} \mathbf{R} & \mathbf{O} \\ \mathbf{O} & \mathbf{R} \end{bmatrix} \begin{bmatrix} \mathbf{A}_{Sj} \\ \mathbf{A}_{Sk} \end{bmatrix} = \begin{bmatrix} \mathbf{S}_{Mjj} & \mathbf{S}_{Mjk} \\ \mathbf{S}_{Mkj} & \mathbf{S}_{Mkk} \end{bmatrix} \begin{bmatrix} \mathbf{R} & \mathbf{O} \\ \mathbf{O} & \mathbf{R} \end{bmatrix} \begin{bmatrix} \mathbf{D}_{Sj} \\ \mathbf{D}_{Sk} \end{bmatrix} \tag{d}$$

To simplify the writing of this equation, let $\mathbf{R}_T$ be the *rotation transformation matrix* for actions and displacements at both ends of the member:

$$\mathbf{R}_T = \begin{bmatrix} \mathbf{R} & \mathbf{O} \\ \mathbf{O} & \mathbf{R} \end{bmatrix} \tag{4-47}$$

Equation (d) then may be rewritten as follows:

$$\mathbf{R}_T \mathbf{A}_S = \mathbf{S}_M \mathbf{R}_T \mathbf{D}_S \tag{e}$$

The vectors $\mathbf{A}_S$ and $\mathbf{D}_S$ in Eq. (e) consist of the actions and displacements at the ends of the member in the directions of structure axes (see Fig. 4-14b).

Premultiplying both sides of Eq. (e) by the inverse of $\mathbf{R}_T$ gives:

$$\mathbf{A}_S = \mathbf{R}_T^{-1} \mathbf{S}_M \mathbf{R}_T \mathbf{D}_S \tag{f}$$

Since the submatrix $\mathbf{R}$ is orthogonal, the matrix $\mathbf{R}_T$ is also orthogonal. This fact can be seen by multiplying $\mathbf{R}_T$ by its transpose, as follows:

$$\mathbf{R}_T^T \mathbf{R}_T = \begin{bmatrix} \mathbf{R}^T & \mathbf{O} \\ \mathbf{O} & \mathbf{R}^T \end{bmatrix} \begin{bmatrix} \mathbf{R} & \mathbf{O} \\ \mathbf{O} & \mathbf{R} \end{bmatrix} = \begin{bmatrix} \mathbf{R}^T\mathbf{R} & \mathbf{O} \\ \mathbf{O} & \mathbf{R}^T\mathbf{R} \end{bmatrix}$$

$$= \begin{bmatrix} \mathbf{R}^{-1}\mathbf{R} & \mathbf{O} \\ \mathbf{O} & \mathbf{R}^{-1}\mathbf{R} \end{bmatrix} = \begin{bmatrix} \mathbf{I} & \mathbf{O} \\ \mathbf{O} & \mathbf{I} \end{bmatrix} = \mathbf{I}$$

Hence, the transpose of $\mathbf{R}_T$ is also the inverse of $\mathbf{R}_T$:

$$\mathbf{R}_T^{-1} = \mathbf{R}_T^T \tag{4-48}$$

and substitution of Eq. (4-48) into Eq. (f) yields:

$$\mathbf{A}_S = \mathbf{R}_T^T \mathbf{S}_M \mathbf{R}_T \mathbf{D}_S \tag{4-49}$$

Since the action-displacement equation that relates the actions $\mathbf{A}_S$ and displacements $\mathbf{D}_S$ is

$$\mathbf{A}_S = \mathbf{S}_{MS} \mathbf{D}_S \tag{4-50}$$

in which $\mathbf{S}_{MS}$ is the member stiffness matrix for structure axes, it is readily seen by comparing Eqs. (4-49) and (4-50) that

$$\mathbf{S}_{\mathrm{MS}} = \mathbf{R}_{\mathrm{T}}^{\mathrm{T}}\mathbf{S}_{\mathrm{M}}\mathbf{R}_{\mathrm{T}} \tag{4-51}$$

Evaluation of the matrix $\mathbf{S}_{\mathrm{MS}}$ from this equation is performed as follows:

$$\mathbf{S}_{\mathrm{MS}} = \left[\begin{array}{cc|cc} C_X & -C_Y & 0 & 0 \\ C_Y & C_X & 0 & 0 \\ \hline 0 & 0 & C_X & -C_Y \\ 0 & 0 & C_Y & C_X \end{array}\right]$$

$$\times \left[\begin{array}{cc|cc} \dfrac{EA_X}{L} & 0 & \dfrac{-EA_X}{L} & 0 \\ 0 & 0 & 0 & 0 \\ \hline \dfrac{-EA_X}{L} & 0 & \dfrac{EA_X}{L} & 0 \\ 0 & 0 & 0 & 0 \end{array}\right] \left[\begin{array}{cc|cc} C_X & C_Y & 0 & 0 \\ -C_Y & C_X & 0 & 0 \\ \hline 0 & 0 & C_X & C_Y \\ 0 & 0 & -C_Y & C_X \end{array}\right]$$

When this matrix multiplication is executed, the result is the matrix $\mathbf{S}_{\mathrm{MS}}$ (see Table 4-15) which was previously obtained by the method of direct formulation.

In addition to the transformation of the member stiffness matrix from member axes to structure axes, the rotation of axes concept can also be used for other purposes in the stiffness method of analysis. One important application arises in the construction of the equivalent load vector $\mathbf{A}_{\mathrm{E}}$ from elements of the matrix $\mathbf{A}_{\mathrm{ML}}$. Contributions to the former array (the elements of which are in the directions of structure axes) from the latter array (which has elements in the directions of member axes) may be obtained by the following transformation:

$$\mathbf{A}_{\mathrm{MS}\,i} = \mathbf{R}_{\mathrm{T}i}^{\mathrm{T}}\mathbf{A}_{\mathrm{ML}\,i} \tag{4-52}$$

In this expression the vector $\mathbf{A}_{\mathrm{MS}\,i}$ represents fixed-end actions in the directions of structure axes, whereas the vector $\mathbf{A}_{\mathrm{ML}\,i}$ is the i-th column of the matrix $\mathbf{A}_{\mathrm{ML}}$. In expanded form, Eq. (4-52) becomes

$$\begin{bmatrix} (A_{MS})_{1,i} \\ (A_{MS})_{2,i} \\ (A_{MS})_{3,i} \\ (A_{MS})_{4,i} \end{bmatrix} = \begin{bmatrix} C_{Xi} & -C_{Yi} & 0 & 0 \\ C_{Yi} & C_{Xi} & 0 & 0 \\ 0 & 0 & C_{Xi} & -C_{Yi} \\ 0 & 0 & C_{Yi} & C_{Xi} \end{bmatrix} \begin{bmatrix} (A_{ML})_{1,i} \\ (A_{ML})_{2,i} \\ (A_{ML})_{3,i} \\ (A_{ML})_{4,i} \end{bmatrix}$$

$$= \begin{bmatrix} C_{Xi}(A_{ML})_{1,i} - C_{Yi}(A_{ML})_{2,i} \\ C_{Yi}(A_{ML})_{1,i} + C_{Xi}(A_{ML})_{2,i} \\ C_{Xi}(A_{ML})_{3,i} - C_{Yi}(A_{ML})_{4,i} \\ C_{Yi}(A_{ML})_{3,i} + C_{Xi}(A_{ML})_{4,i} \end{bmatrix}$$

The resulting terms in $\mathbf{A}_{\mathrm{MS}\,i}$, with the signs reversed, represent the incremental portions of $\mathbf{A}_{\mathrm{E}}$ given previously in Eqs. (4-33).

Another significant application of the rotation of axes concept appears in the computation of final member end-actions. This computation consists of the superposition of the initial actions in member i and the effects of joint displacements. This superposition procedure is expressed by Eq. (4-5), which is repeated below:

$$\mathbf{A}_{\mathrm{M}i} = \mathbf{A}_{\mathrm{ML}i} + \mathbf{S}_{\mathrm{M}i}\mathbf{D}_{\mathrm{M}i} \tag{4-5}$$

repeated

The vector $\mathbf{D}_{\mathrm{M}i}$ in this equation must be ascertained from the vector of joint displacements $\mathbf{D}_{\mathrm{J}}$. The latter displacements are in the directions of structure axes, but the former displacements are in the directions of member axes. Therefore, the vector $\mathbf{D}_{\mathrm{M}i}$ may be obtained by the following transformation:

$$\mathbf{D}_{\mathrm{M}i} = \mathbf{R}_{\mathrm{T}i}\mathbf{D}_{\mathrm{J}i} \tag{4-53}$$

in which $\mathbf{D}_{\mathrm{J}i}$ is the vector of joint displacements for the ends of member i. Substitution of Eq. (4-53) into Eq. (4-5) produces the following expression:

$$\mathbf{A}_{\mathrm{M}i} = \mathbf{A}_{\mathrm{ML}i} + \mathbf{S}_{\mathrm{M}i}\mathbf{R}_{\mathrm{T}i}\mathbf{D}_{\mathrm{J}i} \tag{4-54}$$

Equation (4-54) may also be written in the expanded form:

$$\begin{bmatrix} (A_M)_{1,i} \\ (A_M)_{2,i} \\ (A_M)_{3,i} \\ (A_M)_{4,i} \end{bmatrix} = \begin{bmatrix} (A_{ML})_{1,i} \\ (A_{ML})_{2,i} \\ (A_{ML})_{3,i} \\ (A_{ML})_{4,i} \end{bmatrix}$$

$$+ \frac{EA_{Xi}}{L_i}\begin{bmatrix} 1 & 0 & -1 & 0 \\ 0 & 0 & 0 & 0 \\ -1 & 0 & 1 & 0 \\ 0 & 0 & 0 & 0 \end{bmatrix}\begin{bmatrix} C_{Xi} & C_{Yi} & 0 & 0 \\ -C_{Yi} & C_{Xi} & 0 & 0 \\ 0 & 0 & C_{Xi} & C_{Yi} \\ 0 & 0 & -C_{Yi} & C_{Xi} \end{bmatrix}\begin{bmatrix} (D_J)_{j1} \\ (D_J)_{j2} \\ (D_J)_{k1} \\ (D_J)_{k2} \end{bmatrix}$$

In this expression the subscripts $j1$, $j2$, $k1$, and $k2$ carry the definitions given previously by Eqs. (4-27). When the matrix multiplications indicated above are performed, the resulting four equations are the same as Eqs. (4-34) obtained in Art. 4.11 by the method of direct formulation.

In summary, the concept of rotation of axes has several useful applications in the stiffness method of analysis. The member stiffness matrix for members axes $\mathbf{S}_{\mathrm{M}i}$ may be transformed into the member stiffness matrix for structure axes $\mathbf{S}_{\mathrm{MS}i}$ by means of Eq. (4-51). In addition, the contributions $\mathbf{A}_{\mathrm{MS}i}$ to the equivalent load vector $\mathbf{A}_{\mathrm{E}}$ from a given member may be evaluated conveniently by rotation of axes, as shown by Eq. (4-52). Finally, the member end-actions can be obtained by the rotation of axes formulation given by Eq. (4-54).

It will be seen later that the matrix equations given above for rotation of axes in plane trusses can be generalized and applied to more complicated types of structures.

4.15 Rotation of Axes in Three Dimensions. Consider the action A shown in three dimensions in Fig. 4-26. The two sets of orthogonal axes x_S, y_S, z_S and x_M, y_M, z_M are analogous to the two sets of axes in the two-

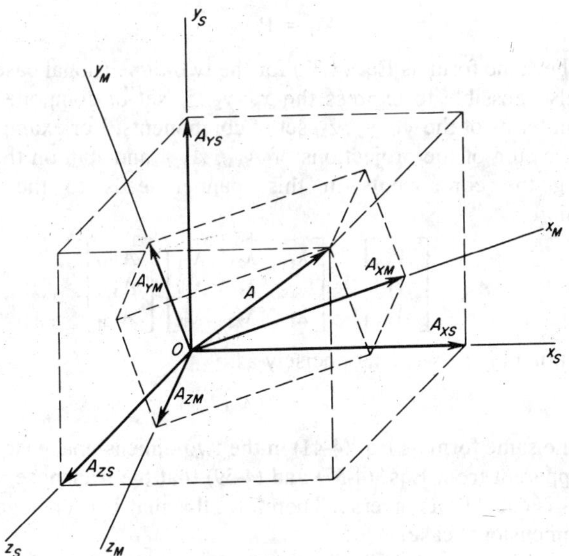

Fig. 4-26. Rotation of axes in three dimensions.

dimensional case in Fig. 4-24. Let the direction cosines of the x_M axis with respect to the x_S, y_S, z_S axes be λ_{11}, λ_{12}, and λ_{13}. They are defined to be the cosines of the angles between the axis x_M and the three axes x_S, y_S, and z_S, respectively. Also, let the direction cosines for the y_M axis be λ_{21}, λ_{22}, and λ_{23}, and let those for the z_M axis be λ_{31}, λ_{32}, and λ_{33}.

The action A may be represented by a set of three orthogonal components A_{XS}, A_{YS}, and A_{ZS} in the x_S, y_S, and z_S directions, respectively, as shown in Fig. 4-26. Alternatively, this action may be represented by a second set of components A_{XM}, A_{YM}, and A_{ZM} in the x_M, y_M, and z_M directions; these are also indicated in the figure. The latter components may be related to the former as in the two-dimensional case. For example, A_{XM} is equal to the sum of the projections of A_{XS}, A_{YS}, and A_{ZS} on the x_M axis. The components A_{YM} and A_{ZM} can be expressed in a similar manner, and the following relationship results:

$$\mathbf{A_M} = \begin{bmatrix} A_{XM} \\ A_{YM} \\ A_{ZM} \end{bmatrix} = \begin{bmatrix} \lambda_{11} & \lambda_{12} & \lambda_{13} \\ \lambda_{21} & \lambda_{22} & \lambda_{23} \\ \lambda_{31} & \lambda_{32} & \lambda_{33} \end{bmatrix} \begin{bmatrix} A_{XS} \\ A_{YS} \\ A_{ZS} \end{bmatrix} \tag{4-55}$$

If a three-dimensional rotation matrix $\mathbf{R}$ is defined as follows,

$$\mathbf{R} = \begin{bmatrix} \lambda_{11} & \lambda_{12} & \lambda_{13} \\ \lambda_{21} & \lambda_{22} & \lambda_{23} \\ \lambda_{31} & \lambda_{32} & \lambda_{33} \end{bmatrix} \tag{4-56}$$

then Eq. (4-55) can be written in the form

$$\mathbf{A}_{\mathrm{M}} = \mathbf{R}\mathbf{A}_{\mathrm{S}} \tag{4-57}$$

which is the same form as Eq. (4-37) for the two-dimensional case.

It is also possible to express the x_S, y_S, z_S set of components of the action A in terms of the x_M, y_M, z_M set of components. For example, A_{XS} is equal to the sum of the projections of A_{XM}, A_{YM}, and A_{ZM} on the x_S axis. Expressing the components in this manner leads to the following relationship:

$$\mathbf{A}_{\mathrm{S}} = \begin{bmatrix} A_{XS} \\ A_{YS} \\ A_{ZS} \end{bmatrix} = \begin{bmatrix} \lambda_{11} & \lambda_{21} & \lambda_{31} \\ \lambda_{12} & \lambda_{22} & \lambda_{32} \\ \lambda_{13} & \lambda_{23} & \lambda_{33} \end{bmatrix} \begin{bmatrix} A_{XM} \\ A_{YM} \\ A_{ZM} \end{bmatrix} \tag{4-58}$$

This equation may be written concisely as:

$$\mathbf{A}_{\mathrm{S}} = \mathbf{R}^{\mathrm{T}}\mathbf{A}_{\mathrm{M}} \tag{4-59}$$

which is the same form as Eq. (4-41) in the two-dimensional case.

It is apparent from Eqs. (4-57) and (4-59) that the transpose of the 3 × 3 matrix **R** is equal to its inverse. Therefore, this matrix is orthogonal, as in the two-dimensional case.

Relationships corresponding to the above equations, which are derived for actions, hold also for the transformation of small displacements in three dimensions. Thus, the components $\mathbf{D}_{\mathrm{M}}$ of a displacement D in the x_M, y_M, z_M directions may be expressed in terms of the components $\mathbf{D}_{\mathrm{S}}$ in the x_S, y_S, z_S directions as follows:

$$\mathbf{D}_{\mathrm{M}} = \mathbf{R}\mathbf{D}_{\mathrm{S}} \tag{4-60}$$

Also, the vector $\mathbf{D}_{\mathrm{S}}$ may be expressed in terms of the vector $\mathbf{D}_{\mathrm{M}}$ by the relationship:

$$\mathbf{D}_{\mathrm{S}} = \mathbf{R}^{\mathrm{T}}\mathbf{D}_{\mathrm{M}} \tag{4-61}$$

Equations (4-60) and (4-61) are of the same form as Eqs. (4-43) and (4-44) for the two-dimensional case. Thus, the relationships for rotation of axes in three dimensions are completely analogous to those for two dimensions, except that the matrix **R** is of order 3 × 3 instead of 2 × 2, and the vectors are of order 3 × 1 instead of 2 × 1.

4.16 Plane Frame Member Stiffnesses. In preparation for the analysis of plane frames, the member stiffness matrix for a typical plane frame member is developed in this article. The matrix is first formulated with respect to member axes and then transformed to structure axes by the method of rotation of axes.

Figure 4-27a shows a typical member i within a plane frame. The joints at the ends of the member are denoted j and k as in previous structures. The orthogonal set of axes x, y, and z shown in Fig. 4-27 are reference axes for the structure. The plane frame lies in the x-y plane, which is assumed to be a principal plane of bending for all of the members. The members of

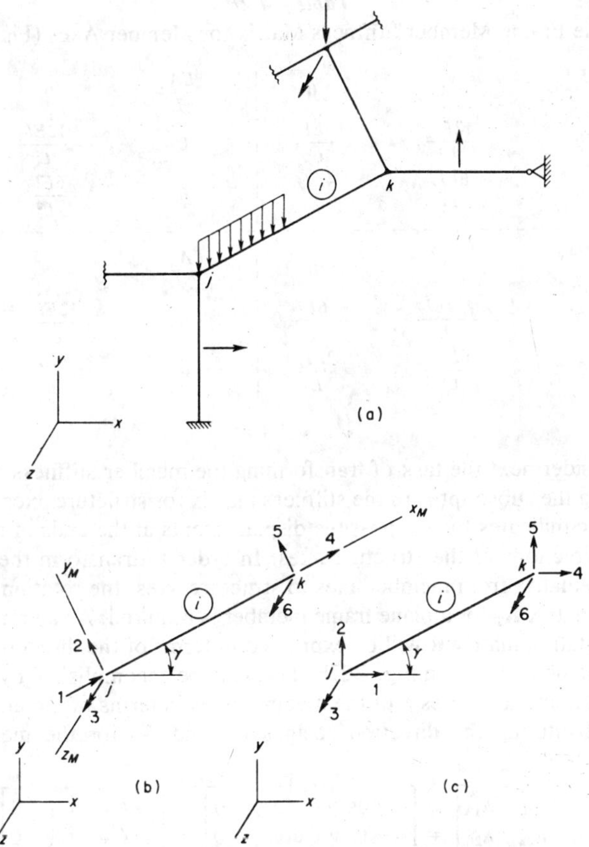

Fig. 4-27. Numbering system for a plane frame member.

the frame are assumed to be rigidly connected, and the significant displacements of the joints consist of translations in the x-y plane and rotations in the z sense.

The possible displacements of the ends of a typical member i are indicated in Fig. 4-27b for member-oriented axes x_M, y_M, and z_M. The member axes are rotated from the structure axes about the z_M axis through the angle γ. The six end-displacements, shown in their positive senses, consist of translations in the x_M and y_M directions and a rotation in the z_M (or z) sense at the ends j and k, respectively. If unit displacements of these types are induced at each end of the member one at a time, the resulting restraint actions will constitute the elements of the member stiffness matrix $\mathbf{S}_{Mi}$ for member axes. These restraint actions may be drawn from cases (1), (2), (6), (7), (8), and (12) of Fig. 4-2 in Art. 4.3. The resulting 6×6 member stiffness matrix for member axes is given in Table 4-26.

Table 4-26

Plane Frame Member Stiffness Matrix for Member Axes (Fig. 4-27b)

$$
\mathbf{S}_{Mi} = \left[\begin{array}{ccc|ccc}
\frac{EA_X}{L} & 0 & 0 & -\frac{EA_X}{L} & 0 & 0 \\
0 & \frac{12EI_Z}{L^3} & \frac{6EI_Z}{L^2} & 0 & -\frac{12EI_Z}{L^3} & \frac{6EI_Z}{L^2} \\
0 & \frac{6EI_Z}{L^2} & \frac{4EI_Z}{L} & 0 & -\frac{6EI_Z}{L^2} & \frac{2EI_Z}{L} \\
\hline
-\frac{EA_X}{L} & 0 & 0 & \frac{EA_X}{L} & 0 & 0 \\
0 & -\frac{12EI_Z}{L^3} & -\frac{6EI_Z}{L^2} & 0 & \frac{12EI_Z}{L^3} & -\frac{6EI_Z}{L^2} \\
0 & \frac{6EI_Z}{L^2} & \frac{2EI_Z}{L} & 0 & -\frac{6EI_Z}{L^2} & \frac{4EI_Z}{L}
\end{array}\right]
$$

Consider next the task of transforming the member stiffness matrix $\mathbf{S}_M$ (omitting the subscript i) to the stiffness matrix for structure axes $\mathbf{S}_{MS}$. Figure 4-27c indicates the six possible displacements at the ends of member i in the directions of the structure axes. In order to transform the member stiffness matrix from member axes to structure axes, the rotation transformation matrix $\mathbf{R}_T$ for a plane frame member is required. As a first step, the 3×3 rotation matrix $\mathbf{R}$ will be expressed in terms of the direction cosines of the member shown in Fig. 4-27b. This may be accomplished by expressing the direction cosines λ of the member axes in terms of the angle γ and then substituting the direction cosines C_X and C_Y for the member, as follows:

$$
\mathbf{R} = \begin{bmatrix} \lambda_{11} & \lambda_{12} & \lambda_{13} \\ \lambda_{21} & \lambda_{22} & \lambda_{23} \\ \lambda_{31} & \lambda_{32} & \lambda_{33} \end{bmatrix} = \begin{bmatrix} \cos\gamma & \sin\gamma & 0 \\ -\sin\gamma & \cos\gamma & 0 \\ 0 & 0 & 1 \end{bmatrix} = \begin{bmatrix} C_X & C_Y & 0 \\ -C_Y & C_X & 0 \\ 0 & 0 & 1 \end{bmatrix} \tag{4-62}
$$

The rotation transformation matrix $\mathbf{R}_T$ for a plane frame member can be shown to take the same form as Eq. (4-47):

$$
\mathbf{R}_T = \begin{bmatrix} \mathbf{R} & \mathbf{O} \\ \mathbf{O} & \mathbf{R} \end{bmatrix} \tag{4-63}
$$

In Eq. (4-63) the submatrix $\mathbf{R}$ is the 3×3 rotation matrix given by Eq. (4-62).

Having the rotation transformation matrix on hand, one may then calculate the member stiffness matrix for structure axes using the type of operation shown previously by Eq. (4-51):

$$
\mathbf{S}_{MS} = \mathbf{R}_T^T \mathbf{S}_M \mathbf{R}_T \tag{4-64}
$$

in which the matrix $\mathbf{R}_T$ is given by Eq. (4-63). The member stiffness matrix which results from this transformation is presented in Table 4-27. This matrix will be used in the analysis of plane frames in the next article.

Table 4-27

Plane Frame Member Stiffness Matrix for Structure Axes (Fig. 4-27c)

$$
\mathbf{S}_{MSi} = \left[\begin{array}{ccc|ccc}
\frac{EA_X}{L}C_X^2 + \frac{12EI_Z}{L^3}C_Y^2 & \left(\frac{EA_X}{L} - \frac{12EI_Z}{L^3}\right)C_XC_Y & -\frac{6EI_Z}{L^2}C_Y & -\left(\frac{EA_X}{L}C_X^2 + \frac{12EI_Z}{L^3}C_Y^2\right) & -\left(\frac{EA_X}{L} - \frac{12EI_Z}{L^3}\right)C_XC_Y & -\frac{6EI_Z}{L^2}C_Y \\
\left(\frac{EA_X}{L} - \frac{12EI_Z}{L^3}\right)C_XC_Y & \frac{EA_X}{L}C_Y^2 + \frac{12EI_Z}{L^3}C_X^2 & \frac{6EI_Z}{L^2}C_X & -\left(\frac{EA_X}{L} - \frac{12EI_Z}{L^3}\right)C_XC_Y & -\left(\frac{EA_X}{L}C_Y^2 + \frac{12EI_Z}{L^3}C_X^2\right) & \frac{6EI_Z}{L^2}C_X \\
-\frac{6EI_Z}{L^2}C_Y & \frac{6EI_Z}{L^2}C_X & \frac{4EI_Z}{L} & \frac{6EI_Z}{L^2}C_Y & -\frac{6EI_Z}{L^2}C_X & \frac{2EI_Z}{L} \\
\hline
-\left(\frac{EA_X}{L}C_X^2 + \frac{12EI_Z}{L^3}C_Y^2\right) & -\left(\frac{EA_X}{L} - \frac{12EI_Z}{L^3}\right)C_XC_Y & \frac{6EI_Z}{L^2}C_Y & \frac{EA_X}{L}C_X^2 + \frac{12EI_Z}{L^3}C_Y^2 & \left(\frac{EA_X}{L} - \frac{12EI_Z}{L^3}\right)C_XC_Y & \frac{6EI_Z}{L^2}C_Y \\
-\left(\frac{EA_X}{L} - \frac{12EI_Z}{L^3}\right)C_XC_Y & -\left(\frac{EA_X}{L}C_Y^2 + \frac{12EI_Z}{L^3}C_X^2\right) & -\frac{6EI_Z}{L^2}C_X & \left(\frac{EA_X}{L} - \frac{12EI_Z}{L^3}\right)C_XC_Y & \frac{EA_X}{L}C_Y^2 + \frac{12EI_Z}{L^3}C_X^2 & -\frac{6EI_Z}{L^2}C_X \\
-\frac{6EI_Z}{L^2}C_Y & \frac{6EI_Z}{L^2}C_X & \frac{2EI_Z}{L} & \frac{6EI_Z}{L^2}C_Y & -\frac{6EI_Z}{L^2}C_X & \frac{4EI_Z}{L}
\end{array}\right]
$$

4.17 Analysis of Plane Frames. In this article a procedure is given for analyzing plane frames of the type illustrated in Fig. 4-27a. Actions applied to such a frame are assumed to be forces in the plane of the structure (the x-y plane) or moment vectors normal to the plane.

As a preliminary step in the analysis, the members and joints of the structure must be numbered. The numbering techniques described in Art. 4.11 for plane trusses may also be applied to plane frames. The joints are numbered consecutively 1 through n_j, and the members are numbered consecutively 1 through m. The sequence of numbering is arbitrary, but each member and each joint must have a number.

Since both axial* and flexural deformations will be taken into account in the analyses of plane frames, the possibility exists for three independent displacements at each joint. These displacements are taken to be the translations of the joint in the x and y directions and the rotation in the z sense. Thus, the possible displacements at a joint j may be designated as follows:

$$
\begin{aligned}
3j - 2 &= \text{index for translation in the } x \text{ direction} \\
3j - 1 &= \text{index for translation in the } y \text{ direction} \\
3j &= \text{index for rotation in the } z \text{ sense}
\end{aligned}
$$

In addition, the number of degrees of freedom n in a plane frame is calculated from the number of joints n_j and the number of restraints n_r by the following expression:

$$n = 3n_j - n_r \tag{4-65}$$

A particular member i in a plane frame will have joint numbers j and k at its ends, as shown in Fig. 4-28. Indexes for the possible displacements of the joints associated with this member are also shown in Fig. 4-28. These displacement indexes are calculated as follows:

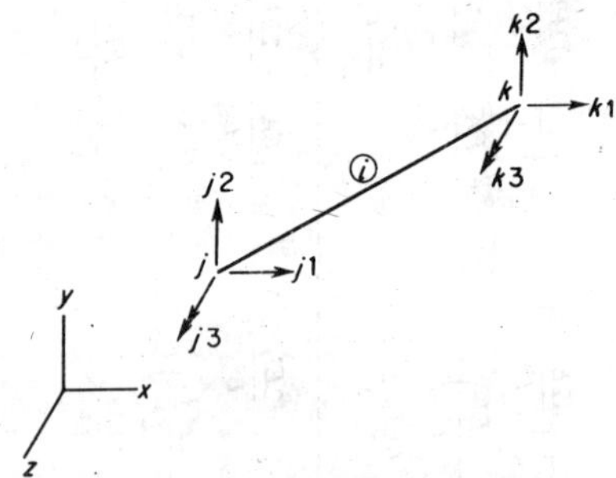

Fig. 4-28. End-displacements for plane frame member.

*If axial deformations in members are to be omitted from the analysis, *constraints* must be introduced to prevent length changes of members. See Greene, B. E., "Application of Generalized Constraints in the Stiffness Method of Structural Analysis", *AIAA Jour.*, Vol. 4, No. 9, Sept. 1966, pp. 1531–1537.

$$j1 = 3j - 2 \qquad j2 = 3j - 1 \qquad j3 = 3j$$
$$k1 = 3k - 2 \qquad k2 = 3k - 1 \qquad k3 = 3k \tag{4-66}$$

In order to construct the joint stiffness matrix in an orderly fashion, the following procedure is recommended. First, the 6×6 stiffness matrix $\mathbf{S}_{\mathrm{MS}\,i}$ for structure axes is generated for the i-th member in the frame (see Art. 4.16, Table 4-27). Member i contributes to the stiffnesses of joints j and k at the ends of the member. Therefore, appropriate elements from the matrix $\mathbf{S}_{\mathrm{MS}\,i}$ for this member may be transferred to the over-all joint stiffness matrix $\mathbf{S}_{\mathrm{J}}$ through an organized handling of subscripts. The first column in the matrix $\mathbf{S}_{\mathrm{MS}\,i}$ consists of restraint actions at j and k due to a unit translation at the j end of member i in the x direction (index $j1$). This column is transferred to the matrix $\mathbf{S}_{\mathrm{J}}$ as follows:

$$\begin{aligned}
(S_J)_{j1,j1} &= \Sigma\, S_{MS} + (S_{MS11})_i \\
(S_J)_{j2,j1} &= \Sigma\, S_{MS} + (S_{MS21})_i \\
(S_J)_{j3,j1} &= \Sigma\, S_{MS} + (S_{MS31})_i \\
(S_J)_{k1,j1} &= (S_{MS41})_i \\
(S_J)_{k2,j1} &= (S_{MS51})_i \\
(S_J)_{k3,j1} &= (S_{MS61})_i
\end{aligned} \tag{4-67}$$

In these equations the first three stiffness coefficients consist of the sums of contributions from all members which frame into joint j, including member i. The last three stiffnesses involve contributions from member i only.

Expressions similar to Eqs. (4-67) may also be written for a unit translation of joint j in the y direction (index $j2$):

$$\begin{aligned}
(S_J)_{j1,j2} &= \Sigma\, S_{MS} + (S_{MS12})_i \\
\cdots \quad &\quad \cdots \qquad \cdots \\
(S_J)_{k3,j2} &= (S_{MS62})_i
\end{aligned} \tag{4-68}$$

Thus, the elements of the second column of $\mathbf{S}_{\mathrm{MS}}$ for member i are transferred as contributions to the matrix $\mathbf{S}_{\mathrm{J}}$.

Similarly, for a unit rotation of joint j in the z sense (index $j3$), the expressions for transferring the third column of $\mathbf{S}_{\mathrm{MS}}$ are:

$$\begin{aligned}
(S_J)_{j1,j3} &= \Sigma\, S_{MS} + (S_{MS13})_i \\
\cdots \quad &\quad \cdots \qquad \cdots \\
(S_J)_{k3,j3} &= (S_{MS63})_i
\end{aligned} \tag{4-69}$$

Expressions for transferring the fourth column of $\mathbf{S}_{\mathrm{MS}}$ to the matrix $\mathbf{S}_{\mathrm{J}}$ are similar to the above equations, except that the first three stiffnesses consist of contributions from member i only and the last three are sums of contributions from all members framing into joint k. Thus, for a unit translation of joint k in the x direction (index $k1$), the expressions are as follows:

$$\begin{aligned}
(S_J)_{j1,k1} &= (S_{MS14})_i \\
\cdots \quad &\quad \cdots \qquad \cdots \\
(S_J)_{k3,k1} &= \Sigma\, S_{MS} + (S_{MS64})_i
\end{aligned} \tag{4-70}$$

Similarly, for a unit translation of joint k in the y direction (index $k2$), the expressions for transferring the fifth column of $\mathbf{S}_{MS}$ are:

$$\begin{aligned}(S_J)_{j1,k2} &= \qquad (S_{MS15})_i \\ \cdots \quad & \cdots \qquad \cdots \\ (S_J)_{k3,k2} &= \Sigma S_{MS} + (S_{MS65})_i\end{aligned} \tag{4-71}$$

Finally, for a unit rotation of joint k in the z sense (index $k3$), the following equations apply:

$$\begin{aligned}(S_J)_{j1,k3} &= \qquad (S_{MS16})_i \\ \cdots \quad & \cdots \qquad \cdots \\ (S_J)_{k3,k3} &= \Sigma S_{MS} + (S_{MS66})_i\end{aligned} \tag{4-72}$$

In the process of transferring elements from the member stiffness matrix $\mathbf{S}_{MS}$ to the over-all joint stiffness matrix $\mathbf{S}_J$, as described above, there is no attempt to take advantage of the symmetry which exists in the matrices. Recognition of this symmetry allows short cuts in the transfer procedure which can readily be made.

Construction of the complete matrix $\mathbf{S}_J$ consists of generating and transferring the matrix $\mathbf{S}_{MS}$ for all members (1 through m) of the structure. After the matrix $\mathbf{S}_J$ is generated, it must be rearranged if necessary into the form given by Eq. (4-8).

In the next phase of the analysis, vectors associated with loads on the frame are formed. External actions applied at joints constitute the vector $\mathbf{A}_J$. Figure 4-29 shows the actions at a typical joint k in a plane frame. The action $(A_J)_{3k-2}$ is the x component of the force applied at k, $(A_J)_{3k-1}$ is the y component of the applied force, and $(A_J)_{3k}$ represents a couple in the z sense applied at the joint. Thus, the vector $\mathbf{A}_J$ will take the following form:

$$\begin{aligned}\mathbf{A}_J = \{&(A_J)_1, (A_J)_2, (A_J)_3, \ldots, (A_J)_{3k-2}, (A_J)_{3k-1}, (A_J)_{3k}, \ldots, \\ &(A_J)_{3nj-2}, (A_J)_{3nj-1}, (A_J)_{3nj}\}\end{aligned} \tag{4-73}$$

Figure 4-30b delineates the actions at the ends of a restrained plane frame member due to loads. End-actions for member i, with respect to member-oriented axes, are defined as follows:

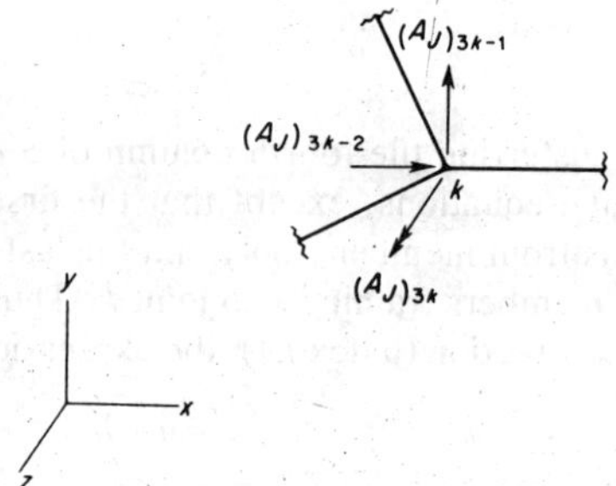

Fig. 4-29. Joint loads for a plane frame.

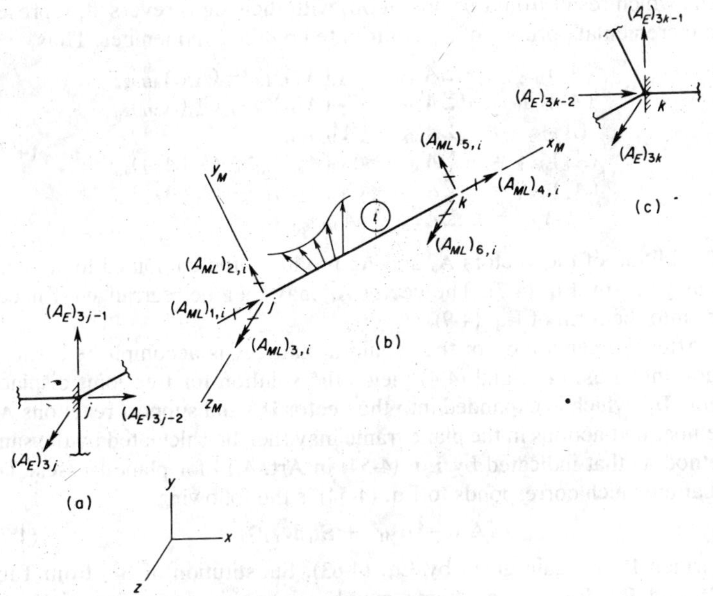

Fig. 4-30. Loads on a plane frame member.

$(A_{ML})_{1,i}$ = force in the x_M direction at the j end
$(A_{ML})_{2,i}$ = force in the y_M direction at the j end
$(A_{ML})_{3,i}$ = moment in the z_M sense at the j end
$(A_{ML})_{4,i}$ = force in the x_M direction at the k end
$(A_{ML})_{5,i}$ = force in the y_M direction at the k end
$(A_{ML})_{6,i}$ = moment in the z_M sense at the k end

These end-actions may be obtained from Appendix B for particular loading conditions. The matrix $\mathbf{A}_{ML}$ is an array of order $6 \times m$, in which each column consists of the elements listed above for a given member; thus:

$$\mathbf{A}_{ML} = \begin{bmatrix} (A_{ML})_{1,1} & \cdots & (A_{ML})_{1,i} & \cdots & (A_{ML})_{1,m} \\ (A_{ML})_{2,1} & \cdots & (A_{ML})_{2,i} & \cdots & (A_{ML})_{2,m} \\ \cdots & \cdots & \cdots & \cdots & \cdots \\ (A_{ML})_{6,1} & \cdots & (A_{ML})_{6,i} & \cdots & (A_{ML})_{6,m} \end{bmatrix} \tag{4-74}$$

The construction of the equivalent load vector $\mathbf{A}_E$ may be executed by the method of rotation of axes. This method was demonstrated previously for plane trusses in Art. 4.14 (see Eq. 4-52). Figures 4-30a and c show the equivalent loads at j and k which receive contributions from member i. Fixed-end actions $\mathbf{A}_{MSi}$, in the directions of structure axes, can be computed using the relationship

$$\mathbf{A}_{MSi} = \mathbf{R}_{Ti}^{T}\mathbf{A}_{MLi} \tag{4-75}$$

in which $\mathbf{R}_{Ti}^{T}$ is the transpose of the matrix given by Eq. (4-63). The expressions which result from this operation, with their signs reversed, represent the incremental portions of $\mathbf{A}_\mathrm{E}$ contributed by the i-th member. Thus,

$$\begin{aligned}
(A_E)_{3j-2} &= -\Sigma A_{MS} - C_{Xi}(A_{ML})_{1,i} + C_{Yi}(A_{ML})_{2,i} \\
(A_E)_{3j-1} &= -\Sigma A_{MS} - C_{Yi}(A_{ML})_{1,i} - C_{Xi}(A_{ML})_{2,i} \\
(A_E)_{3j} &= -\Sigma A_{MS} - (A_{ML})_{3,i} \\
(A_E)_{3k-2} &= -\Sigma A_{MS} - C_{Xi}(A_{ML})_{4,i} + C_{Yi}(A_{ML})_{5,i} \\
(A_E)_{3k-1} &= -\Sigma A_{MS} - C_{Yi}(A_{ML})_{4,i} - C_{Xi}(A_{ML})_{5,i} \\
(A_E)_{3k} &= -\Sigma A_{MS} - (A_{ML})_{6,i}
\end{aligned} \tag{4-76}$$

Addition of the vectors $\mathbf{A}_\mathrm{J}$ and $\mathbf{A}_\mathrm{E}$ produces the combined load vector $\mathbf{A}_\mathrm{C}$ as given by Eq. (4-7). The vector $\mathbf{A}_\mathrm{C}$ may then be rearranged if necessary into the form of Eq. (4-9).

After the generation of the required matrices is accomplished, substitution into Eqs. (4-3) and (4-4) yields the solution for free joint displacements $\mathbf{D}_\mathrm{F}$ (which is expanded into the vector $\mathbf{D}_\mathrm{J}$) and support reactions $\mathbf{A}_\mathrm{R}$. Member end-actions in the plane frame may then be calculated by the same method as that indicated by Eq. (4-54) in Art. 4.14 for plane trusses. The equation which corresponds to Eq. (4-54) is the following:

$$\mathbf{A}_{\mathrm{M}i} = \mathbf{A}_{\mathrm{ML}i} + \mathbf{S}_{\mathrm{M}i}\mathbf{R}_{\mathrm{T}i}\mathbf{D}_{\mathrm{J}i} \tag{4-77}$$

in which $\mathbf{R}_\mathrm{T}$ is again given by Eq. (4-63). Substitution of $\mathbf{S}_{\mathrm{M}i}$ from Table 4-26 and $\mathbf{R}_{\mathrm{T}i}$ for a plane frame member into this expression yields the following:

$$\begin{aligned}
(A_M)_{1,i} &= (A_{ML})_{1,i} + \frac{EA_{Xi}}{L_i}\{[(D_J)_{j1} - (D_J)_{k1}]C_{Xi} \\
&\quad + [(D_J)_{j2} - (D_J)_{k2}]C_{Yi}\} \\
(A_M)_{2,i} &= (A_{ML})_{2,i} - \frac{12EI_{Zi}}{L_i^3}\{[(D_J)_{j1} - (D_J)_{k1}]C_{Yi} \\
&\quad - [(D_J)_{j2} - (D_J)_{k2}]C_{Xi}\} + \frac{6EI_{Zi}}{L_i^2}[(D_J)_{j3} + (D_J)_{k3}] \\
(A_M)_{3,i} &= (A_{ML})_{3,i} + \frac{6EI_{Zi}}{L_i^2}\{-[(D_J)_{j1} - (D_J)_{k1}]C_{Yi} \\
&\quad + [(D_J)_{j2} - (D_J)_{k2}]C_{Xi}\} + \frac{4EI_{Zi}}{L_i}\left[(D_J)_{j3} + \frac{1}{2}(D_J)_{k3}\right] \\
(A_M)_{4,i} &= (A_{ML})_{4,i} - \frac{EA_{Xi}}{L_i}\{[(D_J)_{j1} - (D_J)_{k1}]C_{Xi} \\
&\quad + [(D_J)_{j2} - (D_J)_{k2}]C_{Yi}\} \\
(A_M)_{5,i} &= (A_{ML})_{5,i} + \frac{12EI_{Zi}}{L_i^3}\{[(D_J)_{j1} - (D_J)_{k1}]C_{Yi} \\
&\quad - [(D_J)_{j2} - (D_J)_{k2}]C_{Xi}\} - \frac{6EI_{Zi}}{L_i^2}[(D_J)_{j3} + (D_J)_{k3}] \\
(A_M)_{6,i} &= (A_{ML})_{6,i} + \frac{6EI_{Zi}}{L_i^2}\{-[(D_J)_{j1} - (D_J)_{k1}]C_{Yi} \\
&\quad + [(D_J)_{j2} - (D_J)_{k2}]C_{X}\} + \frac{4EI_{Zi}}{L_i}\left[\frac{1}{2}(D_J)_{j3} + (D_J)_{k3}\right]
\end{aligned} \tag{4-78}$$

In Eqs. (4-78) the subscripts $j1$, $j2$, $j3$, $k1$, $k2$, and $k3$ carry the meanings defined previously in Eqs. (4-66).

The next article contains an example analysis of a plane frame by the method described above, and a computer program for plane frames appears in Art. 5.8.

4.18 Example. Figure 4-31a shows a plane frame having two members, three joints, six restraints, and three degrees of freedom. This frame is to be analyzed by the methods given in the previous article. For this purpose, assume that the cross-sectional area A_X and the moment of inertia I_Z are constant throughout the structure. Assume also that the parameters in the problem have the following numerical values:

$$E = 10{,}000 \text{ ksi} \quad L = 100 \text{ in.} \quad I_Z = 1000 \text{ in.}^4 \quad P = 10 \text{ kips} \quad A_X = 10 \text{ in.}^2$$

Units of kips, inches, and radians are used throughout the analysis.*

A numbering system for members, joints, and displacements is given in Fig. 4-31b, which shows the restrained structure. The sequence for numbering the joints is selected in such a manner that the matrices in the analysis do not require rearrangement.

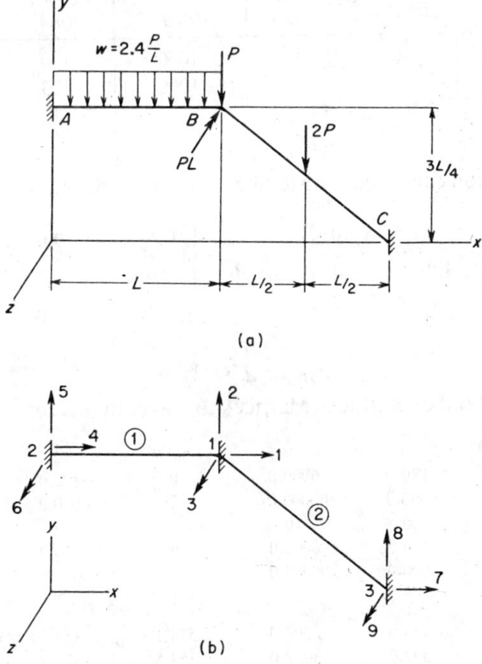

Fig. 4-31. Example (plane frame).

*Units for this example are in the United States (US) customary system. However, about half of the problems for Art. 4.18 (given at the end of the chapter) are in the international (SI) system of units.

Table 4-28
Joint Information for Frame of Fig. 4-31

Joint	Coordinates (in.)		Restraint List		
	x	*y*	*x*	*y*	*z*
1	100	75	0	0	0
2	0	75	1	1	1
3	200	0	1	1	1

Joint information is summarized in Table 4-28, which contains the joint numbers, joint coordinates, and the conditions of restraint. The member information for the frame is presented in Table 4-29. Substitution of the direction cosines for the members into Eq. (4-62) results in the rotation matrices $\mathbf{R}_1$ and $\mathbf{R}_2$ shown in Table 4-30.

Table 4-29
Member Information for Frame of Fig. 4-31

Member	Joint *j*	Joint *k*	Area (in.^{2})	Moment of Inertia (in.^{4})	Length (in.)	Direction Cosines	
						C_X	C_Y
1	2	1	10	1000	100	1.0	0
2	1	3	10	1000	125	0.8	−0.6

Table 4-30
Rotation Matrices for Members of Frame of Fig. 4-31

$$\mathbf{R}_1 = \begin{bmatrix} 1.0 & 0 & 0 \\ 0 & 1.0 & 0 \\ 0 & 0 & 1.0 \end{bmatrix} \qquad \mathbf{R}_2 = \begin{bmatrix} 0.8 & -0.6 & 0 \\ 0.6 & 0.8 & 0 \\ 0 & 0 & 1.0 \end{bmatrix}$$

Table 4-31
Member Stiffness Matrices for Structure Axes

$$\mathbf{S}_{MS1} = \begin{bmatrix} 1000.0 & 0 & 0 & -1000.0 & 0 & 0 \\ 0 & 120.0 & 6000.0 & 0 & -120.0 & 6000.0 \\ 0 & 6000.0 & 400000.0 & 0 & -6000.0 & 200000.0 \\ -1000.0 & 0 & 0 & 1000.0 & 0 & 0 \\ 0 & -120.0 & -6000.0 & 0 & 120.0 & -6000.0 \\ 0 & 6000.0 & 200000.0 & 0 & -6000.0 & 400000.0 \end{bmatrix} \begin{matrix} 4 \\ 5 \\ 6 \\ 1 \\ 2 \\ 3 \end{matrix}$$
$$\begin{matrix} 4 & 5 & 6 & 1 & 2 & 3 \end{matrix}$$

$$\mathbf{S}_{MS2} = \begin{bmatrix} 534.1 & -354.5 & 2304.0 & -534.1 & 354.5 & 2304.0 \\ -354.5 & 327.3 & 3072.0 & 354.5 & -327.3 & 3072.0 \\ 2304.0 & 3072.0 & 320000.0 & -2304.0 & -3072.0 & 160000.0 \\ -534.1 & 354.5 & -2304.0 & 534.1 & -354.5 & -2304.0 \\ 354.5 & -327.3 & -3072.0 & -354.5 & 327.3 & -3072.0 \\ 2304.0 & 3072.0 & 160000.0 & -2304.0 & -3072.0 & 320000.0 \end{bmatrix} \begin{matrix} 1 \\ 2 \\ 3 \\ 7 \\ 8 \\ 9 \end{matrix}$$
$$\begin{matrix} 1 & 2 & 3 & 7 & 8 & 9 \end{matrix}$$

In Eqs. (4-78) the subscripts $j1$, $j2$, $j3$, $k1$, $k2$, and $k3$ carry the meanings defined previously in Eqs. (4-66).

The next article contains an example analysis of a plane frame by the method described above, and a computer program for plane frames appears in Art. 5.8.

4.18 Example. Figure 4-31a shows a plane frame having two members, three joints, six restraints, and three degrees of freedom. This frame is to be analyzed by the methods given in the previous article. For this purpose, assume that the cross-sectional area A_X and the moment of inertia I_Z are constant throughout the structure. Assume also that the parameters in the problem have the following numerical values:

$$E = 10{,}000 \text{ ksi} \quad L = 100 \text{ in.} \quad I_Z = 1000 \text{ in.}^4 \quad P = 10 \text{ kips} \quad A_X = 10 \text{ in.}^2$$

Units of kips, inches, and radians are used throughout the analysis.*

A numbering system for members, joints, and displacements is given in Fig. 4-31b, which shows the restrained structure. The sequence for numbering the joints is selected in such a manner that the matrices in the analysis do not require rearrangement.

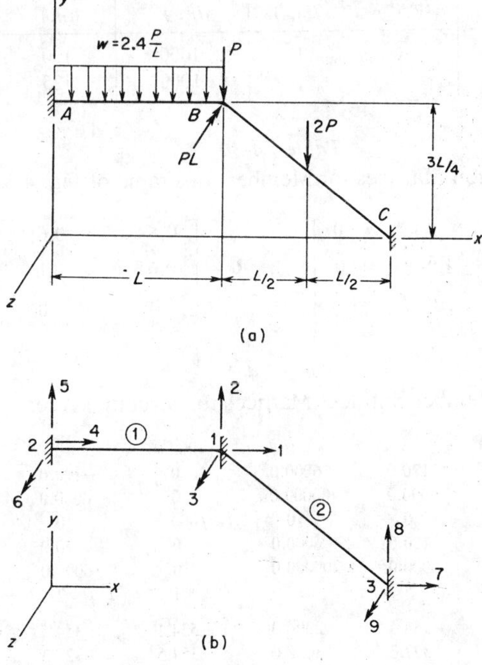

Fig. 4-31. Example (plane frame).

*Units for this example are in the United States (US) customary system. However, about half of the problems for Art. 4.18 (given at the end of the chapter) are in the international (SI) system of units.

Table 4-28
Joint Information for Frame of Fig. 4-31

Joint	Coordinates (in.)		Restraint List		
	x	y	x	y	z
1	100	75	0	0	0
2	0	75	1	1	1
3	200	0	1	1	1

Joint information is summarized in Table 4-28, which contains the joint numbers, joint coordinates, and the conditions of restraint. The member information for the frame is presented in Table 4-29. Substitution of the direction cosines for the members into Eq. (4-62) results in the rotation matrices $\mathbf{R}_1$ and $\mathbf{R}_2$ shown in Table 4-30.

Table 4-29
Member Information for Frame of Fig. 4-31

Member	Joint j	Joint k	Area (in.^{2})	Moment of Inertia (in.^{4})	Length (in.)	Direction Cosines	
						C_X	C_Y
1	2	1	10	1000	100	1.0	0
2	1	3	10	1000	125	0.8	−0.6

Table 4-30
Rotation Matrices for Members of Frame of Fig. 4-31

$$\mathbf{R}_1 = \begin{bmatrix} 1.0 & 0 & 0 \\ 0 & 1.0 & 0 \\ 0 & 0 & 1.0 \end{bmatrix} \qquad \mathbf{R}_2 = \begin{bmatrix} 0.8 & -0.6 & 0 \\ 0.6 & 0.8 & 0 \\ 0 & 0 & 1.0 \end{bmatrix}$$

Table 4-31
Member Stiffness Matrices for Structure Axes

$$\mathbf{S}_{MS1} = \left[\begin{array}{cccccc} 1000.0 & 0 & 0 & -1000.0 & 0 & 0 \\ 0 & 120.0 & 6000.0 & 0 & -120.0 & 6000.0 \\ 0 & 6000.0 & 400000.0 & 0 & -6000.0 & 200000.0 \\ -1000.0 & 0 & 0 & 1000.0 & 0 & 0 \\ 0 & -120.0 & -6000.0 & 0 & 120.0 & -6000.0 \\ 0 & 6000.0 & 200000.0 & 0 & -6000.0 & 400000.0 \end{array}\right] \begin{array}{c} 4 \\ 5 \\ 6 \\ 1 \\ 2 \\ 3 \end{array}$$
$$\begin{array}{cccccc} 4 & 5 & 6 & 1 & 2 & 3 \end{array}$$

$$\mathbf{S}_{MS2} = \left[\begin{array}{cccccc} 534.1 & -354.5 & 2304.0 & -534.1 & 354.5 & 2304.0 \\ -354.5 & 327.3 & 3072.0 & 354.5 & -327.3 & 3072.0 \\ 2304.0 & 3072.0 & 320000.0 & -2304.0 & -3072.0 & 160000.0 \\ -534.1 & 354.5 & -2304.0 & 534.1 & -354.5 & -2304.0 \\ 354.5 & -327.3 & -3072.0 & -354.5 & 327.3 & -3072.0 \\ 2304.0 & 3072.0 & 160000.0 & -2304.0 & -3072.0 & 320000.0 \end{array}\right] \begin{array}{c} 1 \\ 2 \\ 3 \\ 7 \\ 8 \\ 9 \end{array}$$
$$\begin{array}{cccccc} 1 & 2 & 3 & 7 & 8 & 9 \end{array}$$

Table 4-32

Joint Stiffness Matrix for Frame of Fig. 4-31

$$
\mathbf{S}_J = \left[\begin{array}{ccc|ccc|ccc}
1534.1 & -354.5 & 2304.0 & -1000.0 & 0 & 0 & -534.1 & 354.5 & 2304.0 \\
-354.5 & 447.3 & -2928.0 & 0 & -120.0 & -6000.0 & 354.5 & -327.3 & 3072.0 \\
2304.0 & -2928.0 & 720000.0 & 0 & 6000.0 & 200000.0 & -2304.0 & -3072.0 & 160000.0 \\
\hline
-1000.0 & 0 & 0 & 1000.0 & 0 & 0 & 0 & 0 & 0 \\
0 & -120.0 & 6000.0 & 0 & 120.0 & 6000.0 & 0 & 0 & 0 \\
0 & -6000.0 & 200000.0 & 0 & 6000.0 & 400000.0 & 0 & 0 & 0 \\
-534.1 & 354.5 & -2304.0 & 0 & 0 & 0 & 534.1 & -354.5 & -2304.0 \\
354.5 & -327.3 & -3072.0 & 0 & 0 & 0 & -354.5 & 327.3 & -3072.0 \\
2304.0 & 3072.0 & 160000.0 & 0 & 0 & 0 & -2304.0 & -3072.0 & 320000.0
\end{array}\right]
$$

In preparation for generating the over-all joint stiffness matrix $\mathbf{S_J}$, the member stiffness matrices are calculated. This may be done for each member by first generating the member stiffness matrix $\mathbf{S}_{\mathrm{M}i}$ for member axes (see Table 4-26, Art. 4.16) and then calculating the matrix $\mathbf{S}_{\mathrm{MS}i}$ for structure axes by the rotation transformation of Eq. (4-64). For this purpose the matrix $\mathbf{R}_{\mathrm{T}i}$ for each member is composed as shown by Eq. (4-63). Alternatively, the matrix $\mathbf{S}_{\mathrm{MS}i}$ may be calculated directly for each member using Table 4-27. The resulting matrices are given in Table 4-31. The indexes $j1$ through $k3$ (computed by Eqs. 4-66) are also indicated in Table 4-31 down the right-hand side and across the bottom of each matrix $\mathbf{S}_{\mathrm{MS}i}$. These indexes may be used as a guide for the purpose of transferring elements to the matrix $\mathbf{S_J}$. After the transferring process is accomplished, the over-all joint stiffness matrix which results is shown in Table 4-32. This matrix is partitioned in the usual manner, thereby isolating the 3 × 3 stiffness matrix $\mathbf{S_{FF}}$. The inverse of the matrix $\mathbf{S_{FF}}$ is given in Table 4-33.

Table 4-33

Inverse of Stiffness Matrix

$$\mathbf{S}_{\mathrm{FF}}^{-1} = \begin{bmatrix} 798.0 & 632.5 & 0.01877 \\ 632.5 & 2798.0 & 9.355 \\ 0.01877 & 9.355 & 1.426 \end{bmatrix} \times 10^{-6}$$

Next, the load information is processed, beginning with the joint loads shown in Table 4-34. The actions in this table are placed in the vector $\mathbf{A_J}$ as indicated by Eq. (4-73):

$$\mathbf{A_J} = \{0, -10, -1000, 0, 0, 0, 0, 0, 0\}$$

Table 4-34

Actions Applied at Joints

Joint	*Force in x Direction (kips)*	*Force in y Direction (kips)*	*Couple in z Sense (kip-in.)*
1	0	−10	−1000
2	0	0	0
3	0	0	0

In addition, the actions in Table 4-35 are placed in the matrix $\mathbf{A_{ML}}$ and then transferred to the vector of equivalent joint loads $\mathbf{A_E}$ in accordance with Eqs. (4-76). The following vector results:

$$\mathbf{A_E} = \{0, -22, -50, 0, -12, -200, 0, -10, 250\}$$

Adding the vectors $\mathbf{A_J}$ and $\mathbf{A_E}$ produces the combined load vector $\mathbf{A_C}$ as follows:

$$\mathbf{A_C} = \{0, -32, -1050, 0, -12, -200, 0, -10, 250\}$$

Table 4-35

Actions at Ends of Restrained Members Due to Loads

Member	$(A_{ML})_{1,i}$ (kips)	$(A_{ML})_{2,i}$ (kips)	$(A_{ML})_{3,i}$ (kip-in.)	$(A_{ML})_{4,i}$ (kips)	$(A_{ML})_{5,i}$ (kips)	$(A_{ML})_{6,i}$ (kip-in.)
1	0	12	200	0	12	−200
2	−6	8	250	−6	8	−250

The first three elements of this vector constitute the vector $\mathbf{A}_{FC}$:

$$\mathbf{A}_{FC} = \{0, -32, -1050\}$$

and the last six elements are the vector $\mathbf{A}_{RC}$:

$$\mathbf{A}_{RC} = \{0, -12, -200, 0, -10, 250\}$$

Having all of the required matrices on hand, one may complete the solution by calculating the free joint displacements $\mathbf{D}_F$ by Eq. (4-3) with the following result:

$$\mathbf{D}_F = \mathbf{S}_{FF}^{-1}\mathbf{A}_{FC} = \{-0.02026, -0.09936, -0.001797\}$$

The first two elements in the vector $\mathbf{D}_F$ are the translations (inches) in the x and y directions at joint 1, and the last element is the rotation (radians) of the joint in the z sense.

The vector $\mathbf{D}_J$ for this structure consists of the vector $\mathbf{D}_F$ in the first part and zeros in the latter part:

$$\mathbf{D}_J = \{-0.02026, -0.09936, -0.001797, 0, 0, 0, 0, 0, 0\}$$

In the next step the support reactions are computed, using Eq. (4-4) with the matrix $\mathbf{S}_{RF}$ obtained from the lower left-hand portion of Table 4-32:

$$\mathbf{A}_R = -\mathbf{A}_{RC} + \mathbf{S}_{RF}\mathbf{D}_F = \{20.26, 13.14, 436.6, -20.26, 40.86, -889.5\}$$

The elements in the vector $\mathbf{A}_R$ consist of the forces (kips) in the x and y directions and the moments (kip-inches) in the z sense at points 2 and 3, respectively, in the frame.

As the final step in the analysis, the member end-actions $\mathbf{A}_{Mi}$ are calculated, using either Eq. (4-77) or Eqs. (4-78). The results of these calculations are given in Table 4-36, which completes the analysis of the plane frame structure.

Table 4-36

Final Member End-Actions

Member	$(A_M)_{1,i}$ (kips)	$(A_M)_{2,i}$ (kips)	$(A_M)_{3,i}$ (kip-in.)	$(A_M)_{4,i}$ (kips)	$(A_M)_{5,i}$ (kips)	$(A_M)_{6,i}$ (kip-in.)
1	20.26	13.14	436.6	−20.26	10.86	−322.9
2	28.72	−4.53	−677.1	−40.73	20.53	−889.5

4.19 Grid Member Stiffnesses. A grid structure resembles a plane frame in several respects. All of the members and joints lie in the same plane, and the members are assumed to be rigidly connected at the joints (see Fig. 4-32a). Flexural effects tend to predominate in the analysis of both types of structures, with the effects of torsion often being secondary in the grid analysis and axial effects often being secondary in the plane frame analysis. The most important difference between a plane frame and a grid is that forces applied to the former are assumed to lie in its own plane, whereas the forces applied to the latter are normal to its own plane. In addition, moments applied to a plane frame have their vectors normal to the plane of the structure, whereas moment vectors applied to a grid are assumed to lie in the plane of the structure. Both structures could be called plane frames, and the difference between them could be denoted by stating the nature of the loading system. Furthermore, if the applied loads were to have general orientations in space, the analysis of the structure could be divided into two parts. In the first part, the frame would be analyzed for the components of forces in the plane of the structure and moments normal to the plane; and the second part would consist of analyzing for the components of forces normal to the plane and moments in the plane. Superposition of these two analyses would then produce the total solution of the problem. Such a structure might be considered as a special case of a space frame in which all of the members and joints lie in a common plane.

In analyzing a grid structure, the coordinate axes will be taken as shown

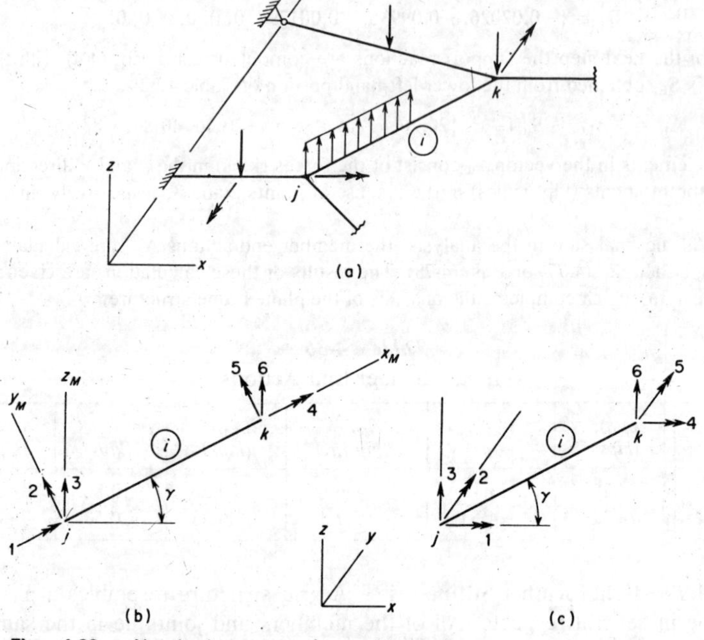

Fig. 4-32. Numbering system for a grid member.

in Fig. 4-32a. The structure lies in the x-y plane, and all applied forces act parallel to the z axis. Loads in the form of moments have their vectors in the x-y plane. The figure shows a typical member i framing into joints j and k. The significant displacements of the joints are rotations in the x and y senses and translations in the z direction. The six possible displacements at the ends of member i in the directions of structure axes are shown in Fig. 4-32c. The numbering system for these displacements is in the order mentioned above. The reason for numbering the rotations before the translation at each joint is to maintain a parallelism with the analysis of a plane frame, as can be seen by comparing the computer programs for plane frames and grids (see Arts. 5.8 and 5.9).

Figure 4-32b depicts the member i in conjunction with a set of member-oriented axes x_M, y_M, and z_M. These axes are rotated from the structure axes about the z_M axis through the angle γ. The x_M-z_M plane for each member in the grid is assumed to be a plane of symmetry (and hence a principal plane of bending). The possible displacements of the ends of member i in the directions of member axes are also indicated in Fig. 4-32b. The six end-displacements, shown in their positive senses, consist of rotations in the x_M and y_M senses and a translation in the z_M (or z) direction at the ends j and k, respectively. Unit displacements of these six types may be induced at the ends of the member one at a time for the purpose of developing the member stiffness matrix $\mathbf{S}_{Mi}$ for member axes. This matrix is of order 6×6, and it may be obtained from cases (3), (4), (5), (9), (10), and (11) of Fig. 4-2 in Art. 4.3. The matrix which results is shown in Table 4-37.

Transformation of the member stiffness matrix from member axes (Fig. 4-32b) to structure axes (Fig. 4-32c) follows the same pattern as that for a

Table 4-37

Grid Member Stiffness Matrix for Member Axes (Fig. 4-32b)

$$
\mathbf{S}_{Mi} = \left[\begin{array}{ccc|ccc}
\frac{GI_X}{L} & 0 & 0 & -\frac{GI_X}{L} & 0 & 0 \\
0 & \frac{4EI_Y}{L} & -\frac{6EI_Y}{L^2} & 0 & \frac{2EI_Y}{L} & \frac{6EI_Y}{L^2} \\
0 & -\frac{6EI_Y}{L^2} & \frac{12EI_Y}{L^3} & 0 & -\frac{6EI_Y}{L^2} & -\frac{12EI_Y}{L^3} \\
\hline
-\frac{GI_X}{L} & 0 & 0 & \frac{GI_X}{L} & 0 & 0 \\
0 & \frac{2EI_Y}{L} & -\frac{6EI_Y}{L^2} & 0 & \frac{4EI_Y}{L} & \frac{6EI_Y}{L^2} \\
0 & \frac{6EI_Y}{L^2} & -\frac{12EI_Y}{L^3} & 0 & \frac{6EI_Y}{L^2} & \frac{12EI_Y}{L^3}
\end{array}\right]
$$

***Table* 4-38**

Grid Member Stiffness Matrix for Structure Axes (Fig. 4-32c)

$$
\mathbf{S}_{MSi} = \left[\begin{array}{ccc|ccc}
\frac{GI_X}{L} C_X^2 + \frac{4EI_Y}{L} C_Y^2 & \left(\frac{GI_X}{L} - \frac{4EI_Y}{L}\right) C_X C_Y & \frac{6EI_Y}{L^2} C_Y & -\frac{GI_X}{L} C_X^2 + \frac{2EI_Y}{L} C_Y^2 & -\left(\frac{GI_X}{L} + \frac{2EI_Y}{L}\right) C_X C_Y & -\frac{6EI_Y}{L^2} C_Y \\
\left(\frac{GI_X}{L} - \frac{4EI_Y}{L}\right) C_X C_Y & \frac{GI_X}{L} C_Y^2 + \frac{4EI_Y}{L} C_X^2 & -\frac{6EI_Y}{L^2} C_X & -\left(\frac{GI_X}{L} + \frac{2EI_Y}{L}\right) C_X C_Y & -\frac{GI_X}{L} C_Y^2 + \frac{2EI_Y}{L} C_X^2 & \frac{6EI_Y}{L^2} C_X \\
\frac{6EI_Y}{L^2} C_Y & -\frac{6EI_Y}{L^2} C_X & \frac{12EI_Y}{L^3} & \frac{6EI_Y}{L^2} C_Y & -\frac{6EI_Y}{L^2} C_X & -\frac{12EI_Y}{L^3} \\
\hline
-\frac{GI_X}{L} C_X^2 + \frac{2EI_Y}{L} C_Y^2 & -\left(\frac{GI_X}{L} + \frac{2EI_Y}{L}\right) C_X C_Y & \frac{6EI_Y}{L^2} C_Y & \frac{GI_X}{L} C_X^2 + \frac{4EI_Y}{L} C_Y^2 & \left(\frac{GI_X}{L} - \frac{4EI_Y}{L}\right) C_X C_Y & -\frac{6EI_Y}{L^2} C_Y \\
-\left(\frac{GI_X}{L} + \frac{2EI_Y}{L}\right) C_X C_Y & -\frac{GI_X}{L} C_Y^2 + \frac{2EI_Y}{L} C_X^2 & -\frac{6EI_Y}{L^2} C_X & \left(\frac{GI_X}{L} - \frac{4EI_Y}{L}\right) C_X C_Y & \frac{GI_X}{L} C_Y^2 + \frac{4EI_Y}{L} C_X^2 & \frac{6EI_Y}{L^2} C_X \\
-\frac{6EI_Y}{L^2} C_Y & \frac{6EI_Y}{L^2} C_X & -\frac{12EI_Y}{L^3} & -\frac{6EI_Y}{L^2} C_Y & \frac{6EI_Y}{L^2} C_X & \frac{12EI_Y}{L^3}
\end{array}\right]
$$

plane frame (see Art. 4.16). The rotation transformation matrix, which involves a rotation about the z_M axis, is exactly the same for both a plane frame member and a grid member because of the choice of numbering system mentioned above. This agreement between the two cases can be seen physically by comparing the orientation of the grid member in Fig. 4-32 with that of the plane frame member in Fig. 4-27. Thus, Eqs. (4-62) through (4-64) in Art. 4.16 may be applied to a grid member as well as a plane frame member. Substitution of the matrix $\mathbf{S}_{Mi}$ from Table 4-37 into Eq. (4-64) results in the member stiffness matrix $\mathbf{S}_{MSi}$ for structure axes. This matrix is given in Table 4-38 and will be used subsequently in the analysis of grid structures.

4.20 Analysis of Grids. The first step in the analysis of a grid is to number the joints and members. This is done in the same manner as for a plane truss or plane frame. As mentioned in the preceding article, there are three possible displacements at each joint. These are the joint rotations in the x and y senses and the joint translation in the z direction. Thus, the possible displacements at a joint j may be denoted by the following:

$$\begin{aligned}3j-2 &= \text{index for rotation in the } x \text{ sense}\\ 3j-1 &= \text{index for rotation in the } y \text{ sense}\\ 3j \quad &= \text{index for translation in the } z \text{ direction}\end{aligned}$$

Note that these indexes are the same as those for the plane frame (see Art. 4.17) but that the meanings are different in the grid structure. Because there are three possible displacements at each joint in both a grid and a plane frame, the number of degrees of freedom in a grid may be determined using Eq. (4-65) in Art. 4.17. Moreover, the indexes for the possible displacements of the two joints associated with member i may be calculated according to Eqs. (4-66) for either type of structure. These indexes are indicated for a grid member in Fig. 4-33. Since several of the steps in the analysis of a grid are symbolically the same as in the analysis of a plane frame, it is sufficient in much of the following discussion merely to refer to the plane frame analysis described in Art. 4.17.

The analysis of the restrained grid for joint stiffnesses follows the same

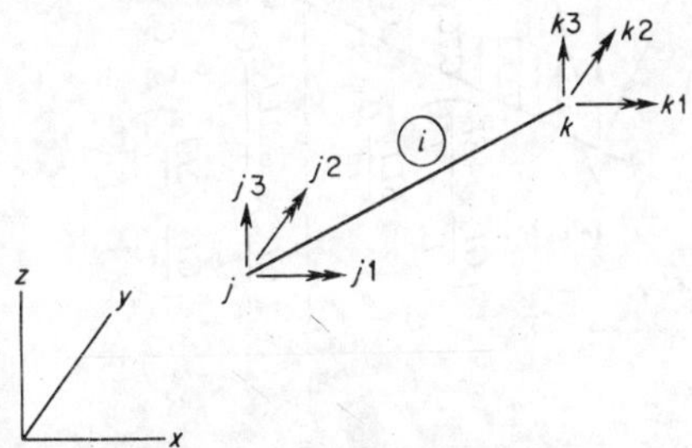

Fig. 4-33. End-displacements for grid member.

Fig. 4-34. Joint loads for a grid.

pattern as in the analysis of a plane frame. The 6 × 6 member stiffness matrix $\mathbf{S}_{\mathrm{MS}i}$ for each member is generated (see Table 4-38), and the elements of this array are transferred systematically to the joint stiffness matrix $\mathbf{S}_{\mathrm{J}}$. Equations (4-67) through (4-72) serve this purpose for a grid as well as for a plane frame.

The analysis of the restrained structure subjected to loads is also analogous to the plane frame, except that the types of actions involved are not the same. Consider first the vector of actions $\mathbf{A}_{\mathrm{J}}$ applied at joints. Figure 4-34 shows the actions which may be imposed at a typical joint k in a grid. The action $(A_J)_{3k-2}$ is the x component of a moment vector applied at k, $(A_J)_{3k-1}$ is the y component of the moment vector, and $(A_J)_{3k}$ represents a force in the z direction applied at the joint. Thus, the vector $\mathbf{A}_{\mathrm{J}}$ may be formed in the sequence denoted by Eq. (4-73).

Actions $\mathbf{A}_{\mathrm{ML}i}$ at the ends of a restrained grid member (due to loads) appear in Fig. 4-35b. The end-actions for the i-th member, with respect to member axes, are defined as follows:

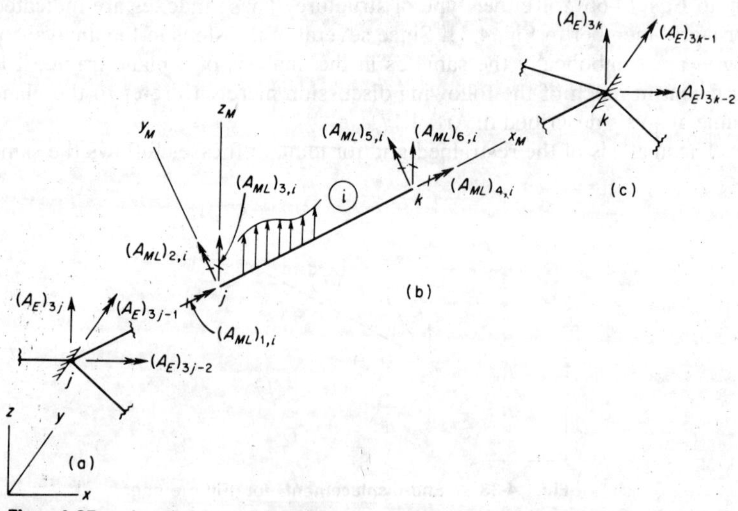

Fig. 4-35. Loads on a grid member.

$(A_{ML})_{1,i}$ = moment in the x_M sense at the j end
$(A_{ML})_{2,i}$ = moment in the y_M sense at the j end
$(A_{ML})_{3,i}$ = force in the z_M direction at the j end
$(A_{ML})_{4,i}$ = moment in the x_M sense at the k end
$(A_{ML})_{5,i}$ = moment in the y_M sense at the k end
$(A_{ML})_{6,i}$ = force in the z_M direction at the k end

The matrix $\mathbf{A}_{\mathrm{ML}}$ may be formulated as a rectangular array of order $6 \times m$ of the type given by Eq. (4-74).

Construction of the equivalent load vector $\mathbf{A}_{\mathrm{E}}$ from the matrix $\mathbf{A}_{\mathrm{ML}}$ is executed as described in Art. 4.17 for plane frames. Figures 4-35a and 4-35c show the equivalent loads at joints j and k which receive contributions from member i. Equations (4-76) may be used for the purpose of computing these actions.

The calculation of displacements, reactions, and end-actions for a grid structure follows the same steps given previously for a plane frame (see Art. 4.17). Joint displacements $\mathbf{D}_{\mathrm{F}}$ are computed, using Eq. (4-3), and then expanded into the vector $\mathbf{D}_{\mathrm{J}}$. Support reactions are calculated, using Eq. (4-4), and member end-actions are determined, using Eq. (4-77). In the latter equation the matrix $\mathbf{R}_{\mathrm{T}i}$ is given by Eq. (4-63), and the matrix $\mathbf{S}_{\mathrm{M}i}$ is obtained from Table 4-37. Substitution of these matrices into Eq. (4-77) produces the following expressions for end-actions:

$$
\begin{aligned}
(A_M)_{1,i} &= (A_{ML})_{1,i} + \frac{GI_{Xi}}{L_i}\{[(D_J)_{j1} - (D_J)_{k1}]C_{Xi} \\
&\quad + [(D_J)_{j2} - (D_J)_{k2}]C_{Yi}\} \\
(A_M)_{2,i} &= (A_{ML})_{2,i} + \frac{4EI_{Yi}}{L_i}\left\{-\left[(D_J)_{j1} + \frac{1}{2}(D_J)_{k1}\right]C_{Yi}\right. \\
&\quad \left.+\left[(D_J)_{j2} + \frac{1}{2}(D_J)_{k2}\right]C_{Xi}\right\} - \frac{6EI_{Yi}}{L_i^2}[(D_J)_{j3} - (D_J)_{k3}] \\
(A_M)_{3,i} &= (A_{ML})_{3,i} + \frac{6EI_{Yi}}{L_i^2}\{[(D_J)_{j1} + (D_J)_{k1}]C_{Yi} \\
&\quad - [(D_J)_{j2} + (D_J)_{k2}]C_{Xi}\} + \frac{12EI_{Yi}}{L_i^3}[(D_J)_{j3} - (D_J)_{k3}] \\
(A_M)_{4,i} &= (A_{ML})_{4,i} - \frac{GI_{Xi}}{L_i}\{[(D_J)_{j1} - (D_J)_{k1}]C_{Xi} \\
&\quad + [(D_J)_{j2} - (D_J)_{k2}]C_{Yi}\} \\
(A_M)_{5,i} &= (A_{ML})_{5,i} + \frac{4EI_{Yi}}{L_i}\left\{-\left[\frac{1}{2}(D_J)_{j1} + (D_J)_{k1}\right]C_{Yi}\right. \\
&\quad \left.+\left[\frac{1}{2}(D_J)_{j2} + (D_J)_{k2}\right]C_{Xi}\right\} - \frac{6EI_{Yi}}{L_i^2}[(D_J)_{j3} - (D_J)_{k3}] \\
(A_M)_{6,i} &= (A_{ML})_{6,i} - \frac{6EI_{Yi}}{L_i^2}\{[(D_J)_{j1} + (D_J)_{k1}]C_{Yi} \\
&\quad - [(D_J)_{j2} + (D_J)_{k2}]C_{Xi}\} - \frac{12EI_{Yi}}{L_i^3}[(D_J)_{j3} - (D_J)_{k3}]
\end{aligned}
\tag{4-79}
$$

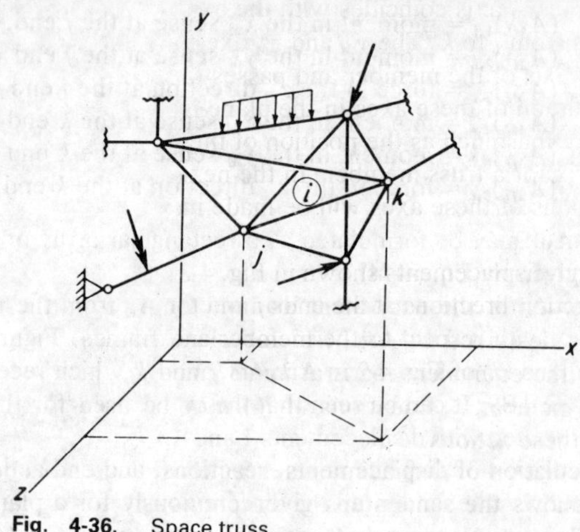

Fig. 4-36. Space truss.

A computer program for the analysis of grids by the method described above is presented in Art. 5.9. Numerical examples of the analysis of grid structures also appear in that article.

4.21 Space Truss Member Stiffnesses. Figure 4-36 shows a portion of a space truss structure in conjunction with a set of structure axes x, y, and z. A typical member i, framing into joints j and k, is indicated in the figure. All joints in the space truss are assumed to be universal hinges. Because of this idealization, rotations of the ends of the members are considered to be immaterial to the analysis. The significant joint displacements are translations, and these translations may be expressed conveniently by their components in the x, y, and z directions.

The possible displacements at the ends of a typical member i are indicated in Fig. 4-37a for member-oriented axes and in Fig. 4-37b for the structure axes. As usual, the member axes in Fig. 4-37a are arranged in such a

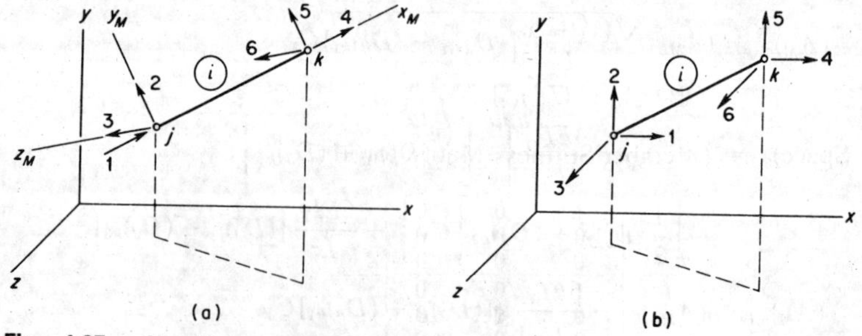

Fig. 4-37. Numbering system for a space truss member.

manner that the x_M axis coincides with the axis of the member and has its positive sense from j to k. The y_M and z_M axes lie in a plane that is perpendicular to the axis of the member and passes through end j. However, the precise orientation of these axes in the plane is immaterial at this stage of the discussion, inasmuch as the position of these axes has no effect on the stiffness matrix for a truss member. In the next article, a particular choice for the directions of these axes will be made in order to facilitate the handling of load data.

The six end-displacements shown in Fig. 4-37a consist of translations in the x_M, y_M, and z_M directions at the ends j and k, respectively. The member stiffness matrix with respect to the member axes may be readily deduced from cases (1) and (7) of Fig. 4-2 in Art. 4.3, and the resulting 6×6 matrix is given in Table 4-39. It can be seen that the only nonzero elements in this matrix are associated with displacements 1 and 4, which are in the direction of the x_M axis (Fig. 4-37a). Thus, as mentioned above, the stiffness matrix for member axes is independent of the directions selected for the y_M and z_M axes.

In order to transform the member stiffness matrix from member axes to structure axes, the rotation transformation matrix $\mathbf{R}_T$ for a space truss member is required. This matrix takes the same form as that given by Eq. (4-63) for plane frames, and the transformation of $\mathbf{S}_M$ to $\mathbf{S}_{MS}$ is shown in Eq. (4-64). The 3×3 rotation matrix $\mathbf{R}$ required for $\mathbf{R}_T$ is explained in the following discussion (omitting the subscript i).

The general form of the rotation matrix $\mathbf{R}$ is given by Eq. (4-56). The three elements λ_{11}, λ_{12} and λ_{13} in the first row of $\mathbf{R}$ are the direction cosines for the x_M axis with respect to the structure axes. Therefore, these three elements are the same as the direction cosines for the member itself ($\lambda_{11} = C_X$, $\lambda_{12} = C_Y$, $\lambda_{13} = C_Z$) and can be found from the coordinates of the ends of the member:

$$C_X = \frac{x_k - x_j}{L} \qquad C_Y = \frac{y_k - y_j}{L} \qquad C_Z = \frac{z_k - z_j}{L} \tag{4-80}$$

The length L of the member may also be computed from the coordinates of its end points:

$$L = \sqrt{(x_k - x_j)^2 + (y_k - y_j)^2 + (z_k - z_j)^2} \tag{4-81}$$

Table 4-39

Space Truss Member Stiffness Matrix for Member Axes (Fig.4-37a)

$$\mathbf{S}_{Mi} = \frac{EA_X}{L}\left[\begin{array}{ccc|ccc} 1 & 0 & 0 & -1 & 0 & 0 \\ 0 & 0 & 0 & 0 & 0 & 0 \\ 0 & 0 & 0 & 0 & 0 & 0 \\ \hline -1 & 0 & 0 & 1 & 0 & 0 \\ 0 & 0 & 0 & 0 & 0 & 0 \\ 0 & 0 & 0 & 0 & 0 & 0 \end{array}\right]$$

The elements in the last two rows of **R** (the direction cosines for the y_M and z_M axes, respectively) can be left in an indefinite form, so that the rotation matrix becomes:

$$\mathbf{R} = \begin{bmatrix} C_X & C_Y & C_Z \\ \lambda_{21} & \lambda_{22} & \lambda_{23} \\ \lambda_{31} & \lambda_{32} & \lambda_{33} \end{bmatrix} \tag{4-82}$$

If this rotation matrix is substituted into Eq. (4-63) for $\mathbf{R}_T$ and then both $\mathbf{R}_T$ and $\mathbf{S}_M$ (see Table 4-39) are substituted into Eq. (4-64), the result will be the 6×6 member stiffness matrix $\mathbf{S}_{MS}$ for structure axes. This matrix, which contains terms involving only the direction cosines for the member itself, is given in Table 4-40.

Table 4-40

Space Truss Member Stiffness Matrix for Structure Axes (Fig. 4-37b)

$$\mathbf{S}_{MSi} = \frac{EA_X}{L}\left[\begin{array}{ccc|ccc} C_X^2 & C_YC_X & C_ZC_X & -C_X^2 & -C_YC_X & -C_ZC_X \\ C_XC_Y & C_Y^2 & C_ZC_Y & -C_XC_Y & -C_Y^2 & -C_ZC_Y \\ C_XC_Z & C_YC_Z & C_Z^2 & -C_XC_Z & -C_YC_Z & -C_Z^2 \\ \hline -C_X^2 & -C_YC_X & -C_ZC_X & C_X^2 & C_YC_X & C_ZC_X \\ -C_XC_Y & -C_Y^2 & -C_ZC_Y & C_XC_Y & C_Y^2 & C_ZC_Y \\ -C_XC_Z & -C_YC_Z & -C_Z^2 & C_XC_Z & C_YC_Z & C_Z^2 \end{array}\right]$$

4.22 Selection of Space Truss Member Axes. For the purpose of handling loads acting directly on the members of a space truss, it is necessary to select a specific orientation for all three member axes x_M, y_M, and z_M. Since the x_M axis has already been selected as the axis of the member (see Fig. 4-37a), it remains only to establish directions for y_M and z_M. For this purpose, the typical member i is shown again in Fig. 4-38, with axes

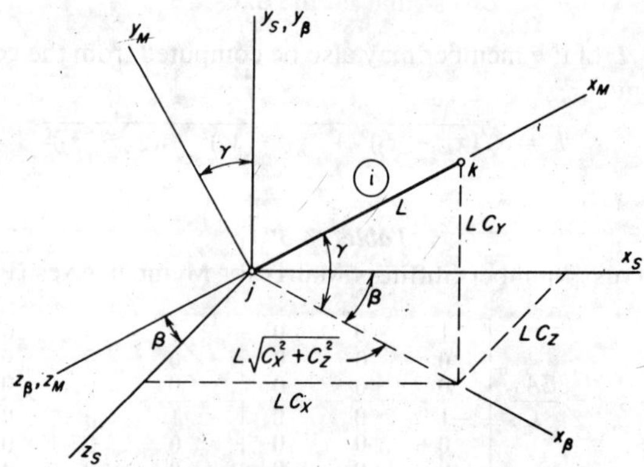

Fig. 4-38. Rotation of axes for a space truss member.

x_S, y_S, and z_S taken parallel to the structural axes. While many choices are possible for the directions of the y_M and z_M axes, a convenient one is to take the z_M axis as being horizontal (that is, lying in the x_S-z_S plane), as shown in the figure. It follows that the y_M axis is located in a vertical plane passing through the x_M and y_S axes.

When the member axes are specified in the manner just described, there is no ambiguity about their orientations except in the case of a vertical member. When the member is vertical, it follows automatically that z_M will be in a horizontal plane; but its position in that plane is not defined uniquely. To overcome this difficulty, the additional specification will be made that the z_M axis is always taken to be along the z_S axis if the member is vertical. The two possibilities for this occurrence are shown in Figs. 4-39a and 4-39b.

In order to perform axis transformations, the rotation matrix **R** (see Eq. 4-82) is required. The direction cosines for the y_M and z_M axes (the last two rows of **R**) can be found directly from Fig. 4-38 by geometrical considerations. However, an alternate approach involves successive rotations of axes. In using the latter method, the transformation from the structure axes (see Fig. 4-38) to the member axes may be considered to take place in two steps. The first of these is a rotation through an angle β about the y_S axis. This rotation places the x axis in the position denoted as x_β, which is the intersection of the x_S-z_S plane and x_M-y_S plane. Also, this rotation places the z_M axis in its final position at the angle β with the z_S axis. The second step in the transformation consists of a rotation through an angle γ about the z_M axis. This rotation places the x_M and y_M axes in their final positions, as shown in the figure.

The rotation matrix **R** which is used in transforming from the structure to the member axes can be developed following the two steps described above. Consider first the rotation about the y_S axis through the angle β. The 3 × 3 rotation matrix $\mathbf{R}_\beta$ for this transformation consists of the direction cosines of the β-axes (that is, the axes x_β, y_β, z_β) with respect to the structure axes (x_S, y_S, z_S):

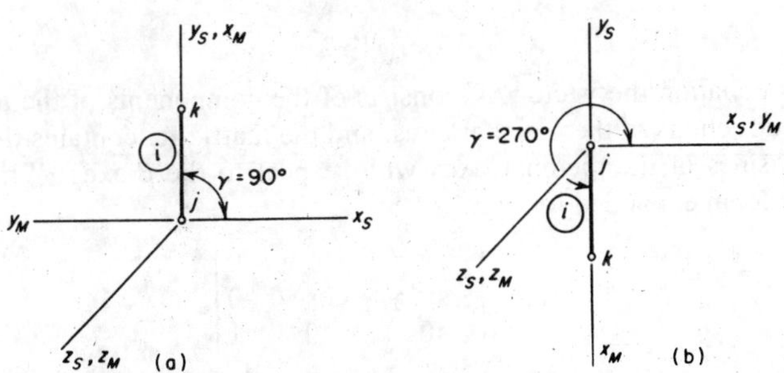

Fig. 4-39. Rotation of axes for a vertical space truss member.

$$\mathbf{R}_\beta = \begin{bmatrix} \cos\beta & 0 & \sin\beta \\ 0 & 1 & 0 \\ -\sin\beta & 0 & \cos\beta \end{bmatrix} \tag{a}$$

The functions cos β and sin β may be expressed in terms of the direction cosines of the member i by referring to the geometry of Fig. 4-38. Thus, these quantities become

$$\cos\beta = \frac{C_X}{C_{XZ}} \qquad \sin\beta = \frac{C_Z}{C_{XZ}} \tag{b}$$

in which $C_{XZ} = \sqrt{C_X^2 + C_Z^2}$. Substitution of these expressions into Eq. (a) gives the matrix $\mathbf{R}_\beta$ in terms of the direction cosines:

$$\mathbf{R}_\beta = \begin{bmatrix} \dfrac{C_X}{C_{XZ}} & 0 & \dfrac{C_Z}{C_{XZ}} \\ 0 & 1 & 0 \\ \dfrac{-C_Z}{C_{XZ}} & 0 & \dfrac{C_X}{C_{XZ}} \end{bmatrix} \tag{4-83}$$

The physical significance of the matrix $\mathbf{R}_\beta$ is that it can be used to relate two alternative orthogonal sets of components of a vector (action or small displacement) in the directions of the β-axes and the structure axes. For example, consider an action A that can be represented by either the vector $\mathbf{A}_\beta$ (consisting of components in the directions of the β axes) or the vector $\mathbf{A}_S$ (consisting of components in the directions of the structure axes). The transformation of the latter vector into the former is performed by the method of Art. 4.15, as follows:

$$\mathbf{A}_\beta = \mathbf{R}_\beta \mathbf{A}_S \tag{c}$$

where $\mathbf{R}_\beta$ is given by Eq. (4-83).

The second rotation about the axis z_M through the angle γ may be handled in a similar manner. In this step a rotation matrix $\mathbf{R}_\gamma$ again relates two orthogonal sets of components of the same vector for two sets of axes. In the case of an action A, the transformation is

$$\mathbf{A}_M = \mathbf{R}_\gamma \mathbf{A}_\beta \tag{d}$$

In this equation the vector $\mathbf{A}_M$ consists of the components of the action A in the directions of the member axes, and the matrix $\mathbf{R}_\gamma$ contains the direction cosines of the member axes with respect to the β-axes. Writing the matrix $\mathbf{R}_\gamma$ in expanded form gives:

$$\mathbf{R}_\gamma = \begin{bmatrix} \cos\gamma & \sin\gamma & 0 \\ -\sin\gamma & \cos\gamma & 0 \\ 0 & 0 & 1 \end{bmatrix} \tag{e}$$

The functions cos γ and sin γ may be expressed in terms of the direction cosines of the member as follows (see Fig. 4-38):

$$\cos \gamma = C_{XZ} \qquad \sin \gamma = C_Y \tag{f}$$

Substitution of these expressions into Eq. (e) yields the matrix $\mathbf{R}_\gamma$ in the following form:

$$\mathbf{R}_\gamma = \begin{bmatrix} C_{XZ} & C_Y & 0 \\ -C_Y & C_{XZ} & 0 \\ 0 & 0 & 1 \end{bmatrix} \tag{4-84}$$

Now that the two separate rotations have been expressed in matrix form, the single transformation matrix $\mathbf{R}$ from the structure axes to the member axes can also be obtained. Again considering the case of an action A, it can be seen that the vector $\mathbf{A}_\mathrm{M}$ may be expressed in terms of the vector $\mathbf{A}_\mathrm{S}$ by substituting $\mathbf{A}_\beta$ (see Eq. c) into Eq. (d):

$$\mathbf{A}_\mathrm{M} = \mathbf{R}_\gamma \mathbf{R}_\beta \mathbf{A}_\mathrm{S} \tag{g}$$

Comparing Eq. (g) with Eq. (4-57) in Art. 4.15 shows that the desired rotation matrix $\mathbf{R}$ consists of the product of $\mathbf{R}_\gamma$ and $\mathbf{R}_\beta$. Thus,

$$\mathbf{R} = \mathbf{R}_\gamma \mathbf{R}_\beta \tag{4-85}$$

If Eqs. (4-83) and (4-84) are substituted into Eq. (4-85), the following matrix results:

$$\mathbf{R} = \begin{bmatrix} C_X & C_Y & C_Z \\ \dfrac{-C_X C_Y}{C_{XZ}} & C_{XZ} & \dfrac{-C_Y C_Z}{C_{XZ}} \\ \dfrac{-C_Z}{C_{XZ}} & 0 & \dfrac{C_X}{C_{XZ}} \end{bmatrix} \tag{4-86}$$

This is the rotation matrix $\mathbf{R}$ for a space truss member. It may be used whenever actions or displacements are to be transformed between the member axes and the structure axes.

The preceding rotation matrix $\mathbf{R}$ is valid for all positions of the member i except when it is vertical. In either of the cases shown in Fig. 4-39a or Fig. 4-39b, the direction cosines of the member axes with respect to the structure axes can be determined by inspection. Thus, the rotation matrix is seen to be

$$\mathbf{R}_\mathrm{vert} = \begin{bmatrix} 0 & C_Y & 0 \\ -C_Y & 0 & 0 \\ 0 & 0 & 1 \end{bmatrix} \tag{4-87}$$

This expression for $\mathbf{R}$ is valid for both cases shown in Figs. 4-39a and b. All that is necessary is to substitute for the direction cosine C_Y its appropriate value, which is 1 for the member in Fig. 4-39a and -1 for the member in Fig. 4-39b.

Substitution of the appropriate rotation matrix $\mathbf{R}$ (either Eq. 4-86 or

4-87) into Eq. (4-63) produces the desired rotation transformation matrix $\mathbf{R}_T$ for the set of member-oriented axes specified in this article.

4.23 Analysis of Space Trusses. The analysis of space truss structures is presented in this article in a manner analogous to the previous discussions for other types of structures. Loads on a space truss structure usually consist of concentrated forces applied at the joints, but in some instances actions of a more general nature may be applied to the individual members, as indicated in Fig. 4-36. As with previous structures, members are numbered 1 through m, and joints are numbered 1 through n_j.

Only axial deformations are taken into account in the analysis of a space truss, but there exists the possibility of three independent displacements at each joint. The possible displacements at a joint j in the structure are signified by the following indexes:

$3j - 2$ = index for translation in the x direction
$3j - 1$ = index for translation in the y direction
$3j$ = index for translation in the z direction

The number of degrees of freedom in a space truss may be calculated using Eq. (4-65) in Art. 4.17. In addition, the indexes for the possible displacements of the ends of member i may be determined using Eqs. (4-66). These indexes are indicated in Fig. 4-40 for a space truss member.

The analysis of space trusses is symbolically similar to the analyses of plane frames (Art. 4.17) and grids (Art. 4.20). The reason for the parallelism is that all three of these types of structures have three possible displacements per joint. The similarity begins in the first phase of the analysis, in which the 6×6 member stiffness matrix $\mathbf{S}_{MSi}$ (see Art. 4.21 for space truss member stiffnesses) is generated for each member in sequence. The elements of each of these member arrays are transferred systematically to the joint stiffness matrix $\mathbf{S}_J$ in the manner represented by Eqs. (4-67) through (4-72).

In the next phase of the analysis of a space truss, loads applied to the structure are processed. Figure 4-41 shows the actions applied at a typical

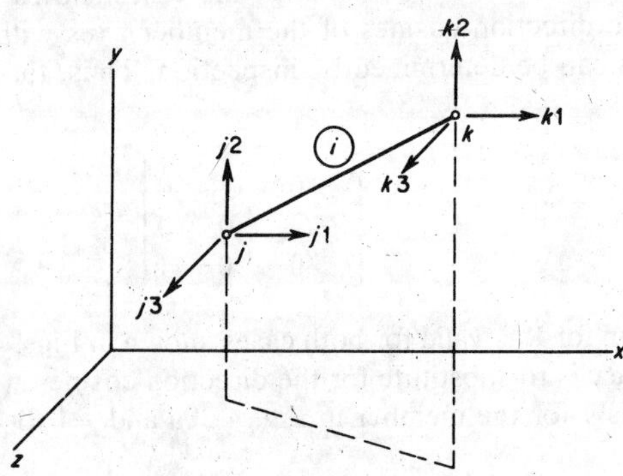

Fig. 4-40. End-displacements for space truss member.

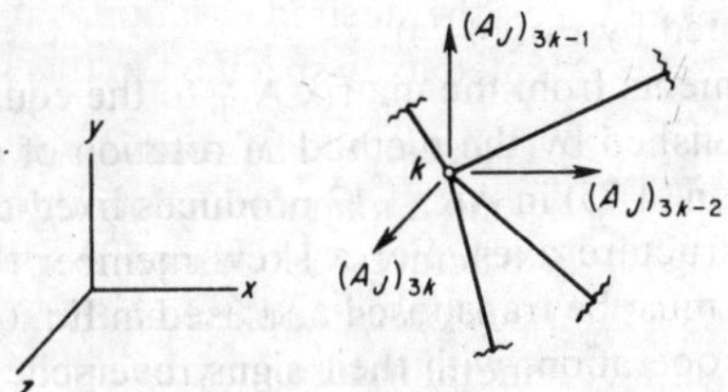

Fig. 4-41. Joint loads for a space truss.

joint k in a space truss. The actions $(A_J)_{3k-2}$, $(A_J)_{3k-1}$, and $(A_J)_{3k}$ are the x, y, and z components of the concentrated force applied at the joint. These actions are placed in the vector $\mathbf{A}_J$, which takes the form given by Eq. (4-73).

Consider next the matrix of actions $\mathbf{A}_{ML}$ at the ends of restrained members due to loads. In a truss structure these actions are determined with the ends of members restrained against translation but not against rotation (see Table B-5 of Appendix B). Figure 4-42b shows such a member i and the six end-actions caused by loading applied along its length. These end-actions are computed as the following:

$$
\begin{aligned}
(A_{ML})_{1,i} &= \text{force in the } x_M \text{ direction at the } j \text{ end} \\
(A_{ML})_{2,i} &= \text{force in the } y_M \text{ direction at the } j \text{ end} \\
(A_{ML})_{3,i} &= \text{force in the } z_M \text{ direction at the } j \text{ end} \\
(A_{ML})_{4,i} &= \text{force in the } x_M \text{ direction at the } k \text{ end} \\
(A_{ML})_{5,i} &= \text{force in the } y_M \text{ direction at the } k \text{ end} \\
(A_{ML})_{6,i} &= \text{force in the } z_M \text{ direction at the } k \text{ end}
\end{aligned}
$$

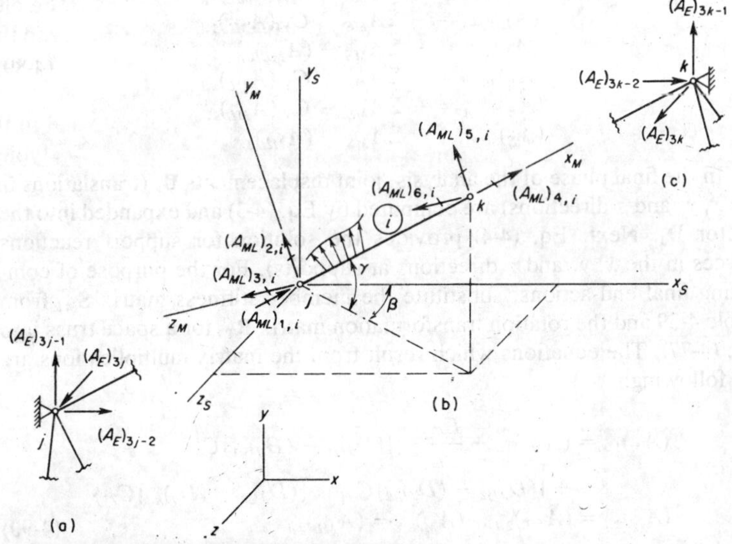

Fig. 4-42. Loads on a space truss member.

The matrix $\mathbf{A}_{ML}$ is conveniently formulated as a rectangular array of order $6 \times m$, as represented by Eq. (4-74).

Transfer of elements from the matrix $\mathbf{A}_{ML}$ to the equivalent load vector $\mathbf{A}_E$ may be accomplished by the method of rotation of axes, as explained previously. Equation (4-75) in Art. 4.17 produces fixed-end actions $\mathbf{A}_{MSi}$ in the directions of structure axes. For a skew member the rotation matrix given by Eq. (4-86) must be transposed and used in Eq. (4-75). Expressions resulting from that operation, with their signs reversed, become the incremental portions of $\mathbf{A}_E$ contributed by the i-th member, as follows:

$$
\begin{aligned}
(A_E)_{3j-2} &= -\Sigma A_{MS} - C_{Xi}(A_{ML})_{1,i} + \frac{C_{Xi}C_{Yi}}{C_{XZi}}(A_{ML})_{2,i} + \frac{C_{Zi}}{C_{XZi}}(A_{ML})_{3,i} \\
(A_E)_{3j-1} &= -\Sigma A_{MS} - C_{Yi}(A_{ML})_{1,i} - C_{XZi}(A_{ML})_{2,i} \\
(A_E)_{3j} &= -\Sigma A_{MS} - C_{Zi}(A_{ML})_{1,i} + \frac{C_{Yi}C_{Zi}}{C_{XZi}}(A_{ML})_{2,i} - \frac{C_{Xi}}{C_{XZi}}(A_{ML})_{3,i} \\
(A_E)_{3k-2} &= -\Sigma A_{MS} - C_{Xi}(A_{ML})_{4,i} + \frac{C_{Xi}C_{Yi}}{C_{XZi}}(A_{ML})_{5,i} + \frac{C_{Zi}}{C_{XZi}}(A_{ML})_{6,i} \\
(A_E)_{3k-1} &= -\Sigma A_{MS} - C_{Yi}(A_{ML})_{4,i} - C_{XZi}(A_{ML})_{5,i} \\
(A_E)_{3k} &= -\Sigma A_{MS} - C_{Zi}(A_{ML})_{4,i} + \frac{C_{Yi}C_{Zi}}{C_{XZi}}(A_{ML})_{5,i} - \frac{C_{Xi}}{C_{XZi}}(A_{ML})_{6,i}
\end{aligned}
\tag{4-88}
$$

Equivalent joint loads receiving the above contributions appear in Figs. 4-42a and 4-42c.

On the other hand, if the member is vertical, the transpose of the rotation matrix given by Eq. (4-87) must be used instead. In this case Eqs. (4-88) for the incremental portions of $\mathbf{A}_E$ simplify to

$$
\begin{aligned}
(A_E)_{3j-2} &= -\Sigma A_{MS} + C_{Yi}(A_{ML})_{2,i} \\
(A_E)_{3j-1} &= -\Sigma A_{MS} - C_{Yi}(A_{ML})_{1,i} \\
(A_E)_{3j} &= -\Sigma A_{MS} - (A_{ML})_{3,i} \\
(A_E)_{3k-2} &= -\Sigma A_{MS} + C_{Yi}(A_{ML})_{5,i} \\
(A_E)_{3k-1} &= -\Sigma A_{MS} - C_{Yi}(A_{ML})_{4,i} \\
(A_E)_{3k} &= -\Sigma A_{MS} - (A_{ML})_{6,i}
\end{aligned}
\tag{4-89}
$$

In the final phase of the analysis, joint displacements $\mathbf{D}_F$ (translations in the x, y, and z directions) are computed by Eq. (4-3) and expanded into the vector $\mathbf{D}_J$. Next, Eq. (4-4) provides the solution for support reactions (forces in the x, y, and z directions at supports). For the purpose of computing final end-actions, substitute the member stiffness matrix $\mathbf{S}_{Mi}$ from Table 4-39 and the rotation transformation matrix $\mathbf{R}_{Ti}$ for a space truss into Eq. (4-77). The equations which result from the matrix multiplications are the following:

$$
\begin{aligned}
(A_M)_{1,i} &= (A_{ML})_{1,i} + \frac{EA_{Xi}}{L_i}\{[(D_J)_{j1} - (D_J)_{k1}]C_{Xi} \\
&\quad + [(D_J)_{j2} - (D_J)_{k2}]C_{Yi} + [(D_J)_{j3} - (D_J)_{k3}]C_{Zi}\} \\
(A_M)_{2,i} &= (A_{ML})_{2,i}; \quad (A_M)_{3,i} = (A_{ML})_{3,i}
\end{aligned}
\tag{4-90}
$$

$$(A_M)_{4,i} = (A_{ML})_{4,i} - \frac{EA_{Xi}}{L_i}\{[(D_J)_{j1} - (D_J)_{k1}]C_{Xi} + [(D_J)_{j2} - (D_J)_{k2}]C_{Yi} + [(D_J)_{j3} - (D_J)_{k3}]C_{Zi}\}$$

$$(A_M)_{5,i} = (A_{ML})_{5,i}; \quad (A_M)_{6,i} = (A_{ML})_{6,i}$$

This set of equations is valid for a member having any orientation, including a vertical member. In the latter case, the direction cosines C_{Xi} and C_{Zi} appearing in the equations will have the value zero.

A computer program for the analysis of space trusses is given in Art. 5.10. Numerical examples may also be found in that article.

4.24 Space Frame Member Stiffnesses. Figure 4-43 shows a portion of a space frame structure and a set of reference axes x, y, and z. A typical member i having positive direction cosines is indicated in the figure with joints at its ends denoted by j and k. The members of the frame are assumed to be rigidly connected at the joints, and each joint that is not restrained is assumed to translate and rotate in a completely general manner in space. Thus, all possible types of joint displacements must be considered and, for convenience, they are taken to be the translations and rotations in the directions of the x, y, and z axes (six possible displacements per joint).

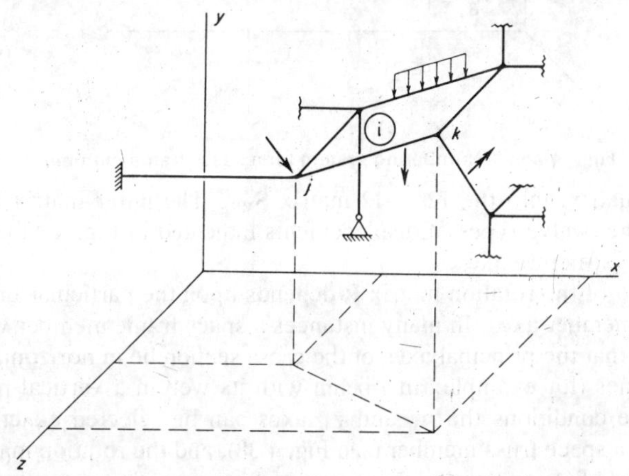

Fig. 4-43. Space frame.

The twelve possible displacements of the two ends of a member were discussed previously in Art. 4.3 (see Fig. 4-1) for member-oriented axes and are indicated again in Fig. 4-44a for the typical member i. In this figure, the x_M axis is taken along the axis of the member, in the same manner as for a space truss (compare with Fig. 4-38). The y_M and z_M axes are selected as the principal axes of the cross section at end j of the member. The complete 12×12 stiffness matrix $\mathbf{S}_{Mi}$ for the member axes was given earlier in Table 4-1. This matrix must be transformed, by means of a rotation trans-

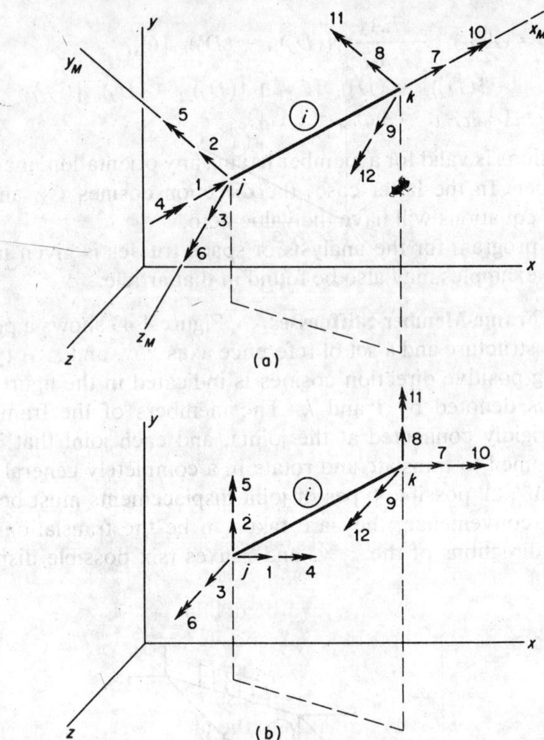

Fig. 4-44. Numbering system for a space frame member.

formation matrix, into the 12 × 12 matrix $\mathbf{S}_{MSi}$. The latter matrix corresponds to the twelve types of displacements indicated in Fig. 4-44b in the directions of structure axes.

The form of the rotation matrix **R** depends upon the particular orientation of the member axes. In many instances a space frame member will be oriented so that the principal axes of the cross section lie in horizontal and vertical planes (for example, an I-beam with its web in a vertical plane). Under these conditions the y_M and z_M axes can be selected exactly the same as for a space truss member (see Fig. 4-38), and the rotation matrix **R** given in Eq. (4-86) can be used for the space frame member as well.

There are other instances in which a space frame member has two axes of symmetry in the cross section and the same moment of inertia about each axis (for example, a circular or square member, either tubular or solid). In such cases the y_M and z_M axes can again be selected as described in the preceding paragraph. This choice can be made because all axes in the cross section are principal axes, and any pair of axes can be selected for y_M and z_M.

However, a space frame member may have its principal axes y_M and z_M in general directions, as indicated in Fig. 4-44a. There are various ways in

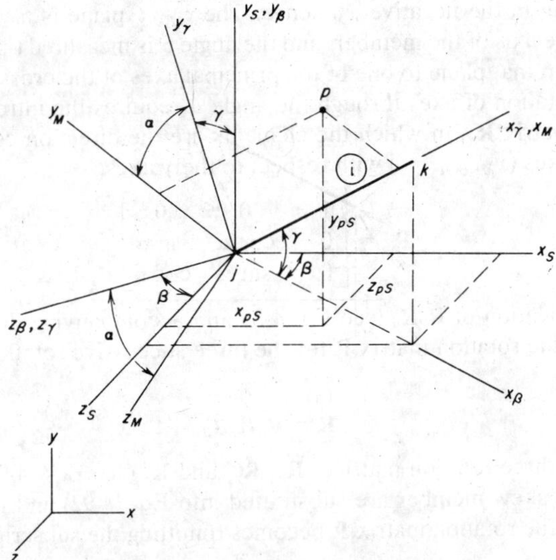

Fig. 4-45. Rotation of axes for a space frame member.

which the orientation of these axes can be defined, and two methods will be described. The first method involves specifying the orientation of the principal axes by means of an angle of rotation about the x_M axis. In order to visualize clearly how such an angle is measured, consider the three successive rotations from the structure axes to the meauber axes shown in Fig. 4-45. The first two rotations through the angles β and γ (about the y_S and z_β axes, respectively) are exactly the same as shown in Fig. 4-38 for the special case described above. The third transformation consists of a rotation through the angle α about the x_M axis, causing the y_M and z_M axes to coincide with the principal axes of thc cross scction. This last rotation is also indicated in Fig. 4-46, which shows a cross sectional view of the mem-

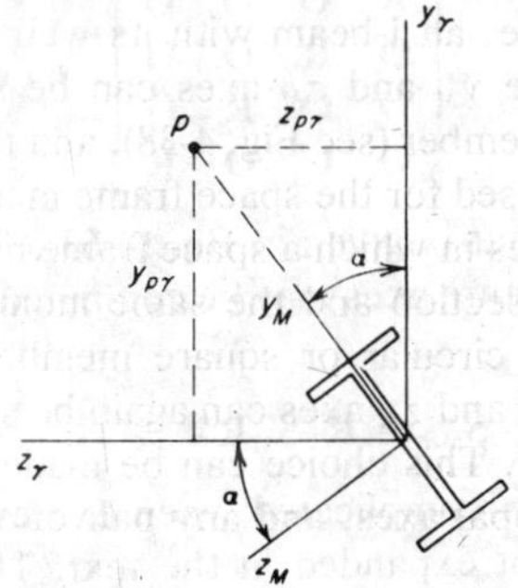

Fig. 4-46. Rotation about x_M axis.

ber looking in the negative x_M sense. The x_M-y_γ plane is a vertical plane through the axis of the member, and the angle α is measured (in the positive sense) from that plane to one of the principal axes of the cross section

The rotation of axes through the angle α requires the introduction of a rotation matrix $\mathbf{R}_\alpha$, in which the elements are the direction cosines of the member axes (x_M, y_M, z_M) with respect to the γ axes:

$$\mathbf{R}_\alpha = \begin{bmatrix} 1 & 0 & 0 \\ 0 & \cos\alpha & \sin\alpha \\ 0 & -\sin\alpha & \cos\alpha \end{bmatrix} \tag{4-91}$$

Premultiplication of $\mathbf{R}_\gamma \mathbf{R}_\beta$ (see Eq. 4-85 and accompanying discussion) by $\mathbf{R}_\alpha$ yields the rotation matrix $\mathbf{R}$ for the three successive rotations shown in Fig. 4-45:

$$\mathbf{R} = \mathbf{R}_\alpha \mathbf{R}_\gamma \mathbf{R}_\beta \tag{4-92}$$

When the three rotation matrices $\mathbf{R}_\alpha$, $\mathbf{R}_\gamma$, and $\mathbf{R}_\beta$ (see Eqs. 4-91, 4-84, and 4-83) for a skew member are substituted into Eq. (4-92) and multiplied as indicated, the rotation matrix $\mathbf{R}$ becomes (omitting the subscript i):

$$\mathbf{R} = \begin{bmatrix} C_X & C_Y & C_Z \\ \dfrac{-C_X C_Y \cos\alpha - C_Z \sin\alpha}{C_{XZ}} & C_{XZ}\cos\alpha & \dfrac{-C_Y C_Z \cos\alpha + C_X \sin\alpha}{C_{XZ}} \\ \dfrac{C_X C_Y \sin\alpha - C_Z \cos\alpha}{C_{XZ}} & -C_{XZ}\sin\alpha & \dfrac{C_Y C_Z \sin\alpha + C_X \cos\alpha}{C_{XZ}} \end{bmatrix} \tag{4-93}$$

This rotation matrix is expressed in terms of the direction cosines of the member (which are readily computed from the coordinates of the joints) and the angle α, which must be given as part of the description of the structure itself. Note that if α is equal to zero, the matrix $\mathbf{R}$ reduces to the form given previously for a space truss member (Eq. 4-86). The special case of a vertical member will be discussed later.

The rotation transformation matrix $\mathbf{R}_\text{T}$ for a space frame member can be shown to take the following form:

$$\mathbf{R}_\text{T} = \begin{bmatrix} \mathbf{R} & \mathbf{O} & \mathbf{O} & \mathbf{O} \\ \mathbf{O} & \mathbf{R} & \mathbf{O} & \mathbf{O} \\ \mathbf{O} & \mathbf{O} & \mathbf{R} & \mathbf{O} \\ \mathbf{O} & \mathbf{O} & \mathbf{O} & \mathbf{R} \end{bmatrix} \tag{4-94}$$

which is analogous to the matrix given in Eq. (4-63). Finally, the member stiffness matrix $\mathbf{S}_\text{MS}$ for structure axes may be computed by the usual matrix multiplications:

$$\mathbf{S}_\text{MS} = \mathbf{R}_\text{T}^\text{T} \mathbf{S}_\text{M} \mathbf{R}_\text{T} \tag{4-95}$$

The resulting matrix is quite complicated when expressed in literal form, and for that reason it is not expanded in the text. The fact that such a lengthy set of relationships can be represented so concisely by Eq. (4-95) is one of the principal advantages of matrix methods in the analysis of structures.

In some structures the orientation of a particular member may be such that the angle α which specifies the location of the principal axes for that member may not be readily available. In that event, a different technique for describing their location can be used. A suitable method is to give the coordinates of a point that lies in one of the principal planes of the member but is not on the axis of the member itself. This point and the x_M axis will define without ambiguity a plane in space, and that plane can be taken as the x_M-y_M plane. All that is necessary is to obtain expressions for the angle of rotation α, which appears in the rotation matrix **R** (Eq. 4-93), in terms of the coordinates of the given point and the coordinates of the ends of the member itself. Once this is accomplished, it becomes possible for the structural analyst to describe the position of the principal axes of a member either by giving the angle α directly or by giving the coordinates of a suitable point.

An arbitrary point p in the x_M-y_M plane is shown in Fig. 4-45. The x, y, and z coordinates of this point (denoted as x_p, y_p, and z_p) are assumed to be given. Since the structure axes x_S, y_S, and z_S have their origin at end j of the member, the coordinates of point p with respect to the structure axes (denoted as x_{pS}, y_{pS}, and z_{pS}) are

$$x_{pS} = x_p - x_j \qquad y_{pS} = y_p - y_j \qquad z_{pS} = z_p - z_j \tag{4-96}$$

in which x_j, y_j, and z_j are the coordinates of end j of the member.

The point p is also shown in Fig. 4-46 in conjunction with the rotation angle α. The coordinates of p with respect to the axes x_γ, y_γ, and z_γ can be obtained in terms of those for the axes x_S, y_S, and z_S by a rotation of axes through the angles β and γ. Let the coordinates of point p with respect to the γ-axes be $x_{p\gamma}$, $y_{p\gamma}$, and $z_{p\gamma}$ (the latter two coordinates are shown positive in Fig. 4-46). Then the desired transformation of coordinates becomes the following:

$$\begin{bmatrix} x_{p\gamma} \\ y_{p\gamma} \\ z_{p\gamma} \end{bmatrix} = \mathbf{R}_\gamma \mathbf{R}_\beta \begin{bmatrix} x_{pS} \\ y_{pS} \\ z_{pS} \end{bmatrix} \tag{a}$$

Substitution of $\mathbf{R}_\gamma \mathbf{R}_\beta$ from Eq. (4-86) into Eq. (a) yields the following expressions:

$$\begin{aligned} x_{p\gamma} &= C_X x_{pS} + C_Y y_{pS} + C_Z z_{pS} \\ y_{p\gamma} &= -\frac{C_X C_Y}{C_{XZ}} x_{pS} + C_{XZ} y_{pS} - \frac{C_Y C_Z}{C_{XZ}} z_{pS} \\ z_{p\gamma} &= -\frac{C_Z}{C_{XZ}} x_{pS} + \frac{C_X}{C_{XZ}} z_{pS} \end{aligned} \tag{4-97}$$

These equations give the coordinates of point p with respect to the γ axes.

From the geometry of Fig. 4-46 the following expressions for the sine and cosine of the angle α are obtained:

$$\sin\alpha = \frac{z_{p\gamma}}{\sqrt{y_{p\gamma}^2 + z_{p\gamma}^2}} \qquad \cos\alpha = \frac{y_{p\gamma}}{\sqrt{y_{p\gamma}^2 + z_{p\gamma}^2}} \tag{4-98}$$

With Eqs. (4-96), (4-97), and (4-98) the quantities $\sin\alpha$ and $\cos\alpha$ can be computed from the coordinates of point p and then substituted into the rotation matrix **R** (Eq. 4-93). This substitution produces the rotation matrix **R** in a form which involves only the direction cosines of the member itself and the coordinates of point p.*

The preceding discussion dealt with a member that was not vertical. As shown in Art. 4.22 for a space truss member, a vertical member is a special case that must be treated separately. The rotation matrix **R** given previously for a vertical space truss member (see Eq. 4-87) can also be used for certain types of vertical space frame members. The only requirement is that one of the principal axes of the cross section of the member be in the direction of the z_S axis. In some cases the member will be oriented in such a manner that this requirement is met (for example, an I-beam with its web in the x_S-y_S plane). The requirement is also satisfied by a member having a cross section that is either circular or square (solid or tubular). In these cases all axes of the cross section are principal axes, and hence one of them can arbitrarily be selected in the z_S direction.

In a more general case of a vertical member, the principal axes of the cross section will be rotated about the vertical axis so that they form an angle α with the directions of the structure axes. The orientation of such a vertical member can best be visualized by considering the successive rotations of axes shown in Fig. 4-47. There is no rotation through the angle β (about the y_S axis) in this case. Instead, the first rotation is through the angle γ, which may be either 90° or 270° (see Figs. 4-47a and 4-47b, respectively). The second rotation is through the angle α and about the x_M axis. The rotation matrix for either of the cases shown in the figure can be obtained by inspection. It consists of the direction cosines of the x_M, y_M, and z_M axes with respect to the structure axes:

$$\mathbf{R}_{\text{vert}} = \begin{bmatrix} 0 & C_Y & 0 \\ -C_Y\cos\alpha & 0 & \sin\alpha \\ C_Y\sin\alpha & 0 & \cos\alpha \end{bmatrix} \tag{4-99}$$

This expression is valid for both orientations shown in Figs. 4-47a and b, provided the appropriate value of C_Y is substituted (+1 in the first case and

*The rotation matrix **R** can also be derived from properties of the vector (or cross) product, as follows: $\mathbf{e}_{zM} = \mathbf{e}_{xM} \times \mathbf{e}_{jp}/\sqrt{e_{xM}^2 + e_{jp}^2}$; $\mathbf{e}_{yM} = \mathbf{e}_{zM} \times \mathbf{e}_{xM}$. (In these expressions the symbol **e** represents a unit vector in the direction indicated by its subscripts.) Such vector operations appear simple because of efficient notation, but their expansions involve the same terms used in the more informative rotation of axes technique described herein.

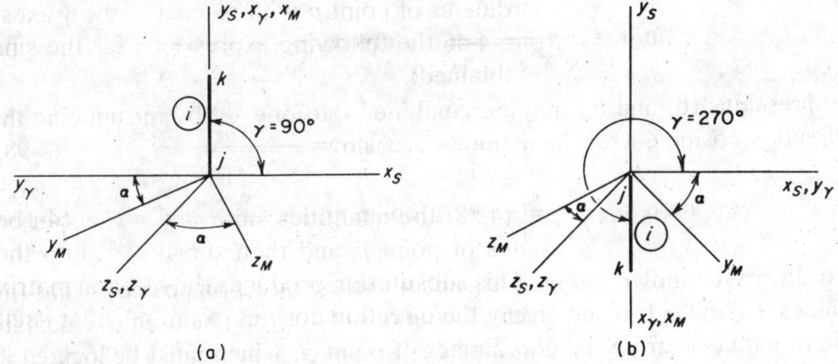

Fig. 4-47. Rotation of axes for a vertical space frame member.

−1 in the second case). It should be noted also that if α is equal to zero, the rotation matrix given in Eq. (4-99) reduces to the rotation matrix for a vertical space truss member (Eq. 4-87).

In those cases where it is desired to begin with the coordinates of a point p that is known to lie in a principal plane, it is possible to calculate sin α and cos α for use in Eq. (4-99) directly from the coordinates of the point. Figure 4-48a shows a vertical member with the lower end designated as joint j and the upper end designated as joint k. Also shown in the figure is the point p. When this point is located so that its coordinates with respect to the S-axes are positive, the angle α will be between 90° and 180°. The sine and cosine of this angle are as follows:

$$\sin \alpha = \frac{z_{pS}}{\sqrt{x_{pS}^2 + z_{pS}^2}} \qquad \cos \alpha = -\frac{x_{pS}}{\sqrt{x_{pS}^2 + z_{pS}^2}} \tag{b}$$

On the other hand, Fig. 4-48b represents a vertical member with the lower end designated as joint k and the upper end designated as joint j. In this case the angle α is between 0° and 90° (when p has positive coordinates), and sin α and cos α are given by the following expressions:

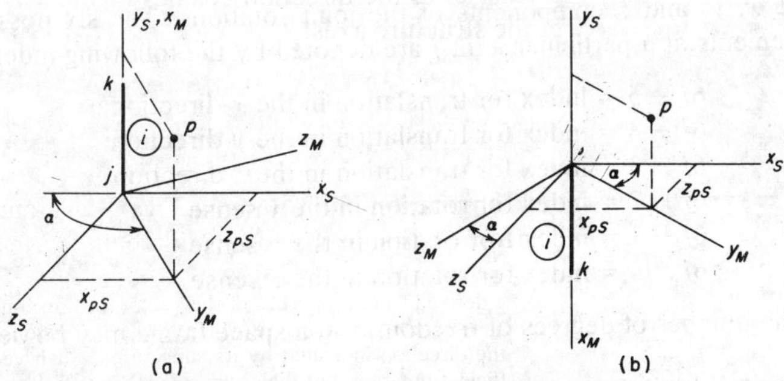

Fig. 4-48. Use of point p for a vertical space frame member.

$$\sin\alpha = \frac{z_{pS}}{\sqrt{x_{pS}^2 + z_{pS}^2}} \qquad \cos\alpha = \frac{x_{pS}}{\sqrt{x_{pS}^2 + z_{pS}^2}} \tag{c}$$

Expressions (b) and (c) may be combined into one set by introducing the direction cosine C_Y for the member, as follows:

$$\sin\alpha = \frac{z_{pS}}{\sqrt{x_{pS}^2 + z_{pS}^2}} \qquad \cos\alpha = \frac{-x_{pS}}{\sqrt{x_{pS}^2 + z_{pS}^2}}\,C_Y \tag{4-100}$$

For the cases shown in Fig. 4-48a and 4-48b the direction cosine C_Y has the values $+1$ and -1, respectively. Equations (4-100) can be used to calculate $\sin\alpha$ and $\cos\alpha$ from the coordinates of point p, which must be located in a principal plane of the member. These functions can then be substituted into the rotation matrix **R** (Eq. 4-99).

In summary, the member stiffness matrix $\mathbf{S}_{Mi}$ for a space frame member is first obtained for the member axes using Table 4-1 in Art. 4.3. Next, a rotation matrix **R** is constructed in a form which depends upon the case under consideration, using either the angle α or the coordinates of a point p to identify a principal plane. The matrix **R** has the form given in either Eq. (4-93) or Eq. (4-99), depending upon whether the member is inclined or vertical. In both of these cases, the angle α may be zero, which means that **R** reduces to one of the forms given previously for a space truss member (see Eqs. 4-86 and 4-87). In all cases, the rotation transformation matrix $\mathbf{R}_T$ is formed according to Eq. (4-94), and the member stiffness matrix for structure axes is computed by Eq. (4-95).

4.25 Analysis of Space Frames. The space frame shown in Fig. 4-43 contains rigidly connected members oriented in a general manner in space. The loads on the structure may be of any type and orientation. The numbering system to be adopted for members and joints is the same as for structures discussed previously.

Axial, flexural, and torsional deformations will be considered in the analysis of space frames.* The unknown displacements at the joints consist of six types, namely, the x, y, and z components of the joint translations and the x, y, and z components of the joint rotations. The six possible displacements at a particular joint j are denoted by the following indexes:

$6j - 5$ = index for translation in the x direction
$6j - 4$ = index for translation in the y direction
$6j - 3$ = index for translation in the z direction
$6j - 2$ = index for rotation in the x sense
$6j - 1$ = index for rotation in the y sense
$6j$ = index for rotation in the z sense

Also, the number of degrees of freedom n in a space frame may be deter-

*To omit axial deformations, constraints against length changes of members must be introduced. See the paper by Greene referenced previously in Art. 4.17.

mined from the number of joints n_j and the number of restraints n_r by the following expression:

$$n = 6n_j - n_r \tag{4-101}$$

A member i in a space frame will have joint numbers j and k at its ends, as shown in Fig. 4-49. The twelve possible displacements of the joints associated with this member are also indicated in Fig. 4-49. These displacements are indexed as follows:

$$\begin{aligned} j1 &= 6j - 5 \quad & j2 &= 6j - 4 \quad & j3 &= 6j - 3 \\ j4 &= 6j - 2 \quad & j5 &= 6j - 1 \quad & j6 &= 6j \\ k1 &= 6k - 5 \quad & k2 &= 6k - 4 \quad & k3 &= 6k - 3 \\ k4 &= 6k - 2 \quad & k5 &= 6k - 1 \quad & k6 &= 6k \end{aligned} \tag{4-102}$$

Construction of the joint stiffness matrix follows the same general pattern as with plane frames (see Art. 4.17), except that the process is complicated by the fact that more terms are involved. The 12 × 12 stiffness matrix $\mathbf{S}_{\mathrm{MS}\,i}$ is generated for each member in the frame (see Art. 4.24), and its contributions to the stiffnesses of joints j and k are assessed. For example, the first column in the matrix $\mathbf{S}_{\mathrm{MS}\,i}$ contributes to the joint stiffness matrix $\mathbf{S}_{\mathrm{J}}$ as follows:

$$\begin{aligned} (S_J)_{j1,j1} &= \Sigma S_{MS} + (S_{MS1,1})_i \quad & (S_J)_{k1,j1} &= (S_{MS7,1})_i \\ (S_J)_{j2,j1} &= \Sigma S_{MS} + (S_{MS2,1})_i \quad & (S_J)_{k2,j1} &= (S_{MS8,1})_i \\ (S_J)_{j3,j1} &= \Sigma S_{MS} + (S_{MS3,1})_i \quad & (S_J)_{k3,j1} &= (S_{MS9,1})_i \\ (S_J)_{j4,j1} &= \Sigma S_{MS} + (S_{MS4,1})_i \quad & (S_J)_{k4,j1} &= (S_{MS10,1})_i \\ (S_J)_{j5,j1} &= \Sigma S_{MS} + (S_{MS5,1})_i \quad & (S_J)_{k5,j1} &= (S_{MS11,1})_i \\ (S_J)_{j6,j1} &= \Sigma S_{MS} + (S_{MS6,1})_i \quad & (S_J)_{k6,j1} &= (S_{MS12,1})_i \end{aligned} \tag{4-103}$$

Eleven other sets of expressions similar to Eqs. (4-103) may be written to make a total of twelve sets of equations. Each set involves transferring

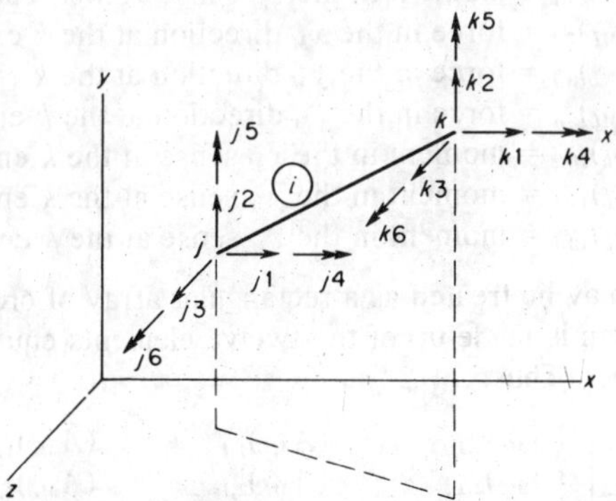

Fig. 4-49. End-displacements for space frame member.

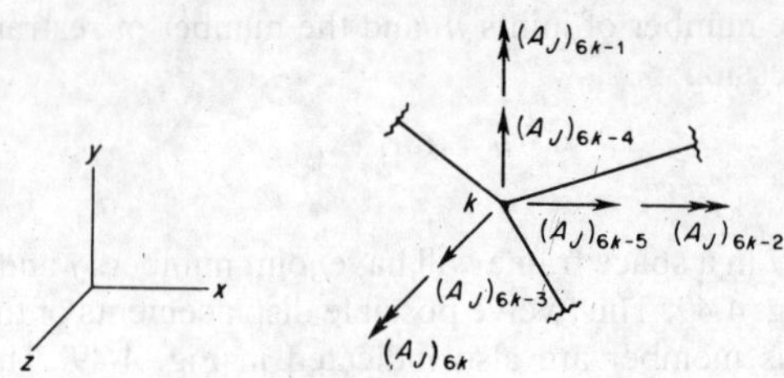

Fig. 4-50. Joint loads for a space frame.

elements from a given column in the matrix $\mathbf{S}_{MSi}$ to the appropriate locations in the matrix $\mathbf{S}_J$.

Actions applied at a typical joint k are shown in Fig. 4-50. In the figure, the actions $(A_J)_{6k-5}$, $(A_J)_{6k-4}$, and $(A_J)_{6k-3}$ are the x, y, and z components of the concentrated force applied at the joint. In addition, the actions $(A_J)_{6k-2}$, $(A_J)_{6k-1}$, and $(A_J)_{6k}$ are the x, y, and z components of the moment vector acting at the joint. These actions are placed in the vector of applied actions $\mathbf{A}_J$, which assumes the following form:

$$\begin{aligned}\mathbf{A}_J = \{&(A_J)_1, (A_J)_2, (A_J)_3, (A_J)_4, (A_J)_5, (A_J)_6, \\ &\ldots, (A_J)_{6k-5}, (A_J)_{6k-4}, (A_J)_{6k-3}, (A_J)_{6k-2}, (A_J)_{6k-1}, (A_J)_{6k}; \\ &\ldots, (A_J)_{6nj-5}, (A_J)_{6nj-4}, (A_J)_{6nj-3}, (A_J)_{6nj-2}, (A_J)_{6nj-1}, (A_J)_{6nj}\}\end{aligned} \tag{4-104}$$

Figure 4-51b shows the actions at the ends of a restrained space frame member i due to loads on the member itself. The end-actions $\mathbf{A}_{ML}$ for member-oriented axes are defined as follows:

$$\begin{aligned}
(A_{ML})_{1,i} &= \text{force in the } x_M \text{ direction at the } j \text{ end}\\
(A_{ML})_{2,i} &= \text{force in the } y_M \text{ direction at the } j \text{ end}\\
(A_{ML})_{3,i} &= \text{force in the } z_M \text{ direction at the } j \text{ end}\\
(A_{ML})_{4,i} &= \text{moment in the } x_M \text{ sense at the } j \text{ end}\\
(A_{ML})_{5,i} &= \text{moment in the } y_M \text{ sense at the } j \text{ end}\\
(A_{ML})_{6,i} &= \text{moment in the } z_M \text{ sense at the } j \text{ end}\\
(A_{ML})_{7,i} &= \text{force in the } x_M \text{ direction at the } k \text{ end}\\
(A_{ML})_{8,i} &= \text{force in the } y_M \text{ direction at the } k \text{ end}\\
(A_{ML})_{9,i} &= \text{force in the } z_M \text{ direction at the } k \text{ end}\\
(A_{ML})_{10,i} &= \text{moment in the } x_M \text{ sense at the } k \text{ end}\\
(A_{ML})_{11,i} &= \text{moment in the } y_M \text{ sense at the } k \text{ end}\\
(A_{ML})_{12,i} &= \text{moment in the } z_M \text{ sense at the } k \text{ end}
\end{aligned}$$

The matrix $\mathbf{A}_{ML}$ may be treated as a rectangular array of order $12 \times m$, in which each column is made up of the twelve elements enumerated above for a given member. Thus,

$$\mathbf{A}_{ML} = \begin{bmatrix} (A_{ML})_{1,1} & \cdots & (A_{ML})_{1,i} & \cdots & (A_{ML})_{1,m} \\ (A_{ML})_{2,1} & \cdots & (A_{ML})_{2,i} & \cdots & (A_{ML})_{2,m} \\ \cdots & \cdots & \cdots & \cdots & \cdots \\ (A_{ML})_{12,1} & \cdots & (A_{ML})_{12,i} & \cdots & (A_{ML})_{12,m} \end{bmatrix} \tag{4-105}$$

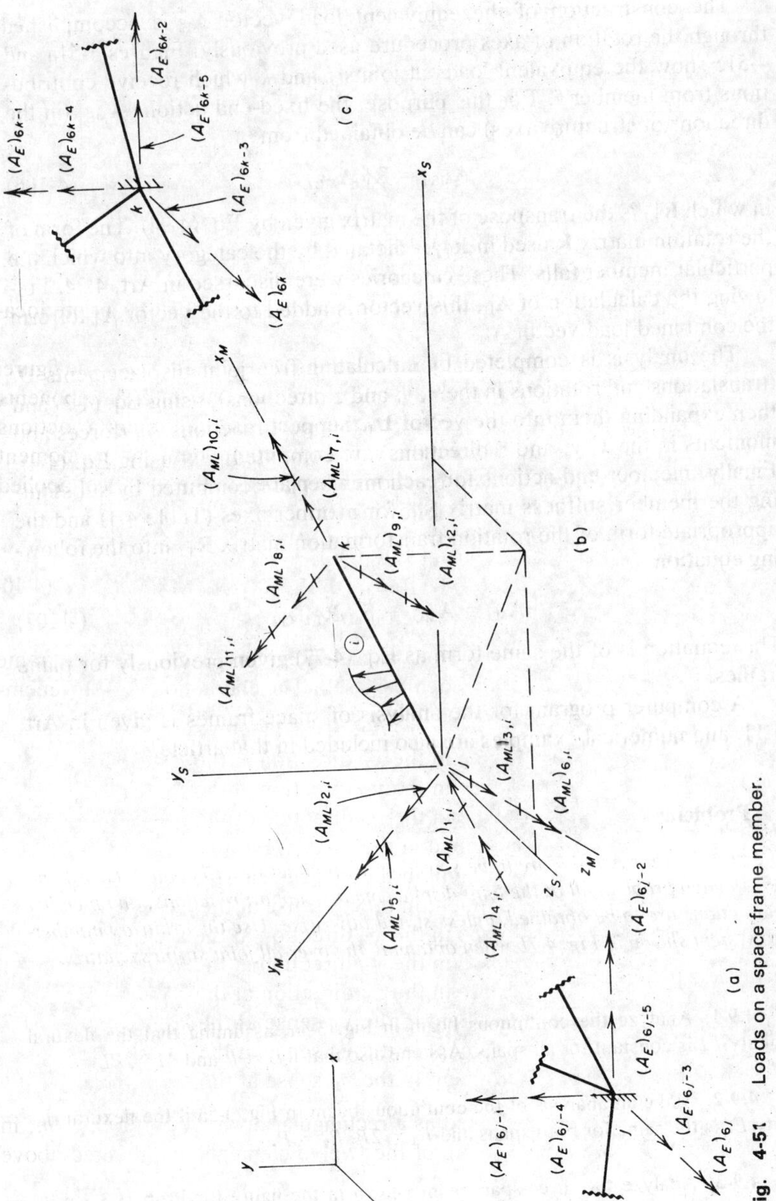

Fig. 4-51 Loads on a space frame member.

The construction of the equivalent load vector $\mathbf{A}_E$ is accomplished through the rotation of axes procedure used previously. Figures 4-51a and 4-51c show the equivalent loads at joints j and k which receive contributions from member i. For this purpose, the fixed-end actions $\mathbf{A}_{MS\,i}$ (in the directions of structure axes) can be obtained from

$$\mathbf{A}_{MS\,i} = \mathbf{R}^T_{T\,i}\mathbf{A}_{ML\,i} \tag{4-106}$$

in which $\mathbf{R}^T_{T\,i}$ is the transpose of the matrix given by Eq. (4-94). The form of the rotation matrix $\mathbf{R}$ used in $\mathbf{R}_{T\,i}$ is dictated by the category into which the particular member falls. These categories were discussed in Art. 4.24. Following the calculation of $\mathbf{A}_E$, this vector is added to the vector $\mathbf{A}_J$ to form the combined load vector $\mathbf{A}_C$.

The analysis is completed by calculating free joint displacements $\mathbf{D}_F$ (translations and rotations in the x, y, and z directions), using Eq. (4-3) and then expanding them into the vector $\mathbf{D}_J$. Support reactions $\mathbf{A}_R$ (forces and moments in the x, y, and z directions) are computed next using Eq. (4-4). Finally, member end-actions for each member are computed by substituting the member stiffness matrix $\mathbf{S}_{M\,i}$ for member axes (Table 4-1) and the appropriate form of the rotation transformation matrix $\mathbf{R}_{T\,i}$ into the following equation:

$$\mathbf{A}_{M\,i} = \mathbf{A}_{ML\,i} + \mathbf{S}_{M\,i}\mathbf{R}_{T\,i}\mathbf{D}_{J\,i} \tag{4-107}$$

This equation is of the same form as Eq. (4-77) given previously for plane frames.

A computer program for the analysis of space frames is given in Art. 5.11, and numerical examples are also included in that article.

Problems

The problems for Art. 4.9 are to be solved in the manner described in Arts. 4.8 and 4.9. In each problem all of the joint displacements, support reactions, and member end-actions are to be obtained, unless stated otherwise. Use the arbitrary numbering system shown in Fig. 4-7b when obtaining the over-all joint stiffness matrix.

4.9-1. Analyze the continuous beam in Fig. 3-4a, assuming that the flexural rigidity EI is constant for all spans. Assume also that $wL = P$ and $M = PL$.

4.9-2. Make an analysis of the continuous beam in Fig. 3-5a if the flexural rigidity EI is the same for both spans and $P_1 = 2P$, $P_2 = P$.

4.9-3. Analyze the three-span beam shown in the figure for Prob. 3.3-7 if $L_1 = L_3 = L$, $L_2 = 2L$, $P_1 = P_2 = P_3 = P$, $M = PL$, and $wL = P$. The flexural rigidity for members AB and CD is EI and for member BC is $2EI$.

4.9-4. Analyze the beam shown in the figure for Prob. 3.3-14 assuming that $EI_1 = 2EI$ and $EI_2 = EI$.

4.9-5. Analyze the overhanging beam shown in the figure, taking points A, B, and C as joints. The beam has constant flexural rigidity EI.

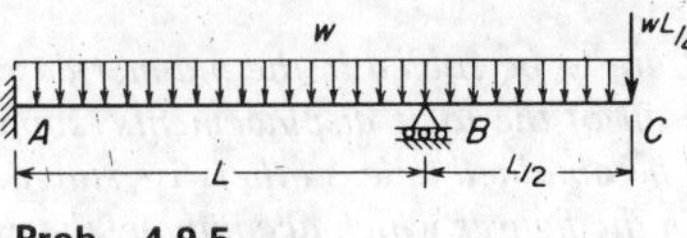

Prob. 4.9-5.

4.9-6. Analyze the three-span beam having constant flexural rigidity EI (see figure).

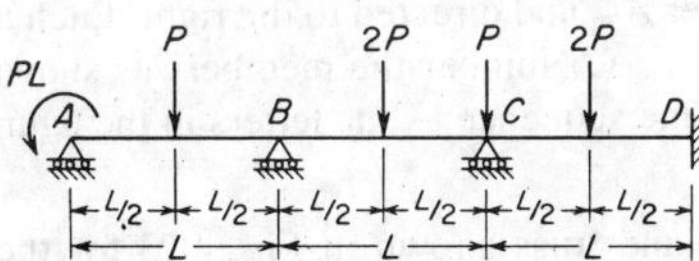

Prob. 4.9-6.

4.9-7. Analyze the beam shown in the figure, taking points A, B, C, and D as joints. The segment AB has a flexural rigidity of $2EI$, and the portion from B to D has a constant flexural rigidity of EI.

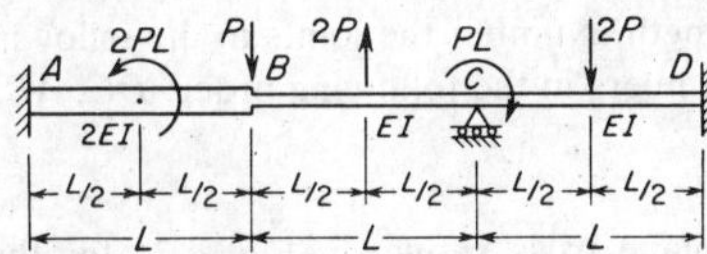

Prob. 4.9-7.

4.9-8. Obtain the joint stiffness matrix $\mathbf{S}_J$ for the beam shown in Prob. 3.3-8; also, rearrange and partition $\mathbf{S}_J$ into the form given by Eq. (4-8). Assume that the beam has constant EI.

4.9-9. Obtain the joint stiffness matrix $\mathbf{S}_J$ for the beam shown in Prob. 3.3-11, assuming that EI is constant. Also, rearrange and partition the matrix into the form given by Eq. (4-8).

4.9-10. Find the joint stiffness matrix $\mathbf{S}_J$ for the beam of Prob. 3.3-12, assuming that the flexural rigidity of the middle span is twice that of the end spans. Also, rearrange and partition the matrix into the form given by Eq. (4-8).

4.9-11. Find the joint stiffness matrix $\mathbf{S}_J$ for a continuous beam on five simple supports having four identical spans, each of length L. Also, rearrange and partition the matrix into the form given by Eq. (4-8). Assume that EI is constant for all spans.

4.9-12. Obtain the joint stiffness matrix $\mathbf{S}_J$ for the beam shown in the figure.

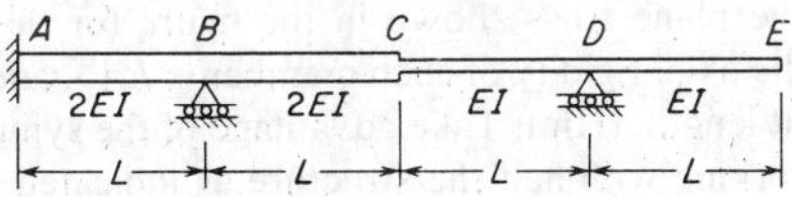

Prob. 4.9-12.

Also, rearrange and partition the matrix into the form given by Eq. (4-8). The beam has a flexural rigidity of $2EI$ from A to C and EI from C to E.

The problems for Art. 4.12 are to be solved in the manner described in Arts. 4.11 and 4.12. In each problem all of the joint displacements, support reactions, and member end-actions are to be obtained unless otherwise stated. Use the arbitrary numbering systems shown in the figures which accompany the problems.

4.12-1. Analyze the plane truss structure in the figure for Prob. 3.3-16 under the effect of the vertical load P shown in the figure plus a horizontal load of $2P$ applied at the mid-height of member AC and directed to the right. Each member of the truss has the same axial rigidity EA_X. Number the members as shown in the figure, and number the joints in the same sequence as the letters in the figure.

4.12-2. Analyze the plane truss shown in Fig. 4-23 for the loads shown, but with a hinge support which prevents translation at point B (joint 2). Otherwise, the data are the same as in the example problem.

4.12-3. Analyze the plane truss shown in the figure for Prob. 3.3-20 due to the weights of the members. Each member has the same axial rigidity EA_X and the same weight w per unit length. Number the joints in the following sequence: D, C, A, and B. Number the members in the following order: DC, AB, AD, BC, DB, and AC.

4.12-4. Analyze the plane truss shown in Fig. 4-23 for the effect of its own weight, assuming a roller support exists at joint B which prevents translation in the x direction. Assume that the weight of each member is w per unit length. Otherwise, the structural data are the same as in the example problem.

4.12-5. Analyze the plane truss for the loads shown (see figure). All members have the same axial rigidity EA_X.

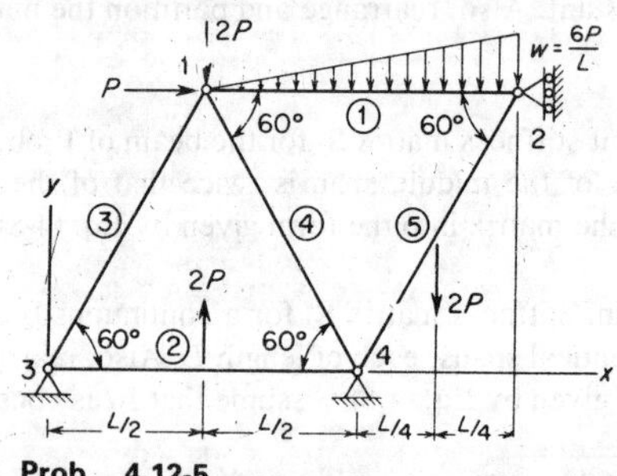

Prob. 4.12-5.

4.12-6. Analyze the plane truss shown in the figure for the effect of its own weight. Assume that the axial rigidity of each member is EA_X, and that each has the same weight w per unit length. (Hint: Take advantage of the symmetry of the structure and loading by working with half the structure as indicated in the second part of the figure. Note that both the area and the weight of the middle bar must be halved.)

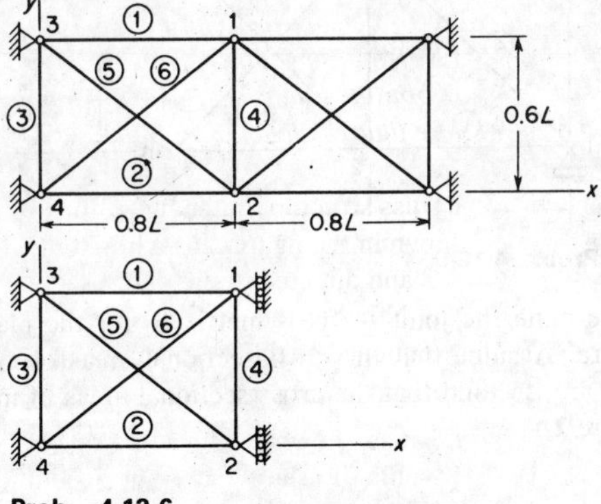

Prob. 4.12-6.

4.12-7. Obtain the joint stiffness matrix $\mathbf{S}_J$ for the plane truss shown in the figure. Assume that all members have the same axial rigidity EA_X.

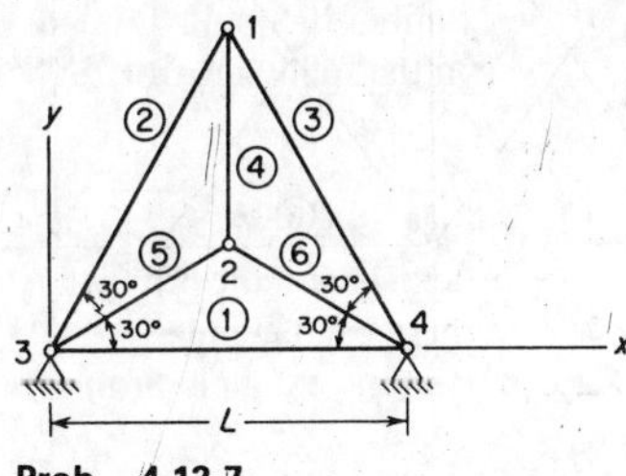

Prob. 4.12-7.

4.12-8. Determine the joint stiffness matrix $\mathbf{S}_J$ for the plane truss shown in the figure. Assume that all members have the same axial rigidity EA_X.

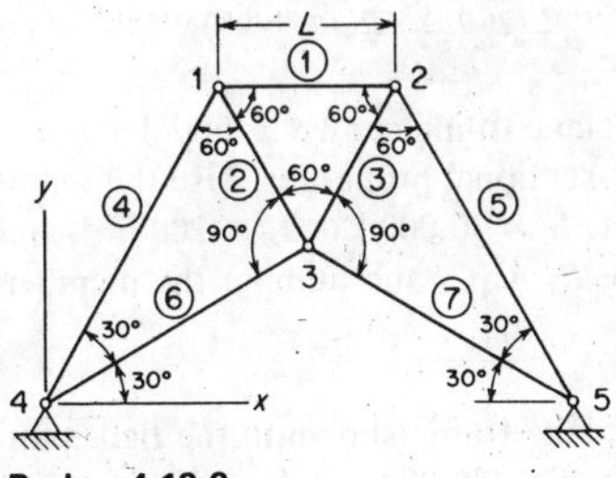

Prob. 4.12-8.

4.12-9. Obtain the joint stiffness matrix $\mathbf{S}_J$ for the plane truss structure shown in the figure. Rearrange and partition the matrix into the form given by Eq. (4-8). Assume that the horizontal and vertical members have cross-sectional areas A_X, and that the diagonal members have cross-sectional areas $2A_X$.

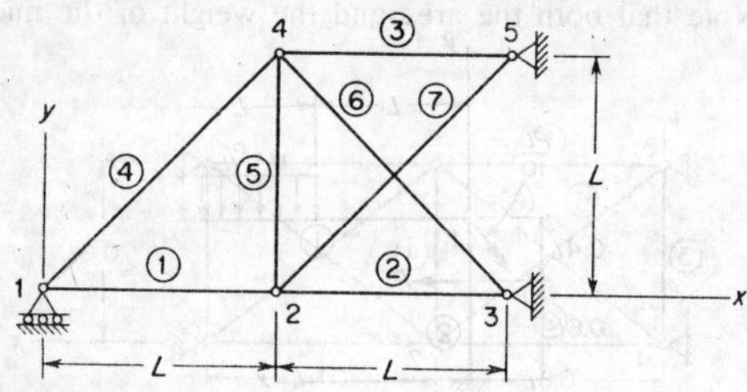

Prob. 4.12-9.

4.12-10. Determine the joint stiffness matrix S_J for the plane truss structure given in the figure. Assume that the cross-sectional areas of members numbered one through five are A_X, and that the cross-sectional areas of members numbered six through ten are $2A_X$.

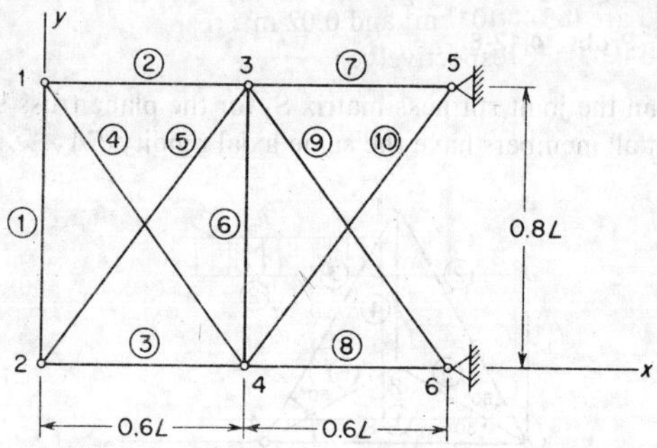

Prob. 4.12-10.

The problems for Art. 4.18 are to be solved in the manner described in Arts. 4.17 and 4.18. In each problem all of the joint displacements, support reactions, and member end-actions are to be obtained unless otherwise stated. Use the numbering systems shown in the figures which accompany the problems. In the problems with US units, convert to kips and inches where necessary.

4.18-1. Analyze the plane frame shown in Fig. 3-10a, assuming that both members have the same cross-sectional properties. Use the following numerical data: $P = 10$ kips, $L = H = 12$ ft, $E = 30{,}000$ ksi, $I_Z = 200$ in.^{4}, and $A_X = 10$ in.^{2} Number the joints in the sequence *B, A, C*, and number the members in the sequence *AB, BC*.

4.18-2. Analyze the plane frame shown in the figure for Prob. 3.3-25 if $P = 6$ kips, $L = 24$ ft, $H = 16$ ft, $E = 30{,}000$ ksi, $I_Z = 350$ in.^{4}, and $A_X = 16$ in.^{2} Number the joints in the sequence *B, A, C, D*, and number the members in the sequence *AB, BC, DB*.

4.18-3. Analyze the plane frame shown in the figure, assuming that both members have the same cross-sectional properties. Use the following data: $P = 50$ kN, $L = 6$ m, $E = 200 \times 10^6$ kN/m^2, $I_Z = 1 \times 10^{-3}$ m^4, and $A_X = 0.02$ m^2.

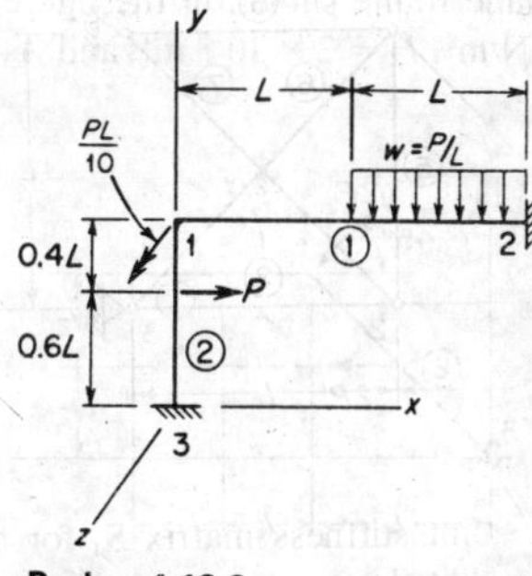

Prob. 4.18-3.

4.18-4. Analyze the plane frame shown in the figure, assuming that $P = 40$ kN, $L = 2$ m, and $E = 70 \times 10^6$ kN/m². The moment of inertia and cross-sectional area of members 1 and 3 are 0.5×10^{-3} m⁴ and 0.02 m², respectively, and for member 2 are 1×10^{-3} m⁴ and 0.03 m², respectively.

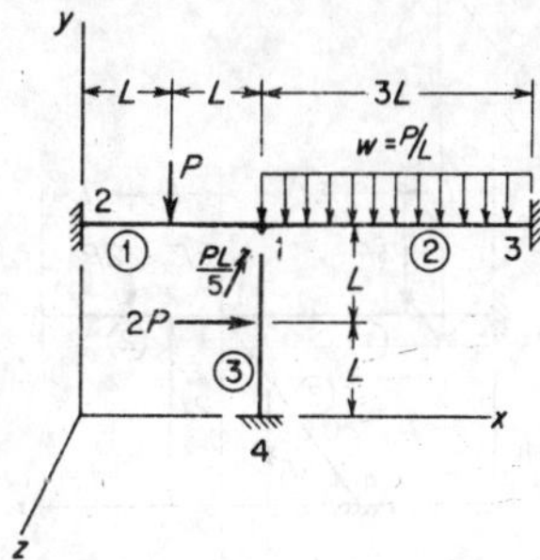

Prob. 4.18-4.

4.18-5. Analyze the plane frame shown in figure (a) for the effect of the loads shown. Assume that all members have $E = 10{,}500$ ksi, $I_Z = 26$ in.⁴, $A_X = 8$ in.², and also assume that $L = 60$ in. and $P = 2000$ lb. Take advantage of the symmetry of the structure and loading by analyzing the structure shown in figure (b). Note that the area and moment of inertia of member 1 must be halved. The imaginary restraint at joint 1 prevents both x translation and z rotation.

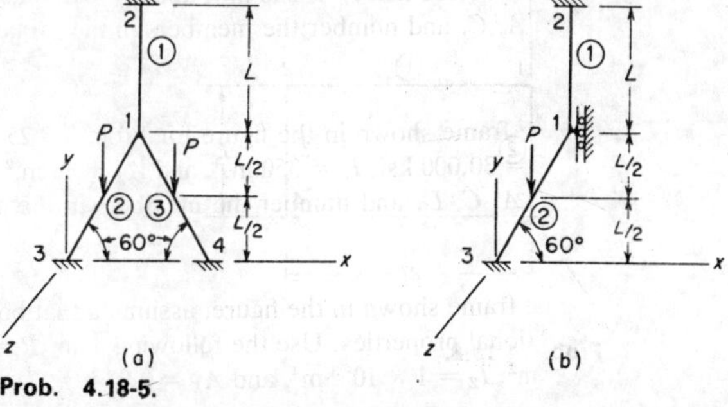

Prob. 4.18-5.

4.18-6. Analyze the plane frame shown in the figure, assuming that (for both members) $E = 200 \times 10^6$ kN/m², $I_Z = 2 \times 10^{-3}$ m⁴, and $A_X = 0.04$ m². Also, assume $L = 3$ m and $P = 30$ kN.

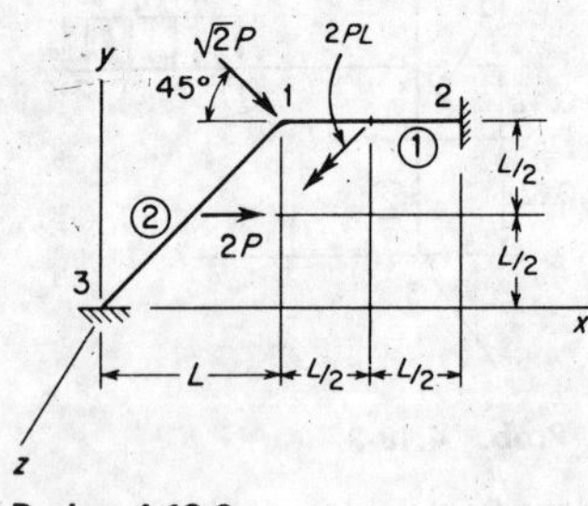

Prob. 4.18-6.

4.18-7. The plane frame shown in the figure is to be analyzed on the basis of the following numerical data: $E = 200 \times 10^6$ kN/m², $P = 10$ kN, $L = 2$ m; for members 1 and 2, $I_Z = 1.5 \times 10^{-3}$ m⁴, $A_X = 0.01$ m²; for member 3, $I_Z = 1.8 \times 10^{-3}$ m⁴, $A_X = 0.03$ m².

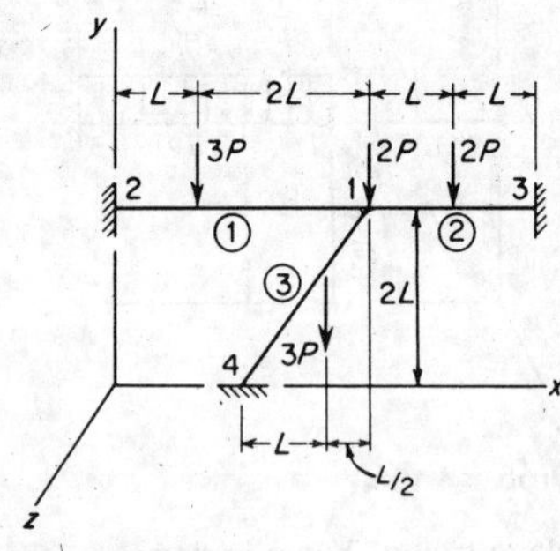

Prob. 4.18-7.

4.18-8. Determine the joint stiffness matrix $\mathbf{S}_J$ for the plane frame shown in the figure if all members have the same cross-sectional properties. Assume the following data: $E = 30{,}000$ ksi, $L = 20$ ft, $I_Z = 1600$ in.⁴, $A_X = 24$ in.²

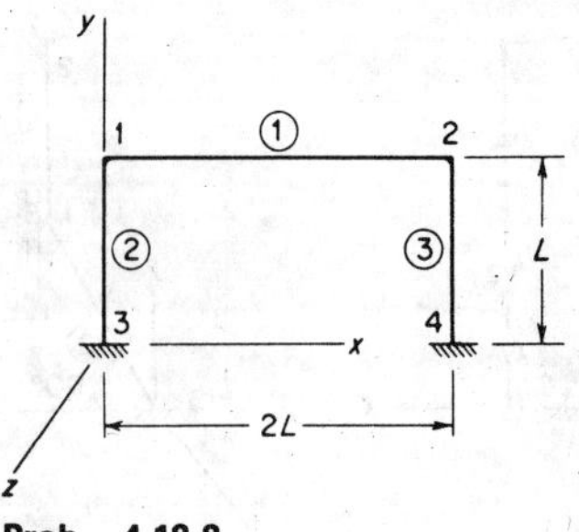

Prob. 4.18-8.

4.18-9. Repeat the preceding problem for the plane frame shown in the figure. Also, rearrange and partition the matrix into the form given by Eq. (4-8).

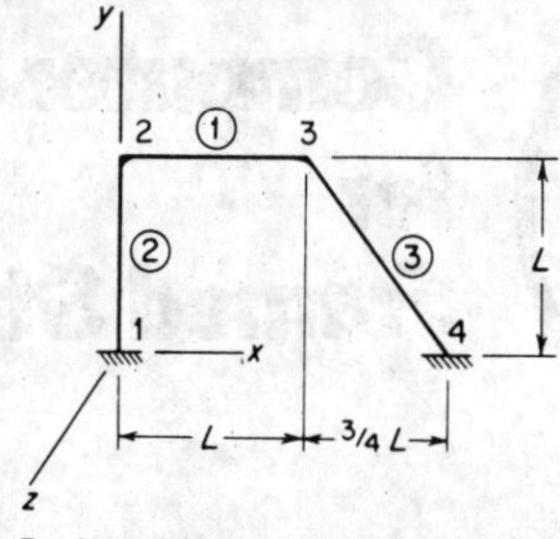

Prob. 4.18-9.

Computer Programs for Framed Structures

5.1 Introduction. This chapter contains flow charts of computer programs for analyzing the six basic types of framed structures by the direct stiffness method. The flow charts are sufficiently detailed so that persons who are familiar with the elements of computer programming can write their own programs for structural analysis.

All of the structures to be analyzed by the programs of this chapter are assumed to consist of straight, prismatic members. The material properties for a given structure are taken to be constant throughout the structure. Only the effects of loads are considered, and no other influences, such as temperature changes, are taken directly into account. However, such influences can usually be handled by converting them into equivalent joint loads, as described in Chapter 6. The programs are designed to handle in a single computer run any number of structures as well as any number of loading systems for the same structure.

Article 5.2 contains a brief description of the essential features of computer languages, with emphasis on FORTRAN and flow chart symbols. Program notation, preparation of data, and a description of program organization are given in Arts. 5.3 through 5.5. Flow charts of the programs for the six types of framed structures are presented in Arts. 5.6 through 5.11. Each of these articles also contains examples to demonstrate the preparation of input data and the form of output results from a digital computer (for both customary US units and SI units). Finally, a combined program for analyzing all types of framed structures is discussed in Art. 5.12.

5.2 FORTRAN Programming and Flow Charts. It is assumed that the reader is already acquainted with FORTRAN or some other algorithmic programming language. Some programming experience with elementary problems, not necessarily related to structural engineering, is also desirable. If the reader has this rudimentary programming ability as well as a fundamental understanding of the material in Chapter 4, there should be no great difficulty in implementing the programs in this chapter. The programs are presented in the form of annotated flow charts that are oriented toward the features of FORTRAN IV.* These FORTRAN-oriented charts are sufficiently detailed so that every step in the calculations is covered. Furthermore, a person who wishes to write these programs in some computer language other than FORTRAN should be able to do so quite easily by going step-by-step through the flow charts.

Some of the essential features of an algorithmic language (such as FORTRAN), as well as the symbols used in the flow charts, are summarized in this article. The discussion is not intended to be complete, and it covers only those features that are required for the structural programs.

Numbers in a computer program may be either *decimal numbers* or *integer numbers*. The decimal numbers are also referred to as *real numbers* or *floating-point numbers*. Integer numbers are whole numbers that do not include a decimal point. Both types will be used in the programs, but they must be distinguished from one another because arithmetic operations are not always the same for the two types of numbers. Most of the calculations in the programs are made with decimal numbers. Integer numbers are used primarily as indexes (or subscripts).

Identifiers are names for variables, constants, or other entities used in a program. For example, the symbol E is used as an identifier for the modulus of elasticity of the material. In general, the symbols used previously in Chapter 4 are adopted in the programs as identifiers, although some minor modifications are required. All alphabetical characters will be written as capital letters, and all symbols must be on the same line. For example, the symbol for the number of joints n_j becomes the identifier NJ in a computer program. Identifiers of this type require only one storage location in the memory of a computer and are said to represent *simple variables*.

There are other variables, called *subscripted variables*, which require more than one storage location. Thus, an identifier for a subscripted variable represents an array of numbers, as in the case of a vector or a matrix. The location of a particular element in the array is determined by the subscripts following the symbol. These subscripts are enclosed in parentheses; for example, an element of the vector $\mathbf{A}_E$, such as A_{E4}, becomes AE(4) in a computer program. Similarly, an element of the matrix $\mathbf{A}_{ML}$, such as A_{ML23}, becomes AML(2,3). It is necessary to reserve in advance a block of storage locations in the computer memory for each subscripted variable, as explained later.

Operators used in computer programming are of several types, and those used in the structural programs are summarized in Table 5-1. *Arithmetic operators* are used to signify the operations of addition, subtraction, multiplication, division, and exponentiation. Note that multiplication is indicated by an asterisk, as in FORTRAN; thus, the product of B times C is represented in a program as B*C. Since all symbols must be on the same

*For instance, see Hammond, R. H., Rogers, W. B., and Houck, B., Jr., *Introduction to FORTRAN IV,* 2nd ed., McGraw-Hill, 1978; or McCracken, D. D., *A Guide to FORTRAN IV Programming,* 2nd ed., Wiley, New York, 1972.

Table 5-1
Operators Used in Computer Programs

Operator	*Symbol*
Arithmetic Operators:	
Addition	+
Subtraction	–
Multiplication	*
Division	/
Exponentiation	**
Replacement Operator	=
Relational Operators:	
Less than	$<$
Less or equal	$\leqslant$
Equal	=
Greater or equal	$\geqslant$
Greater than	$>$
Not equal	$\neq$

line, exponentiation is denoted by two asterisks followed by the exponent. For example, A^4 is represented in a program by A**4.

The *replacement operator* is an equal sign, as in FORTRAN, which represents the operation of replacing an existing number in a storage location with a new number. The new number is then said to be the current value of the variable.

The *relational operators* are also listed in Table 5-1. These consist of the relations less than, less or equal, equal, greater or equal, greater than, and not equal. All of the relational operators are represented in the table by the usual algebraic symbols. These symbols are not the same as their counterparts in FORTRAN, which are .LT., .LE., .EQ., .GE., .GT., and .NE. However, the former are preferred to the latter for use in flow charts because they are more general.

Computer programs consist of *declarations* and *statements*. The purpose of a declaration may be to define the properties of one or more identifiers used in a program. For example, declarations are used to specify which identifiers represent integer numbers and which represent decimal numbers (*type declarations*). The former are declared to be identifiers of integer type, and the latter are declared to be identifiers of floating (or real) type. In FORTRAN identifiers beginning with the letters I, J, K, L, M, and N are implied to be of type integer unless otherwise specified. Another important kind of declaration specifies the amounts of storage to be reserved for arrays in the program (*array declarations*). Declarations are also used to specify the items in sets of input or output data, as well as the details of format for the output of information (*format declarations*).

Statements, on the other hand, define operations to be performed by the computer. The order in which statements are written is an important feature in computer programming because instructions are executed one-by-one in the sequence in which they appear in the program, unless a specific command causes control to pass to another statement that is not in sequence. A statement may be preceded by a *label*, which must be an integer number in FORTRAN. Statement labels are useful for proceeding from one statement to another in a sequence other than that in which they are listed.

The types of statements used in the programs in this book are summarized in Table 5-2. Flow chart symbols denoting these statements also appear in the table. Each of the statements listed in Table 5-2 is described briefly in the following discussion.

The *input statement* causes information to be transferred into the mem-

Table 5-2

Statements Used in Computer Programs

Type of Statement	Flow Chart Symbol
(a) Input	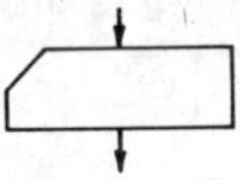
(b) Output	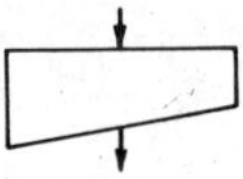
(c) Assignment	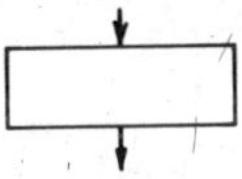
(d) Unconditional control	
(e) Conditional control	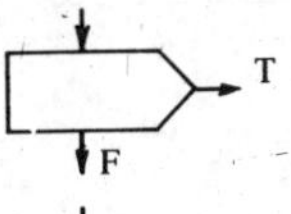 T F
(f) Continue	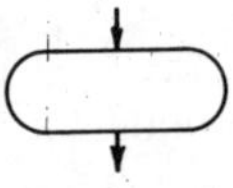
(g) Iterative control	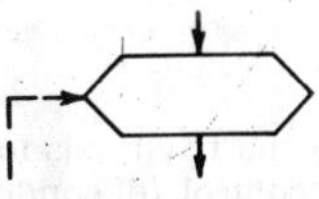

ory unit of the computer from some external source, such as a terminal, data cards, or magnetic tape. An input statement is represented in a flow chart by a box in the shape of a data card (see Table 5-2). As an example, suppose that the numbers A, B, and C are to be brought into the computer as input data. This may be represented in the flow chart as shown in Fig. 5-1a. (The arrows denote the flow of steps in a chart.)

The *output statement* causes information in the computer to be transferred outside and communicated to the programmer by means of a printer, a cathode ray tube (CRT), a card punch, or other device. An output statement appears in a flow chart as a box with a diagonal line at the base, as shown in Table 5-2. Figure 5-1b demonstrates the appearance in a flow chart of an output statement that causes the numbers A, B, and C to be printed (or otherwise communicated to the programmer).

The *assignment statement* causes the current value of a variable to be replaced with a new value. For example, the statement

$$A = B*C$$

means that B*C is to be evaluated and put into the storage location for the

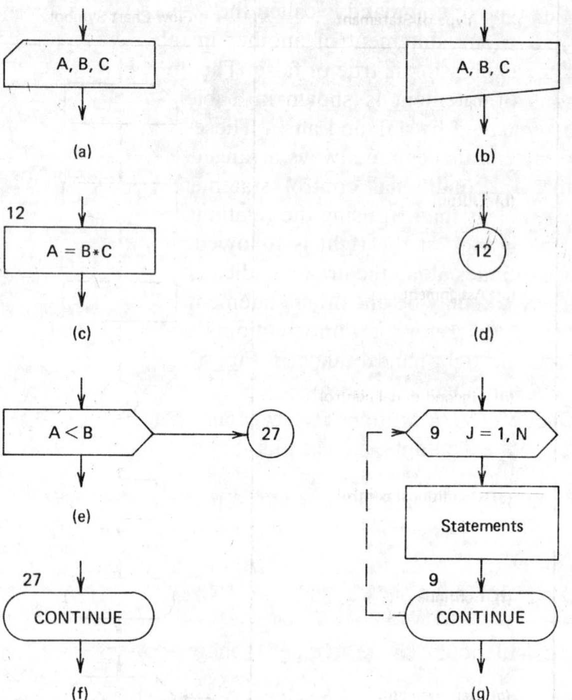

Fig. 5-1. Flow chart symbols for statements: (a) input, (b) output, (c) assignment, (d) unconditional control, (e) conditional control, (f) continue, and (g) iterative control.

variable A. Thus, the current value of A is replaced by the product of B times C. The replacement operator (an equal sign) is an integral part of the assignment statement. In a flow chart the assignment statement is enclosed in a rectangle (see Table 5-2). The statement described above is shown in Fig. 5-1c as it would appear in a flow chart. The manner of showing a statement label in a flow chart is also illustrated in Fig. 5-1c. The label is written at the upper left-hand corner of the rectangle, and in this example the label is assumed to be the number 12.

The *unconditional control statement* causes a change in the order of execution of statements in a program. This instruction is commonly known as the *GO TO statement*. By means of this command, control may be transferred to any other statement in the program by referring to its statement label. The unconditional control statement is represented in a flow chart by a circle, as shown in Table 5-2. The label of the statement to which control is transferred is indicated inside the circle. For example, in Fig. 5-1d the statement label 12 in the circle indicates that control is to be transferred to the statement in the program having the label 12.

The *conditional control statement* provides the ability to make decisions based upon a test, which may be a relational expression. In FORTRAN this type of command is called the *logical IF statement*. Control is transferred to one statement or another in the program depending upon whether the expression is true or false. The flow chart symbol for a conditional control statement is shown in Table 5-2 with the true and false branches indicated by a T and an F. These letters are not required if the true and false branches are always arranged in the manner shown. Figure 5-1e shows a conditional control statement which tests to determine whether A is less than B, using the relational operator $<$. If the condition is true, the branch at the right is followed; if the condition is false, the program continues along the branch indicated by the lower arrow. In FORTRAN there can only be one other statement on the true branch of a logical IF statement. Therefore, an unconditional control (or GO TO) statement appears on the right-hand branch in Fig. 5-1e, representing typical FORTRAN construction.

The *CONTINUE statement* serves as a spacer in a FORTRAN program. It is not executable, except that it means "go on" (or continue with) the next statement. When labeled, it is frequently used as the object of a GO TO statement, as indicated in Fig. 5-1f.

The *iterative control statement* is a useful device for executing the same statement or set of statements a specific number of times. In FORTRAN this type of command is known as a *DO statement* (or "DO-loop"). It causes an index, such as J, to be set to a specified initial value for the first cycle of calculations. Then the index is incremented a specified amount for subsequent cycles until it reaches or exceeds a final value, which is also specified. The increment for a DO-loop in FORTRAN is implied to be unity, unless otherwise stated. In a flow chart the list of specified parameters is enclosed in a hexagonal figure of the type shown in Table 5-2. The dashed line and arrowhead at the left of the symbol indicate that control is returned to this statement in cyclic fashion.

Figure 5-1g shows an iterative control statement that causes other statements within the rectangular box to be executed N times. The iterative control statement contains the following indexes:

J = 1,N

This expression means that the quantity J takes the value 1 for the first execution and then is incremented by an implied value of 1 for subsequent executions until it is finally set equal to N, and the statements are executed for the last time. If the increment is not equal to 1, its value must be specified as a third parameter (following N) in the list. Note the use of the CONTINUE statement in Fig. 5-1g to delineate the scope of the DO-loop.

As an example, suppose that A, B, and C are all subscripted variables of N elements each. If the corresponding elements of B and C are to be multiplied and assigned to A, the statement in the rectangular box would be

A(J) = B(J)*C(J)

After N repetitions of the calculation, all elements of the arrays have been processed, and control is transferred to the next statement in the program.

It is often necessary to evaluate a *function* in the course of executing a given statement. For example, the assignment statement

A = B*COS(C)

involves the evaluation of the cosine of C. In this example COS is the function designator, and its use in a program implies that a specific set of operations is to be performed automatically, resulting in the evaluation of the cosine of C. Certain frequently used functions such as sine, cosine, square root, logarithm, absolute value, etc. (designated as SIN, COS, SQRT, LOG, ABS, etc.) are incorporated into algorithmic languages such as FORTRAN for the convenience of the programmer.

Another important feature of algorithmic languages enables the programmer to use a series of statements as a *subprogram* (or *subroutine* in FORTRAN). A subprogram is a block of statements that may be used at several points in the same program or in different programs without alteration. If a program is constructed using subprograms as building blocks, it is said to be *modular*. The programs for structural analysis presented in this chapter are constructed in this manner.

The block of FORTRAN coding constituting a subroutine is treated as a declaration (called a *subroutine declaration*) that contains both a heading and a body. The heading includes the *subroutine name*, to which may be appended a list of *formal parameters*, or *arguments*, enclosed in parentheses. These parameters represent quantities from the main program that are used in the subroutine. As an example of a subroutine name, consider the operation of multiplying a vector A (of M elements) by a scalar B to produce the vector C. A possible name for this subroutine would be:

SUBROUTINE VSCALE(M,A,B,C)

An important characteristic of a subroutine is that notation in the subroutine declaration has no connection with notation in the main program. When a program makes use of a subroutine, the formal parameters are replaced by *actual parameters* being used in the main program. Insertion of a *subroutine statement* (or *subroutine call*) into the program causes the steps in the subroutine to be executed. A subroutine call has the same format as a subroutine declaration, but in this case the list of actual parameters is used. For example, if a vector D (of N elements) in the program is to be multiplied by a scalar E to produce the vector F, the subroutine call becomes:

CALL VSCALE(N,D,E,F)

Note that the list of actual parameters must contain all of the identifiers to be substituted for the corresponding items in the list of formal parameters found in the subroutine declaration. Both the number and the sequence of actual parameters must agree with the list of formal parameters.

If the parameters are numerous, it is more convenient to list them in a *COMMON declaration** in both the subroutine and the main program. In this case the parameters are omitted from the subroutine name and the subroutine call, which would both become simply VSCALE for the example. If the main program calls various subroutines of this nature, all COMMON statements must have the same number and sequence of parameters.

Although the foregoing discussion of some of the features of an algorithmic language is brief and incomplete, its purpose has been to explain the conventions used in the FORTRAN-oriented flow charts of this book. With these features in mind, the programmer can follow the charts and write programs for structural analysis.

5.3 Program Notation. The notation used in the computer programs for framed structures is summarized in this article. For convenience, items are listed in alphabetical order. A symbol in the list followed by parentheses ( ) denotes a subscripted variable with a single subscript (or a vector). A symbol followed by parentheses enclosing a comma (,) denotes a subscripted variable with two subscripts (or a matrix). The type of each variable (integer or decimal) is implied by the default rule of FORTRAN. That is, variables having names that begin with the letters I, J, K, L, M, and N are taken to be integers, while all others are assumed to be decimals.

In conjunction with a computer program, symbols such as A_{ML} and AML may be used interchangeably. It is understood that the symbol A_{ML}

*See McCracken (referenced earlier).

represents a matrix and that the identifier AML stands for its counterpart in the program.

Identifiers Used in Computer Programs

Identifier	*Definition*
AC( )	Combined joint loads $\mathbf{A}_C$ (in directions of structure axes)
AE( )	Equivalent joint loads $\mathbf{A}_E$ (in directions of structure axes)
AJ( )	Actions $\mathbf{A}_J$ (loads) applied at joints (in directions of structure axes)
AM( )	Final member end-actions $\mathbf{A}_M$
AMD( )	Actions $\mathbf{A}_{MD}$ at ends of member due to joint displacements
AML(,)	Actions $\mathbf{A}_{ML}$ at ends of restrained members (in directions of member axes) due to loads
AR ( )	Support reactions $\mathbf{A}_R$ (in directions of structure axes)
AX( )	Cross-sectional area A_X of member
CX, CY, CZ	x, y, and z direction cosines of member (C_X, C_Y, C_Z)
CXZ	Identifier for expression CXZ = SQRT(CX*CX + CZ*CZ)
DF( )	Free joint displacements $\mathbf{D}_F$ (in directions of structure axes)
DJ( )	Joint displacements $\mathbf{D}_J$ for all joints (in directions of structure axes)
E	Modulus of elasticity E for tension or compression
EL( )	Length L of element (member)
G	Modulus of elasticity G for shear
I	Member index i
IA	Identifier used to indicate whether or not the angle α is zero
ID( )	Displacement indexes for joints
IM( )	Displacement indexes for a member
IR, IC	Row and column indexes
ISN	Structure number
ITS	Type of structure
J, K	Joint indexes
J1, . . . , J6	Indexes for displacements at j end of member
JE, KE	Indexes for expanded vectors
JJ( )	Designation for j end of member (joint j)
JK( )	Designation for k end of member (joint k)
JRL( )	Joint restraint list
K1, . . ., K6	Indexes for displacements at k end of member
LML( )	Loaded member list
LN	Loading number
M	Number of members
MD	Number of displacement coordinates for a member (MD = 2 NDJ)
N	Number of degrees of freedom
NB	Semi-band width of stiffness matrix
ND	Number of displacement coordinates for all joints (ND = NDJ*NJ)
NDJ	Number of displacements per joint
NJ	Number of joints
NLJ	Number of loaded joints
NLM	Number of loaded members

Identifier	Definition
NLS	Number of loading systems
NR	Number of support restraints
NRJ	Number of restrained joints
R(,)	Rotation matrix
SCM	Stiffness constant of member
SFF(,)	Stiffness matrix $\mathbf{S}_{FF}$ for free joint displacements
SINA, COSA	Sine and cosine of the angle α ($\sin \alpha$, $\cos \alpha$)
SM(,)	Member stiffness matrix $\mathbf{S}_M$ for member-oriented axes
SMRT(,)	Identifier used for temporary storage of the product $\mathbf{S}_M\mathbf{R}_T$
SMS(,)	Member stiffness matrix $\mathbf{S}_{MS}$ for structure-oriented axes
X( ), Y( ), Z( )	x, y, and z coordinates of joints
XCL, YCL, ZCL	x, y, and z components of length of member
XI( )	Torsion constant of member (I_X)
XP, YP, ZP	x, y, and z coordinates of point p
XPS, YPS, ZPS	x_S, y_S, and z_S coordinates of point p
YI( )	Moment of inertia about the y_M axis (I_Y)
YPG, ZPG	y_γ and z_γ coordinates of point p
ZI( )	Moment of inertia about the z_M axis (I_Z)

5.4 Preparation of Data. The manner in which a program is written depends to some extent upon the form in which the data are supplied. As a preliminary matter, the required input data for all of the programs in this chapter are described in this article. It is assumed that the data are given either as a group of lines typed on a terminal or as a deck of punched cards accompanying the coded program. Thus, the words "line" and "card" will be used interchangeably when discussing data.

The data lines (or cards) for a particular structure are divided into the three categories of *control data, structural data,* and *load data*. Only one set of control and structural data is required for each structure, but there may be any number of sets of load data. The control data consist of three parameters that are most useful when the six special-purpose programs are assembled into a combined program for all types of framed structures. These parameters are the structure number ISN, the type of structure ITS, and the number of loading systems NLS. The structure number ISN will be 1 for the first structure, 2 for the second structure, etc. On the other hand, the type of structure ITS is represented by one of the numbers 1 through 6, as follows:

ITS = 1: Continuous beam	ITS = 4: Grid
ITS = 2: Plane truss	ITS = 5: Space truss
ITS = 3: Plane frame	ITS = 6: Space frame

The number of loading systems NLS indicates the number of sets of load data that accompany a given set of structural data.

The input data required for continuous beams are given in Table 5-3. This table specifies the number of data lines (or cards) required for each category of information as well as the items to be included on the lines. The

Table 5-3
Preparation of Data for Continuous Beams

	Data	*Number of Lines*	*Items on Data Lines*			
	Control Data	1	ISN	ITS	NLS	
Structural Data	a. Structural parameters	1	M	NR	NRJ	E
	b. Member information	M	I	EL(I)	ZI(I)	
	c. Joint restraint list	NRJ	K	JRL(2K-1)	JRL(2K)	
Load Data	a. Load parameters	1	NLJ	NLM		
	b. Actions at joints	NLJ	K	AJ(2K-1)	AJ(2K)	
	c. Actions at ends of restrained members due to loads	NLM	I	AML(1,I) AML(3,I)	AML(2,I) AML(4,I)	

first line in the table under the category of structural data contains the number of members M, the number of restraints NR, the number of restrained joints NRJ, and the elastic modulus E of the material. The reason for including NRJ is that the number of data lines required to fill the joint restraint list (see line 4 of the table) is minimized if the number of restrained joints is known.

Each data line containing member information (line 3 of Table 5-3) includes for one member the member number I, the length EL(I), and the moment of inertia ZI(I) about the z axis. A total of M lines is required for the member data.

Each of the lines in the next series (a total of NRJ lines) contains a joint number K and two code numbers which indicate the conditions of restraint at that joint. The term JRL(2K-1) denotes the restraint against translation in the y direction at joint k, and the term JRL(2K) denotes the restraint against rotation in the z sense at joint k. The convention adopted in this program is the following: If the restraint exists, the integer 1 is assigned as the value of JRL; and if there is no restraint, a value of zero is assigned.

The first line in the load data contains two items. These are the number of loaded joints NLJ and the number of loaded members NLM. One reason for introducing these numbers is that they serve to minimize the amount of subsequent load data required. Another reason is that certain calculations in the program can be bypassed if either NLJ or NLM is equal to zero; for example, if NLM is equal to zero, the calculation of the vector A_E can be omitted.

Each line of joint load data (a total of NLJ lines) contains a joint number K and the two actions AJ(2K-1) and AJ(2K) applied at that joint. These actions are a force in the y direction and a moment in the z sense. Finally, each line for member loads (NLM lines total) contains a member number I and the four fixed-end actions AML for that member. The fixed-end actions consist of a force in the y direction and a moment in the z sense at each end of the member.

Table 5-4 shows the input data required for plane trusses. Since both beams and plane trusses have two possible displacements at each joint, Tables 5-3 and 5-4 are very similar. The structural parameters required for plane trusses are similar to those for beams, except that the number of joints NJ must be specified in addition to the other items. Moreover, a set of data (NJ lines total) is required to specify the coordinates of the joints of a truss. Each line in this set contains a joint number J, the x coordinate X(J) of the joint, and the y coordinate Y(J) of the joint.

On each of the lines containing member information the member number I is listed first, followed by the joint j number JJ(I) and the joint k number JK(I) for the two ends of the member. The choice of which end of the member is to be the j end and which is to be the k end is made arbitrarily by the programmer. The last item on each line is the cross-sectional area AX(I) of the member.

The data for the joint restraint list are similar to those for continuous beams, except that the types of restraints differ. For the plane truss, the terms JRL(2K-1) and JRL(2K) denote the restraints against translations in the x and y directions, respectively, at joint k. Similarly, the load data are symbolically the same as for continuous beams, but the applied actions AJ consist of forces in the x and y directions, and the hinged-end actions AML consist of forces in the x_M and y_M directions at the ends of the members.

Table 5-5 lists the input data required for plane frames. This table contains information similar to that in Table 5-4 for plane trusses, but there are certain differences caused by the fact that plane frames have three possible

Table 5-4

Preparation of Data for Plane Trusses

	Data	Number of Lines	Items on Data Lines			
	Control Data	1	ISN	ITS	NLS	
Structural Data	a. Structural parameters	1	M	NJ	NR NRJ	E
	b. Joint coordinates	NJ	J	X(J) Y(J)		
	c. Member information	M	I	JJ(I) JK(I) AX(I)		
	d. Joint restraint list	NRJ	K	JRL(2K-1)	JRL(2K)	
Load Data	a. Load parameters	1	NLJ	NLM		
	b. Actions at joints	NLJ	K	AJ(2K-1)	AJ(2K)	
	c. Actions at ends of restrained members due to loads	NLM	I	AML(1,I) AML(3,I)	AML(2,I) AML(4,I)	

Table 5-5
Preparation of Data for Plane Frames

	Data	Number of Lines	Items on Data Lines					
	Control Data	1	ISN	ITS	NLS			
Structural Data	a. Structural parameters	1	M	NJ	NR	NRJ	E	
	b. Joint coordinates	NJ	J	X(J)	Y(J)			
	c. Member information	M	I	JJ(I)	JK(I)	AX(I)	ZI(I)	
	d. Joint restraint list	NRJ	K	JRL(3K-2)		JRL(3K-1)	JRL(3K)	
Load Data	a. Load parameters	1	NLJ	NLM				
	b. Actions at joints	NLJ	K	AJ(3K-2)		AJ(3K-1)	AJ(3K)	
	c. Actions at ends of restrained members due to loads	NLM	I	AML(1,I) AML(4,I)		AML(2,I) AML(5,I)	AML(3,I) AML(6,I)	

displacements at each joint. One dissimilarity is that an additional member property ZI (moment of inertia of the cross section) is required on each of the lines containing member information. Another distinction is that each of the lines in the joint restraint series contains (in addition to the joint number K) three code numbers instead of two. The terms JRL(3K-2), JRL(3K-1), and JRL(3K) denote the restraints at joint k against translations in the x and y directions and rotation in the z sense, respectively.

Each line of joint load data contains a joint number K and the three actions AJ applied at that joint. These actions are applied forces in the x and y directions and a moment in the z sense. In addition, each line of member loads contains a member number I and the six actions AML at the ends of the restrained member. These actions consist of forces in the x_M and y_M directions and a moment in the z_M (or z) sense at the ends j and k, respectively.

The input data required for grids, shown in Table 5-6, are nearly the same as for plane frames (see Table 5-5). However, an additional material property (the shear modulus of elasticity G) is required on the first line of structural data in Table 5-6. Also, the torsion constant XI and moment of inertia YI appear on the lines containing member properties instead of the cross-sectional area AX and moment of inertia ZI.

Data for the joint restraint list are similar to those for plane frames, except that the nature of the restraints is different. For grids, the terms JRL(3K-2), JRL(3K-1), and JRL(3K) denote the restraints against rotations in the x and y senses and translation in the z direction, respectively, at joint k.

Similarly, the load data are symbolically the same as for plane frames, but the meanings are different. For a grid structure, the actions AJ applied at joints consist of moments in the x and y senses and a force in the z

Table 5-6
Preparation of Data for Grids

	Data	Number of Lines	Items on Data Lines
	Control Data	1	ISN ITS NLS
Structural Data	a. Structural parameters	1	M NJ NR NRJ E G
	b. Joint coordinates	NJ	J X(J) Y(J)
	c. Member information	M	I JJ(I) JK(I) XI(I) YI(I)
	d. Joint restraint list	NRJ	K JRL(3K-2) JRL(3K-1) JRL(3K)
Load Data	a. Load parameters	1	NLJ NLM
	b. Actions at joints	NLJ	K AJ(3K-2) AJ(3K-1) AJ(3K)
	c. Actions at ends of restrained members due to loads	NLM	I AML(1,I) AML(2,I) AML(3,I) AML(4,I) AML(5,I) AML(6,I)

direction. Finally, the actions AML at the ends of restrained members due to loads are moments in the x_M and y_M senses and a force in the z_M (or z) direction at the ends j and k, respectively.

Table 5-7 shows the input data required for space trusses. This table is similar to Table 5-5 for plane frames, with some exceptions. For example, each of the lines in Table 5-7 containing joint coordinates provides not only the x and y coordinates but also the z coordinate for a joint. On the lines containing member properties, only the cross-sectional area AX is required.

Table 5-7
Preparation of Data for Space Trusses

	Data	Number of Lines	Items on Data Lines
	Control Data	1	ISN ITS NLS
Structural Data	a. Structural parameters	1	M NJ NR NRJ E
	b. Joint coordinates	NJ	J X(J) Y(J) Z(J)
	c. Member information	M	I JJ(I) JK(I) AX(I)
	d. Joint restraint list	NRJ	K JRL(3K-2) JRL(3K-1) JRL(3K)
Load Data	a. Load parameters	1	NLJ NLM
	b. Actions at joints	NLJ	K AJ(3K-2) AJ(3K-1) AJ(3K)
	c. Actions at ends of restrained members due to loads	NLM	I AML(1,I) AML(2,I) AML(3,I) AML(4,I) AML(5,I) AML(6,I)

Data for the joint restraint list are similar to those for plane frames, except that the types of restraints are different. For space trusses, the terms JRL(3K-2), JRL(3K-1), and JRL(3K) denote the restraints at joint k against translations in the x, y, and z directions, respectively.

Similarly, the load data are symbolically the same as for plane frames, but the meanings are different. Actions AJ applied at joints consist of forces in the x, y, and z directions, and end-actions AML consist of forces in the x_M, y_M, and z_M directions at each end of a loaded member.

The input data required for space frames are summarized in Table 5-8, which is also similar to Table 5-5 for plane frames. However, since the space frame is three-dimensional and has six possible displacements per joint, there are more items of data for this type of structure than for plane frames. The first line of structural data in Table 5-8 includes the shear modulus G as well as the modulus E, and each of the lines containing joint coordinates must provide the z coordinate as well as the x and y coordinates of a joint.

Member information must include the cross-sectional properties AX, XI, YI, and ZI. In addition, an identifier IA is provided in order to indicate whether the rotation angle α is zero or not (see Art. 4.24 for the meaning of the angle α). If the angle α is zero for a given member, the value of zero is assigned to IA. If the angle α is not zero, the integer 1 is assigned. This is an arbitrary convention indicating that the member has its principal axes in skew directions. Under these conditions the principal axes are located by

Table 5-8
Preparation of Data for Space Frames

	Data	Number of Lines	Items on Data Lines					
	Control Data	1	ISN	ITS	NLS			
Structural Data	a. Structural parameters	1	M	NJ	NR	NRJ	E	G
	b. Joint coordinates	NJ	J	X(J)	Y(J)	Z(J)		
	c. Member information	M	I	JJ(I)	JK(I)	AX(I)		
				XI(I)	YI(I)	ZI(I)	IA	
	Coordinates of point p (required when IA = 1)		I	XP	YP	ZP		
	d. Joint restraint list	NRJ	K	JRL(6K-5)	JRL(6K-4)	JRL(6K-3)		
				JRL(6K-2)	JRL(6K-1)	JRL(6K)		
Load Data	a. Load parameters	1	NLJ	NLM				
	b. Actions at joints	NLJ	K	AJ(6K-5)	AJ(6K-4)	AJ(6K-3)		
				AJ(6K-2)	AJ(6K-1)	AJ(6K)		
	c. Actions at ends of restrained members due to loads	2NLM	I	AML(1,I)	AML(2,I)	AML(3,I)		
				AML(4,I)	AML(5,I)	AML(6,I)		
				AML(7,I)	AML(8,I)	AML(9,I)		
				AML(10,I)	AML(11,I)	AML(12,I)		

means of the coordinates of a point p in the x_M-y_M plane (see Figs. 4-45 and 4-46) but not on the axis of the member. Therefore, each data line for which IA is equal to 1 must be followed by an additional line on which the member index I and the three coordinates XP, YP, and ZP are given. These coordinates are used to obtain the rotation matrix as described in Art. 4.24.

An alternate method of programming would be to provide for the contingency that the angle α is known, in which case it would not be necessary to give the coordinates of a point p. The actual value of α could be assigned to an identifier named AA, after which the rotation matrix can be calculated directly from Eq. (4-93) or Eq. (4-99).

Each of the lines in the joint restraint series contains a joint number K and six code numbers indicating the conditions of restraint at that joint. The terms JRL(6K-5) through JRL(6K) denote the restraints against translations and rotations in the x, y, and z directions at joint k.

Each line of joint load data contains a joint number K and the six actions applied at that point. These actions are the components of the applied force vector and the applied moment vector in the x, y, and z directions.

In the last group of data, two lines must be provided for each member to which loads are applied because the information may not fit on one line (or card). The first line contains a member number I and the six actions at the j end of the restrained member, and the second line provides the six actions at the k end of the member. The actions at each end are the forces and moments in the x_M, y_M, and z_M directions.

Units. A consistent system of units must be used for the input data, which involve units of force and length only. The results of the calculations will have the same units as the input data, with the addition that all angles will automatically be in radians. One example of a consistent US system of units is to give loads in kips, lengths in inches, areas in square inches, modulus of elasticity in kips per square inch, etc. In such a case all of the final results will be in units of kips, inches, and radians. On the other hand, if SI units of kilonewtons and meters are used for the input data, the final results will be in terms of kilonewtons, meters, and radians.

5.5 Description of Programs. The primary objective of this chapter is to present computer programs for individual types of framed structures. Each of these special-purpose programs consists of a main program that uses a series of subprograms to perform the detailed calculations. The main program is designed to handle any number of structures in one run and any number of loading systems for each structure. It calls upon subprograms that accomplish the following steps for a given type of structure:

1. Read and print structural data
   a. Problem identification
   b. Structural parameters
   c. Joint coordinates (except continuous beams)
   d. Member information

   e. Joint restraint list
   f. Joint displacement indexes (computed internally)
2. Construct stiffness matrix
   a. Member stiffnesses
   b. Transfer to joint stiffness matrix
3. Read and print load data
   a. Load parameters
   b. Joint loads
   c. Member loads
4. Construct load vector
   a. Equivalent joint loads
   b. Combined joint loads
5. Calculate and print results
   a. Joint displacements
   b. Member end-actions
   c. Support reactions

To conserve computer time and storage, only the upper band of the stiffness matrix S_{FF} (for free joint displacements) is constructed. This matrix is generated as a rectangular array of size N × NB, where N is the number of degrees of freedom and NB is the semi-band width. Figure 5-2a shows the banded form of S_{FF}, and Fig. 5-2b depicts storage of the upper band as a rectangular matrix. The semi-band width NB for a particular structure is determined by the system for numbering joints. The value of this parameter may be calculated from the formula

$$NB = [NDJ*ABS(JK(I) - JJ(I) + 1) - NSR]_{max} \tag{5-1}$$

In this FORTRAN expression the identifiers JJ(I) and JK(I) denote the joint numbers at the two ends of a member I, and NDJ is the number of displacements per joint. The symbol NSR represents the number of support restraints having displacement indexes within the range of indexes spanned by the term NDJ*ABS(JK(I) − JJ(I) + 1). The member for which the right-hand side of Eq. (5-1) is a maximum determines the semi-band width NB. Since the magnitude of NB dictates the amount of storage required for the upper band portion of S_{FF}, the joints of a structure should be numbered in a sequence which minimizes that parameter. If the term NSR is equal to zero, minimization of the difference JK(I) − JJ(I) serves to minimize NB also.

Generation of the upper band of the stiffness matrix S_{FF} in the form shown by Fig. 5-2b requires modification of the subscripts calculated for the form in Fig. 5-2a, which are

$$ID(J) = J - N1 \tag{5-2a}$$
$$ID(K) = K - N2 \tag{5-2b}$$

These FORTRAN expressions for row and column indexes of a typical

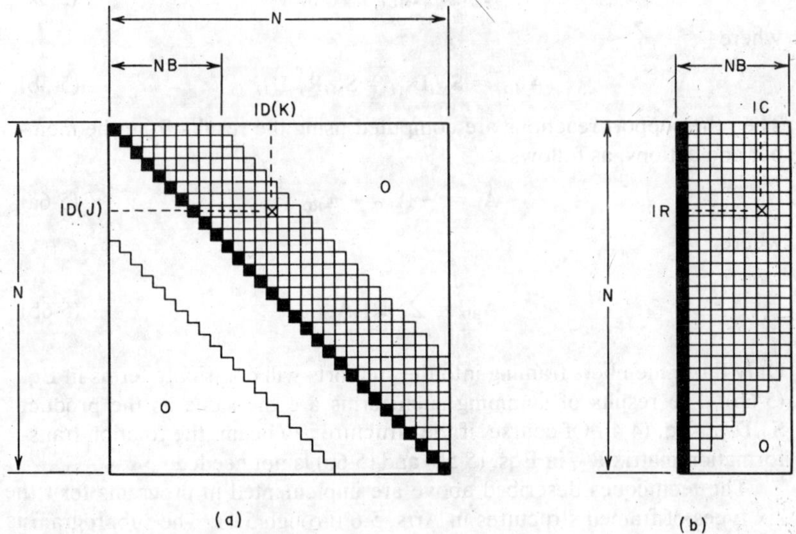

Fig. 5-2. Stiffness matrix $\mathbf{S}_{FF}$: (a) usual form of storage and (b) storage of upper band as a rectangular matrix.

element in $\mathbf{S}_{FF}$ are drawn from Eq. (a) in Art. 4.6. They produce revised displacement indexes ID(J) and ID(K) from the original indexes J and K by subtracting cumulative numbers of restraints N1 and N2, corresponding to J and K. Then the row and column indexes of the same element appearing in Fig. 5-2b are obtained as

$$\text{IR} = \text{ID(J)} \tag{5-3a}$$

$$\text{IC} = \text{ID(K)} - \text{IR} + 1 \tag{5-3b}$$

The stiffness matrix generated in the form of Fig. 5-2b is factored using the subprogram named BANFAC given in Appendix D. This factorization need be done only once for a given structure. Subsequently, the free joint displacements $\mathbf{D}_F$ are found using the subprogram BANSOL, which is also given in Appendix D. This subprogram must be applied a number of times equal to the number of loading systems NLS. For each loading system the matrix equation to be solved is

$$\mathbf{S}_{FF}\mathbf{D}_F = \mathbf{A}_{FC} \tag{5-4}$$

After the free joint displacements $\mathbf{D}_F$ have been calculated, the member end-actions $\mathbf{A}_{Mi}$ and the support reactions $\mathbf{A}_R$ are also obtained. Because the stiffness submatrix $\mathbf{S}_{RF}$ is not generated, support reactions due to displacements $\mathbf{D}_F$ cannot be calculated by matrix multiplication, as indicated in Eq. (4-4). Instead, the member end-actions $\mathbf{A}_{Mi}$ are first obtained from Eq. (4-5) as

$$\mathbf{A}_{\mathrm{M}i} = \mathbf{A}_{\mathrm{ML}i} + \mathbf{A}_{\mathrm{MD}i} \tag{5-5a}$$

where

$$\mathbf{A}_{\mathrm{MD}i} = \mathbf{S}_{\mathrm{M}i}\mathbf{D}_{\mathrm{M}i} = \mathbf{S}_{\mathrm{M}i}\mathbf{R}_{\mathrm{T}i}\mathbf{D}_{\mathrm{J}i} \tag{5-5b}$$

Then the support reactions are computed using the results from the member end-actions, as follows:

$$\mathbf{A}_{\mathrm{R}} = -\mathbf{A}_{\mathrm{RC}} + \mathbf{A}_{\mathrm{RD}} \tag{5-6a}$$

where

$$\mathbf{A}_{\mathrm{RD}} = \sum_{i=1}^{m} \mathbf{R}_{\mathrm{T}i}^{\mathrm{T}}\mathbf{A}_{\mathrm{MD}i} \tag{5-6b}$$

Only those members framing into the supports will contribute terms in Eq. (5-6b). The results of summing such terms are the same as the product $\mathbf{S}_{\mathrm{RF}}\mathbf{D}_{\mathrm{F}}$ in Eq. (4-4). Of course, if the structure is a beam, the rotation transformation matrix $\mathbf{R}_{\mathrm{T}i}$ in Eqs. (5-5b) and (5-6b) is not needed.

The techniques described above are implemented in programs for the six types of framed structures in Arts. 5.6 through 5.11. The subprograms from these six programs may also be used to form a single combined program for all types of framed structures, as shown in Art. 5.12.

5.6 Continuous Beam Program. This article contains flow charts for the main program and subprograms for analyzing continuous beams, as described in Art. 4.8. In addition, numerical examples are given to demonstrate the form of the data and the results obtained from a computer program.

The first part of Flow Chart 5-1 shows the logic of the main program, called CB, for analyzing continuous beams. Comments on the right side of the flow chart are intended to help the reader understand what is meant by the symbols in the boxes. (Double boxes indicate subprograms that are called by the main program.) More than half of the statements in the main program are independent of the type of structure. These items include reading the control data, factoring the stiffness matrix, processing multiple loading systems, and solving joint equilibrium equations for each load set. Note that the layout of the program allows any number of structures to be processed in one run, and each structure may be analyzed for a number of loading systems equal to NLS. The upper band of the stiffness matrix $\mathbf{S}_{\mathrm{FF}}$ is generated and factored only once for a given structure, but the solution for $\mathbf{D}_{\mathrm{F}}$ (and other results) is obtained NLS times (for a series of loading systems).

The other items in the main program consist of the five subprogram statements numbered 1 through 5 in the chart. These subprograms pertain specifically to continuous beams and do the following blocks of operations:

1. SDATA1: Reads and prints the structural data
2. STIFF1: Constructs the stiffness matrix
3. LDATA1: Reads and prints the load data
4. LOADS1: Constructs the load vector
5. RESUL1: Calculates and prints the results

Statements in these five subprograms are given in Sections 1.1 through 1.5 of the flow chart.

Example 1. The continuous beam structure analyzed in Art. 4.9 (see Fig. 4-13a) is presented as the first example to demonstrate the computer program. For this purpose the following numerical values (in US units) are assumed:

$E = 10{,}000$ ksi $\quad L = 100$ in. $\quad I_Z = 1000$ in.^{4} $\quad P = 10$ kips

The input data required by the computer program are summarized in Table 5-9,

Table 5-9

Data for Continuous Beam Example 1

Type of Data		*Numerical Values*					
Control Data		1	1	1			
Structural Data	(a)	3	5	3	10000.0		
	(b)	1	100.0		1000.0		
		2	100.0		2000.0		
		3	200.0		2000.0		
	(c)	1	1	1			
		3	1	0			
		4	1	1			
Load Data	(a)	2	3				
	(b)	2	−10.0		1000.0		
		3	−10.0		0.0		
	(c)	1	10.0		250.0	10.0	−250.0
		2	10.0		250.0	10.0	−250.0
		3	10.0		333.333	10.0	−333.333

which conforms to the specifications for continuous beams given previously (see Table 5-3). Results from the computer program for this data are listed in Table 5-10.

Example 2. The continuous beam in Fig. 5-3a has constant flexural rigidity and is to be analyzed for the given loads in three stages, as follows: (1) concentrated loads, (2) distributed loads, and (3) total loads. Numerical values (in SI units) for this problem are:

$E = 200 \times 10^6$ kN/m^2 $L_1 = 6$ m $L_2 = 5$ m $I_Z = 3.6 \times 10^{-3}$ m^4
$P = 50$ kN $w = 30$ kN/m $M = 150$ kN-m

Figure 5-3b shows the numbering system for the restrained structure. Input data for this example are given in Table 5-11; and the results appear in Table 5-12.

Table 5-10
Results for Continuous Beam Example 1

```
STRUCTURE NO.   1 CONTINUOUS BEAM
NUMBER OF LOADING SYSTEMS =  1

STRUCTURAL PARAMETERS
     M     N    NJ    NR   NRJ                E
     3     3     4     5     3          10000.0

MEMBER INFORMATION
MEMBER          EL            ZI
     1    100.000    1000.00000
     2    100 000    2000.00000
     3    200.000    2000.00000

JOINT RESTRAINTS
JOINT   JR1   JR2
    1     1     1
    3     1     0
    4     1     1


LOADING NO.   1
  NLJ  NLM
    2    3

ACTIONS AT JOINTS
JOINT         AJ1         AJ2
    2     -10.000    1000.000
    3     -10.000       0.000

ACTIONS AT ENDS OF RESTRAINED MEMBERS DUE TO LOADS
MEMBER        AML1        AML2        AML3        AML4
     1      10.000     250.000      10.000    -250.000
     2      10.000     250.000      10.000    -250.000
     3      10 000     333.333      10.000    -333.333

JOINT DISPLACEMENTS
JOINT          DJ1           DJ2
    1  .0000E+00    .0000E+00
    2-.1316E+00    .1210E-02
    3  .0000E+00    .8433E-03
    4  .0000E+00    .0000E+00

MEMBER END-ACTIONS
MEMBER         AM1          AM2          AM3          AM4
     1      33.056     1281.746      -13.056     1023.810
     2       3.056      -23.810       16.944     -670.635
     3      12.530      670.635        7.470     -164.682

SUPPORT REACTIONS
JOINT          AR1          AR2
    1       33.056     1281.746
    3       39.474        0.000
    4        7.470     -164.682
```

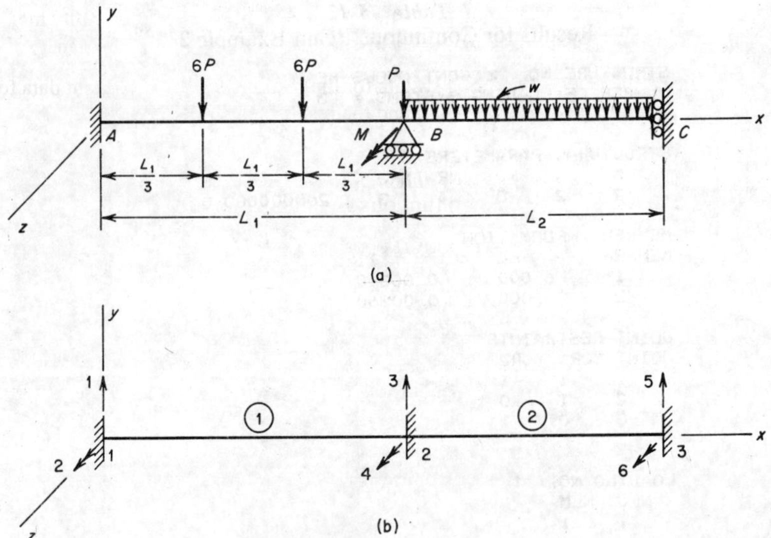

Fig. 5-3. Continuous beam Example 2.

Table 5-11
Data for Continuous Beam Example 2

Type of Data			*Numerical Values*				
Control Data			2	1	3		
Structural Data		(a)	2	4	3	20000000.0	
		(b)	1	6.00	0.0036		
			2	5.00	0.0036		
		(c)	1	1	1		
			2	1	0		
			3	0	1		
Load Data Sets	1	(a)	1	1			
		(b)	2	−50.0	150.0		
		(c)	1	300.0	400.0	300.0	−400.0
	2	(a)	0	1			
		(c)	2	75.0	62.5	75.0	−62.5
	3	(a)	1	2			
		(b)	2	−50.0	150.0		
		(c)	1	300.0	400.0	300.0	−400.0
			2	75.0	62.5	75.0	−62.5

Table 5-12
Results for Continuous Beam Example 2

```
STRUCTURE NO.  2 CONTINUOUS BEAM
NUMBER OF LOADING SYSTEMS =  3

STRUCTURAL PARAMETERS
   M     N    NJ   NR   NRJ                 E
   2     2     3    4     3       200000000.0

MEMBER INFORMATION
MEMBER         EL              ZI
    1       6.000        0.00360
    2       5.000        0.00360

JOINT RESTRAINTS
JOINT   JR1   JR2
    1     1     1
    2     1     0
    3     0     1


LOADING NO.   1
  NLJ  NLM
    1    1

ACTIONS AT JOINTS
JOINT         AJ1          AJ2
    2     -50.000      150.000

ACTIONS AT ENDS OF RESTRAINED MEMBERS DUE TO LOADS
MEMBER       AML1         AML2         AML3         AML4
    1     300.000      400.000      300.000     -400.000

JOINT DISPLACEMENTS
JOINT         DJ1          DJ2
    1 .0000E+00 .0000E+00
    2 .0000E+00 .8814E-03
    3 .2204E-02 .0000E+00

MEMBER END-ACTIONS
MEMBER        AM1          AM2          AM3          AM4
    1     405.769      611.538      194.231       23.077
    2       0.000      126.923        0.000     -126.923

SUPPORT REACTIONS
JOINT         AR1          AR2
    1     405.769      611.538
    2     244.231        0.000
    3       0.000     -126.923

LOADING NO.   2
  NLJ  NLM
    0    1

ACTIONS AT ENDS OF RESTRAINED MEMBERS DUE TO LOADS
MEMBER       AML1         AML2         AML3         AML4
    2      75.000       62.500       75.000      -62.500

JOINT DISPLACEMENTS
JOINT         DJ1          DJ2
    1 .0000E+00 .0000E+00
    2 .0000E+00-.4006E-03
    3-.2087E-02 .0000E+00
```

```
MEMBER END-ACTIONS
MEMBER       AM1        AM2        AM3        AM4
   1     -48.077    -96.154    348.077   -592.308
   2     150.000    192.308      0.000    182.692

SUPPORT REACTIONS
JOINT        AR1        AR2
   1     -48.077    -96.154
   2     198.077      0.000
   3       0.000    182.692


LOADING NO.   3
  NLJ  NLM
   1    2

ACTIONS AT JOINTS
JOINT        AJ1        AJ2
   2     -50.000    150.000

ACTIONS AT ENDS OF RESTRAINED MEMBERS DUE TO LOADS
MEMBER      AML1       AML2       AML3       AML4
   1     300.000    400.000    300.000   -400.000
   2      75.000     62.500     75.000    -62.500

JOINT DISPLACEMENTS
JOINT         DJ1         DJ2
   1 .0000E+00 .0000E+00
   2 .0000E+00 .4808E-03
   3 .1169E-03 .0000E+00

MEMBER END-ACTIONS
MEMBER       AM1        AM2        AM3        AM4
   1     357.692    515.385    242.308   -169.231
   2     150.000    319.231     -0.000     55.769

SUPPORT REACTIONS
JOINT        AR1        AR2
   1     357.692    515.385
   2     442.308      0.000
   3       0.000     55.769
```

Flow Chart 5-1

Main Program CB for Continuous Beams

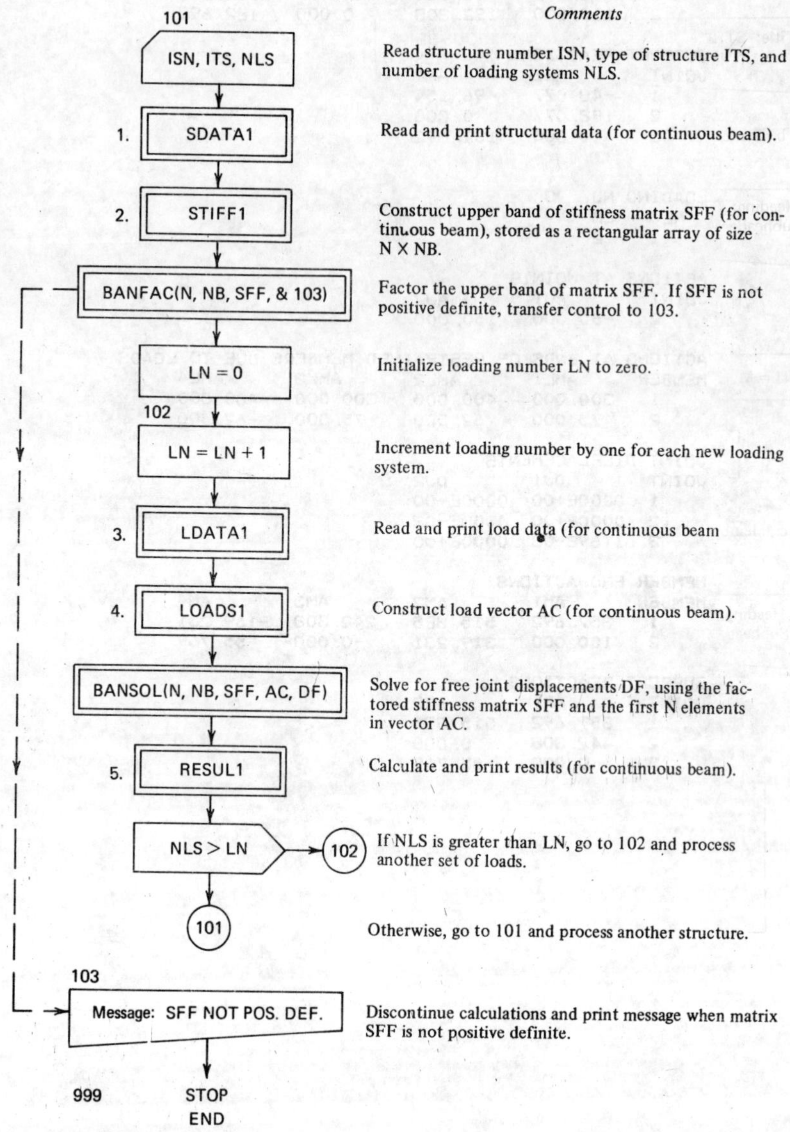

1.1 Subprogram SDATA1 for Program CB

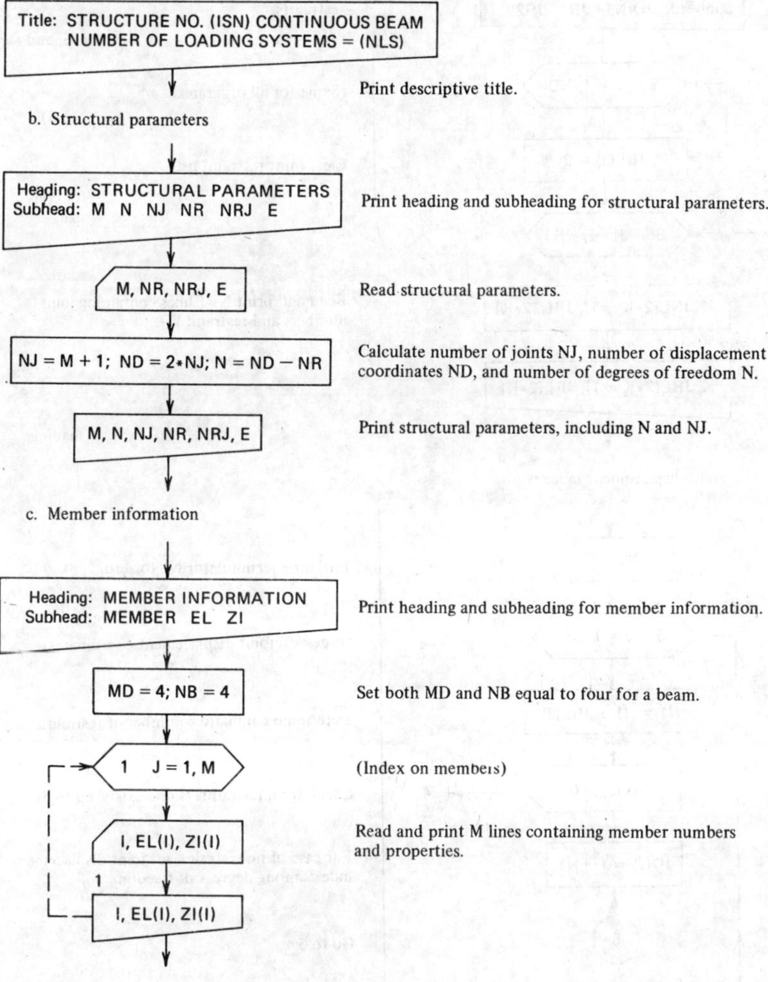

d. Joint restraints

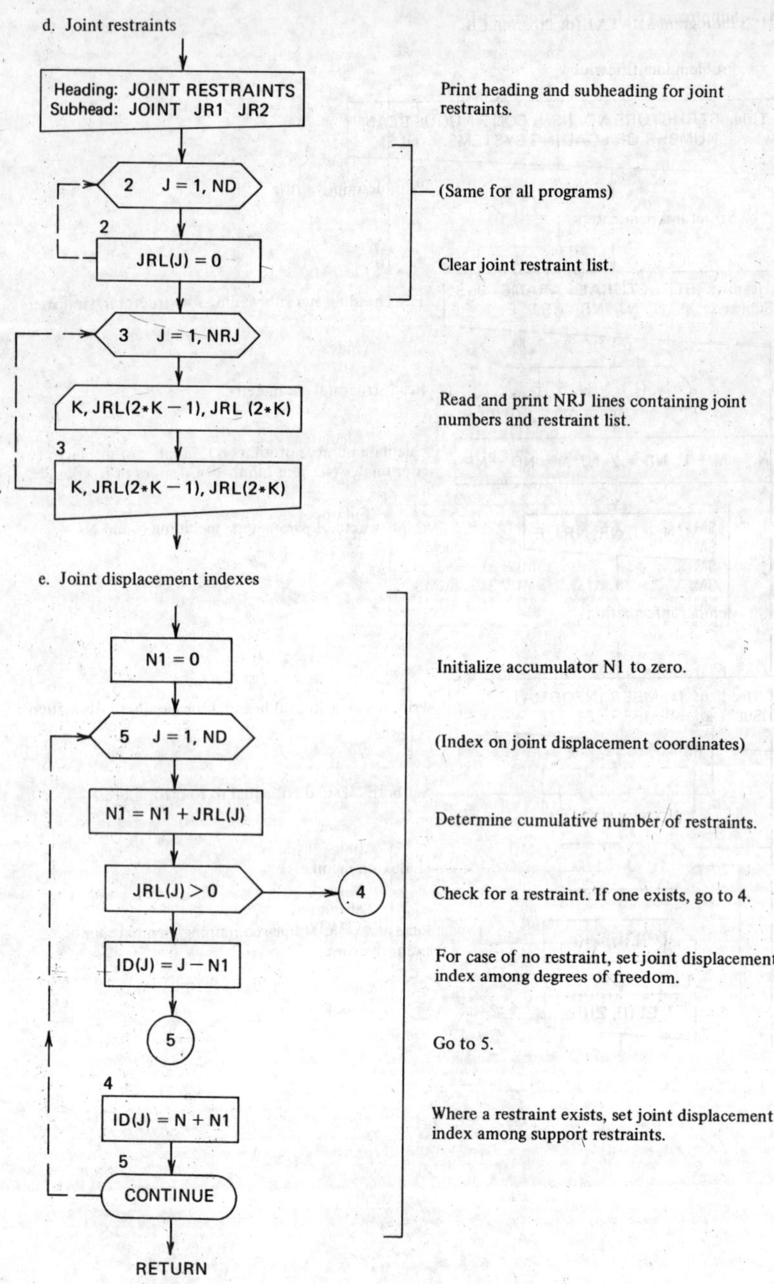

1.2 Subprogram STIFF1 for Program CB

a. Member stiffnesses

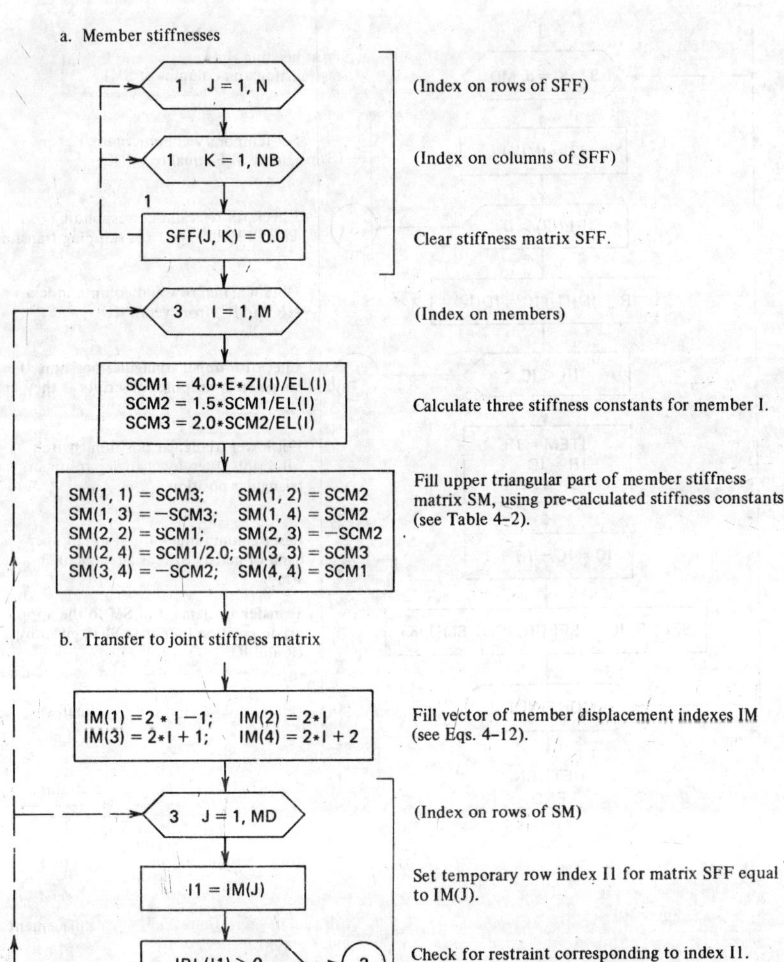

b. Transfer to joint stiffness matrix

1.2b (cont.)

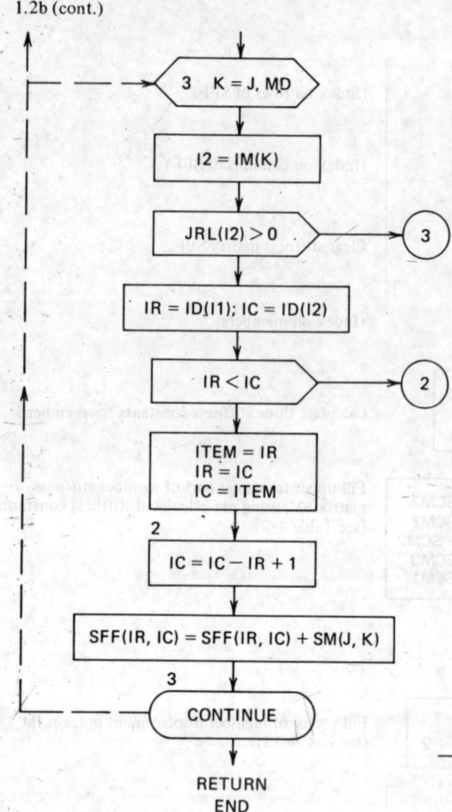

(Index on columns of SM)

Set temporary column index I2 for matrix SFF equal to IM(K).

Check for restraint corresponding to index I2. If so, go to 3, skipping transfer.

Obtain actual row and column indexes (IR and IC) from vector ID.

Check for upper triangular position. If so, go to 2, keeping subscripts as they are.

For lower triangular position, switch subscripts, thereby changing to upper triangular position.

Reset column index for storing upper band of matrix SFF as a rectangular array.

Transfer an element of SM to the appropriate position in matrix SFF, given by IR and IC.

1.3 Subprogram LDATA1 for Program CB

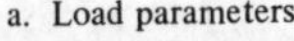
a. Load parameters

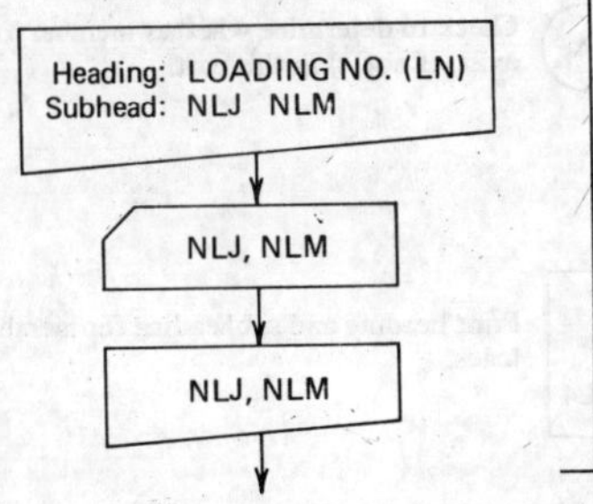

Print heading and subheading for load parameters.

Read and print number of loaded joints and number of loaded members.

b. Joint loads

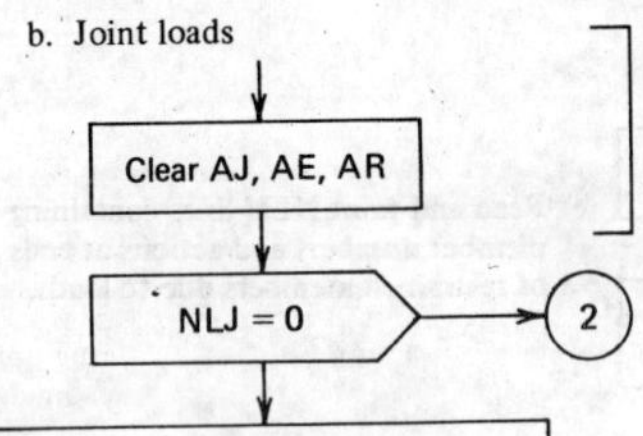

Check to determine whether joint loads exist. If not, skip this part.

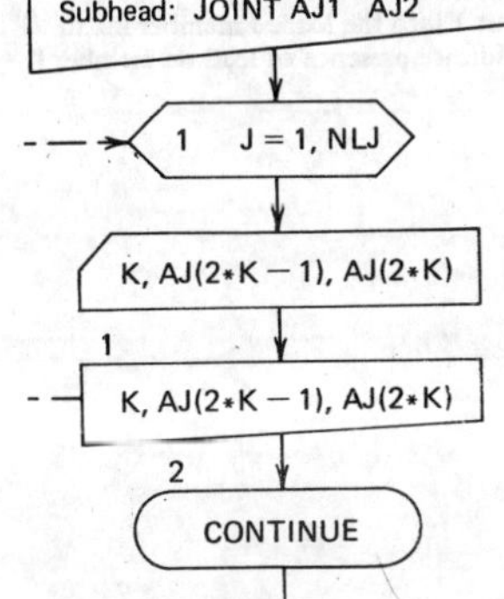

Print heading and subheading for joint loads.

Read and print NLJ lines containing joint numbers and joint loads.

c. Member loads

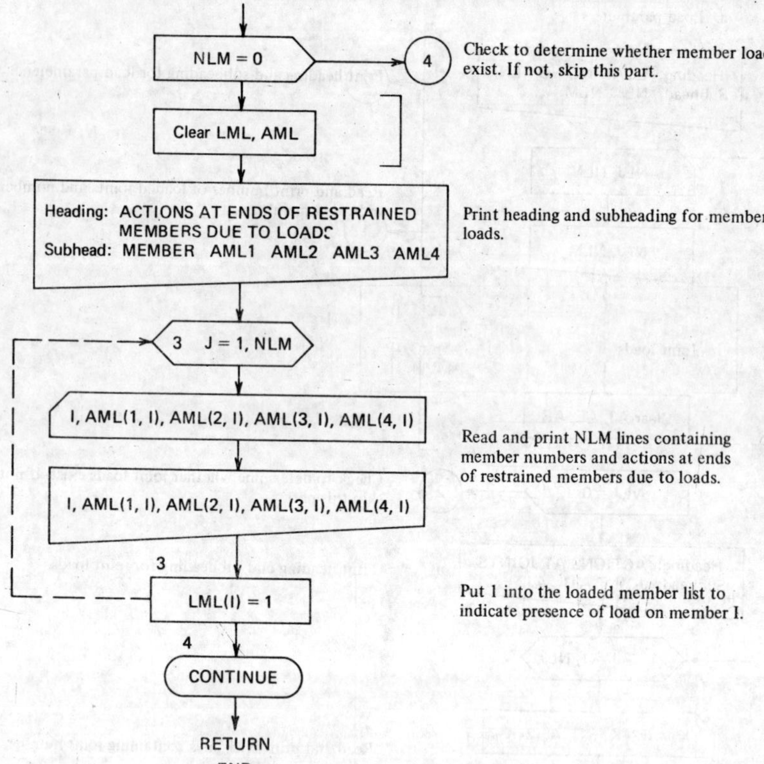

1.4 Subprogram LOADS1 for Program CB

a. Equivalent joint loads

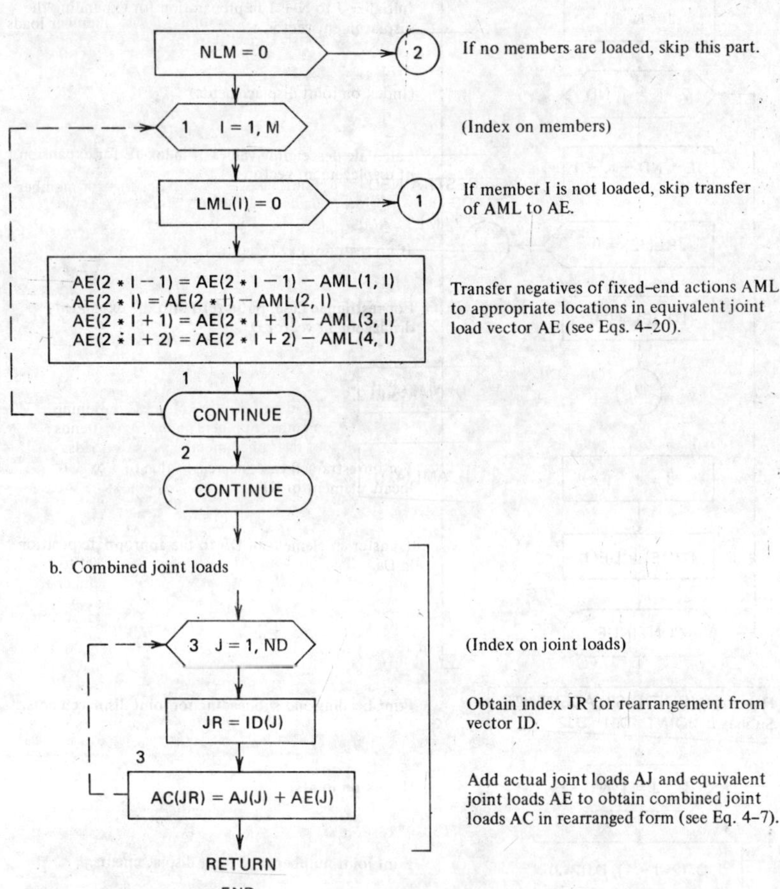

1.5 Subprogram RESUL1 for Program CB

a. Joint displacements

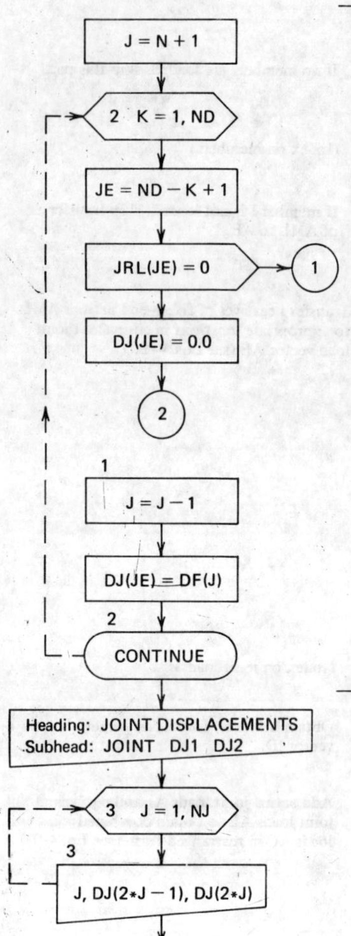

Initialize J to N + 1 in preparation for expanding the displacement vector.

(Index on joint displacements)

Calculate descending values of index JE for expansion of displacement vector.

If no restraint exists, go to 1.

For restrained case, put a zero into the expanded displacement vector DJ.

Go to 2.

For unrestrained case, decrease value of J by 1 to locate element of DF.

Transfer an element of DF to the appropriate position in DJ.

Print heading and subheading for joint displacements.

(Index on joints)

Print joint numbers and joint displacements.

b. Member end–actions

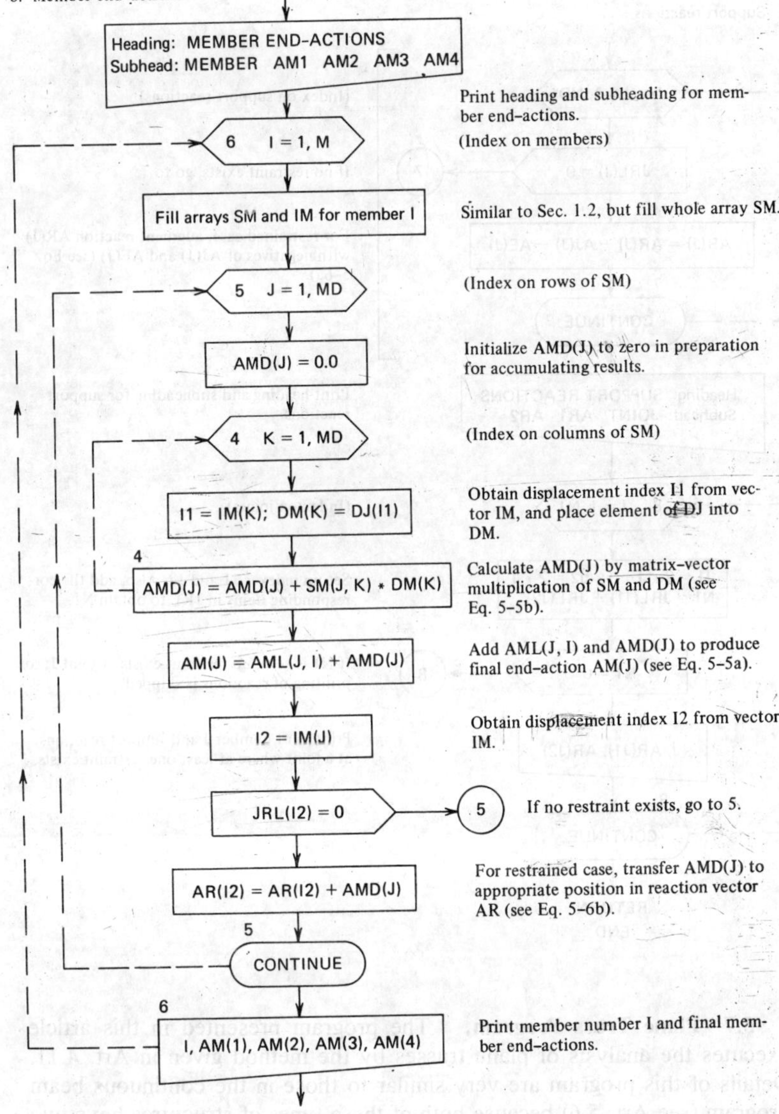

c. Support reactions

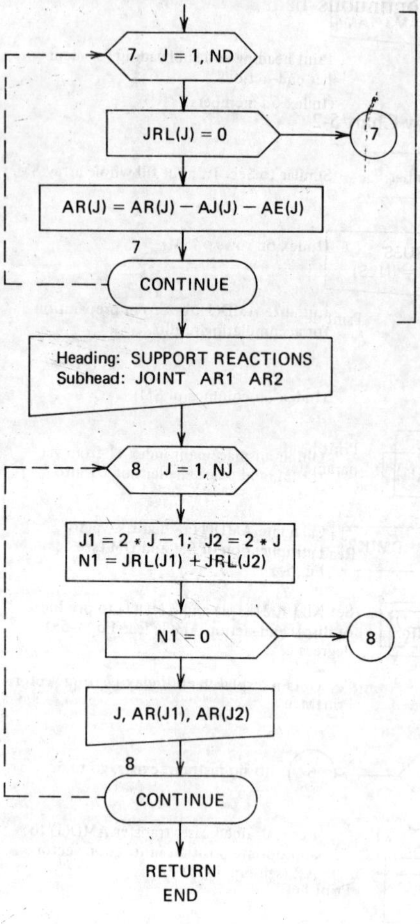

Flow chart	Explanation
7 J = 1, ND	(Index on support reactions)
JRL(J) = 0 → 7	If no restraint exists, go to 7.
AR(J) = AR(J) − AJ(J) − AE(J)	For restrained case, augment reaction AR(J) with negatives of AJ(J) and AE(J) (see Eq. 5–6a).
7 CONTINUE	
Heading: SUPPORT REACTIONS Subhead: JOINT AR1 AR2	Print heading and subheading for support reactions.
8 J = 1, NJ	(Index on joints)
J1 = 2 * J − 1; J2 = 2 * J N1 = JRL(J1) + JRL(J2)	Set up indexes J1 and J2. Also, add the corresponding items in JRL to obtain N1.
N1 = 0 → 8	If N1 is zero, no restraint exists at joint J; so printing of reactions is skipped.
J, AR(J1), AR(J2)	Print joint number J and support reactions at a joint where at least one restraint exists.
8 CONTINUE	
RETURN END	

5.7 Plane Truss Program. The program presented in this article executes the analysis of plane trusses by the method given in Art. 4.11. Details of this program are very similar to those in the continuous beam program (see Art. 5.6) because both of these types of structures have two possible displacements at each joint. Therefore, in order to avoid repetition, the emphasis in this article is upon those parts of the program that are different from the continuous beam program.

The main program for Program PT (plane trusses) is similar to that for Program CB (see Flow Chart 5-1). However, the five subprograms pertaining specifically to plane trusses are named SDATA2, STIFF2, LDATA2,

LOADS2, and RESUL2. Statements in these subprograms that are different from their counterparts for continuous beams appear in Flow Chart 5-2.

Flow Chart 5-2

2.1 Subprogram SDATA2 for Program PT

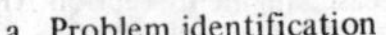

a. Problem identification

Title: STRUCTURE NO. (ISN) PLANE TRUSS
NUMBER OF LOADING SYSTEMS = (NLS)

Print descriptive title.

b. Structural parameters

Heading: STRUCTURAL PARAMETERS
Subhead: M N NJ NR NRJ E

Print heading and subheading for structural parameters.

M, NJ, NR, NRJ, E

Read structural parameters.

NDJ = 2; ND = NDJ * NJ; N = ND − NR

Set NDJ for a plane truss. Then calculate number of displacement coordinates ND and number of degrees of freedom N.

M, N, NJ, NR, NRJ, E

Print structural parameters, including N.

c. Joint coordinates

Heading: JOINT COORDINATES
Subhead: JOINT X Y

Print heading and subheading for joint coordinates.

6 K = 1, NJ

(Index on joints)

J, X(J), Y(J)

6

J, X(J), Y(J)

Read and print NJ lines containing joint numbers and x and y coordinates.

d. Member information

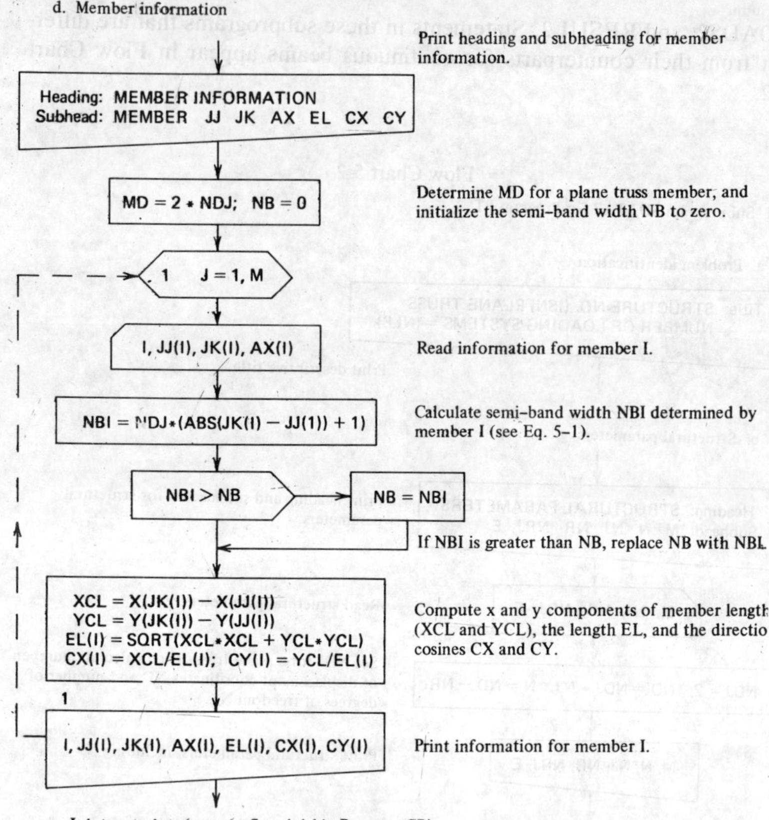

e. Joint restraints (same as Sec. 1.1d in Program CB)

f. Displacement indexes (same as Sec. 1.1e in Program CB)

2.2 Subprogram STIFF2 for Program PT

a. Member stiffnesses

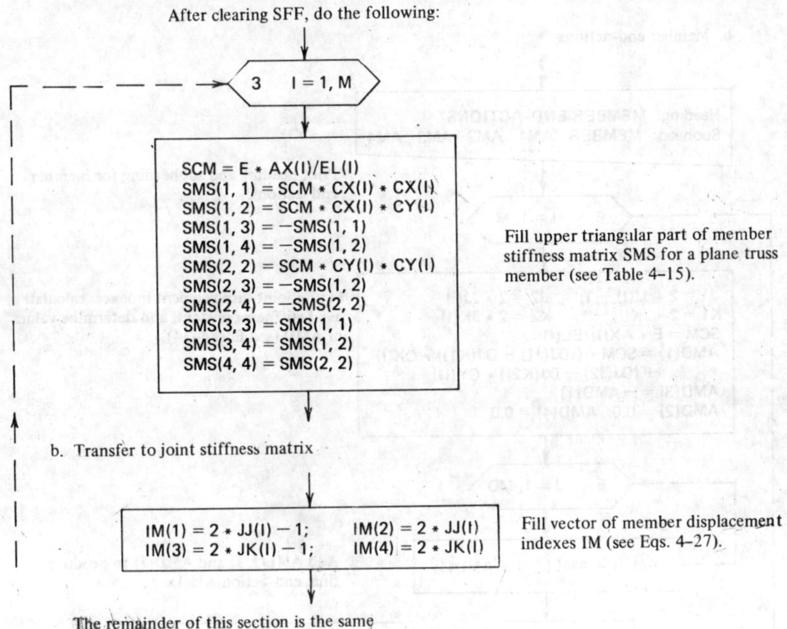

The remainder of this section is the same as Sec. 1.2b of Program CB, except that SMS replaces SM.

2.3 Subprogram LDATA2 for Program PT

This subprogram is the same as LDATA1 for Program CB.

2.4 Subprogram LOADS2 for Program PT

This subprogram is similar to LOADS1 for Program CB, but in Sec. 2.4a the equivalent joint loads AE are computed as follows:

Transfer contributions of pinned–end actions AML to appropriate locations in equivalent load vector AE (see Eqs. 4–33).

```
J1 = 2 * JJ(I) − 1;     J2 = 2 * JJ(I)
K1 = 2 * JK(I) − 1;     K2 = 2 * JK(I)
AE(J1) = AE(J1) − CX(I) * AML(1, I) + CY(I) * AML(2, I)
AE(J2) = AE(J2) − CY(I) * AML(1, I) − CX(I) * AML(2, I)
AE(K1) = AE(K1) − CX(I) * AML(3, I) + CY(I) * AML(4, I)
AE(K2) = AE(K2) − CY(I) * AML(3, I) − CX(I) * AML(4, I)
```

2.5 Subprogram RESUL2 for Program PT

Sections 2.5a and 2.5c in this subprogram are the same as in RESUL1 for Program CB. However, Sec. 2.5b is as follows:

b. Member end-actions

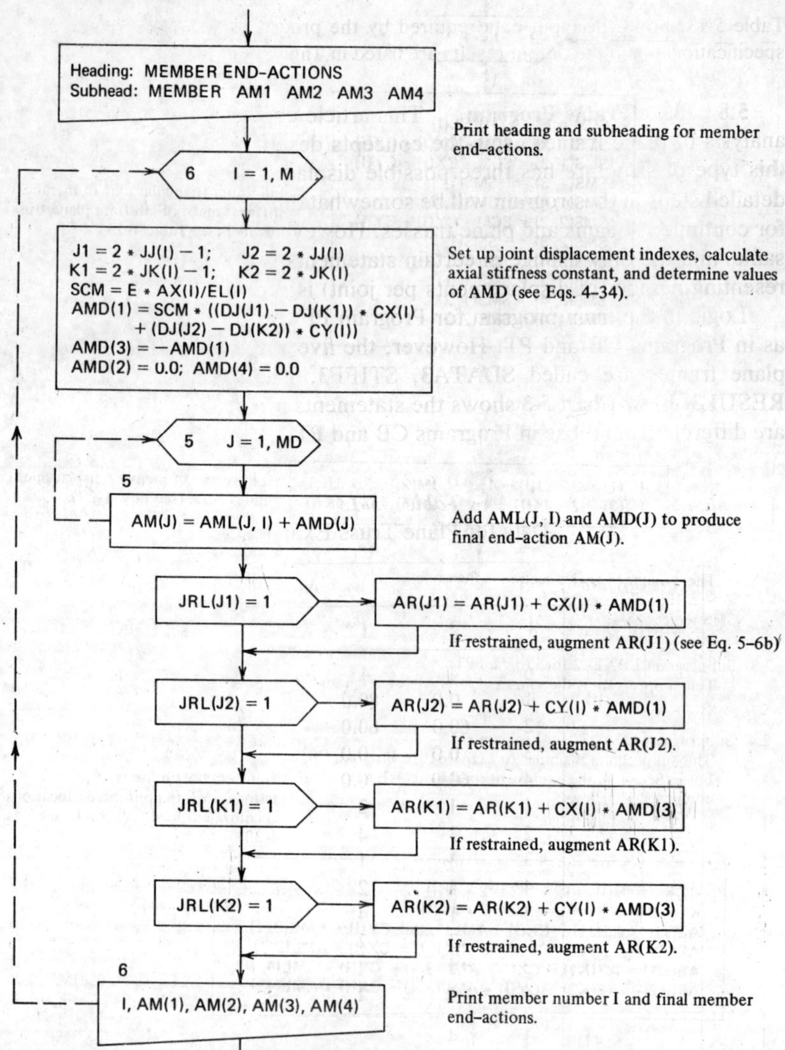

Example. To demonstrate the use of Program PT, the plane truss structure in Art. 4.12 (see Fig. 4-23) will be analyzed again. The following numerical values (in US units) are assumed for this problem:

$$E = 10{,}000 \text{ ksi} \qquad L = 100 \text{ in.} \qquad A_X = 10 \text{ in.}^2 \qquad P = 10 \text{ kips}$$

Table 5-13 shows the input data required by the program (see Table 5-4 for data specifications), and the output results are listed in Table 5-14.

5.8 Plane Frame Program. This article contains a program for the analysis of plane frames using the concepts described in Art. 4.17. Since this type of structure has three possible displacements at each joint, the detailed steps in the program will be somewhat different from the programs for continuous beams and plane trusses. However, the methodology is the same, and the only change in certain statements is that the integer 2 (representing number of displacements per joint) is replaced by the integer 3.

Logic in the main program for Program PF (plane frames) is the same as in Programs CB and PT. However, the five subprograms dealing with plane frames are called SDATA3, STIFF3, LDATA3, LOADS3, and RESUL3. Flow Chart 5-3 shows the statements in these subprograms that are different from those in Programs CB and PT.

Table 5-13
Data for Plane Truss Example

Type of Data		***Numerical Values***				
Control Data		1	2	1		
Structural Data	(a)	6	4	4	2	10000.0
	(b)	1	0.0	80.0		
		2	60.0	80.0		
		3	0.0	0.0		
		4	60.0	0.0		
	(c)	1	1	2	6.0	
		2	3	4	6.0	
		3	3	1	8.0	
		4	4	2	8.0	
		5	1	4	10.0	
		6	3	2	10.0	
	(d)	3	1	1		
		4	1	1		
Load Data	(a)	1	4			
	(b)	2	20.0	10.0		
	(c)	1	0.0	-20.0	0.0	20.0
		2	0.0	10.0	0.0	10.0
		3	0.0	10.0	0.0	10.0
		4	5.0	5.0	5.0	5.0

Table 5-14
Results for Plane Truss Example

STRUCTURE NO. 1 PLANE TRUSS

NUMBER OF LOADING SYSTEMS = 1

STRUCTURAL PARAMETERS

M	N	NJ	NR	NRJ	E
6	4	4	4	2	10000.0

JOINT COORDINATES

JOINT	X	Y
1	0.000	80.000
2	60.000	80.000
3	0.000	0.000
4	60.000	0.000

MEMBER INFORMATION

MEMBER	JJ	JK	AX	EL	CX	CY
1	1	2	6.000	60.000	1.000	0.000
2	3	4	6.000	60.000	1.000	0.000
3	3	1	8.000	80.000	0.000	1.000
4	4	2	8.000	80.000	0.000	1.000
5	1	4	10.000	100.000	0.600	-0.800
6	3	2	10.000	100.000	0.600	0.800

JOINT RESTRAINTS

JOINT	JR1	JR2
3	1	1
4	1	1

LOADING NO. 1

NLJ	NLM
1	4

ACTIONS AT JOINTS

JOINT	AJ1	AJ2
2	20 000	10.000

ACTIONS AT ENDS OF RESTRAINED MEMBERS DUE TO LOADS

MEMBER	AML1	AML2	AML3	AML4
1	0.000	-20.000	0.000	20.000
2	0.000	10.000	0.000	10.000
3	0.000	10.000	0.000	10.000
4	5.000	5.000	5.000	5.000

JOINT DISPLACEMENTS

JOINT	DJ1	DJ2
1	.1000E+00	.4147E-01
2	.1061E+00	-.4020E-01
3	.0000E+00	.0000E+00
4	.0000E+00	.0000E+00

MEMBER END-ACTIONS

MEMBER	AM1	AM2	AM3	AM4
1	-6.099	-20.000	6.099	20.000
2	0.000	10.000	0.000	10.000
3	-41.465	10.000	41.465	10.000
4	45.201	5.000	-35.201	5.000
5	26.832	0.000	-26.832	0.000
6	-31.502	0.000	31.502	0.000

SUPPORT REACTIONS

JOINT	AR1	AR2
3	-28.901	-56.667
4	-21.099	76.667

Flow Chart 5-3

3.1 Subprogram SDATA3 for Program PF

a. Problem identification

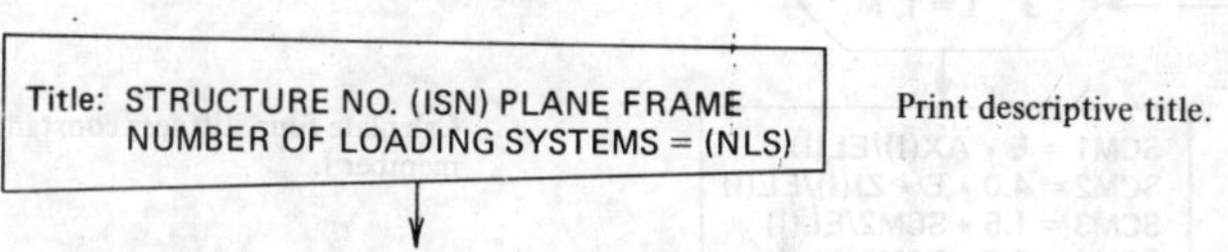

b. Structural parameters (same as Sec. 2.1b in Program PT, except that NDJ = 3)

c. Joint coordinates (same as Sec. 2.1c in Program PT)

d. Member information (same as Sec. 2.1d in Program PT, except that ZI is also read and printed)

e. Joint restraints

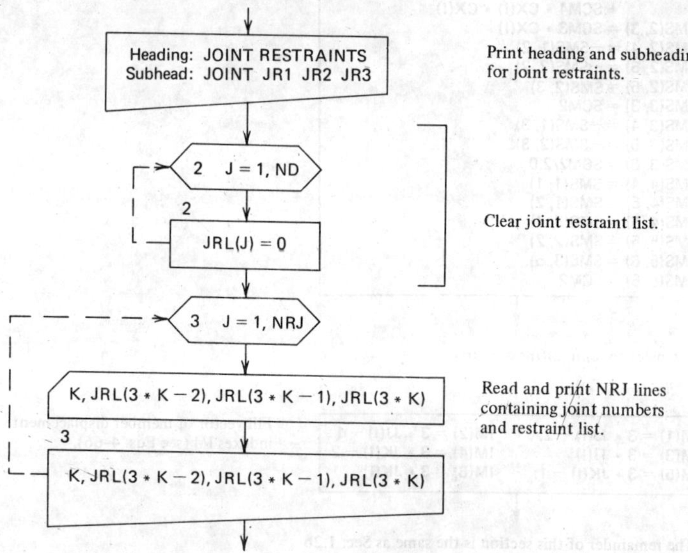

f. Displacement indexes (same as Sec. 1.1e in Program CB)

3.2 Subprogram STIFF3 for Program PF

a. Member stiffnesses

After clearing SFF, do the following:

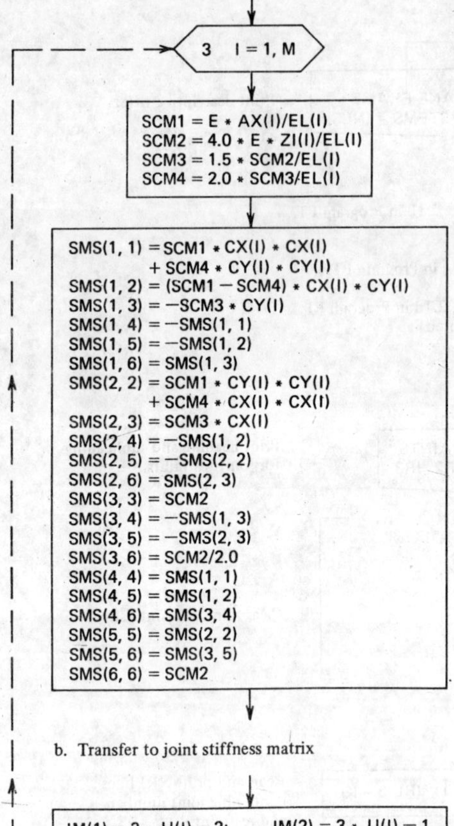

Calculate four stiffness constants for member I.

Fill upper triangular part of member stiffness matrix SMS for a plane frame member (see Table 4–27).

b. Transfer to joint stiffness matrix

```
IM(1) = 3 * JJ(I) − 2;     IM(2) = 3 * JJ(I) − 1
IM(3) = 3 * JJ(I);         IM(4) = 3 * JK(I) − 2
IM(5) = 3 * JK(I) − 1;     IM(6) = 3 * JK(I)
```

Fill vector of member displacement indexes IM (see Eqs. 4–66).

The remainder of this section is the same as Sec. 1.2b of Program CB, except that SMS replaces SM.

3.3 Subprogram LDATA3 for Program PF

a. Load parameters (same as Sec. 1.3a in Program CB)

b. Joint loads

If NLJ ≠ 0, do the following:

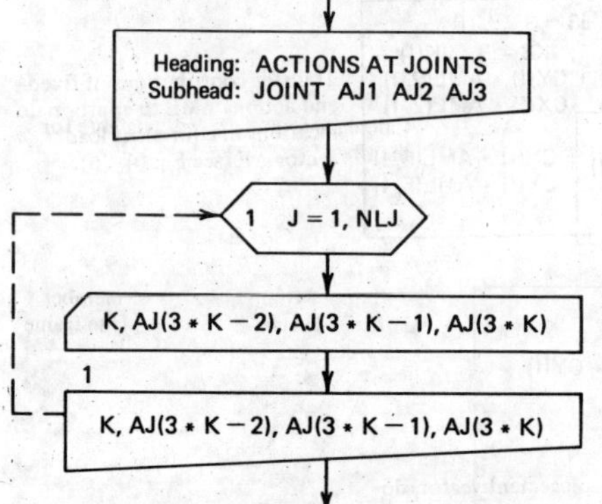

Print heading and subheading for joint loads.

Read and print NLJ lines containing joint numbers and joint loads.

c. Member loads

After clearing LML and AML, do the following:

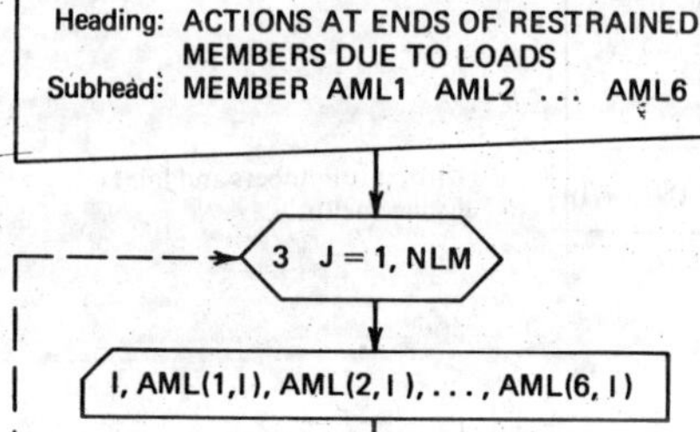

Print heading and subheading for member loads.

Read and print NLM lines containing member numbers and actions at ends of restrained members due to loads.

3.4 Subprogram LOADS3 for Program PF

This subprogram is similar to LOADS1 for Program CB, but in Sec. 3.4a the equivalent joint loads are computed as follows:

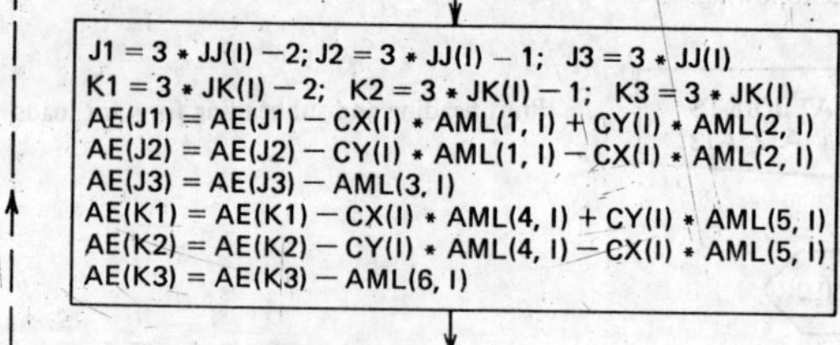

Transfer contributions of fixed-end actions AML to appropriate locations in equivalent load vector AE (see Eqs. 4–76).

3.5 Subprogram RESUL3 for Program PF

a. Joint displacements

After expanding the joint displacement vector, do the following:

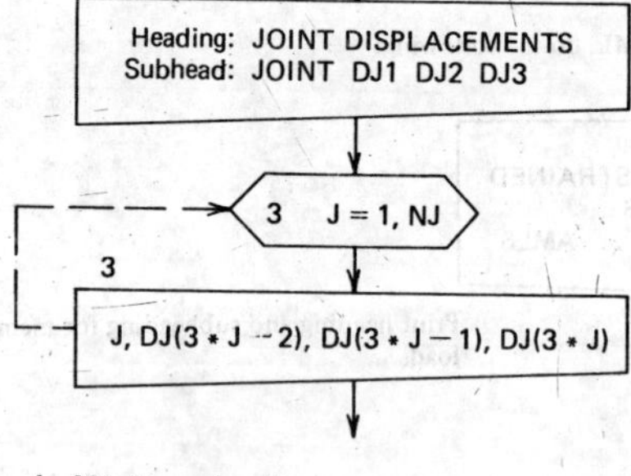

Print heading and subheading for joint displacements.

Print joint numbers and joint displacements.

b. Member end-actions

Heading: MEMBER END-ACTIONS
Subhead: MEMBER AM1 AM2 . . . AM6

Print heading and subheading for member end-actions.

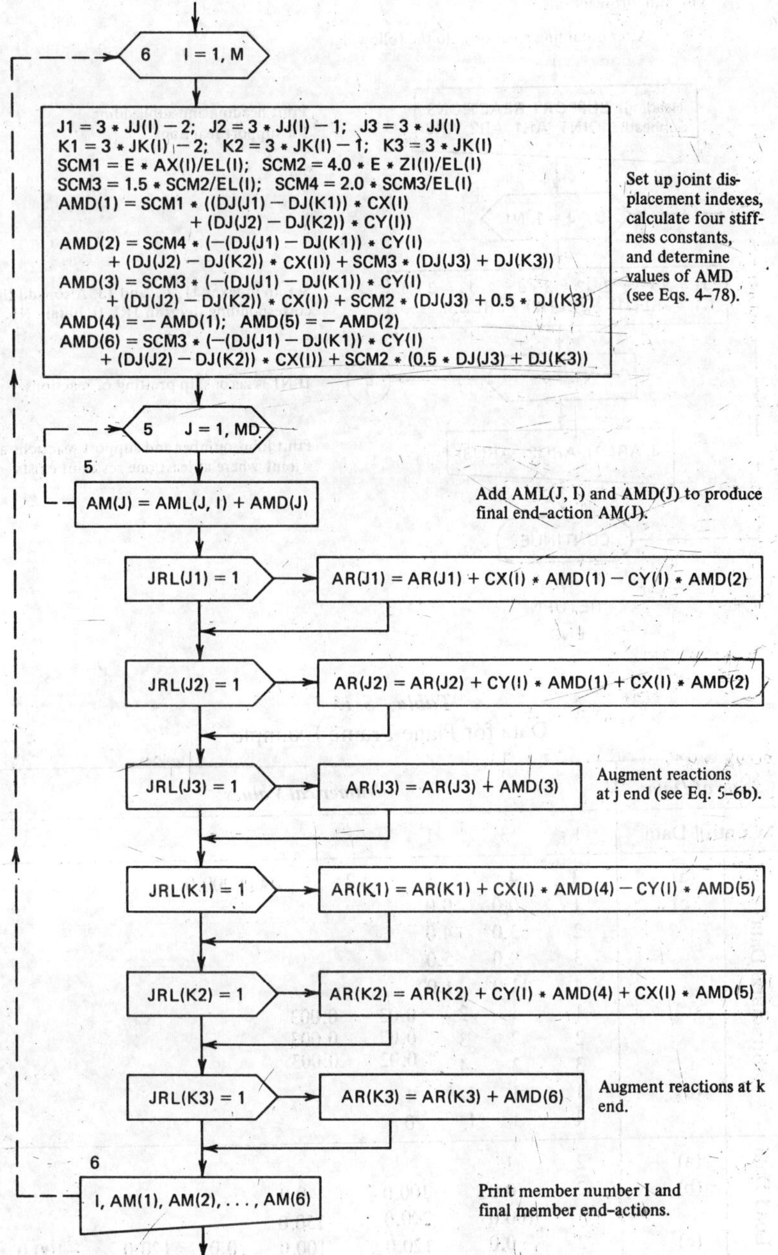
6 I = 1, M
J1 = 3 * JJ(I) − 2; J2 = 3 * JJ(I) − 1; J3 = 3 * JJ(I)
K1 = 3 * JK(I) − 2; K2 = 3 * JK(I) − 1; K3 = 3 * JK(I)
SCM1 = E * AX(I)/EL(I); SCM2 = 4.0 * E * ZI(I)/EL(I)
SCM3 = 1.5 * SCM2/EL(I); SCM4 = 2.0 * SCM3/EL(I)
AMD(1) = SCM1 * ((DJ(J1) − DJ(K1)) * CX(I)
+ (DJ(J2) − DJ(K2)) * CY(I))
AMD(2) = SCM4 * (−(DJ(J1) − DJ(K1)) * CY(I)
+ (DJ(J2) − DJ(K2)) * CX(I)) + SCM3 * (DJ(J3) + DJ(K3))
AMD(3) = SCM3 * (−(DJ(J1) − DJ(K1)) * CY(I)
+ (DJ(J2) − DJ(K2)) * CX(I)) + SCM2 * (DJ(J3) + 0.5 * DJ(K3))
AMD(4) = − AMD(1); AMD(5) = − AMD(2)
AMD(6) = SCM3 * (−(DJ(J1) − DJ(K1)) * CY(I)
+ (DJ(J2) − DJ(K2)) * CX(I)) + SCM2 * (0.5 * DJ(J3) + DJ(K3))
Set up joint displacement indexes, calculate four stiffness constants, and determine values of AMD (see Eqs. 4–78).
5 J = 1, MD
5
AM(J) = AML(J, I) + AMD(J)
Add AML(J, I) and AMD(J) to produce final end-action AM(J).
JRL(J1) = 1
AR(J1) = AR(J1) + CX(I) * AMD(1) − CY(I) * AMD(2)
JRL(J2) = 1
AR(J2) = AR(J2) + CY(I) * AMD(1) + CX(I) * AMD(2)
JRL(J3) = 1
AR(J3) = AR(J3) + AMD(3)
Augment reactions at j end (see Eq. 5–6b).
JRL(K1) = 1
AR(K1) = AR(K1) + CX(I) * AMD(4) − CY(I) * AMD(5)
JRL(K2) = 1
AR(K2) = AR(K2) + CY(I) * AMD(4) + CX(I) * AMD(5)
JRL(K3) = 1
AR(K3) = AR(K3) + AMD(6)
Augment reactions at k end.
6
I, AM(1), AM(2), . . . , AM(6)
Print member number I and final member end-actions.

c. Support reactions

After obtaining reactions, do the following:

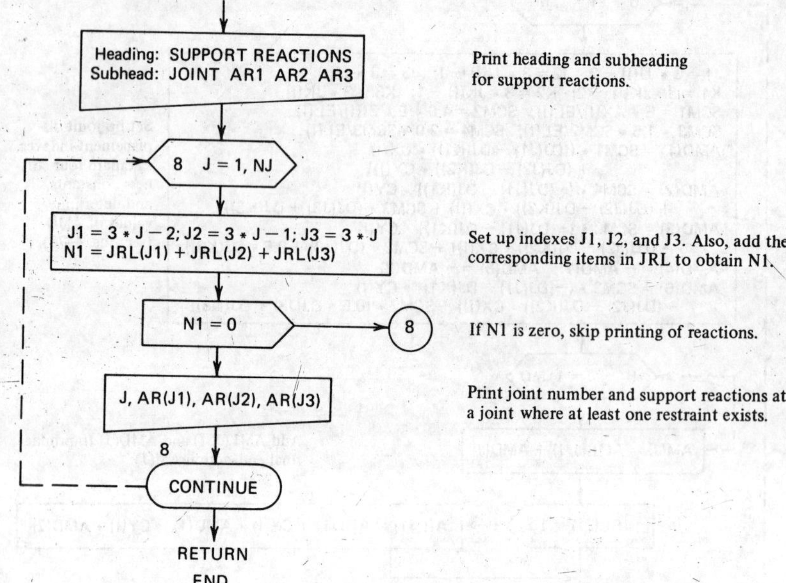

Table 5-15
Data for Plane Frame Example

Type of Data		*Numerical Values*						
Control Data		1	3	1				
Structural Data	(a)	3	4	5	2	70000000.0		
	(b)	1	0.0	0.0				
		2	3.0	4.0				
		3	9.0	5.0				
		4	11.0	1.0				
	(c)	1	1	2	0.02	0.003		
		2	2	3	0.02	0.003		
		3	3	4	0.02	0.003		
	(d)	1	1	1	1			
		4	1	1	0			
Load Data	(a)	2	1					
	(b)	2	0.0	−100.0	0.0			
		3	100.0	200.0	−150.0			
	(c)	2	0.0	120.0	100.0	0.0	120.0	−100.0

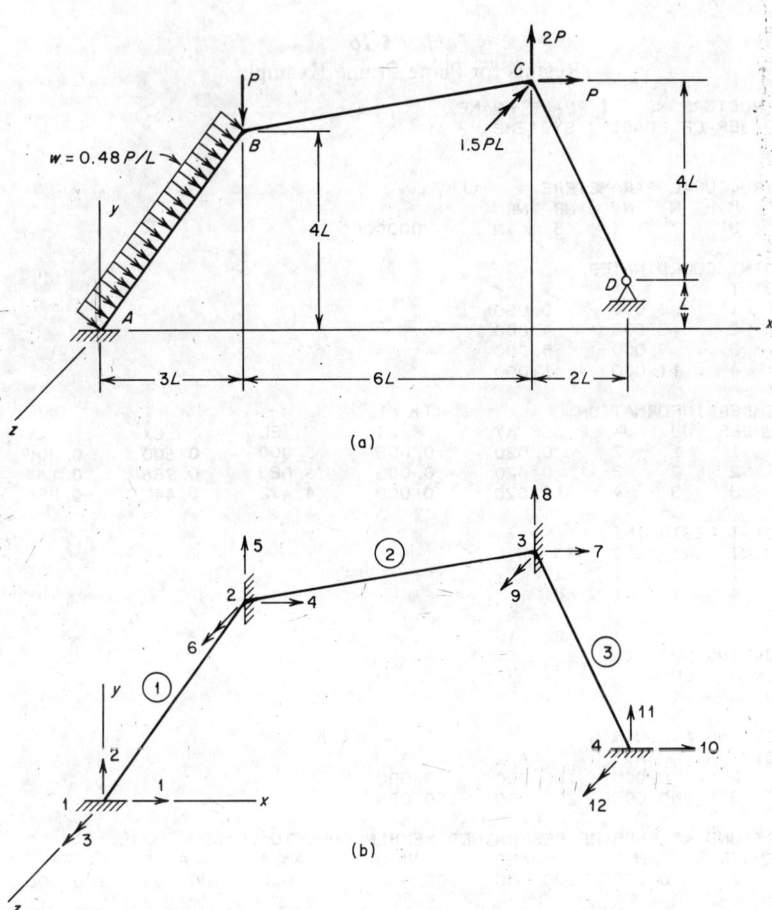

Fig. 5-4. Plane frame example.

Example. Figure 5-4a shows a plane frame consisting of prismatic members that all have the same cross-sectional properties. A numbering system for the restrained structure is given in Fig. 5-4b. Numerical values (in SI units) for this problem are:

$$E = 70 \times 10^6 \text{ kN/m}^2 \quad L = 1 \text{ m} \quad A_X = 0.02 \text{ m}^2$$
$$I_Z = 3 \times 10^{-3} \text{m}^4 \quad P = 100 \text{ kN}$$

Table 5-15 contains the input data required by the computer program (see Table 5-5 for the form of plane frame data), and results from the program are given in Table 5-16.

Table 5-16
Results for Plane Frame Example

```
STRUCTURE NO.   1 PLANE FRAME
NUMBER OF LOADING SYSTEMS =  1

STRUCTURAL PARAMETERS        O.K.
    M     N    NJ    NR   NRJ                E
    3     7     4     5     2       70000000.0

JOINT COORDINATES
JOINT           X           Y
    1       0.000       0.000
    2       3.000       4.000
    3       9.000       5.000
    4      11.000       1.000

MEMBER INFORMATION                 O.K.
MEMBER   JJ   JK           AX          ZI          EL          CX          CY
    1     1    2        0.020       0.003       5.000       0.600       0.800
    2     2    3        0.020       0.003       6.083       0.986       0.164
    3     3    4        0.020       0.003       4.472       0.447      -0.894

JOINT RESTRAINTS
JOINT   JR1   JR2   JR3
    1     1     1     1
    4     1     1     0


LOADING NO.   1
  NLJ  NLM
    2    1

ACTIONS AT JOINTS                            O.K.
JOINT         AJ1          AJ2          AJ3
    2       0.000     -100.000        0.000
    3     100.000      200.000     -150.000

ACTIONS AT ENDS OF RESTRAINED MEMBERS DUE TO LOADS      O.K.
MEMBER        AML1         AML2         AML3         AML4         AML5         AML6
    1        0.000      120.000      100.000        0.000      120.000     -100.000

JOINT DISPLACEMENTS
JOINT         DJ1          DJ2          DJ3
    1 .0000E+00 .0000E+00 .0000E+00
    2 .7368E-02-.5192E-02-.1335E-03
    3 .6410E-02 .3421E-02-.2685E-03
    4 .0000E+00 .0000E+00-.2302E-02

MEMBER END-ACTIONS
MEMBER         AM1          AM2          AM3          AM4          AM5          AM6
    1      -74.807      294.903      542.865       74.807      -54.903      331.648
    2     -108.461     -110.577     -331.648      108.461      110.577     -340.968
    3      -60.373       42.702      190.968       60.373      -42.702       -0.000

SUPPORT REACTIONS
JOINT         AR1          AR2          AR3
    1     -280.806      117.096      542.865
    4      -11.194      -73.096        0.000
```

5.9 Grid Program. A program for the analysis of grid structures is given in this article, using the method described earlier in Art. 4.20. Since the analytical procedure for grids is symbolically similar to that for plane frames, Program GR (for grids) is not much different from Program PF in the preceding article. The five subprograms pertaining to grids are named SDATA4, STIFF4, LDATA4, LOADS4, and RESUL4. Statements in these subprograms that differ from those in Program PF are indicated in Flow Chart 5-4.

Example 1. Figure 5-5a shows a grid structure having geometrical properties and loads similar to the plane frame problem in Art. 4.18 (see Fig. 4-31a). Assume that the cross-sectional properties of both members of the grid are the same and that the following numerical values (in US units) apply:

$$E = 10{,}000 \text{ ksi} \quad G = 4{,}000 \text{ ksi} \quad L = 100 \text{ in.}$$
$$I_X = 1000 \text{ in.}^4 \quad I_Y = 1000 \text{ in.}^4 \quad P = 10 \text{ kips}$$

A numbering system is given in Fig. 5-5b, which shows the restrained structure. Numerical values to be supplied as data are shown in Table 5-17 (see Table 5-6 for data specifications), and the output of the computer program is given in Table 5-18.

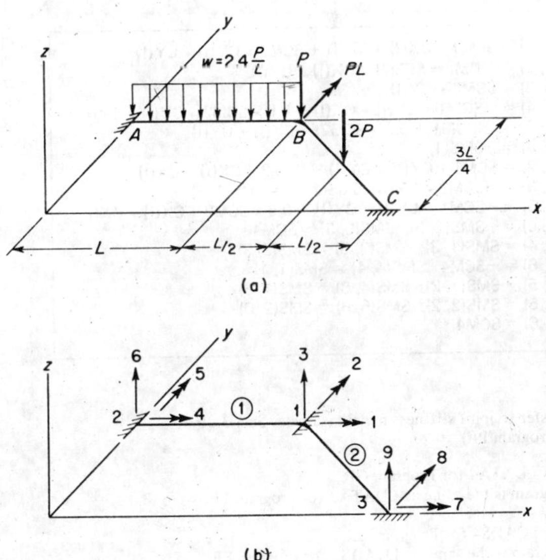

Fig. 5-5. Grid Example 1.

Flow Chart 5-4

4.1 Subprogram SDATA4 for Program GR

a. Problem identification

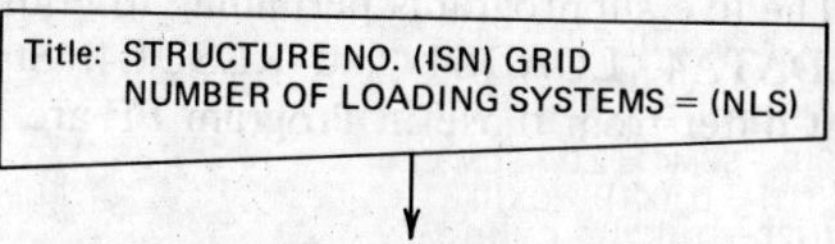

Print title.

The remainder of this subprogram is similar to Subprogram SDATA3 in Program PF; but the shear modulus G is added, and the member properties AX and ZI are replaced by XI and YI.

4.2 Subprogram STIFF4 for Program GR

a. Member stiffnesses

After clearing SFF, do the following:

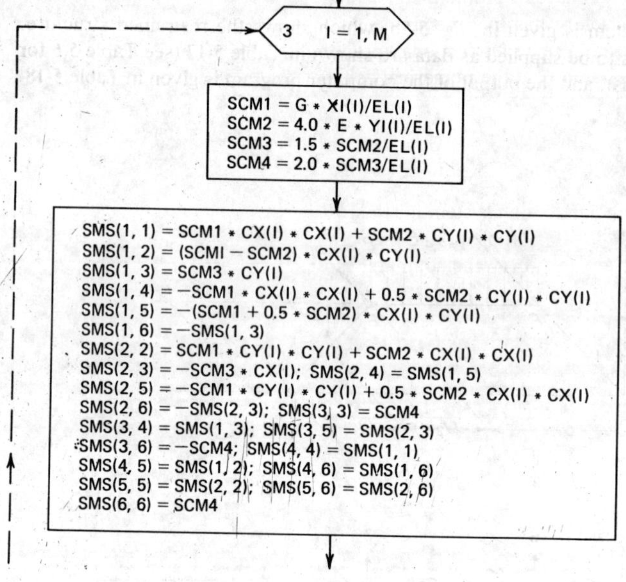

Calculate four stiffness constants for member I.

Fill upper triangular part of member stiffness matrix SMS for a grid member (see Table 4-38).

b. Transfer to joint stiffness matrix (same as Sec. 3.2b in Program PF)

4.3 Subprogram LDATA4 for Program GR
This subprogram is the same as LDATA3 for Program PF.

4.4 Subprogram LOADS4 for Program GR
This subprogram is the same as LOADS3 for Program PF.

4.5 Subprogram RESUL4 for Program GR
This subprogram is similar to RESUL3 for Program PF, but in Sec. 4.5b the member end-actions AMD are computed as follows:

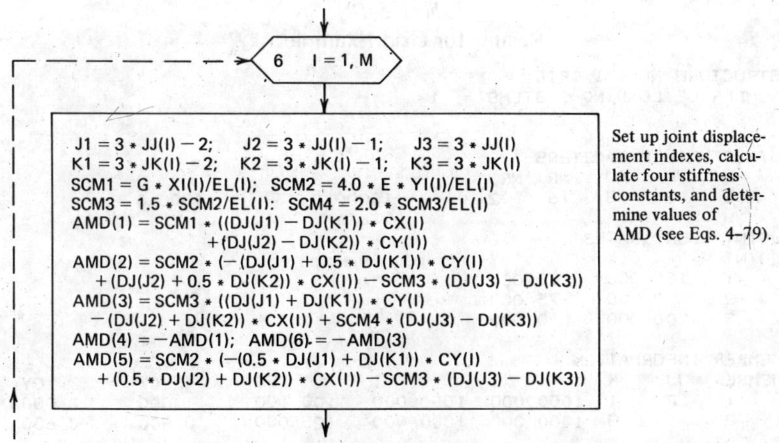

Example 2. A second problem consisting of a rectangular grid is shown in Fig. 5-6a. This structure has five members, six joints, twelve restraints, and six degrees of freedom. The cross-sectional properties of all of the members are assumed to be the same, and the numerical constants (in SI units) for the problem are the following:

$$E = 200 \times 10^6 \text{ kN/m}^2 \qquad G = 80 \times 10^6 \text{ kN/m}^2 \qquad L = 2 \text{ m}$$
$$I_X = 2 \times 10^{-3} \text{ m}^4 \qquad I_Y = 1 \times 10^{-3} \text{ m}^4 \qquad P = 50 \text{ kN}$$

Figure 5-6b shows a numbering system for the restrained structure. Table 5-19 contains the input data for this problem, and the solution is given in Table 5-20, which shows only the final results from the computer program.

Table 5-17
Data for Grid Example 1

Type of Data		***Numerical Values***							
Control Data		*1*	*4*	*1*					
Structural Data	(a)	2	3	6	2	10000.0	4000.0		
	(b)	1	100.0	75.0					
		2	0.0	75.0					
		3	200.0	0.0					
	(c)	1	2	1	1000.0	1000.0			
		2	1	3	1000.0	1000.0			
	(d)	2	1	1	1				
		3	1	1	1				
Load Data	(a)	1	2						
	(b)	1	0.0	1000.0	−10.0				
	(c)	1	0.0	−200.0	12.0	0.0	200.0	12.0	
		2	0.0	−312.5	10.0	0.0	312.5	10.0	

Table 5-18
Results for Grid Example 1

```
STRUCTURE NO.   1 GRID
NUMBER OF LOADING SYSTEMS =   1

STRUCTURAL PARAMETERS
   M    N   NJ   NR  NRJ             E                G
   2    3    3    6    2       10000.0           4000.0

JOINT COORDINATES
JOINT          X            Y
   1    100.000       75.000
   2      0.000       75.000
   3    200.000        0.000

MEMBER INFORMATION
MEMBER   JJ   JK        XI         YI          EL           CX          CY
   1      2    1  1000.000   1000.000     100.000        1.000       0.000
   2      1    3  1000.000   1000.000     125.000        0.800      -0.600

JOINT RESTRAINTS
JOINT   JR1   JR2   JR3
   2     1     1     1
   3     1     1     1

LOADING NO.   1
  NLJ  NLM
   1    2

ACTIONS AT JOINTS
JOINT        AJ1          AJ2         AJ3
   1       0.000     1000.000     -10.000

ACTIONS AT ENDS OF RESTRAINED MEMBERS DUE TO LOADS
MEMBER      AML1         AML2         AML3        AML4         AML5         AML6
   1       0.000     -200.000       12.000       0.000      200.000       12.000
   2       0.000     -312.500       10.000       0.000      312.500       10.000

JOINT DISPLACEMENTS
JOINT         DJ1         DJ2         DJ3
   1-.7599E-02 .5095E-02-.3551E+00
   2 .0000E+00 .0000E+00 .0000E+00
   3 .0000E+00 .0000E+00 .0000E+00

MEMBER END-ACTIONS
MEMBER        AM1          AM2          AM3          AM4          AM5          AM6
   1     303.942    -1311.474       24.040     -303.942      107.504       -0.040
   2    -292.344      896.362       -9.960      292.344     1598.676       29.960

SUPPORT REACTIONS
JOINT         AR1          AR2          AR3
   2     303.942    -1311.474       24.040
   3    1193.081     1103.534       29.960
```

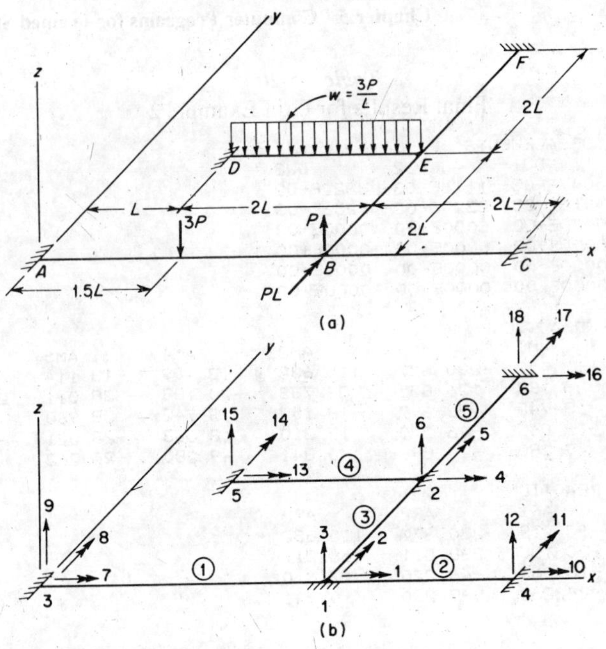

Fig. 5-6. Grid Example 2.

Table 5-19

Data for Grid Example 2

Type of Data		***Numerical Values***						
Control Data		2	4	1				
Structural Data	(a)	5	6	12	4	200000000.0	80000000.0	
	(b)	1	6.0	0.0				
		2	6.0	4.0				
		3	0.0	0.0				
		4	10.0	0.0				
		5	2.0	4.0				
		6	6.0	8.0				
	(c)	1	3	1	0.002	0.001		
		2	1	4	0.002	0.001		
		3	1	2	0.002	0.001		
		4	5	2	0.002	0.001		
		5	2	6	0.002	0.001		
	(d)	3	1	1	1			
		4	1	1	1			
		5	1	1	1			
		6	1	1	1			
Load Data	(a)	1	2					
	(b)	1	0.0	100.0	50.0			
	(c)	1	0.0	−112.5	75.0	0.0	112.5	75.0
		4	0.0	−100.0	150.0	0.0	100.0	150.0

Table 5-20
Final Results for Grid Example 2

JOINT DISPLACEMENTS

JOINT	DJ1	DJ2	DJ3
1	.5047E-03	-.1116E-03	-.3322E-02
2	.4516E-03	.2320E-03	-.9256E-03
3	.0000E+00	.0000E+00	.0000E+00
4	.0000E+00	.0000E+00	.0000E+00
5	.0000E+00	.0000E+00	.0000E+00
6	.0000E+00	.0000E+00	.0000E+00

MEMBER END-ACTIONS

MEMBER	AM1	AM2	AM3	AM4	AM5	AM6
1	-13.459	-230.677	115.632	13.459	-13.114	34.368
2	20.188	126.855	33.783	-20.188	338.011	266.217
3	-13.742	33.647	-18.152	13.742	38.960	18.152
4	-18.063	-46.220	17.310	18.063	-23.021	-17.310
5	9.280	-20.896	-0.841	-9.280	24.262	0.841

SUPPORT REACTIONS

JOINT	AR1	AR2	AR3
3	-13.459	-230.677	115.632
4	-20.188	338.011	266.217
5	-18.063	-46.220	17.310
6	-24.262	-9.280	0.841

5.10 Space Truss Program. The program in this article performs the analysis of space trusses by the method explained in Art. 4.23. The steps in this program are similar to those in the plane frame and grid programs (Arts. 5.8 and 5.9) because all three types of structures have three possible displacements at each joint. The space truss program also bears a great similarity to the plane truss program, but the three-dimensional nature of space trusses causes some additional complications.

The main program ST for space trusses is similar to that for other types of structures, but in this case the five structure-oriented subprograms are called SDATA5, STIFF5, LDATA5, LOADS5, and RESUL5. Flow Chart 5-5 shows how statements in these subprograms differ from those for the other types of structures considered previously.

Example *1.* Figure 5-7a shows a space truss having seven members, five joints, nine restraints (at three pinned supports), and six degrees of freedom. The cross-sectional areas of all members are assumed to be the same, and the following numerical values (in US units) apply:

$$E = 30{,}000 \text{ ksi} \qquad L = 120 \text{ in.} \qquad A_X = 10 \text{ in.}^2 \qquad P = 20 \text{ kips}$$

Figure 5-7b shows the restrained structure and an appropriate numbering system. The input data required for this problem are listed in Table 5-21 (see Table 5-7 for the form of data), and the results from the computer program are given in Table 5-22.

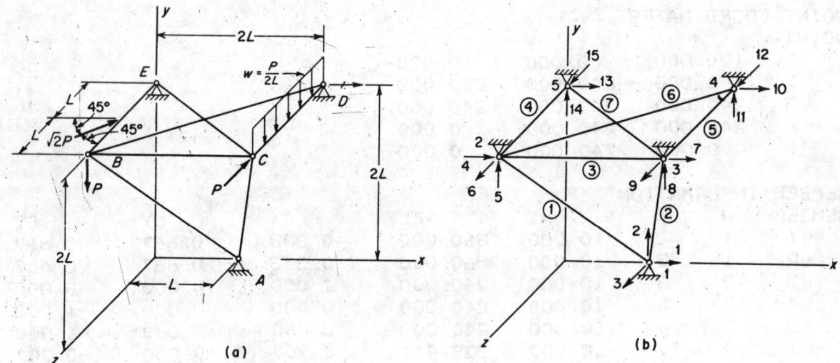

Fig. 5-7. Space truss Example 1.

Table 5-21

Data for Space Truss Example 1

Type of Data		*Numerical Values*						
Control Data		1	5	1				
Structural Data	(a)	7	5	9	3	30000.0		
	(b)	1	120.0	0.0	0.0			
		2	0.0	240.0	240.0			
		3	240.0	240.0	240.0			
		4	240.0	240.0	0.0			
		5	0.0	240.0	0.0			
	(c)	1	1	2	10.0			
		2	1	3	10.0			
		3	2	3	10.0			
		4	2	5	10.0			
		5	3	4	10.0			
		6	2	4	10.0			
		7	3	5	10.0			
	(d)	1	1	1	1			
		4	1	1	1			
		5	1	1	1			
Load Data	(a)	2	2					
	(b)	2	0.0	−20.0	0.0			
		3	0.0	0.0	−20.0			
	(c)	4	−10.0	0.0	−10.0	−10.0	0.0	−10.0
		5	0.0	10.0	0.0	0.0	10.0	0.0

Table 5-22

Results for Space Truss Example 1

```
STRUCTURE NO.   1 SPACE TRUSS
NUMBER OF LOADING SYSTEMS =   1


STRUCTURAL PARAMETERS
     M     N    NJ    NR   NRJ                 E
     7     6     5     9     3           30000.0

JOINT COORDINATES
JOINT              X              Y              Z
    1        120.000          0.000          0.000
    2          0.000        240.000        240.000
    3        240.000        240.000        240.000
    4        240.000        240.000          0.000
    5          0.000        240.000          0.000

MEMBER INFORMATION
MEMBER   JJ    JK         AX          EL          CX          CY          CZ
    1     1     2     10.000     360.000      -0.333       0.667       0.667
    2     1     3     10.000     360.000       0.333       0.667       0.667
    3     2     3     10.000     240.000       1.000       0.000       0.000
    4     2     5     10.000     240.000       0.000       0.000      -1.000
    5     3     4     10.000     240.000       0.000       0.000      -1.000
    6     2     4     10.000     339.411       0.707       0.000      -0.707
    7     3     5     10.000     339.411      -0.707       0.000      -0.707

JOINT RESTRAINTS
JOINT   JR1   JR2   JR3
    1     1     1     1
    4     1     1     1
    5     1     1     1


LOADING NO.   1
  NLJ  NLM
    2    2

ACTIONS AT JOINTS
JOINT          AJ1          AJ2          AJ3
    2        0.000      -20.000        0.000
    3        0.000        0.000      -20.000

ACTIONS AT ENDS OF RESTRAINED MEMBERS DUE TO LOADS
MEMBER        AML1         AML2         AML3         AML4         AML5         AML6
    4      -10.000        0.000      -10.000      -10.000        0.000      -10.000
    5        0.000       10.000        0.000        0.000       10.000        0.000

JOINT DISPLACEMENTS
JOINT          DJ1          DJ2          DJ3
    1 .0000E+00 .0000E+00 .0000E+00
    2 .1477E-01-.5538E-01 .9769E-02
    3 .1654E-01-.2504E-01-.1023E-01
    4 .0000E+00 .0000E+00 .0000E+00
    5 .0000E+00 .0000E+00 .0000E+00
```

MEMBER END-ACTIONS

MEMBER	AM1	AM2	AM3	AM4	AM5	AM6
1	30. 000	0. 000	0. 000	-30. 000	0. 000	0. 000
2	15. 000	0. 000	0. 000	-15. 000	0. 000	0. 000
3	-2. 211	0. 000	0. 000	2. 211	0. 000	0. 000
4	-22. 211	0. 000	-10. 000	2. 211	0. 000	-10. 000
5	12. 789	10. 000	0. 000	-12. 789	10. 000	0. 000
6	3. 127	0. 000	0. 000	-3. 127	0. 000	0. 000
7	-3. 944	0. 000	0. 000	3. 944	0. 000	0. 000

SUPPORT REACTIONS

JOINT	AR1	AR2	
1	-5. 000	30. 000	30. 000
4	-2. 211	10. 000	15. 000
5	-12. 789	0. 000	-5. 000

Flow Chart 5-5

5.1 Subprogram SDATA5 for Program ST

a. Problem identification

Title: STRUCTURE NO. (ISN) SPACE TRUSS NUMBER OF LOADING SYSTEMS = (NLS)	Print title.

b. Structural parameters (same as Sec. 2.1b in Program PT, except that NDJ = 3)

c. Joint coordinates (similar to Sec. 2.1c in Program PT, but z coordinates are also required)

d. Member information (similar to Sec. 2.1d in Program PT)

For each member the direction cosines are computed and printed, as follows:

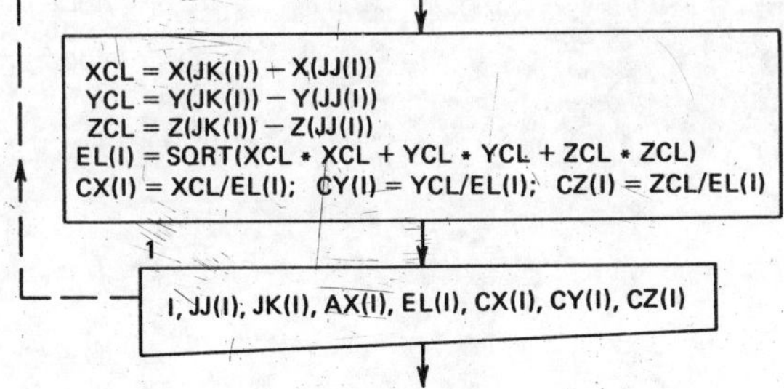

Compute x, y, and z components of the member length (XCL, YCL, and ZCL), the length EL, and the direction cosines CX, CY, and CZ.

Print information for member I.

e. Joint restraints (same as Sec. 3.1e in Program PF)

f. Displacement indexes (same as Sec. 1.1e in Program CB)

5.2 Subprogram STIFF5 for Program ST

a. Member stiffnesses

After clearing SFF, do the following:

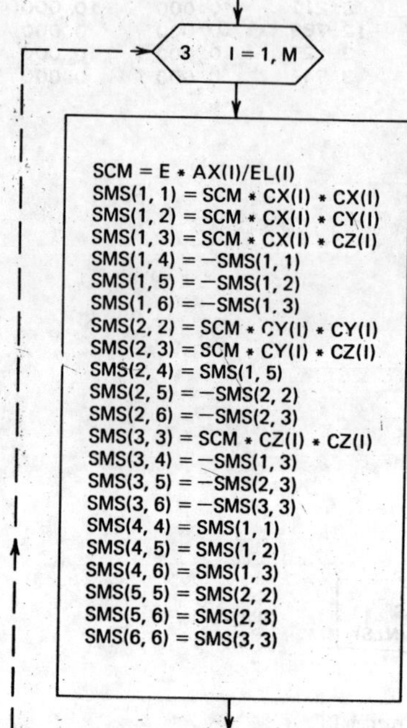

Fill upper triangular part of member stiffness matrix SMS for a space truss member (see Table 4–40).

b. Transfer to joint stiffness matrix (same as Sec. 3.2b in Program PF)

5.3 Subprogram LDATA5

This subprogram is the same as LDATA3 for Program PF.

5.4 Subprogram LOADS5 for Program ST

a. Equivalent joint loads

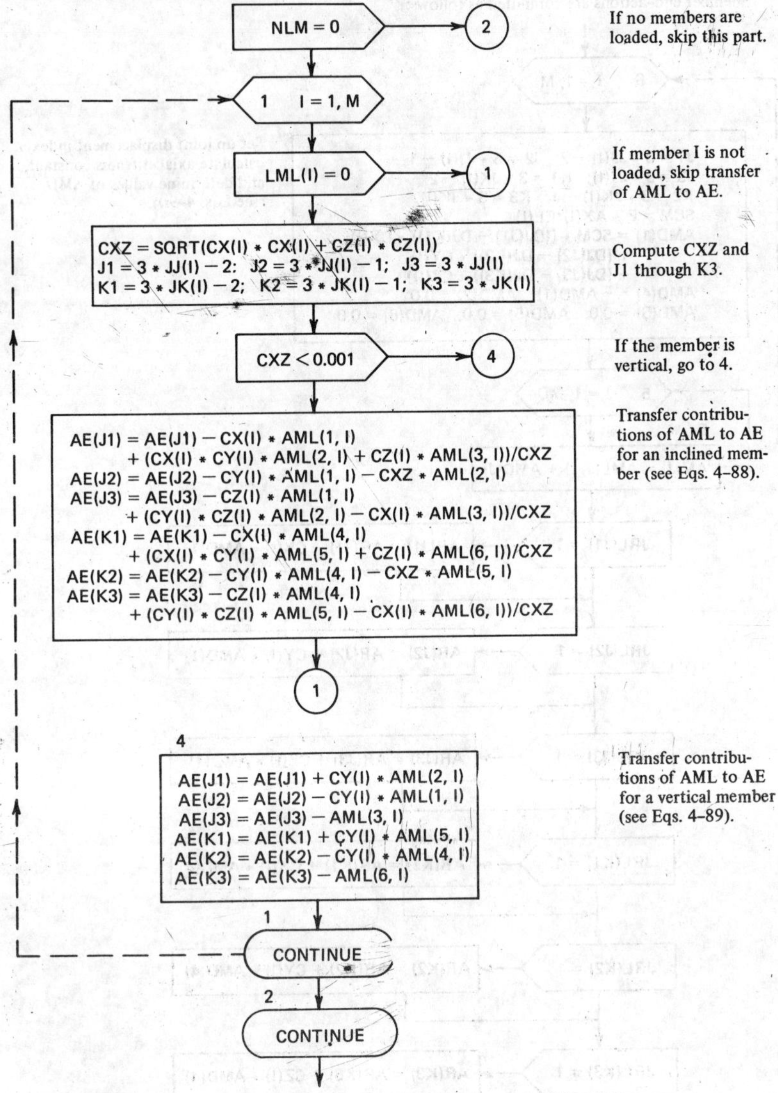

b. Combined joint loads (same as Sec. 1.4b in Program CB)

5.5 Subprogram RESUL5 for Program ST

Sections 5.5a and 5.5c in this subprogram are the same as in RESUL3 for Program PF. However, in Sec. 5.5b the member end-actions are computed as follows:

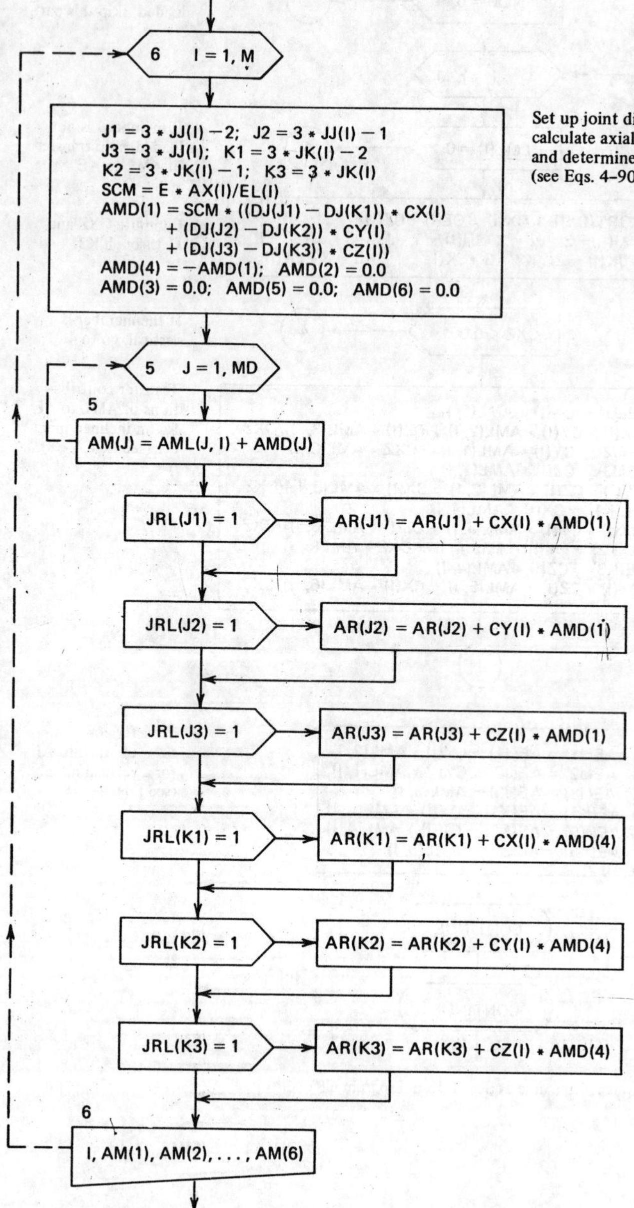

Set up joint displacement indexes, calculate axial stiffness constant, and determine values of AMD (see Eqs. 4–90).

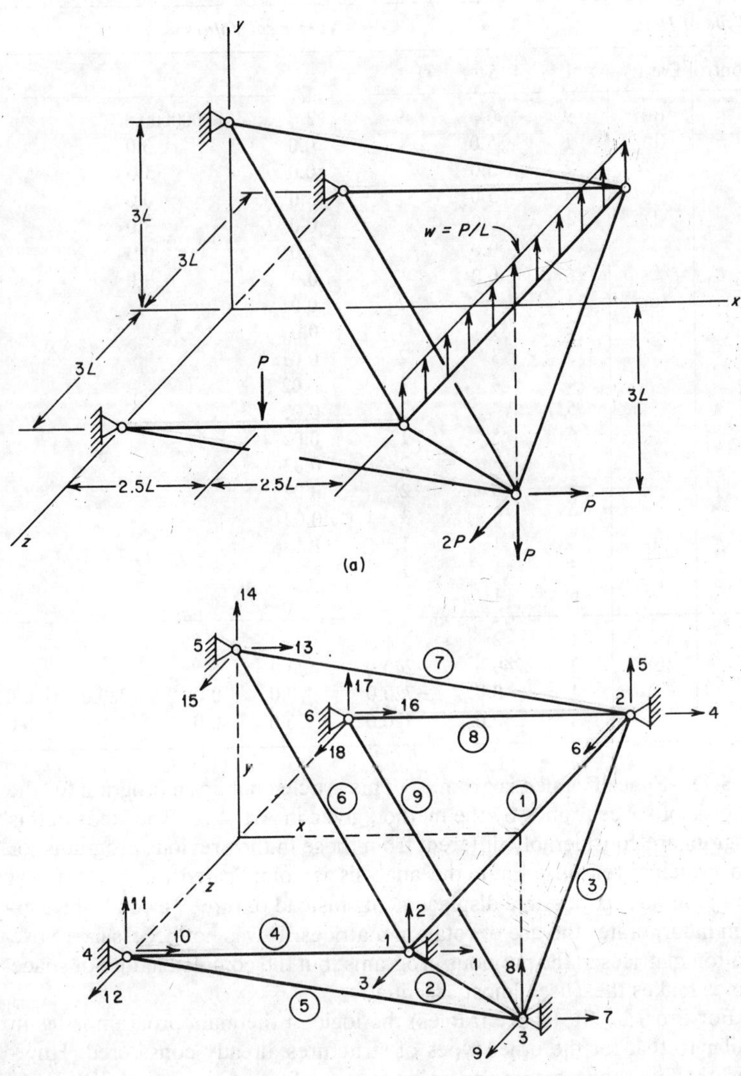

Fig. 5-8. Space truss Example 2.

Example 2. The space truss in Fig. 5-8a has nine members, six joints, nine restraints (at three pinned supports), and nine degrees of freedom. The members numbered 1, 2, and 3 in Fig. 5-8b (the restrained structure) have cross-sectional areas equal to AX, but the other six members have areas equal to 2AX. For this problem the following numerical values (in SI units) are assumed:

$$E = 80 \times 10^6 \text{ kN/m}^2 \qquad L = 1 \text{ m} \qquad A_X = 0.01 \text{ m}^2 \qquad P = 240 \text{ kN}$$

Table 5-23 contains the required input data, and the final results from the computer program appear in Table 5-24.

Table 5-23

Data for Space Truss Example 2

Type of Data		Numerical Values						
Control Data		1	5	1				
Structural Data	(a)	9	6	9	3	80000000.0		
	(b)	1	5.0		0.0	3.0		
		2	5.0		0.0	−3.0		
		3	5.0		−3.0	0.0		
		4	0.0		0.0	3.0		
		5	0.0		3.0	0.0		
		6	0.0		0.0	−3.0		
	(c)	1	1	2	0.01			
		2	1	3	0.01			
		3	3	2	0.01			
		4	4	1	0.02			
		5	4	3	0.02			
		6	5	1	0.02			
		7	5	2	0.02			
		8	6	2	0.02			
		9	6	3	0.02			
	(d)	4	1	1	1			
		5	1	1	1			
		6	1	1	1			
Load Data	(a)	1	2					
	(b)	3	240.0	−240.0	480.0			
	(c)	1	0.0	−720.0	0.0	0.0	−720.0	0.0
		4	0.0	120.0	0.0	0.0	120.0	0.0

5.11 Space Frame Program. This article contains a program for the analysis of space frames by the method given in Art. 4.25. The steps in this program are considerably different from those in the previous programs for two principal reasons. First, the analysis is complicated by the fact that each joint has six possible displacements instead of three; second, the program incorporates the use of rotation matrices. It was not necessary to use rotation matrices in the previous programs, but the complex nature of space frames makes their use almost essential.

For Program SF (space frames) the logic in the main program is again similar to that for the other types of structures already considered. However, the five subprograms pertaining to space frames are named SDATA6, STIFF6, LDATA6, LOADS6, and RESUL6. Flow Chart 5-6 gives the statements in these subprograms that are different from those for the other types of structures.

Example 1. A space frame having three members and four joints is shown in Fig. 5-9a. Inspection of the figure shows that there are twelve degrees of freedom (six at each of the joints B and C) and twelve restraints (six at each of the support points A and D). The joint loads on the frame consist of a force $2P$ in the positive x direction at point B, a force P in the negative y direction at point C, and a moment

Table 5-24
Final Results for Space Truss Example 2

JOINT DISPLACEMENTS

JOINT	DJ1	DJ2	DJ3
1	.3031E-02	.2088E-01	.4436E-02
2	.3344E-02	.1794E-01	.2059E-03
3	.1346E-01	.2102E-01	.4112E-02
4	.0000E+00	.0000E+00	.0000E+00
5	.0000E+00	.0000E+00	.0000E+00
6	.0000E+00	.0000E+00	.0000E+00

MEMBER END-ACTIONS

MEMBER	AM1	AM2	AM3	AM4	AM5	AM6
1	-564.000	-720.000	0.000	564.000	-720.000	0.000
2	-25.456	0.000	0.000	25.456	0.000	0.000
3	-110.309	0.000	0.000	110.309	0.000	0.000
4	-970.000	120.000	0.000	970.000	120.000	0.000
5	301.642	0.000	0.000	-301.642	0.000	0.000
6	1272.143	0.000	0.000	-1272.143	0.000	0.000
7	1403.292	0.000	0.000	-1403.292	0.000	0.000
8	-1070.000	0.000	0.000	1070.000	0.000	0.000
9	-616.399	0.000	0.000	616.399	0.000	0.000

SUPPORT REACTIONS

JOINT	AR1	AR2	
4	-740.000	-18.000	-138.000
5	2040.000	-1224.000	-60.000
6	-1540.000	282.000	-282.000

PL in the negative z sense at C. Member BC is subjected to force $4P$ in the positive z direction applied at the midlength of the member.

The x-y plane is a principal plane for members AB and BC. and member CD has a principal plane parallel to the y axis. Therefore, all members have the angle α equal to zero. It is assumed that all members have the same cross-sectional properties and that the numerical values (in SI units) for this problem are the following:

$$E = 200 \times 10^6 \text{ kN/m}^2 \qquad G = 80 \times 10^6 \text{ kN/m}^2 \qquad L = 3 \text{ m} \qquad A_X = 0.01 \text{ m}^2$$
$$I_X = 2 \times 10^{-3} \text{ m}^4 \qquad I_Y = I_Z = 1 \times 10^{-3} \text{ m}^4 \qquad P = 60 \text{ kN}$$

Figure 5-9b shows the restrained structure with an appropriate numbering system. The data required as input to the computer program are given in Table 5-25 (see Table 5-8 for data specifications). Note that the identifier IA is zero on each of the lines containing member information. Table 5-26 shows the output of the computer program. (Tables on pp. 356–357.)

Example 2. Figure 5-10a shows a space frame having three members, four joints, six degrees of freedom (at joint A), and eighteen restraints (six at each of the points B, C, and D). The frame is loaded at joint A with a force P in the negative z direction and a moment $PL/4$ in the negative x sense. In addition, a force of magnitude P acting in the negative z direction is applied at the middle of member AB, and a force P acting in the negative y direction is applied at the middle of member AD. All members in the frame have a principal plane x_M-y_M that contains the point p indicated in Fig. 5-10a. Point p is located in the plane containing joints B, C, and D (that is, the x-y plane). Any point on the line pA (except point A) would suffice to define the principal planes in this problem. Since a principal plane for member AB is parallel to the y axis, the angle α is taken to be zero for this member. However, for members AC and AD the principal planes are located by giving the coordinates of point p. (Example 2 continued on p. 355.)

Flow Chart 5-6

6.1 Subprogram SDATA6 for Program SF

a. Problem identification

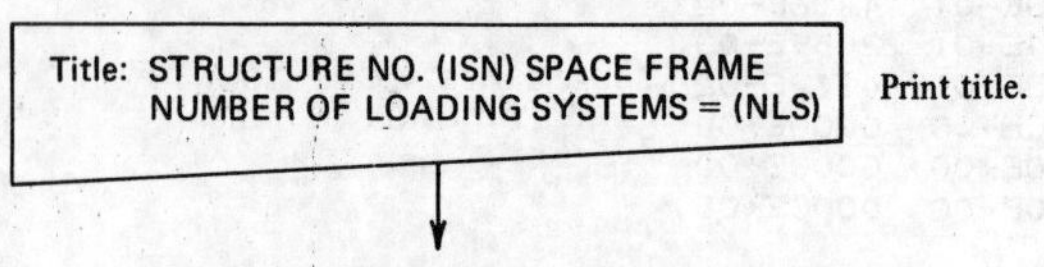

Title: STRUCTURE NO. (ISN) SPACE FRAME
NUMBER OF LOADING SYSTEMS = (NLS)
Print title.

b. Structural parameters (same as Sec. 2.1b in Program PT, except that the shear modulus G is added and NDJ = 6)

c. Joint coordinates (similar to Sec. 2.1c in Program PT, but z coordinates are also required)

d. Member information

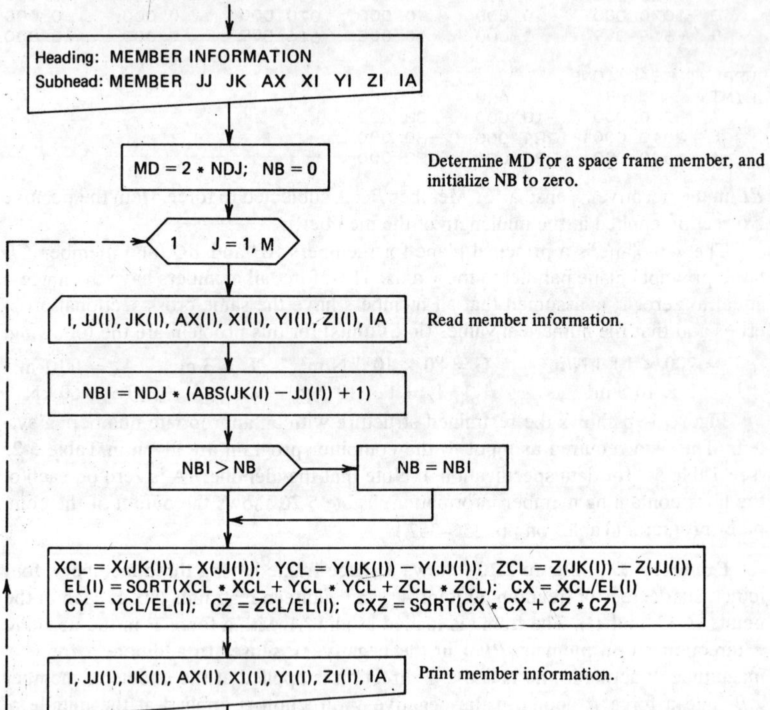

Heading: MEMBER INFORMATION
Subhead: MEMBER JJ JK AX XI YI ZI IA
MD = 2 * NDJ; NB = 0
Determine MD for a space frame member, and initialize NB to zero.
1 J = 1, M
I, JJ(I), JK(I), AX(I), XI(I), YI(I), ZI(I), IA
Read member information.
NBI = NDJ * (ABS(JK(I) − JJ(I)) + 1)
NBI > NB
NB = NBI
XCL = X(JK(I)) − X(JJ(I)); YCL = Y(JK(I)) − Y(JJ(I)); ZCL = Z(JK(I)) − Z(JJ(I))
EL(I) = SQRT(XCL * XCL + YCL * YCL + ZCL * ZCL); CX = XCL/EL(I)
CY = YCL/EL(I); CZ = ZCL/EL(I); CXZ = SQRT(CX * CX + CZ * CZ)
I, JJ(I), JK(I), AX(I), XI(I), YI(I), ZI(I), IA
Print member information.

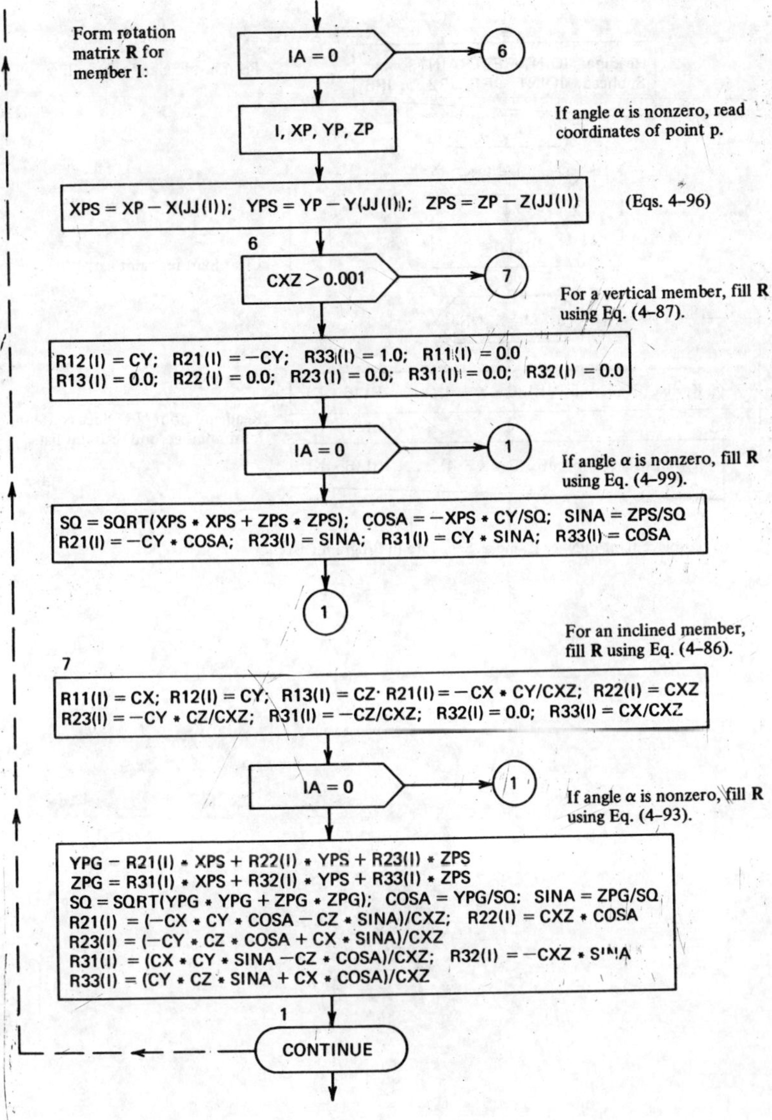
Form rotation
matrix R for
member I:
IA = 0
6
I, XP, YP, ZP
If angle α is nonzero, read coordinates of point p.
XPS = XP − X(JJ(I)); YPS = YP − Y(JJ(I)); ZPS = ZP − Z(JJ(I))
(Eqs. 4–96)
6
CXZ > 0.001
7
For a vertical member, fill R using Eq. (4–87).
R12(I) = CY; R21(I) = −CY; R33(I) = 1.0; R11(I) = 0.0
R13(I) = 0.0; R22(I) = 0.0; R23(I) = 0.0; R31(I) = 0.0; R32(I) = 0.0
IA = 0
1
If angle α is nonzero, fill R using Eq. (4–99).
SQ = SQRT(XPS * XPS + ZPS * ZPS); COSA = −XPS * CY/SQ; SINA = ZPS/SQ
R21(I) = −CY * COSA; R23(I) = SINA; R31(I) = CY * SINA; R33(I) = COSA
1
For an inclined member, fill R using Eq. (4–86).
7
R11(I) = CX; R12(I) = CY; R13(I) = CZ· R21(I) = −CX * CY/CXZ; R22(I) = CXZ
R23(I) = −CY * CZ/CXZ; R31(I) = −CZ/CXZ; R32(I) = 0.0; R33(I) = CX/CXZ
IA = 0
1
If angle α is nonzero, fill R using Eq. (4–93).
YPG − R21(I) * XPS + R22(I) * YPS + R23(I) * ZPS
ZPG = R31(I) * XPS + R32(I) * YPS + R33(I) * ZPS
SQ = SQRT(YPG * YPG + ZPG * ZPG); COSA = YPG/SQ; SINA = ZPG/SQ
R21(I) = (−CX * CY * COSA − CZ * SINA)/CXZ; R22(I) = CXZ * COSA
R23(I) = (−CY * CZ * COSA + CX * SINA)/CXZ
R31(I) = (CX * CY * SINA −CZ * COSA)/CXZ; R32(I) = −CXZ * SINA
R33(I) = (CY * CZ * SINA + CX * COSA)/CXZ
1
CONTINUE

e. Joint restraints

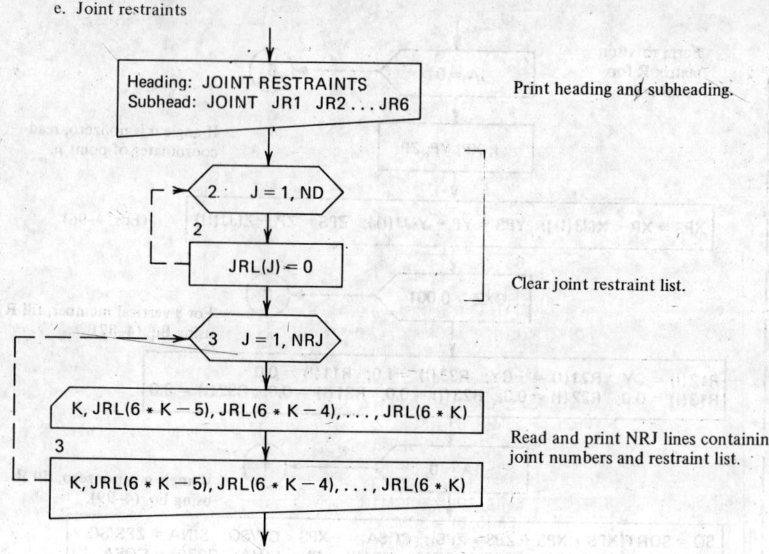

f. Displacement indexes (same as Sec. 1.1e in Program CB)

6.2 Subprogram STIFF6 for Program SF

a. Member stiffnesses

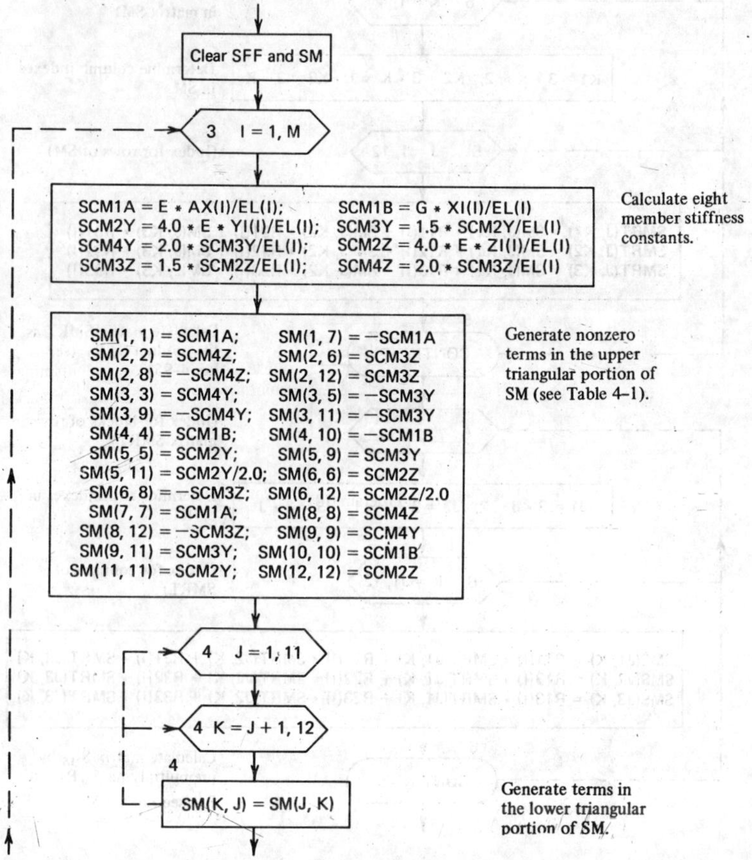

Clear SFF and SM

3 I = 1, M

```
SCM1A = E * AX(I)/EL(I);          SCM1B = G * XI(I)/EL(I)
SCM2Y = 4.0 * E * YI(I)/EL(I);    SCM3Y = 1.5 * SCM2Y/EL(I)
SCM4Y = 2.0 * SCM3Y/EL(I);        SCM2Z = 4.0 * E * ZI(I)/EL(I)
SCM3Z = 1.5 * SCM2Z/EL(I);        SCM4Z = 2.0 * SCM3Z/EL(I)
```

Calculate eight member stiffness constants.

```
SM(1, 1) = SCM1A;        SM(1, 7) = -SCM1A
SM(2, 2) = SCM4Z;        SM(2, 6) = SCM3Z
SM(2, 8) = -SCM4Z;       SM(2, 12) = SCM3Z
SM(3, 3) = SCM4Y;        SM(3, 5) = -SCM3Y
SM(3, 9) = -SCM4Y;       SM(3, 11) = -SCM3Y
SM(4, 4) = SCM1B;        SM(4, 10) = -SCM1B
SM(5, 5) = SCM2Y;        SM(5, 9) = SCM3Y
SM(5, 11) = SCM2Y/2.0;   SM(6, 6) = SCM2Z
SM(6, 8) = -SCM3Z;       SM(6, 12) = SCM2Z/2.0
SM(7, 7) = SCM1A;        SM(8, 8) = SCM4Z
SM(8, 12) = -SCM3Z;      SM(9, 9) = SCM4Y
SM(9, 11) = SCM3Y;       SM(10, 10) = SCM1B
SM(11, 11) = SCM2Y;      SM(12, 12) = SCM2Z
```

Generate nonzero terms in the upper triangular portion of SM (see Table 4–1).

4 J = 1, 11

4 K = J + 1, 12

4

SM(K, J) = SM(J, K)

Generate terms in the lower triangular portion of SM.

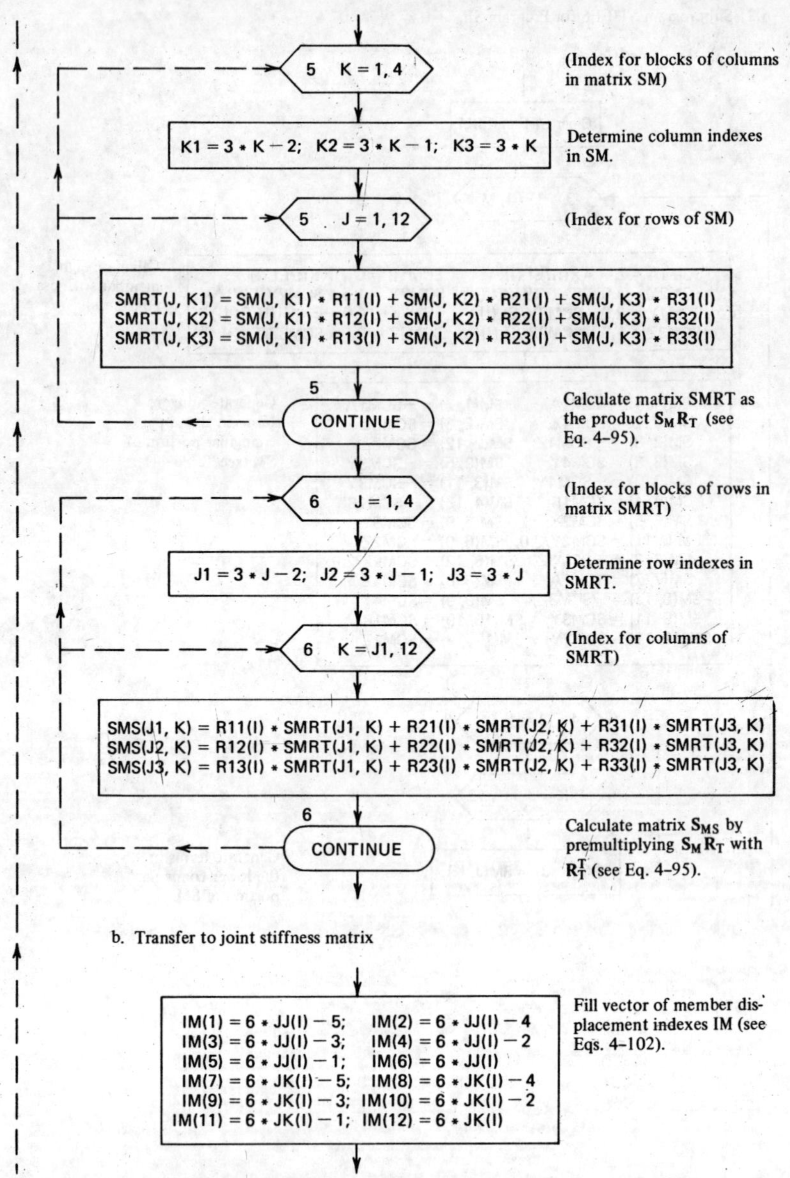

b. Transfer to joint stiffness matrix

```
IM(1) = 6 * JJ(I) − 5;    IM(2) = 6 * JJ(I) − 4
IM(3) = 6 * JJ(I) − 3;    IM(4) = 6 * JJ(I) − 2
IM(5) = 6 * JJ(I) − 1;    IM(6) = 6 * JJ(I)
IM(7) = 6 * JK(I) − 5;    IM(8) = 6 * JK(I) − 4
IM(9) = 6 * JK(I) − 3;   IM(10) = 6 * JK(I) − 2
IM(11) = 6 * JK(I) − 1;  IM(12) = 6 * JK(I)
```

Fill vector of member displacement indexes IM (see Eqs. 4–102).

The remainder of this section is the same as Sec. 1.2b of Program CB, except that SMS replaces SM.

6.3 Subprogram LDATA6 for Program SF

a. Load parameters (same as Sec. 1.3a in Program CB)

b. Joint loads

If NLJ ≠ 0, do the following:

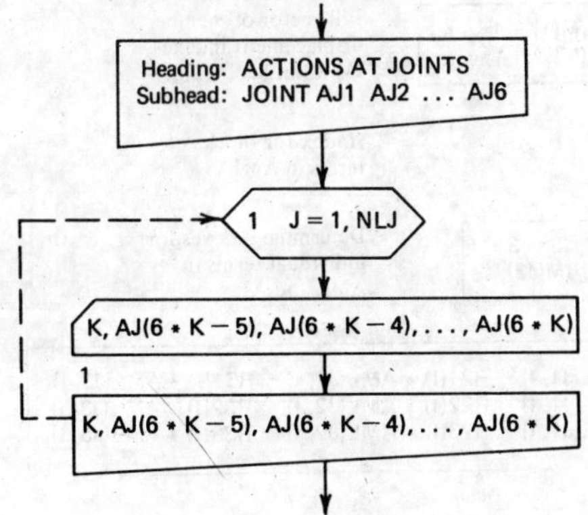

Print heading and subheading for joint loads.

Read and print NLJ lines containing joint numbers and joint loads.

c. Member loads

After clearing LML and AML, do the following:

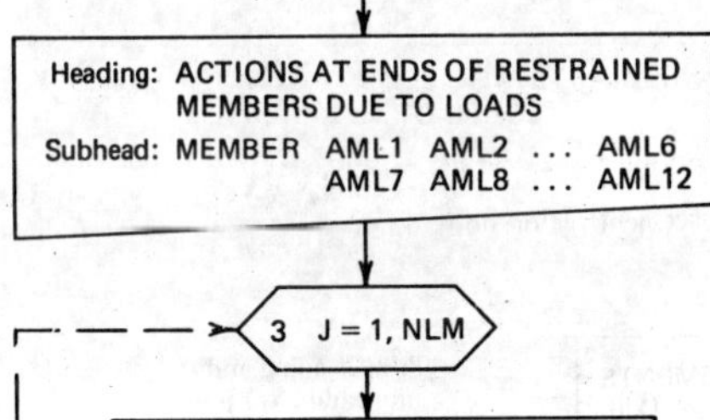

Print heading and subheadings for member loads.

3 J = 1, NLM

I, AML(1, I), AML(2, I), . . . , AML(6, I)
AML(7, I), AML(8, I), . . . , AML(12, I)

I, AML(1, I), AML(2, I), . . . , AML(6, I)
AML(7, I), AML(8, I), . . . , AML(12, I)

3 LML(I) = 1

Read and print 2 (NLM) lines containing member numbers and actions at ends of restrained members due to loads.

6.4 Subprogram LOADS6 for Program SF

This subprogram is similar to LOADS1 for Program CB, but in Sec. 6.4a the equivalent joint loads are computed as follows:

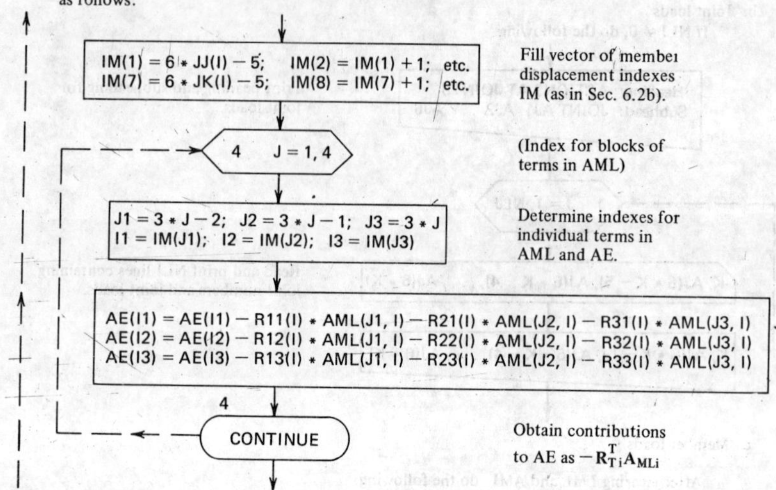

6.5 Subprogram RESUL6 for Program SF

a. Joint displacements

After expanding the joint displacement vector, do the following:

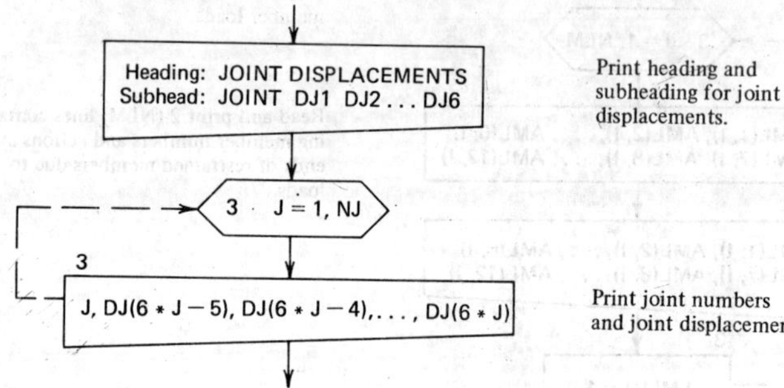

b. Member end-actions

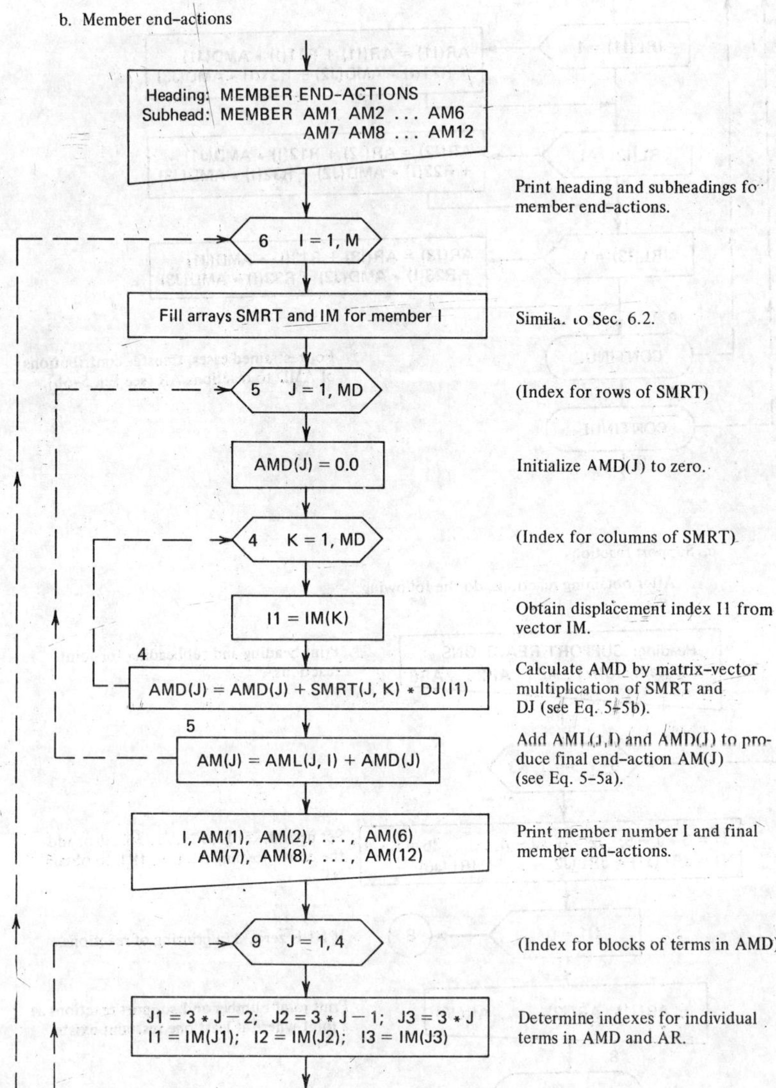

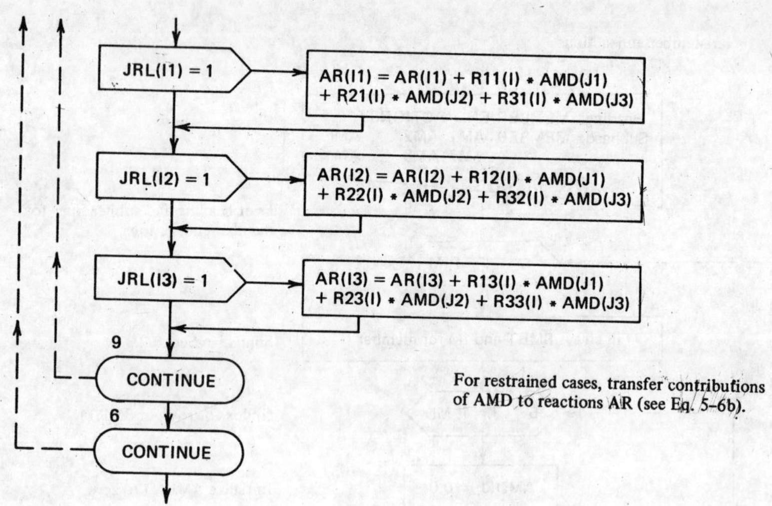

c. Support reactions

After obtaining reactions, do the following:

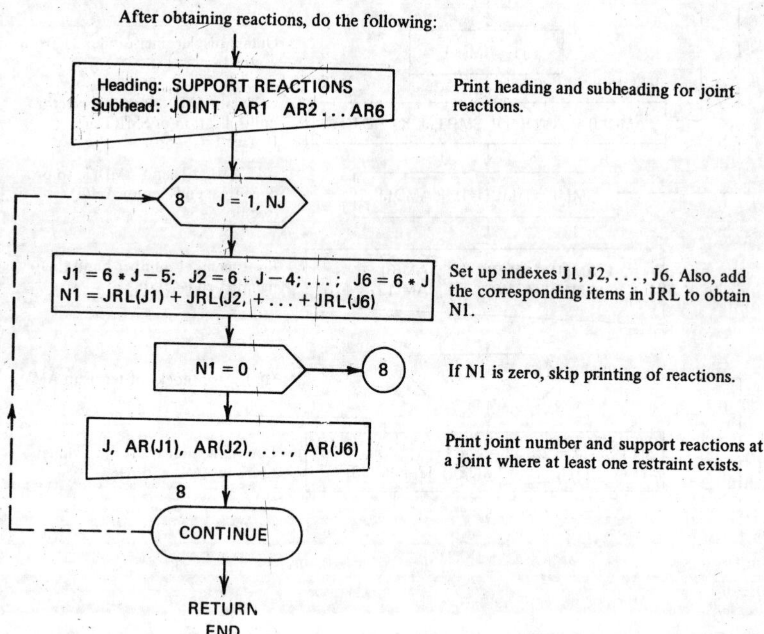

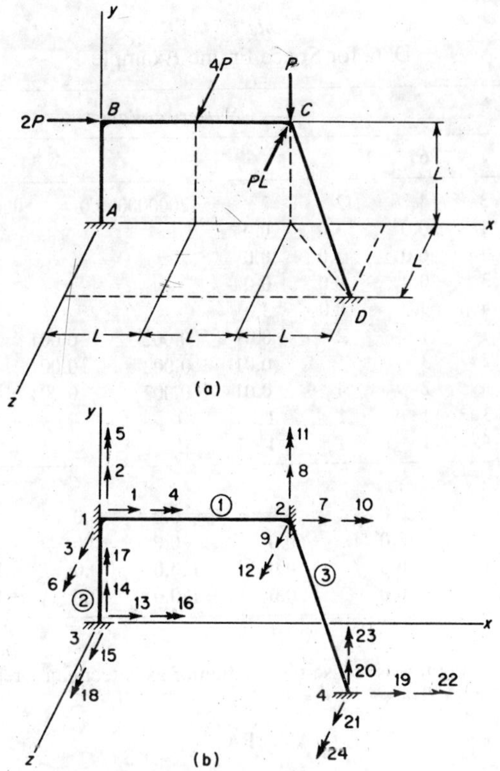

Fig. 5-9. Space frame Example 1.

The cross-sectional properties of all members are assumed to be the same, and the numerical constants in the problem are as follows:

$$E = 10{,}000 \text{ ksi} \qquad G = 4{,}000 \text{ ksi} \qquad L = 96 \text{ in.} \qquad A_X = 9 \text{ in.}^2$$
$$I_X = 64 \text{ in.}^4 \qquad I_Y = 28 \text{ in.}^4 \qquad I_Z = 80 \text{ in.}^4 \qquad P = 5 \text{ kips}$$

Figure 5-10b shows a numbering system for the restrained structure. Input data required for this problem are listed in Table 5-27. Note that the lines containing member information for members 2 and 3 are each followed by a line that gives the coordinates of point *p*.

The calculation of actions $\mathbf{A}_{ML}$ at the ends of the restrained members due to loads (the last four data cards) requires further explanation in this example. The end-actions $\mathbf{A}_{ML1}$ for member 1 are obtained easily because the load *P* applied at the middle lies in a principal plane for the member. Figure 5-11 shows all of the end-actions for this member. Note that the orientation of the member axes in the figure results from the sequence of β and γ rotations depicted in Fig. 4-38 (see Art. 4.22).

On the other hand, the load *P* applied to member 3 does not lie in a principal plane of the member. In order to calculate the end-actions $\mathbf{A}_{ML3}$ for this member, the components of the load *P* in the directions of member axes must be ascertained.

Table 5-25
Data for Space Frame Example 1

Type of Data		*Numerical Values*							
Control Data		1	6	1					
Structural Data	(a)	3	4	12	2	200000000.0	80000000.0		
	(b)	1	0.0	3.0	0.0				
		2	6.0	3.0	0.0				
		3	0.0	0.0	0.0				
		4	9.0	0.0	3.0				
	(c)	1	1	2	0.01	0.002	0.001	0.001	0
		2	3	1	0.01	0.002	0.001	0.001	0
		3	2	4	0.01	0.002	0.001	0.001	0
	(d)	3	1	1	1 1	1	1		
		4	1	1	1 1	1	1		
Load Data	(a)	2	1						
	(b)	1	120.0	0.0	0.0	0.0	0.0	0.0	
		2	0.0	−60.0	0.0	0.0	0.0	−180.0	
	(c)	1	0.0	0.0	−120.0	0.0	180.0	0.0	
			0.0	0.0	−120.0	0.0	−180.0	0.0	

One method for doing this is to use the rotation of axes technique represented by Eq. (4-57) in Art. 4.15. Thus,

$$\mathbf{A}_{\mathrm{M}} = \mathbf{R}\mathbf{A}_{\mathrm{S}} \tag{4-57}$$

repeated

Figure 5-12 shows the member axes for member 3 as well as the load P applied in the negative y direction. The rotation matrix $\mathbf{R}$ for the member axes may be obtained either from the geometry of the figure or from Eq. (4-93) in Art. 4.24 (using the coordinates of point p if desired). This rotation matrix is the following:

$$\mathbf{R} = \begin{bmatrix} \dfrac{16}{\sqrt{481}} & -\dfrac{12}{\sqrt{481}} & -\dfrac{9}{\sqrt{481}} \\ -\dfrac{36}{5\sqrt{481}} & \dfrac{27}{5\sqrt{481}} & -\dfrac{20}{\sqrt{481}} \\ \dfrac{3}{5} & \dfrac{4}{5} & 0 \end{bmatrix} \tag{a}$$

Since the applied load P is 5 kips, the vector $\mathbf{A}_s$ becomes:

$$\mathbf{A}_{\mathrm{S}} = \{0, -5, 0\} \tag{b}$$

Substitution of expressions (a) and (b) into Eq. (4-57) produces the vector $\mathbf{A}_{\mathrm{M}}$, as follows:

$$\mathbf{A}_{\mathrm{M}} = \{60, -27, -4\sqrt{481}\}\big/\sqrt{481} = \{2.736, -1.231, -4.000\} \tag{c}$$

The elements in this vector are the components of the applied load in the directions of member axes. End-actions $\mathbf{A}_{\mathrm{ML3}}$ caused by these forces acting on member 3 are listed in the last two lines of Table 5-27. When the data in Table 5-27 are analyzed by Program SF, the final results obtained are as shown in Table 5-28.

Table 5-26

Results for Space Frame Example 1

```
STRUCTURE NO.  1 SPACE FRAME
NUMBER OF LOADING SYSTEMS =  1

STRUCTURAL PARAMETERS
   M     N    NJ    NR   NRJ            E                G
   3    12     4    12     2   200000000.0       80000000.0

JOINT COORDINATES
JOINT          X          Y          Z
   1       0.000      3.000      0.000
   2       6.000      3.000      0.000
   3       0.000      0.000      0.000
   4       9.000      0.000      3.000

MEMBER INFORMATION
MEMBER  JJ   JK        AX         XI         YI         ZI    IA
   1     1    2     0.010      0.002      0.001      0.001     0
   2     3    1     0.010      0.002      0.001      0.001     0
   3     2    4     0.010      0.002      0.001      0.001     0

JOINT RESTRAINTS
JOINT   JR1   JR2   JR3   JR4   JR5   JR6
   3     1     1     1     1     1     1
   4     1     1     1     1     1     1


LOADING NO.  1
  NLJ  NLM
   2    1

ACTIONS AT JOINTS
JOINT       AJ1        AJ2        AJ3        AJ4        AJ5        AJ6
   1    120.000      0.000      0.000      0.000      0.000      0.000
   2      0.000    -60.000      0.000      0.000      0.000   -180.000

ACTIONS AT ENDS OF RESTRAINED MEMBERS DUE TO LOADS
MEMBER     AML1       AML2       AML3       AML4       AML5       AML6
           AML7       AML8       AML9      AML10      AML11      AML12
   1      0.000      0.000   -120.000      0.000    180.000      0.000
          0.000      0.000   -120.000      0.000   -180.000      0.000

JOINT DISPLACEMENTS
JOINT        DJ1        DJ2        DJ3        DJ4        DJ5        DJ6
    1-.8594E-03  .5776E-04  .5008E-02  .2292E-02-.1620E-02  .6813E-03
    2-.1176E-02  .3253E-02  .5256E-02  .1288E-02  .1721E-02-.7715E-03
    3 .0000E+00  .0000E+00  .0000E+00  .0000E+00  .0000E+00  .0000E+00
    4 .0000E+00  .0000E+00  .0000E+00  .0000E+00  .0000E+00  .0000E+00

MEMBER END-ACTIONS
MEMBER      AM1        AM2        AM3        AM4        AM5        AM6
            AM7        AM8        AM9       AM10       AM11       AM12
   1    105.548    -38.509   -126.013     29.464     86.569    -67.100
       -105.548     38.509   -113.987    -29.464     50.491   -163.954
   2    -38.509     14.452   -126.013     86.569    348.575    -23.744
         38.509    -14.452    126.013    -86.569     29.464     67.100
   3    183.623      9.192      5.967    -21.404     46.703    -32.181
       -183.623     -9.192     -5.967     21.404     77.711     79.946

SUPPORT REACTIONS
JOINT       AR1        AR2        AR3        AR4        AR5        AR6
   3    -14.452    -38.509   -126.013   -348.575     86.569    -23.744
   4   -105.548     98.509   -113.987    -75.898    -75.808     37.163
```

Table 5-27
Data for Space Frame Example 2

Type of Data		*Numerical Values*							
Control Data		2	6	1					
Structural Data	(a)	3	4	18	3	10000.0	4000.0		
	(b)	1	128.0	96.0	72.0				
		2	128.0	192.0	0.0				
		3	0.0	0.0	0.0				
		4	256.0	0.0	0.0				
	(c)	1	1	2	9.0	64.0	28.0	80.0	0
		2	3	1	9.0	64.0	28.0	80.0	1
		2	128.0	96.0	0.0				
		3	1	4	9.0	64.0	28.0	80.0	1
		3	128.0	96.0	0.0				
	(d)	2	1	1	1	1	1	1	
		3	1	1	1	1	1	1	
		4	1	1	1	1	1	1	
Load Data	(a)	1	2						
	(b)	1	0.0	0.0	−5.0	−120.0	0.0	0.0	
	(c)	1	−1.5	2.0	0.0	0.0	0.0	60.0	
			−1.5	2.0	0.0	0.0	0.0	−60.0	
		3	−1.368	0.616	2.0	0.0	−87.727	27.0	
			−1.368	0.616	2.0	0.0	87.727	−27.0	

Table 5-28
Final Results for Space Frame Example 2

```
JOINT DISPLACEMENTS
JOINT      DJ1          DJ2          DJ3          DJ4          DJ5          DJ6
    1-.2176E-03-.4062E-02-.1674E-01-.5202E-02 .1870E-03-.4495E-02
    2 .0000E+00  .0000E+00  .0000E+00  .0000E+00  .0000E+00  .0000E+00
    3 .0000E+00  .0000E+00  .0000E+00  .0000E+00  .0000E+00  .0000E+00
    4 .0000E+00  .0000E+00  .0000E+00  .0000E+00  .0000E+00  .0000E+00

MEMBER END-ACTIONS
MEMBER       AM1          AM2          AM3          AM4          AM5          AM6
             AM7          AM8          AM9          AM10         AM11         AM12
    1      3.594        0.178        0.406        6.073       30.492      -84.001
          -6.594        3.822       -0.406       -6.073       16.233     -134.638
    2      4.744        0.485       -0.131        8.080        7.610       27.616
          -4.744       -0.485        0.131       -8.080       15.390       57.444
    3      3.214        0.178        1.678       -2.995       60.178      -24.964
          -5.950        1.054        2.322        2.995      106.594      -51.865

SUPPORT REACTIONS
JOINT        AR1          AR2          AR3          AR4          AR5          AR6
    2     -0.406       -2.982        7.014     -134.638       14.599       -9.343
    3      3.699        2.610        1.505       -8.177       28.387       -3.624
    4     -3.293        5.372        1.480      -63.927      -16.885      -98.434
```

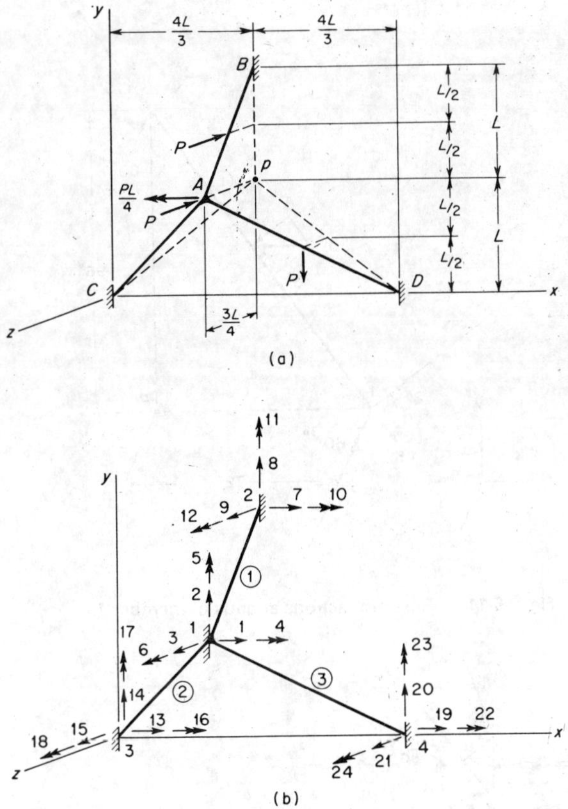

Fig. 5-10. Space frame Example 2.

5.12 Combined Program for Framed Structures. Although the primary objective in this chapter is to present programs for individual types of structures, it is also of interest to combine the work into a larger program for analyzing all types of framed structures. This secondary objective can be fulfilled using the subprograms already discussed for the various types of structures.

Flow Chart 5-7 shows the main program (called FS) for analyzing all types of framed structures. It has the same attributes as the main programs CB, PT, PF, etc., for analyzing continuous beams, plane trusses, plane frames, etc. However, in this case the program must branch in accordance with the type of structure being processed. For this purpose, the five subprograms named SDATA, STIFF, LDATA, LOADS, and RESUL are

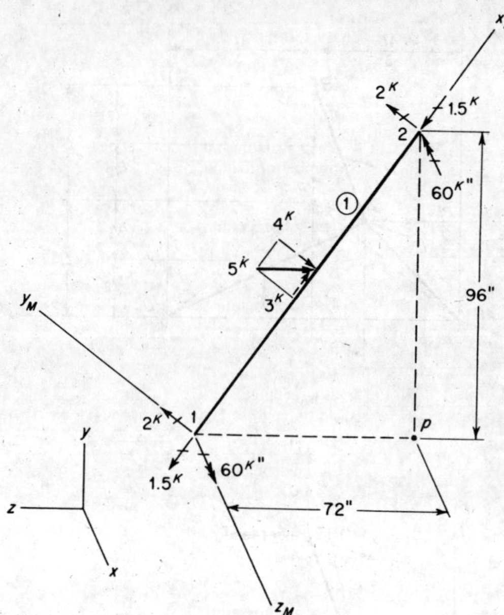

Fig. 5-11. Restraint actions at ends of member 1.

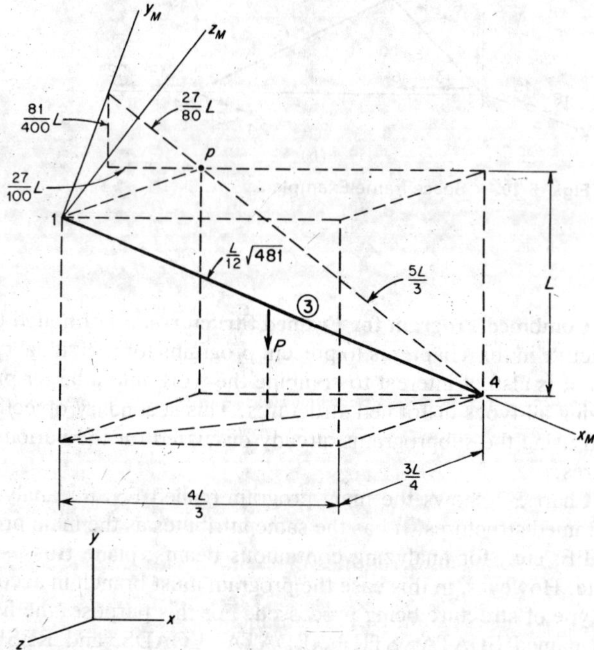

Fig. 5-12. Member axes for member 3.

Flow Chart 5-7

Main Program FS for Framed Structures

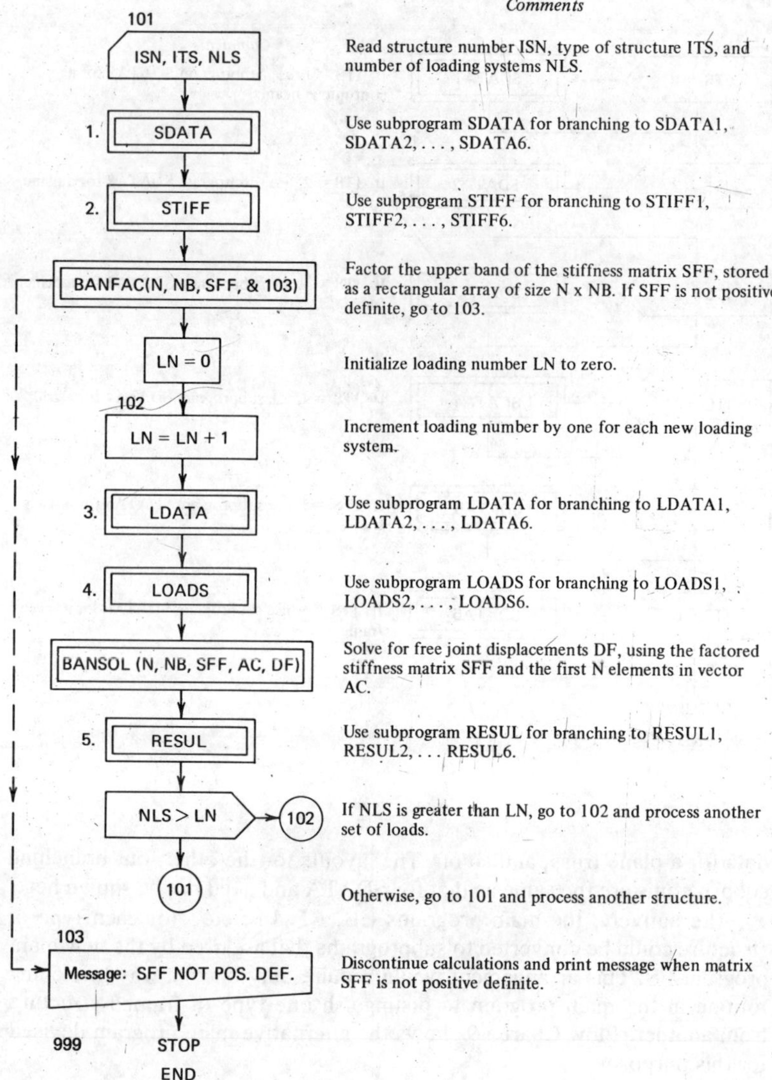

written as six-way branching routines that call upon other subprograms to do the actual calculations.

For example, subprogram SDATA branches as shown in Flow Chart 5-8 and uses subprogram SDATA1 to read and print structural data for a continuous beam. It uses subprogram SDATA2 to read and print structural

Flow Chart 5-8

Subprogram SDATA for Program FS

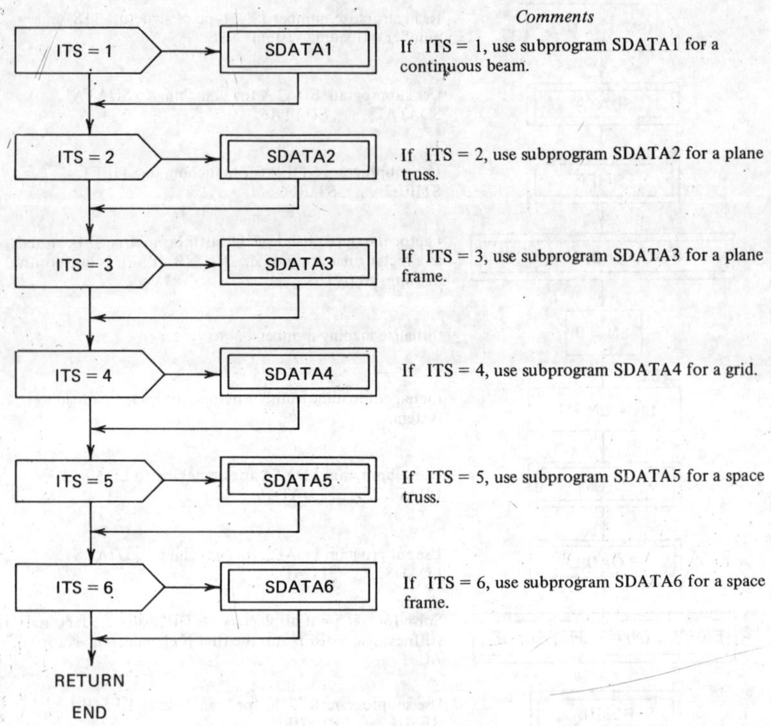

data for a plane truss, and so on. The layouts for the other four branching subprograms are the same as that for SDATA and need not be shown here.

Alternatively, the main programs CB, PT, PF, etc., for each type of structure could be converted to subprograms that are used by the new main program FS. This arrangement would require only one six-way branching routine in the main program to distinguish one type of framed structure from another. Flow Chart 5-9 shows the alternative main program devised for this purpose.

While such combined programs are rather elegant, they are not very suitable for small machines where core storage is limited. The computer code requires space that could otherwise be occupied by data for problems with large numbers of degrees of freedom. In addition, a COMMON declaration designed for the most complicated type of structure (the space

Flow Chart 5-9

Alternative Main Program FS for Framed Structures

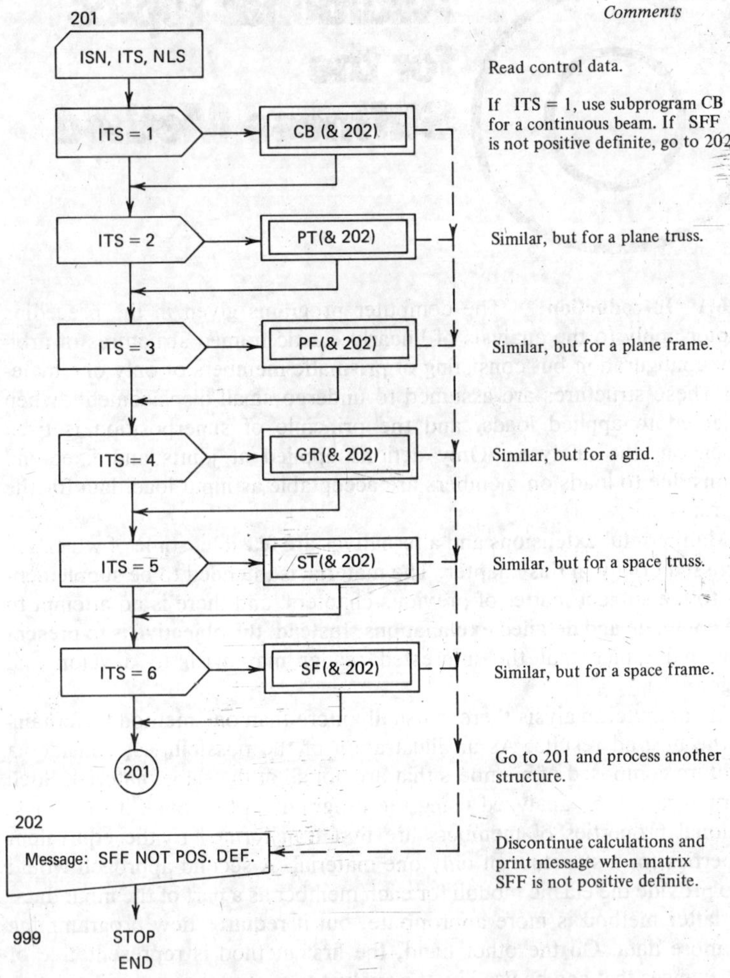

frame) must be used in the main program FS and all of the structure-oriented subprograms, regardless of the type of structure involved. In spite of these drawbacks, readers may wish to code combined programs after they have written all of the subprograms required for the various types of framed structures.

Additional Topics for the Stiffness Method

6.1 Introduction. The computer programs given in the preceding chapter apply to the analysis of linearly elastic framed structures of arbitrary configuration but consisting of prismatic members of only one material. These structures are assumed to undergo small displacements when subjected to applied loads, and the principle of superposition is used throughout the analyses. Only actions applied at joints and fixed-end actions due to loads on members are acceptable as input load data for the programs.

Many useful extensions and alternatives are feasible, some of which are discussed briefly in this chapter. The material is intended to be supplementary to the subject matter of previous chapters, and there is no attempt to give complete and detailed explanations. Instead, the objective is to present additional topics that the interested reader may wish to develop and program.

In computer analysis there is usually more than one method for obtaining the desired results. As an illustration of the possibilities, consider a structure composed of members that are not all of the same material. Such a structure can be analyzed using the programs in this book if the cross-sectional properties of members are first transformed to the equivalent properties for a structure of only one material. A second approach would be to provide the elastic moduli for each member as a part of the input data. The latter method is more appropriate, but it requires new programming and more data. On the other hand, the first method is representative of contrivances that can be used by the analyst to extend the scope of existing programs merely by manipulating the input data.

The first two topics discussed in this chapter (Arts. 6.2 and 6.3) are the *geometric specializations* of rectangular framing and symmetric structures. The next three articles (6.4 through 6.6) deal with *load considerations* not covered by the documented programs. Then several topics relating to *support conditions* are discussed (Arts. 6.7 through 6.9). These topics include support displacements, oblique supports, and elastic supports. The remaining articles in the chapter are devoted to *member characteristics* that alter the stiffnesses and fixed-end actions for individual members. Covered under this category are nonprismatic members, curved members, discontinuities in members, elastic connections, shearing deformations, offset connections, and axial-flexural interactions.

6.2 Rectangular Framing. Many framed structures have geometric properties that allow simplifications in their analyses. For example, continuous beams consist of collinear members, and advantage of this fact is taken in the analysis of such structures. Similarly, plane and space structures with rectangular framing are much easier to analyze than those of skew geometry. The joint coordinates, member orientations, and loads may be specified with a minimum of input data, and no rotation-of-axes transformations are required.

The programs in this book will, of course, accommodate structures with rectangular framing, but the computations will not be particularly efficient. Special-purpose programs should be used for the analysis of structures with special geometry in order to take advantage of the simplifications inherent in such problems.

Figures 6-1 through 6-3 show examples of structures with rectangular framing. These are a plane frame, a grid, and a space frame. Each member in a structure of this type has its axis parallel to one of the coordinate axes for the system. Therefore, the member axes should always be taken parallel to the structure axes, and member stiffness matrices should be generated in a form that permits a direct transfer of terms to the joint stiffness matrix without rotation of axes. For example, both the horizontal and the vertical members in the rectangular plane frame (see Figs. 6-1b and 6-1c) have their member axes x_M, y_M, and z_M parallel to the structure axes x, y, and z. Such axes are also indicated for typical members of the rectangular grid (see Figs. 6-2b and 6-2c) and the rectangular space frame (see Figs. 6-3b, 6-3c, and 6-3d). In the latter type of structure, principal axes of members are usually parallel to the structure axes. However, if this is not the case, a single rotation transformation will be required for each member.

Rectangular frames are most commonly used for multi-story buildings,* in which the horizontal displacements of floor and roof diaphragms are of primary interest. Figure 6-3a shows such diaphragms (or laminae) for a three-story example treated as a space frame. Each lamina is assumed to have infinite rigidity in its own plane, but zero rigidity normal to the plane. Consequently, the joints at a particular framing level are constrained to displace in accordance with the rigid-body motions of the diaphragm at that level. These motions consist of the x and y translations and the z rotation indicated by the numbered arrows at arbitrarily selected reference points p_1, p_2, and p_3, which are considered to be rigidly attached to the laminae

*See "Three-Dimensional Analysis of Tier Buildings," by W. Weaver, Jr. and M. F. Nelson, *Jour. Struc. Div., Proc. ASCE,* Vol. 92, No. ST6, Proc. Paper 5019, Dec. 1966, pp. 385–404.

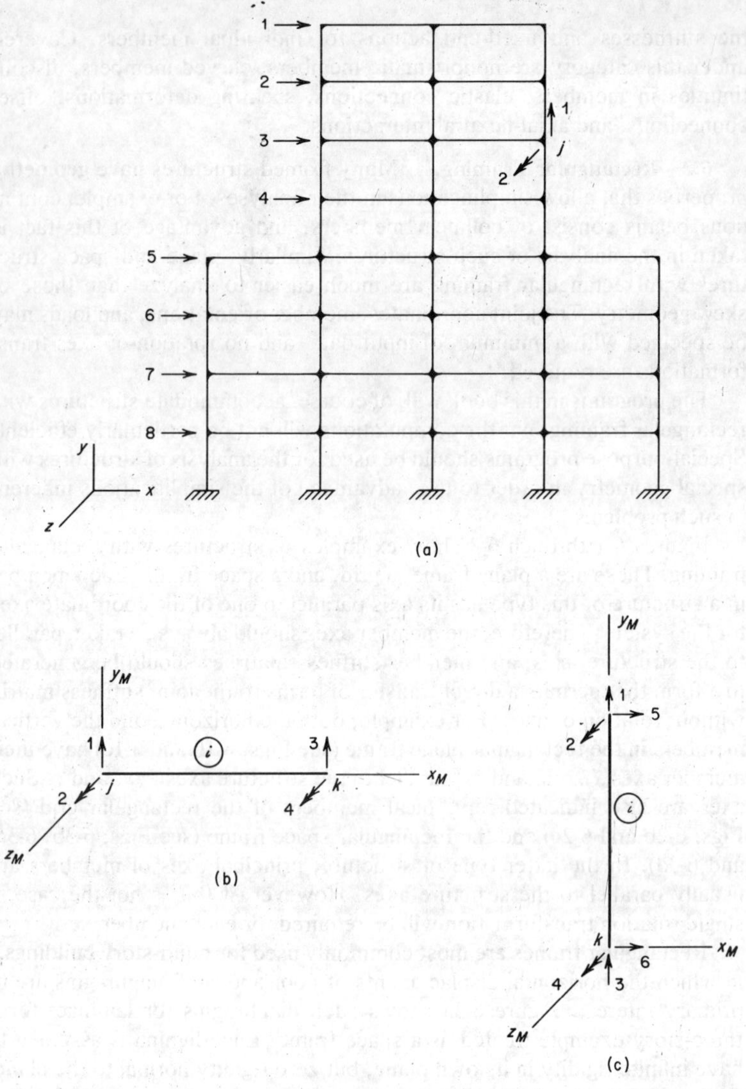

Fig. 6-1. Rectangular plane frame: (a) structure, (b) member parallel to x axis, (c) member parallel to y axis.

in Fig. 6-3a. At each joint there remain three independent displacements, which are the x and y rotations and the z translation. These displacements are indicated by the numbered arrows at the typical joint j at the right-hand side of Fig. 6-3a. Thus, the independent joint translations are the same as those for a grid (see also Fig. 6-2a). If n_j is the number of joints at a given

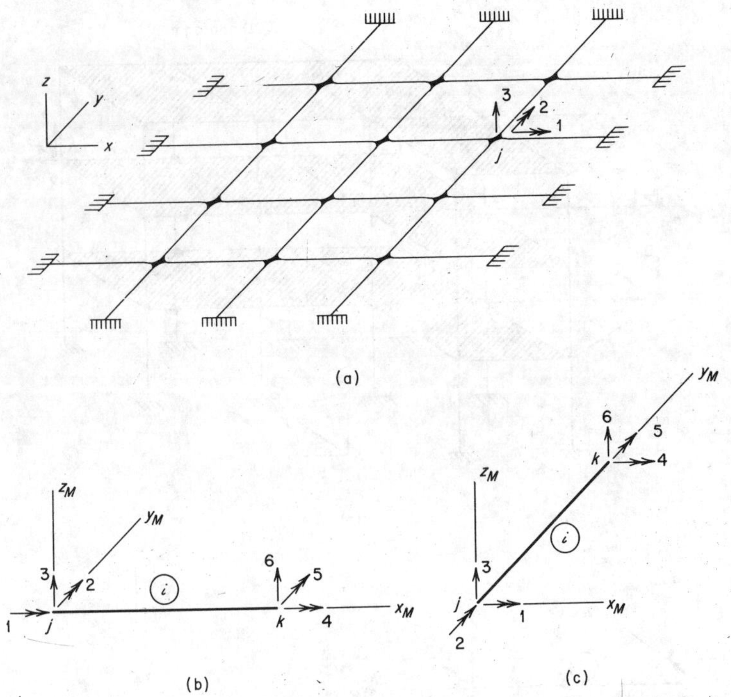

Fig. 6-2. Rectangular grid: (a) structure, (b) member parallel to x axis, (c) member parallel to y axis.

level, the number of independent displacements for that level is $3n_j + 3$ (instead of $6n_j$) due to the presence of the lamina.

Similarly, if axial strains are neglected in the horizontal members (beams) of the plane building frame in Fig. 6-1a, the beams are implied to be infinitely rigid in their axial directions. Consequently, all points at a given level are inherently constrained to translate the same amount in the x direction. That is, each horizontal line of beams in Fig. 6-1a may be considered to be the one-dimensional counterpart of a two-dimensional lamina in Fig. 6-3a. Thus, there is only one independent translation in the x direction at each framing level, as indicated by the numbered arrows at the left side of Fig. 6-1a. In addition, there are two independent displacements at each joint, shown as the y translation and the z rotation at a typical joint j. These joint displacements are the same as those for a beam, so the displacement vectors indicated in Fig. 6-1b imply the generation of only a 4×4 member stiffness matrix for a horizontal member. On the other hand, the vertical member (column) in Fig. 6-1c provides finite terms for a 6×6 member stiffness matrix. Note that the displacement indexes at the ends of the members are numbered in a sequence that takes independent

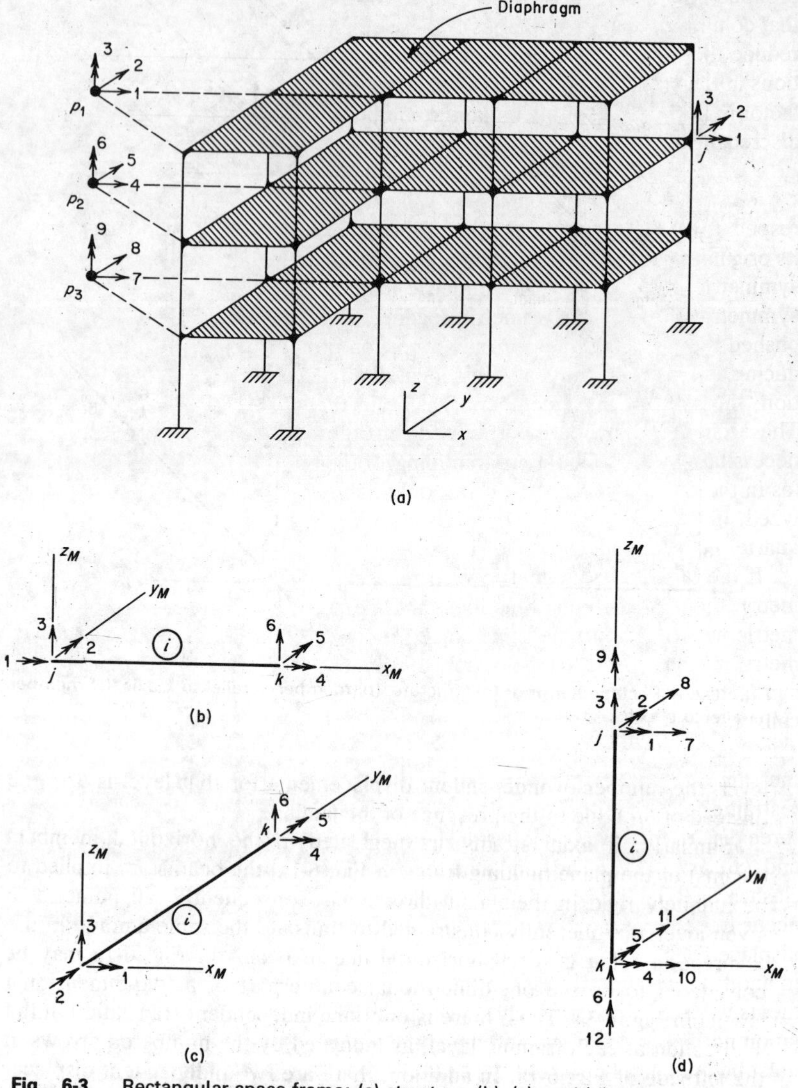

Fig. 6-3. Rectangular space frame: (a) structure, (b), (c), and (d) members parallel to *x, y,* and *z* axes, respectively.

joint displacements before story displacements and proceeds from top-to-bottom and left-to-right in the structure. Similar numbering sequences are also indicated at the ends of the members in Figs. 6-3b, c, and d for the space frame.

The omission of axial strains in the columns of a building frame would produce constraints against vertical translations at every joint (assuming that column lines are continuous). These constraints would serve to further reduce the number of degrees of freedom and allow additional simplifications in the analysis. However, axial strains in columns can be a significant factor in the analysis of tall buildings, and their omission may cause serious discrepancies.

6.3 Symmetric Structures. When a structure is symmetric with respect to one or more planes, the problem may be reduced to a fraction of its original size. This statement applies even when the loading pattern is not symmetric, because any unsymmetric loading can be decomposed into a symmetric and an antisymmetric loading. The reduction may be accomplished by providing data for only a portion of the structure and by introducing artificial restraints at joints located on planes of symmetry. In addition, the properties of members that lie in those planes must be altered. These changes may be incorporated into the structural data and do not necessitate any additional programming. If there is one plane of symmetry (as in the plane frame in Fig. 6-1a), only half of the structure need be analyzed. If two planes of symmetry exist (as in the grid in Fig. 6-2a), only a quarter need be analyzed, and so on.

If the loading is symmetric with respect to a plane of structural symmetry, the deformations, reactions, and member actions will also be symmetric with respect to the same plane. Therefore, joints on a plane of symmetry must be restrained in such a manner that the structure deforms symmetrically with respect to that plane. Figure 6-4a portrays schematically a typical joint k located on a plane of symmetry normal to the x axis. The loading on the structure is assumed to be symmetric with respect to the same plane. Of the six possible displacements shown, three must be restrained in order to produce a symmetric pattern of distortions. These three displacements are the x translation ($k1$) and the y and z rotations ($k5$ and $k6$) at joint k. The fact that they are restrained is indicated by the slashes on their vectors. Other displacements at the joint are allowed to occur freely, and it follows that the reactions corresponding to these free displacements must be zero. Of course, if the plane of symmetry is oriented in another way with respect to the reference axes, the restrained displacements will change accordingly. In general, the component of the joint translation normal to the plane and the components of the rotation in the plane must be prevented.

If the loading is antisymmetric with respect to a plane of structural symmetry, the deformations, reactions, and member actions will also be antisymmetric with respect to the same plane. Therefore, joints on a plane of symmetry must be compelled to displace in an antisymmetric manner. For this purpose, the joint k in Fig. 6-4b is restrained against translations ($k2$ and $k3$) in the plane of symmetry and rotation ($k4$) normal to the plane. Other displacements at the joint are allowed to occur freely, and the cor-

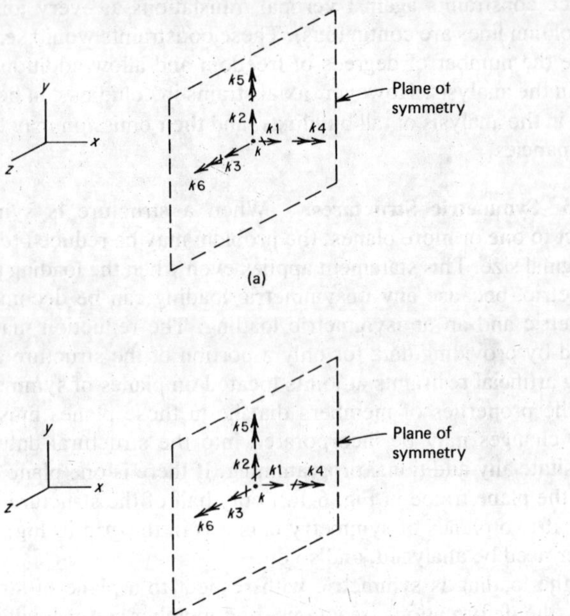

Fig. 6-4. Joint restraints on planes of symmetry: (a) symmetric loading (b) antisymmetric loading.

responding reactions will be zero. This system of restraints guarantees an antisymmetric pattern of distortion.

The rigidities of members lying in a plane of symmetry must be halved in order to divide the structure into two equal parts. In the event that a member lies in two planes of symmetry, its rigidities must be divided by four, and so on. If a member is normal to and bisected by a plane of symmetry, a new joint is introduced at midlength and restrained as described above.

As an example, consider the plane frame in Fig. 6-5a, which is symmetric with respect to the y-z plane. The applied force P represents an unsymmetric loading that can be decomposed into the symmetric and antisymmetric patterns shown in Figs. 6-5b and 6-5e, respectively. An analysis for the symmetric pattern may be carried out using either the left-hand half of the structure (Fig. 6-5c) or the right-hand half (Fig. 6-5d). The left-hand set of structural axes x', y, and z pertain to the former case, and the right-hand set x, y, and z pertain to the latter case. Since the use of right-hand axes is always preferred, the latter case represents the better choice. The artificial restraint conditions at point C preclude translation in the x direction (across the plane of symmetry) and rotation in the z sense (in the plane of symmetry). However, there is no restraint at C against translation in the y

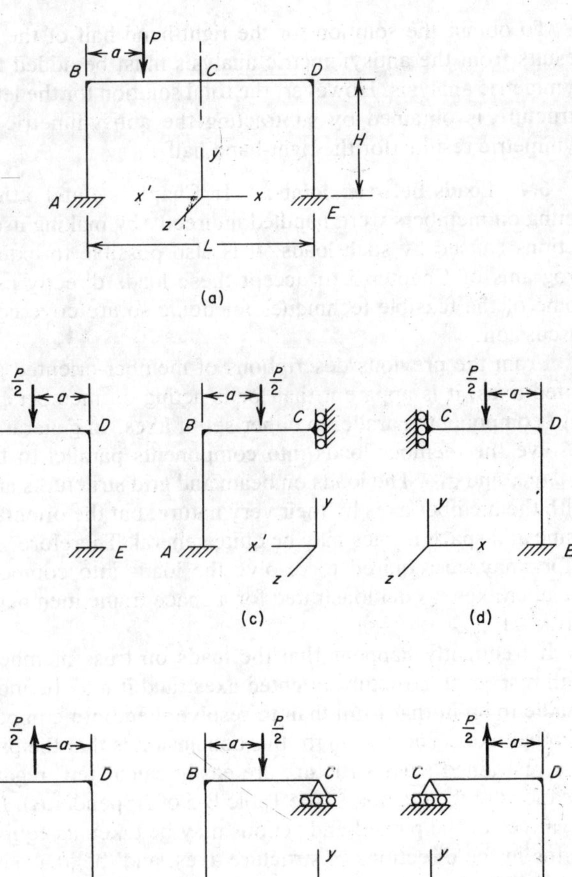

Fig. 6-5. Symmetric plane frame example.

direction (in the plane of symmetry). Therefore, the corresponding reaction in the y direction at point C will be zero in the symmetric analysis.

Similarly, an analysis for the antisymmetric pattern of loading (Fig. 6-5e) could be performed using either half of the structure, as indicated in Figs. 6-5f and 6-5g. Again the right-hand half (Fig. 6-5g) would be preferred because of the orientation of structural axes. In this case the artificial restraint conditions at point C allow the x translation and the z rotation to occur, but not the y translation. Hence, the corresponding reactive force in the x direction and moment in the z sense at point C will be zero in the antisymmetric analysis.

To obtain the solution for the right-hand half of the original problem, results from the antisymmetric analysis must be added to those from the symmetric analysis. However, the total solution for the left-hand half of the structure is obtained by subtracting the antisymmetric results from the symmetric results (for the right-hand half).

6.4 Loads Between Joints. In Chapters 4 and 5 the effects of loads acting on members were handled indirectly by making use of the fixed-end actions caused by such loads. It is also possible to extend the computer programs of Chapter 5 to accept these loads directly as input data, and some of the feasible techniques for doing so are covered in the following discussion.

From the previous descriptions of member-oriented and structure-oriented axes, it is apparent that loads acting on members may be resolved into components parallel to either set of axes. In general, it is desirable to resolve the member loads into components parallel to the member axes (x_M, y_M, and z_M). The loads on beam and grid structures are usually aligned with the member axes by their very nature, but the orientations of loads on plane and space frames may be quite general. Therefore, some preliminary effort may be required to resolve the loads into components parallel to member axes, as demonstrated for a space frame member in Example 2 of Art. 5.11.

It frequently happens that the loads on truss members are expressed with respect to structure-oriented axes, and it may be more convenient to handle them in that form than to resolve them into components parallel to member axes. The reason for this conclusion is that the pinned-end actions in a restrained truss structure are easily calculated, regardless of the orientations of the members (see Table B-5 of Appendix B). In such a case the negatives of the pinned-end actions may be taken as equivalent joint loads acting in the directions of structure axes, and the analysis may be carried out as usual from this point on. The drawback to this approach is the fact that the final member end-actions will not include the initial pinned-end actions A_{ML} in the directions of member axes. Thus, the only member end-actions A_M that are calculated are the axial forces in the members due to the equivalent joint loads. This limitation is not a serious disadvantage in the analysis of trusses, because their principal features relate to the effects of loads (actual or equivalent) at the joints.

Now consider again those structures for which the member loads have been resolved into components parallel to member axes. The task of incorporating the loads directly into the stiffness method of analysis will be facilitated if they are expressed in discrete form (that is, as concentrated forces and moments). Any distributed loadings also can be accommodated by integrating the formulas for A_{ML} due to concentrated forces. If this approach is followed, only a certain number of concentrated loads will be possible at any point on a member. For example, at a point ℓ on a beam there may exist either a concentrated force $A_{\ell 1}$ in the y_M direction or a moment $A_{\ell 2}$ in the z_M sense, as shown in Fig. 6-6a. Each of these loads will cause four fixed-end actions of the types indicated in the figure. Similarly, there are three possible concentrated loads at any point on a plane truss member (see Fig. 6-6b), and four possible pinned-end actions. The most general category of member (a space frame member) is illustrated in Fig. 6-6c. Six load components may exist at any point on such a member, causing twelve types of fixed-end actions, as shown in the figure.

If the loads on a member consist only of concentrated forces and moments as described above, there are several methods by which they may be incorporated directly into the analysis of a structure. The simplest approach is to assume that a joint exists at every point of loading on the member, and then to analyze the structure as if it were subjected to joint loads only. This method is suitable for structures with rigid joints but can-

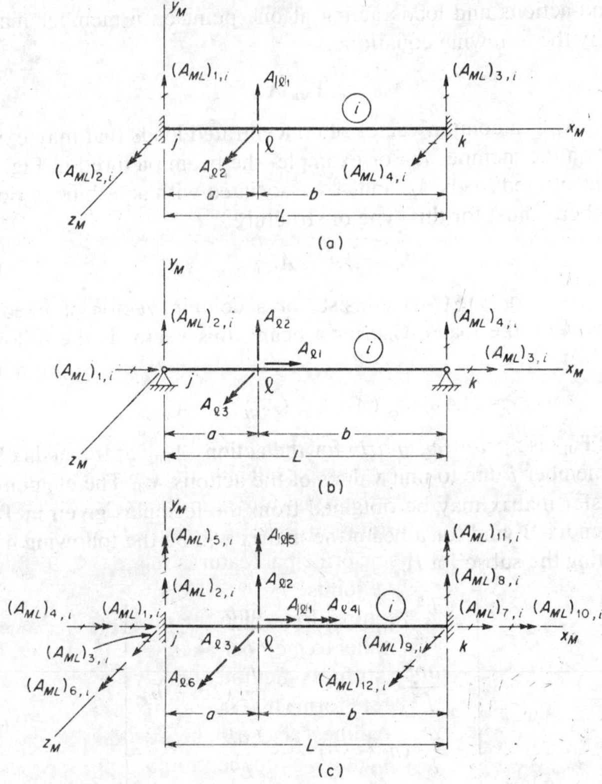

Fig. 6-6. Concentrated loads on members: (a) beam member, (b) plane truss member, and (c) space frame member.

not be adapted easily to trusses; moreover, the loads must be structure-oriented instead of member-oriented in this approach. The principal disadvantage of this method is that the number of joints in the structure may become excessive if there are loads on many members.

A second method consists of analyzing each loaded member as a separate structure, using the stiffness method of analysis, and then incorporating the results of each such subanalysis into the over-all analysis of the structure. For example, the restrained beam in Fig. 6-6a could be analyzed as a structure consisting of two members of lengths a and b, respectively, and having two degrees of freedom at the point of application of the loads. The fixed-end actions A_{ML} for the member i will be the reactions determined in the subanalysis of the beam considered as two members. This line of reasoning may be generalized to include any type of framed structure and any number of load points on a member.

A third approach to the problem involves the use of transfer matrices for fixed-end actions due to unit loads on the members. The relationships between end-actions and loads acting at one point on a member may be expressed by the following equation:

$$\mathbf{A}_{\mathrm{ML}\,i} = \mathbf{T}_{\mathrm{ML}\,i}\mathbf{A}_{\ell i} \tag{6-1}$$

In Eq. (6-1), $\mathbf{A}_{\ell i}$ is a column vector of concentrated loads that may exist at any point ℓ on the member i. For example, the beam pictured in Fig. 6-6a has the concentrated loads $A_{\ell 1}$ and $A_{\ell 2}$ associated with an arbitrary point ℓ on the member. Thus, for this type of structure:

$$\mathbf{A}_{\ell i} = \{A_{\ell 1}, A_{\ell 2}\} \tag{6-2}$$

The matrix $\mathbf{A}_{\mathrm{ML}\,i}$ in Eq. (6-1) consists of a column vector of fixed-end actions caused by the loads $\mathbf{A}_{\ell i}$. For a beam, this vector is the following (see Fig. 6-6a):

$$\mathbf{A}_{\mathrm{ML}\,i} = \{(A_{ML})_{1,i}, (A_{ML})_{2,i}, (A_{ML})_{3,i}, (A_{ML})_{4,i}\} \tag{6-3}$$

The matrix $\mathbf{T}_{\mathrm{ML}\,i}$ is a *transfer matrix* for the actions $\mathbf{A}_{\mathrm{ML}\,i}$ at the ends of the restrained member i due to unit values of the actions $\mathbf{A}_{\ell i}$. The elements of such a transfer matrix may be obtained from the formulas given in Table B-1 of Appendix B, and for a beam the matrix will be the following 4×2 array (omitting the subscript i):

$$\mathbf{T}_{\mathrm{ML}} = \begin{bmatrix} -\dfrac{b^2}{L^3}(3a + b) & \dfrac{6ab}{L^3} \\ -\dfrac{ab^2}{L^2} & \dfrac{b}{L^2}(2a - b) \\ -\dfrac{a^2}{L^3}(a + 3b) & -\dfrac{6ab}{L^3} \\ \dfrac{a^2 b}{L^2} & \dfrac{a}{L^2}(2b - a) \end{bmatrix} \tag{6-4}$$

The vector of member end-actions $\mathbf{A}_{ML}$ for a member with loads acting at several load points can be obtained by applying Eq. (6-4) for every such point and then summing the results. In addition, formulas for fixed-end actions $\mathbf{A}_{ML}$ due to distributed forces can be derived from the terms in the first column ($\mathbf{T}_{ML1}$) of $\mathbf{T}_{ML}$. This is accomplished by replacing a and b with x and $L - x$ and integrating the product of $\mathbf{T}_{ML1}$ and the distributed-load function $f(x)$ over the length to which it is applied. That is,

$$\mathbf{A}_{ML} = \int \mathbf{T}_{ML1} f(x)\, dx \tag{6-5}$$

Alternatively, these formulas can be drawn from sources such as Table B-1 in Appendix B.

Transfer matrices similar to Eq. (6-4) may be developed for all types of framed structures. For a plane truss member subjected to loads as indicated in Fig. 6-6b, the transfer matrix $\mathbf{T}_{ML}$ for pinned-end actions will be of order 4×3, as follows:

$$\mathbf{T}_{ML} = \begin{bmatrix} -\dfrac{b}{L} & 0 & 0 \\ 0 & -\dfrac{b}{L} & \dfrac{1}{L} \\ -\dfrac{a}{L} & 0 & 0 \\ 0 & -\dfrac{a}{L} & -\dfrac{1}{L} \end{bmatrix} \tag{6-6}$$

Similarly, the transfer matrices for the other types of framed structures may be easily derived. Such transfer matrices may be incorporated into a computer program for the purpose of automatically calculating the actions $\mathbf{A}_{ML}$ at the ends of restrained members due to loads at a discrete number of points. After this is done, the calculation of equivalent joint loads and the remaining parts of the analysis proceed as before.

6.5 Automatic Dead Load Analysis. Because structures must carry their own weight, dead load analyses are of great interest to structural engineers. Of course, the programs in this book will produce dead load analyses if the input data includes fixed-end actions due to the weights of individual members. Such load data becomes unwieldy for structures with many members, however, and a less tedious approach is desired. The programs can easily be modified to perform dead load analyses automatically for any type of framed structure. Moreover, the weight of the structure may be handled either separately or in combination with other loading systems.

If cross-sectional areas of members and unit weights of materials are included among the input data, the weights of individual members may be calculated by the computer. This task is especially straightforward for prismatic members. Next, the fixed-end actions A_{ML} are computed by predetermined formulas. For example, the actions at the ends of a prismatic member in a plane frame (see Fig. 6-7) due to its own weight w (per unit of length) are

$$
\begin{aligned}
A_{ML1} &= \frac{wL}{2}C_Y & A_{ML4} &= \frac{wL}{2}C_Y \\
A_{ML2} &= \frac{wL}{2}C_X & A_{ML5} &= \frac{wL}{2}C_X \\
A_{ML3} &= \frac{wL^2}{12}C_X & A_{ML6} &= -\frac{wL^2}{12}C_X
\end{aligned}
\tag{6-7}
$$

Expressions similar to Eqs. (6-7) may be programmed for each type of framed structure. After these terms are generated and placed into the matrix $\mathbf{A}_{ML}$, the calculation of equivalent joint loads and the remaining parts of the analysis proceed as before.

Uniformly distributed live loads applied to members may be treated in a manner similar to that for dead loading. For this purpose, an alternative form of input data is required for loads on members. Each data card must specify the member number, the direction of loading, the intensity of loading, and a code number indicating whether the intensity is for a unit of length along the member or along its projection on a reference plane. Formulas for fixed-end actions due to the former type of loading are analogous to those for dead weight calculations. Fixed-end actions due to loads of the latter type are also easily determined. For example, if the loading in Fig. 6-7 is defined to be of intensity w per unit of length along the x axis, the fixed-end actions are obtained simply by multiplying the formulas in Eqs. (6-7) by the direction cosine C_X of the member.

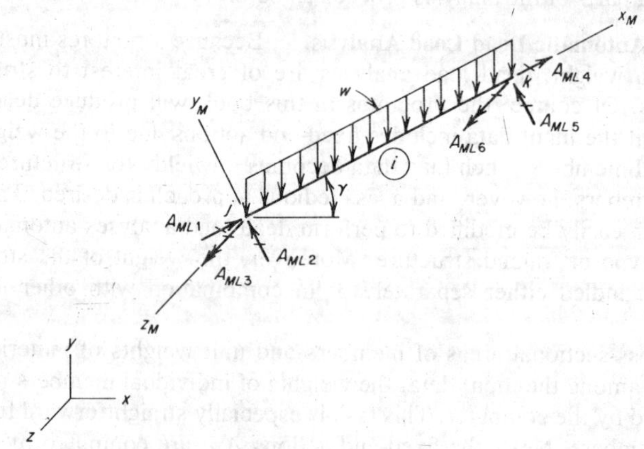

Fig. 6-7. Dead weight on prismatic member in a plane frame.

6.6 Temperature Changes and Prestrains. The effects of temperature changes and prestrains were considered in the hand solutions of Chapters 2 and 3. These effects may be readily incorporated into a computer-oriented analysis by treating them in a manner analogous to the procedure described in Chapter 4 (Art. 4.5) for loads on members, that is, by replacing them with equivalent joint loads. Both temperature changes and prestrain effects will cause fixed-end actions in the members of a restrained structure. Formulas for fixed-end actions of this nature are listed in Tables B-2 and B-3 of Appendix B. Such end-actions may be handled in the same manner as those due to loads on members; in other words, they may be listed in matrices $\mathbf{A}_{MT}$ and $\mathbf{A}_{MP}$, respectively, analogous to the listing of fixed-end actions due to loads in the matrix $\mathbf{A}_{ML}$. Then the sum of all three effects can be combined into a single matrix $\mathbf{A}_{MC}$. Of course, in computer programming it may be desirable to place all terms directly into the matrix $\mathbf{A}_{MC}$, without the necessity of forming separately the matrices $\mathbf{A}_{ML}$, $\mathbf{A}_{MT}$, and $\mathbf{A}_{MP}$. Following this step, the fixed-end actions in the matrix $\mathbf{A}_{MC}$ are converted to equivalent joint loads, and the analysis proceeds as described in Chapters 4 and 5 from then on. All free joint displacements $\mathbf{D}_F$, reactions $\mathbf{A}_R$, and member end-actions $\mathbf{A}_M$ are calculated in exactly the same manner as for the case of loads only.

6.7 Support Displacements. Support displacements consist of known translations or rotations of support restraints. Two alternative methods will be described for including the effects of support displacements in the stiffness method of analysis.

The first approach requires the calculation of the actions at the ends of members in the restrained structure due to the displacements of the supports. (This technique was used in Chapter 3 for support displacements and discussed in Art. 6.6 for temperature and prestrain effects.) These fixed-end actions are then placed into the matrix $\mathbf{A}_{MR}$ (or $\mathbf{A}_{MC}$) and treated in the same manner as those due to loads. The conversion of these quantities to equivalent joint loads and the subsequent analysis then proceed as described in Chapter 4. The advantage of treating support displacements in this member-oriented fashion is that the analysis is cast into the same format as that for other secondary effects.

The second approach involves the generation of the over-all stiffness matrix $\mathbf{S}_J$ appearing in Eq. (3-31) of Chapter 3, which is repeated below:

$$\mathbf{A}_J = \mathbf{S}_J \mathbf{D}_J \tag{3-31}$$

repeated

The expanded form of this equation is

$$\begin{bmatrix} \mathbf{A}_F \\ \mathbf{A}_R \end{bmatrix} = \begin{bmatrix} \mathbf{S}_{FF} & \mathbf{S}_{FR} \\ \mathbf{S}_{RF} & \mathbf{S}_{RR} \end{bmatrix} \begin{bmatrix} \mathbf{D}_F \\ \mathbf{D}_R \end{bmatrix} \tag{3-33}$$

repeated

Writing Eq. (3-33) as two matrix equations produces

$$\mathbf{A}_F = \mathbf{S}_{FF}\mathbf{D}_F + \mathbf{S}_{FR}\mathbf{D}_R \tag{3-34a}$$

repeated

and

$$\mathbf{A}_{\mathrm{R}} = \mathbf{S}_{\mathrm{RF}}\mathbf{D}_{\mathrm{F}} + \mathbf{S}_{\mathrm{RR}}\mathbf{D}_{\mathrm{R}} \tag{3-34b}$$
repeated

Equation (3-34a) may be solved (symbolically) for the free joint displacements:

$$\mathbf{D}_{\mathrm{F}} = \mathbf{S}_{\mathrm{FF}}^{-1}(\mathbf{A}_{\mathrm{F}} - \mathbf{S}_{\mathrm{FR}}\mathbf{D}_{\mathrm{R}}) \tag{3-35}$$
repeated

In this expression it is clearly seen that the second term in the parentheses represents equivalent joint loads corresponding to $\mathbf{D}_{\mathrm{F}}$ due to the support displacements $\mathbf{D}_{\mathrm{R}}$. Furthermore, the last term in Eq. (3-34b) produces the influence of support displacements upon support reactions.

Both of the methods described above for calculating the effects of support displacements may be readily incorporated into a computer program. However, the member end-actions must be computed differently for these two approaches. In the former method the initial member end-actions $\mathbf{A}_{\mathrm{MR}}$ contain the effects of support displacements. Therefore, only the influence of free joint displacements is utilized in the augmenting terms $\mathbf{A}_{\mathrm{MD}}$ (see Eq. 5-5b). In the latter method the initial member end-actions are zero, and the effects of both free displacements and support displacements must be included in the augmenting terms.

6.8 Oblique Supports. The restraint list code used in the computer programs of Chapter 5 will account for the presence or absence of restraints in directions parallel to the structure axes x, y, and z. Thus, for example, a roller support parallel to one of the structure axes can be taken into account. The code, however, will not accommodate oblique roller supports such as those indicated in Fig. 6-8a for a plane truss and Fig. 6-8c for a plane frame. There are several methods by which such support conditions may be included in the analysis of structures by the stiffness method. One possible approach is to rotate the structure axes so that the reference planes are either parallel or perpendicular to the inclined planes. This technique is limited to those cases in which there is only one oblique support or in which all inclines are mutually orthogonal. In those cases where this method is applicable, however, it is likely that the joint coordinates and the orientations of members and loads with respect to the structure axes will become more complicated.

Another method of handling the problem is to replace the actual support by a member having a large cross-sectional area A_X and having its longitudinal axis in the direction normal to the inclined support. Such a substitution is indicated in Fig. 6-8b for the support shown in Fig. 6-8a. A similar substitution is also indicated in Fig. 6-8d for the frame support of Fig. 6-8c. In this latter case the cross-sectional area A_X must be made large and the moment of inertia I_Z must be set equal to zero. Since the cross-sectional areas of the substitute members are large compared to those of the other

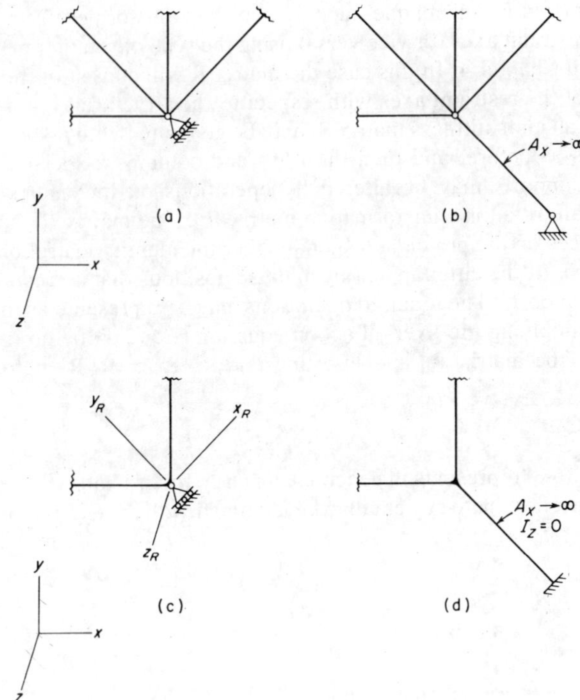

Fig. 6-8. Oblique supports.

members of the structure, their axial changes of length will be negligible in the over-all analysis. Therefore, such members will create essentially the same effects as the rollers on the inclined planes. Also, the axial forces in the substitute members will be approximately the same as the reactions of the roller supports. The length of a substitute member should be at least the same order of magnitude as the other members in the structure in order to insure that the angle of rotation of the member is small. The advantage of this method of handling supports is that the addition of one or two extra members and joints does not require any changes in the programs of Chapter 5.

A third approach to the problem of oblique supports requires a modification of the method for writing action equations in the stiffness method. These equations represent summations of forces and moments in the directions of the structure axes. However, at an oblique support the suitable directions are parallel and perpendicular to the inclined plane. Therefore, the over-all formulation of the action equations for the structure should include some equations that are written for *restraint-oriented axes* instead of structure-oriented axes. A set of restraint-oriented axes x_R, y_R, and z_R is shown in Fig. 6-8c in conjunction with the oblique support. Actions and displacements at an oblique support may be transformed from structure axes to restraint axes (or vice versa) using the type of rotation matrix **R** described in Chapter 4. In this case the matrix **R** will consist of the direction cosines of the restraint axes with respect to the structure axes. Moreover, the over-all joint stiffness matrix $\mathbf{S}_J$ may be generated with respect to structure axes as before, and then the rows and columns associated with the oblique supports may be altered by operations on the matrix using an appropriate rotation transformation matrix $\mathbf{R}_R$. The matrix $\mathbf{R}_R$ contains as submatrices on the principal diagonal either the identity matrix or the rotation matrix **R**, the latter appearing in those positions that correspond to the oblique supports. The required operations may be represented symbolically by premultiplying the over-all action equation (Eq. 3-31 before rearrangement) by the matrix $\mathbf{R}_R$ and inserting $\mathbf{I} = \mathbf{R}_R^{-1}\mathbf{R}_R = \mathbf{R}_R^T\mathbf{R}_R$ before $\mathbf{D}_J$, as follows:

$$\mathbf{R}_R\mathbf{A}_J = \mathbf{R}_R\mathbf{S}_J\mathbf{R}_R^T\mathbf{R}_R\mathbf{D}_J \tag{6-8}$$

Equation (6-8) represents in general a change of coordinates, and quantities in the new coordinate system may be identified with an asterisk and defined as follows:

$$\mathbf{R}_R\mathbf{A}_J = \mathbf{A}_J^* \tag{6-9a}$$

$$\mathbf{R}_R\mathbf{S}_J\mathbf{R}_R^T = \mathbf{S}_J^* \tag{6-9b}$$

$$\mathbf{R}_R\mathbf{D}_J = \mathbf{D}_J^* \tag{6-9c}$$

When Eq. (6-8) is rewritten in these terms, it becomes

$$\mathbf{A}_J^* = \mathbf{S}_J^*\mathbf{D}_J^* \tag{6-10}$$

which has the same form as Eq. (3-31). The solution for displacements and reactions in the new coordinate system may now be carried out as before, but the computation of member end-actions requires that $\mathbf{D}_J^*$ be transformed back to the original coordinates, using Eq. (6-9c). Thus,

$$\mathbf{D}_J = \mathbf{R}_R^{-1}\mathbf{D}_J^* = \mathbf{R}_R^T\mathbf{D}_J^* \tag{6-11}$$

Among the three methods described above for handling oblique supports, the approach using substitute members is the most convenient because it requires no changes in the programs of Chapter 5. Although the method involving rotation of axes is mathematically more elegant, its application involves a substantial increase in the amount of programming effort.

6.9 Elastic Supports. Cases of support restraint conditions may exist that are intermediate between the extremes of zero restraint and full restraint. If such restraints against either translations or rotations are linearly elastic, they may be included easily within the scope of the stiffness method of analysis.

In order to illustrate conditions of elastic support, consider the continuous beam shown in Fig. 6-9. A comparison of this figure with Fig. 4-7a in

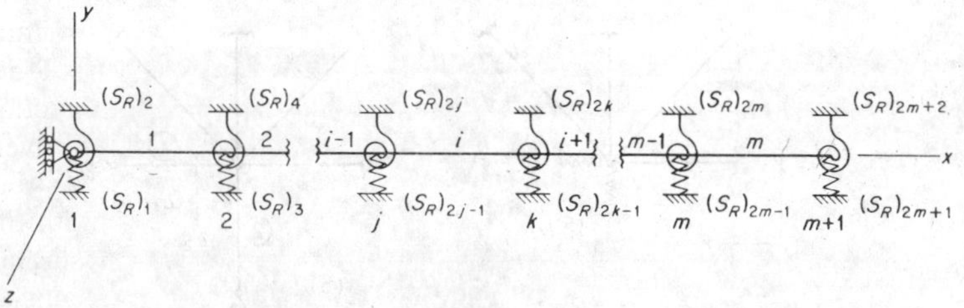

Fig. 6-9. Beam with elastic supports.

Art. 4.8 shows that full restraints and zero restraints at every joint have been replaced by elastic springs having stiffness constants denoted by the symbol S_R. The odd-numbered stiffnesses are restraints against translations in the y direction, and the even-numbered stiffnesses are restraints against rotations in the z sense. If two springs exist at every joint, and if there are m members in the beam, there will be $2(m + 1)$ elastic restraints, and the analysis of the structure by the stiffness method will involve that number of degrees of freedom. The restrained structure for the beam in Fig. 6-9 will be the same as before (see Fig. 4-7b), but the joint stiffness matrix will be altered because of the restraint stiffnesses S_R. If these restraint stiffnesses are added to the diagonal elements of the over-all joint stiffness matrix $\mathbf{S}_J$, the resulting matrix can then be used in the solution for joint displacements as in Art. 4.8. Member end-actions may also be computed as described in that article. On the other hand, the reactions at the elastic supports will be equal to the actions in the elastic springs and may be calculated as the negatives of the products of the spring constants times the joint displacements. Thus,

$$A_R = -S_R D_J \tag{6-12}$$

Although a beam serves as an example in the preceding discussion, this method of handling elastic supports is applicable to all types of framed structures.

An alternative approach to the problem of elastic supports is to substitute equivalent structural members for the restraints. As an example of this technique, consider the problem of handling the joint j of the plane truss shown in Fig. 6-10a. The joint is restrained by two translational springs having stiffnesses $(S_R)_{2j-1}$ and $(S_R)_{2j}$. The two springs can be replaced by the two additional truss members 1 and 2 shown in Fig. 6-10b. These members can be assigned arbitrary lengths L_1 and L_2 (comparable to the lengths of the other members in the truss) and cross-sectional areas A_{X1} and A_{X2}, computed as follows:

$$A_{X1} = \frac{L_1}{E}(S_R)_{2j-1} \qquad A_{X2} = \frac{L_2}{E}(S_R)_{2j}$$

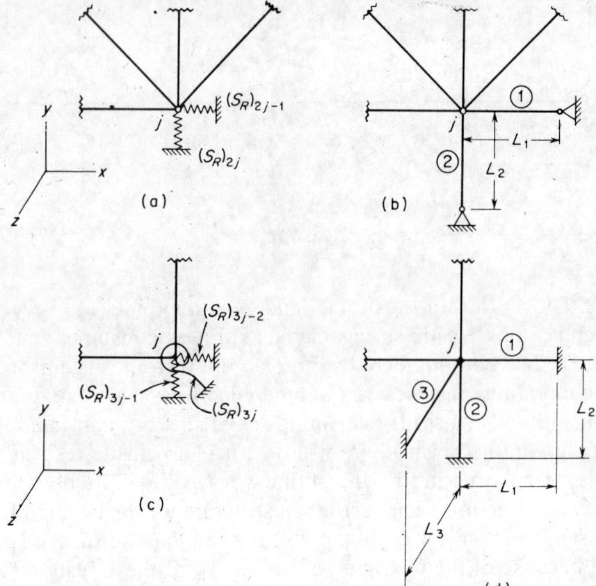

Fig. 6-10. Substitution of members.

The assignment of these cross-sectional areas gives the substitute members axial stiffnesses equal to those of the elastic restraints they replace.

A similar example is illustrated in Figs. 6-10c and 6-10d, which show a joint j in a plane frame. The substitution of members 1 and 2 for the translational restraints (see Fig. 6-10d) is similar to that for the plane truss example, except that the moments of inertia of these members must be set equal to zero. A suitable substitution for the rotational restraint of stiffness $(S_R)_{3j}$ consists of a torsion bar, denoted as member 3 in Fig. 6-10d, parallel to the z axis (and hence perpendicular to the plane of the frame). This member would be assigned an arbitrary length L_3 and a torsion constant I_{X3}, computed as follows:

$$I_{X3} = \frac{L_3}{G}(S_R)_{3j}$$

The difficulty arising in this example is that the plane frame is converted into a space frame by the addition of member 3, and the analysis of the structure must proceed on that basis. Thus, it is seen that there will be some cases in which the replacement method is disadvantageous compared to the first method discussed above, which involves augmenting the joint stiffness matrix directly.

6.10 Translation of Axes. In Chapter 4 the concept of rotation of axes was found to be of great practical value for handling actions and displacements in members that are skewed in space. Similarly, the concept of translation of axes will be useful in subsequent articles of this chapter. By this technique it will be possible to transform a set of actions or displacements from one point to another for parallel sets of axes.

Figure 6-11a shows a force vector $\mathbf{P}_q$ and a moment $\mathbf{M}_q$ at point q and their *static equivalents* ($\mathbf{P}_p$ and $\mathbf{M}_p$) at point p. The actions at p may be computed from those at q by the following expressions:

$$\mathbf{P}_p = \mathbf{P}_q \tag{a}$$

$$\mathbf{M}_p = \mathbf{r}_{pq} \times \mathbf{P}_q + \mathbf{M}_q \tag{b}$$

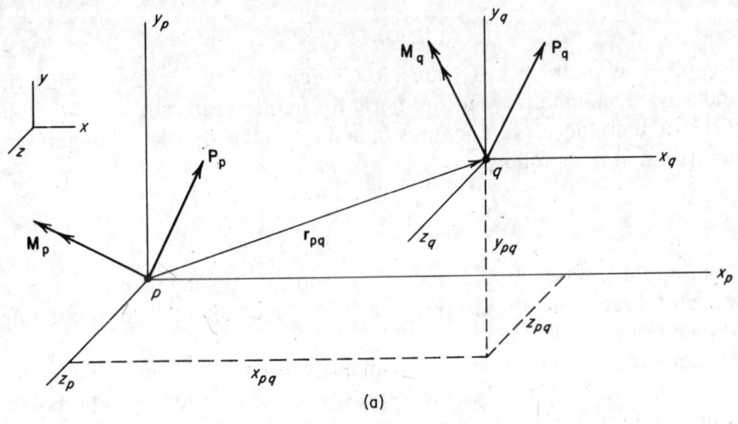

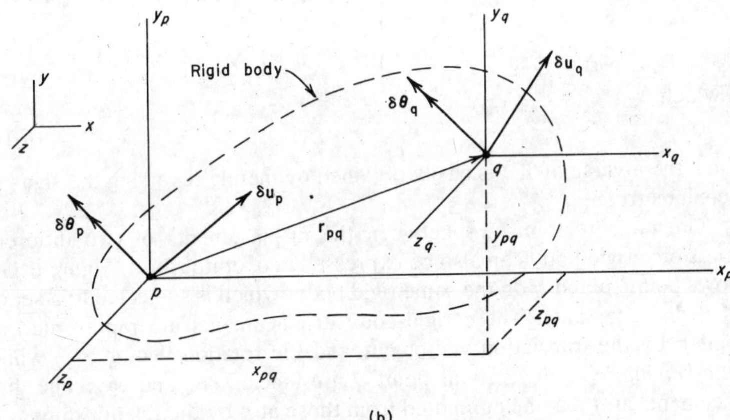

Fig. 6-11. Translation of axes: (a) action and (b) displacements.

As seen in the figure, $\mathbf{r}_{pq}$ is the location vector from p to q, and its components in the x, y, and z directions are $\mathbf{x}_{pq}$, $\mathbf{y}_{pq}$, and $\mathbf{z}_{pq}$. The matrix operation that is equivalent to Eqs. (a) and (b) takes the form

$$\mathbf{A}_p = \mathbf{T}_{pq}\mathbf{A}_q = \begin{bmatrix} \mathbf{I}_3 & \mathbf{O} \\ \mathbf{c}_{pq} & \mathbf{I}_3 \end{bmatrix} \begin{bmatrix} \mathbf{P}_q \\ \mathbf{M}_q \end{bmatrix} \tag{6-13}$$

In this equation the operator $\mathbf{T}_{pq}$ contains a 3×3 submatrix $\mathbf{c}_{pq}$ in the lower left-hand position, which consists of the components of $\mathbf{r}_{pq}$ arranged in a pattern for generating the vector product (or cross product) in Eq. (b). For this purpose, the following skew-symmetric submatrix is required:

$$\mathbf{c}_{pq} = \begin{bmatrix} 0 & -z_{pq} & y_{pq} \\ z_{pq} & 0 & -x_{pq} \\ -y_{pq} & x_{pq} & 0 \end{bmatrix} \tag{6-14}$$

The other submatrices in $\mathbf{T}_{pq}$ are a 3×3 identity matrix $\mathbf{I}_3$ (appearing twice) and a 3×3 null matrix. Furthermore, the action vectors $\mathbf{A}_p$ and $\mathbf{A}_q$ in Eq. (6-13) contain the six components of actions in the x, y, and z directions at points p and q, as follows:

$$\mathbf{A}_p = \{A_{p1}, A_{p2}, \ldots, A_{p6}\} \tag{c}$$
$$\mathbf{A}_q = \{A_{q1}, A_{q2}, \ldots, A_{q6}\} \tag{d}$$

The first three terms in $\mathbf{A}_p$ are the x, y, and z components of the force $\mathbf{P}_p$; the next three terms in $\mathbf{A}_p$ are the x, y, and z components of the moment $\mathbf{M}_p$; and similarly for $\mathbf{A}_q$.

Equation (6-13) may be used in analysis whenever static equivalents are required. The operator $\mathbf{T}_{pq}$ transforms a set of component actions $\mathbf{A}_q$ at point q into a parallel set $\mathbf{A}_p$ at point p, and it is called a *translation-of-axes transformation matrix*. The reverse transformation (from p to q) is

$$\mathbf{A}_q = \mathbf{T}_{pq}^{-1}\mathbf{A}_p = \begin{bmatrix} \mathbf{I}_3 & \mathbf{O} \\ \mathbf{c}_{qp} & \mathbf{I}_3 \end{bmatrix} \begin{bmatrix} \mathbf{P}_p \\ \mathbf{M}_p \end{bmatrix} \tag{6-15}$$

in which

$$\mathbf{c}_{qp} = -\mathbf{c}_{pq} \tag{6-16}$$

Thus, the inverse of $\mathbf{T}_{pq}$ is easily obtained by merely reversing the signs in the submatrix $\mathbf{c}_{pq}$.

Kinematic relationships between the displacements of two different points on a rigid body can also be expressed in operator form. Figure 6-11b shows points p and q on the same rigid body, which is subjected to a set of small displacements. These rigid-body displacements are represented at point p by the translation vector $\delta\mathbf{u}_p$ and the rotation vector $\delta\boldsymbol{\theta}_p$, while their *kinematic equivalents* at point q are labeled $\delta\mathbf{u}_q$ and $\delta\boldsymbol{\theta}_q$. The displacements at q may be computed from those at p by the relationships

$$\delta\mathbf{u}_q = \delta\mathbf{u}_p + \delta\boldsymbol{\theta}_p \times \mathbf{r}_{pq} \tag{e}$$
$$\delta\boldsymbol{\theta}_q = \delta\boldsymbol{\theta}_p \tag{f}$$

In Eq. (e) the cross product can be rewritten as

$$\delta\boldsymbol{\theta}_p \times \mathbf{r}_{pq} = -\mathbf{r}_{pq} \times \delta\boldsymbol{\theta}_p = \mathbf{r}_{qp} \times \delta\boldsymbol{\theta}_p \tag{g}$$

Then the matrix operation that is equivalent to Eqs. (e) and (f) becomes

$$\mathbf{D}_q = \mathbf{T}_{pq}^{\mathrm{T}}\mathbf{D}_p = \begin{bmatrix} \mathbf{I}_3 & \mathbf{c}_{pq}^{\mathrm{T}} \\ \mathbf{O} & \mathbf{I}_3 \end{bmatrix} \begin{bmatrix} \delta\mathbf{u}_p \\ \delta\boldsymbol{\theta}_p \end{bmatrix} \tag{6-17}$$

in which

$$\mathbf{c}_{pq}^{\mathrm{T}} = -\mathbf{c}_{pq} = \mathbf{c}_{qp} = \begin{bmatrix} 0 & z_{pq} & -y_{pq} \\ -z_{pq} & 0 & x_{pq} \\ y_{pq} & -x_{pq} & 0 \end{bmatrix} \tag{6-18}$$

Here it is seen that transposition of $\mathbf{c}_{pq}$ changes its sign, which is an inherent property of a skew-symmetric matrix.

The kinematically equivalent displacement vectors $\mathbf{D}_p$ and $\mathbf{D}_q$ in Eq. (6-17) each contain three components of translation and three of rotation in the x, y, and z directions at points p and q, respectively. In this case the transposed operator $\mathbf{T}_{pq}^{\mathrm{T}}$ converts a set of component displacements $\mathbf{D}_p$ at point p into a parallel set $\mathbf{D}_q$ at point q, assuming that there is no relative displacement of the points. The reverse transformation is

$$\mathbf{D}_p = \mathbf{T}_{pq}^{-\mathrm{T}}\mathbf{D}_q = \begin{bmatrix} \mathbf{I}_3 & \mathbf{c}_{pq} \\ \mathbf{O} & \mathbf{I}_3 \end{bmatrix} \begin{bmatrix} \delta\mathbf{u}_q \\ \delta\boldsymbol{\theta}_q \end{bmatrix} \tag{6-19}$$

Thus, the calculation of displacements at p in terms of those at q involves the transposed inverse of the operator used for obtaining actions at p from those at q (see Eq. 6-13).

6.11 Member Stiffnesses and Fixed-End Actions from Flexibilities. All subsequent articles in this chapter deal with member characteristics that alter the stiffnesses and fixed-end actions for individual members. If the two matrices $\mathbf{S}_{\mathrm{M}i}$ (see Art. 4.3) and $\mathbf{T}_{\mathrm{ML}i}$ (see Art. 6.4) are revised to account for nonprismatic shapes, shearing deformations, etc., the stiffness method of analysis will be otherwise unchanged. In many instances it is easier to derive these matrices from flexibility considerations than to use the direct stiffness approach. For this reason a technique that requires inversion of member flexibilities and translation-of-axes transformations will be described in this article. This procedure can then be applied as desired in subsequent articles to find $\mathbf{S}_{\mathrm{M}i}$ and $\mathbf{T}_{\mathrm{ML}i}$ for members with various characteristics.

Figure 6-12a shows a prismatic space frame member that is fixed at its j end and free at its k end. The 6×6 flexibility matrix $\mathbf{F}_{\mathrm{M}kk}$ for this type of member was derived previously in Chapter 2 (see Eq. 2-23). Inversion of $\mathbf{F}_{\mathrm{M}kk}$ produces the stiffness submatrix $\mathbf{S}_{\mathrm{M}kk}$. Thus,

$$\mathbf{S}_{\mathrm{M}kk} = \mathbf{F}_{\mathrm{M}kk}^{-1} \tag{6-20}$$

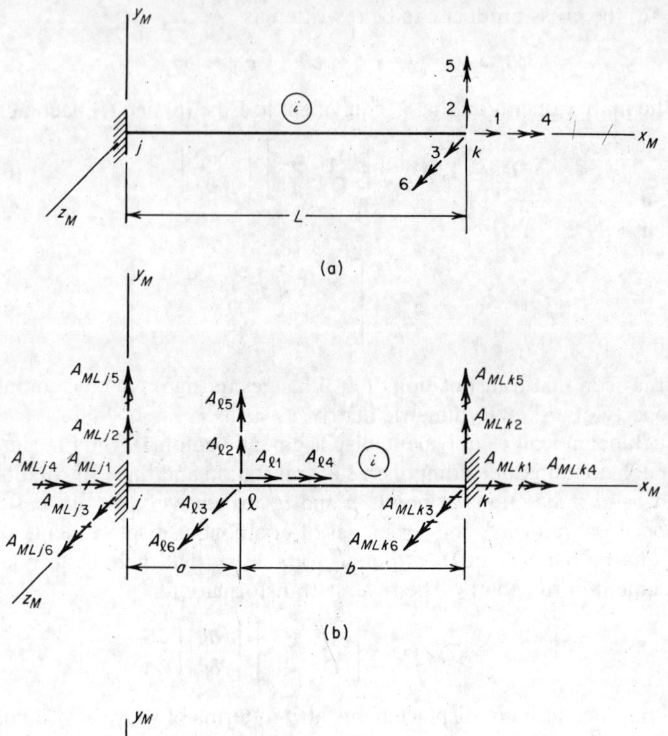

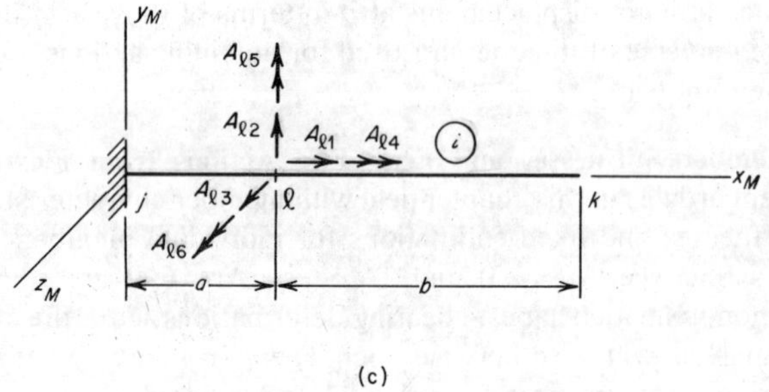

Fig. 6-12. Space frame member.

As discussed in Art 4.3, the complete member stiffness matrix $\mathbf{S}_{\mathrm{M}}$ (omitting the subscript i) has the form

$$\mathbf{S}_{\mathrm{M}} = \begin{bmatrix} \mathbf{S}_{\mathrm{M}jj} & \mathbf{S}_{\mathrm{M}jk} \\ \mathbf{S}_{\mathrm{M}kj} & \mathbf{S}_{\mathrm{M}kk} \end{bmatrix} \tag{6-21}$$

After the 6×6 submatrix $\mathbf{S}_{\mathrm{M}kk}$ has been obtained by inversion, the other three 6×6 submatrices of $\mathbf{S}_{\mathrm{M}}$ can be found by the type of axis transformations described in the preceding article.

Terms in the submatrices of $\mathbf{S}_{\mathrm{M}}$ are restraint actions due to unit displacements, and in particular the terms in submatrix $\mathbf{S}_{\mathrm{M}kk}$ are defined as the restraint actions at the k end of the member due to unit displacements at the same end. Statically equivalent actions at the j end may be computed using the transformation matrix

$$\mathbf{T}_{jk} = \left[\begin{array}{ccc|ccc} 1 & 0 & 0 & 0 & 0 & 0 \\ 0 & 1 & 0 & 0 & 0 & 0 \\ 0 & 0 & 1 & 0 & 0 & 0 \\ \hline 0 & 0 & 0 & 1 & 0 & 0 \\ 0 & 0 & -L & 0 & 1 & 0 \\ 0 & L & 0 & 0 & 0 & 1 \end{array}\right] \tag{6-22}$$

This operator is obtained from Eq. (6-13) by replacing points p and q with points j and k and using member-oriented axes. Then terms in the submatrix $\mathbf{S}_{\mathrm{M}jk}$ can be computed as the *static equilibrants* (equal and opposite to equivalents) of those in $\mathbf{S}_{\mathrm{M}kk}$, as follows:

$$\mathbf{S}_{\mathrm{M}jk} = -\mathbf{T}_{jk}\mathbf{S}_{\mathrm{M}kk} \tag{6-23}$$

The terms in $\mathbf{S}_{\mathrm{M}jk}$ are the restraint actions at j due to unit displacements at k, and they are in equilibrium with the terms in $\mathbf{S}_{\mathrm{M}kk}$.

Because the member stiffness matrix $\mathbf{S}_{\mathrm{M}}$ is symmetric, the submatrix $\mathbf{S}_{\mathrm{M}kj}$ must be equal to the transpose of $\mathbf{S}_{\mathrm{M}jk}$. Thus,

$$\mathbf{S}_{\mathrm{M}kj} = \mathbf{S}_{\mathrm{M}jk}^{\mathrm{T}} = -\mathbf{S}_{\mathrm{M}kk}\mathbf{T}_{jk}^{\mathrm{T}} \tag{6-24}$$

Terms in $\mathbf{S}_{\mathrm{M}kj}$ consist of restraint actions at k due to unit displacements at j.

The remaining submatrix $\mathbf{S}_{\mathrm{M}jj}$ can be found from $\mathbf{S}_{\mathrm{M}kj}$ by using an expression analogous to Eq. (6-23). That is,

$$\mathbf{S}_{\mathrm{M}jj} = -\mathbf{T}_{jk}\mathbf{S}_{\mathrm{M}kj} = \mathbf{T}_{jk}\mathbf{S}_{\mathrm{M}kk}\mathbf{T}_{jk}^{\mathrm{T}} \tag{6-25}$$

The terms in $\mathbf{S}_{\mathrm{M}jj}$ are restraint actions at j due to unit displacements at j, and they are in equilibrium with the terms in $\mathbf{S}_{\mathrm{M}kj}$.

In summary, Eqs. (6-23), (6-24), and (6-25) constitute the relationships for finding the other submatrices of $\mathbf{S}_{\mathrm{M}}$ after the submatrix $\mathbf{S}_{\mathrm{M}kk}$ has been found by inversion of $\mathbf{F}_{\mathrm{M}kk}$. A formulation similar to the above may also be carried out by starting with the j end free and the k end fixed. In this approach the submatrix $\mathbf{S}_{\mathrm{M}jj}$ is determined as the inverse of $\mathbf{F}_{\mathrm{M}jj}$, and a transformation matrix $\mathbf{T}_{kj}$ is used to find the other submatrices of $\mathbf{S}_{\mathrm{M}}$. The operator $\mathbf{T}_{kj}$ required for this purpose is the inverse of matrix $\mathbf{T}_{jk}$ and is obtained by merely changing the signs of the nonzero terms in the lower left-hand portion of $\mathbf{T}_{jk}$.

Furthermore, the expressions developed above for a space frame member can be specialized to any other type of framed structure by removing appropriate rows and columns from the transformation operator. For example, a beam member involves only forces in the y_M direction and moments in the z_M sense. Therefore, only the second and sixth rows and columns of $\mathbf{T}_{jk}$ need be retained, as follows:

$$\mathbf{T}_{jk} = \begin{bmatrix} 1 & 0 \\ L & 1 \end{bmatrix} \tag{6-26}$$

Of course, the submatrices of $\mathbf{S}_{\mathrm{M}}$ for a beam member will all be of size 2×2 instead of 6×6.

It is also possible to develop expressions for the fixed-end actions in matrix $\mathbf{T}_{\mathrm{ML}}$ (see Art. 6.4) from member flexibilities and axis transformations. For this purpose, consider the fixed-end space frame member in Fig. 6-12b, which has six components of actions applied at an intermediate point ℓ. They are

$$\mathbf{A}_{\ell} = \{A_{\ell 1}, A_{\ell 2}, \ldots, A_{\ell 6}\} \tag{a}$$

Due to these applied actions, six fixed-end actions will occur at each end of the member. At the j end the fixed-end actions are

$$\mathbf{A}_{\mathrm{ML}j} = \{A_{MLj1}, A_{MLj2}, \ldots, A_{MLj6}\} \tag{b}$$

and at the k end they are

$$\mathbf{A}_{\mathrm{ML}k} = \{A_{MLk1}, A_{MLk2}, \ldots, A_{MLk6}\} \tag{c}$$

When the fixed-end actions at j and k are distinguished in this manner, the matrix $\mathbf{T}_{\mathrm{ML}}$ can be partitioned as follows:

$$\mathbf{T}_{\mathrm{ML}} = \begin{bmatrix} \mathbf{T}_{\mathrm{ML}j} \\ \mathbf{T}_{\mathrm{ML}k} \end{bmatrix} \tag{6-27}$$

Formulas for the submatrices $\mathbf{T}_{\mathrm{ML}j}$ and $\mathbf{T}_{\mathrm{ML}k}$ will be derived separately in the ensuing discussion.

The fixed-end member in Fig. 6-12b is statically indeterminate to the sixth degree, and the flexibility method of Chapter 2 can be used to find six of the unknown actions. If the fixed-end actions $\mathbf{A}_{\mathrm{ML}k}$ at the k end are chosen as redundants, the released structure is as shown in Fig. 6-12c. Due to the applied loads on the released structure, the displacements at the load point are

$$\mathbf{D}_{\ell} = \mathbf{F}_{a\ell\ell} \mathbf{A}_{\ell} \tag{6-28}$$

In this expression the symbol $\mathbf{F}_{a\ell\ell}$ represents a 6×6 flexibility matrix for the ℓ end of the beam segment of length a. Whereas this segment deforms due to the loads, the other segment (of length b) displaces as a rigid body. Therefore, the displacements at k can be computed as

$$\mathbf{D}_k = \mathbf{T}_{\ell k}^{\mathrm{T}} \mathbf{D}_{\ell} \tag{6-29}$$

which is drawn from Eq. (6-17) in the preceding article. Substitution of Eq. (6-28) into Eq. (6-29) yields

$$\mathbf{D}_k = \mathbf{T}_{\ell k}^{\mathrm{T}} \mathbf{F}_{a\ell\ell} \mathbf{A}_{\ell} \tag{6-30}$$

This vector of displacements corresponds to the redundants and is due to the loads on the released structure. Therefore, it is recognized to be $\mathbf{D}_{QL}$ in the flexibility equation, which is

$$\mathbf{Q} = \mathbf{F}^{-1}(\mathbf{D}_Q - \mathbf{D}_{QL}) \qquad \text{(2-9) repeated}$$

Since the final displacements $\mathbf{D}_Q$ corresponding to the redundants are zero, the solution for the fixed-end actions $\mathbf{A}_{MLk}$ becomes

$$\mathbf{A}_{MLk} = -\mathbf{S}_{Mkk}\mathbf{T}^T_{\ell k}\mathbf{F}_{a\ell\ell}\mathbf{A}_\ell \qquad \text{(6-31)}$$

This formula is obtained from Eq. (2-9) by replacing $\mathbf{Q}$, $\mathbf{F}^{-1}$, and $\mathbf{D}_{QL}$ with $\mathbf{A}_{MLk}$, $\mathbf{S}_{Mkk}$, and $\mathbf{D}_k$ (from Eq. 6-30).

Similarly, a matrix formula for $\mathbf{A}_{MLj}$ can be derived with the flexibility method by choosing the fixed-end actions at j as the redundants and releasing the j end instead of the k end. The resulting expression is

$$\mathbf{A}_{MLj} = -\mathbf{S}_{Mjj}\mathbf{T}^T_{\ell j}\mathbf{F}_{b\ell\ell}\mathbf{A}_\ell \qquad \text{(6-32)}$$

In this formula the symbol $\mathbf{F}_{b\ell\ell}$ represents a 6×6 flexibility matrix for the ℓ end of the beam segment of length b; and the operator $\mathbf{T}^T_{\ell j}$ transforms displacements from ℓ to j. Since the coefficients of $\mathbf{A}_\ell$ in Eqs. (6-31) and (6-32) are $\mathbf{T}_{MLk}$ and $\mathbf{T}_{MLj}$, they can be placed into Eq. (6-27) to produce

$$\mathbf{T}_{ML} = \begin{bmatrix} \mathbf{T}_{MLj} \\ \mathbf{T}_{MLk} \end{bmatrix} = \begin{bmatrix} -\mathbf{S}_{Mjj}\mathbf{T}^T_{\ell j}\mathbf{F}_{b\ell\ell} \\ -\mathbf{S}_{Mkk}\mathbf{T}^T_{\ell k}\mathbf{F}_{a\ell\ell} \end{bmatrix} \qquad \text{(6-33)}$$

From principles of static equilibrium, the following relationships can also be written:

$$\mathbf{A}_{MLj} = -\mathbf{T}_{jk}\mathbf{A}_{MLk} - \mathbf{T}_{j\ell}\mathbf{A}_\ell = -(\mathbf{T}_{jk}\mathbf{T}_{MLk} + \mathbf{T}_{j\ell})\mathbf{A}_\ell \qquad \text{(6-34a)}$$

and

$$\mathbf{A}_{MLk} = -\mathbf{T}_{kj}\mathbf{A}_{MLj} - \mathbf{T}_{k\ell}\mathbf{A}_\ell = -(\mathbf{T}_{kj}\mathbf{T}_{MLj} + \mathbf{T}_{k\ell})\mathbf{A}_\ell \qquad \text{(6-34b)}$$

The coefficients of $\mathbf{A}_\ell$ in the latter form of these expressions are seen to be $\mathbf{T}_{MLj}$ and $\mathbf{T}_{MLk}$. Thus,

$$\mathbf{T}_{MLj} = -(\mathbf{T}_{jk}\mathbf{T}_{MLk} + \mathbf{T}_{j\ell}) \qquad \text{(6-35a)}$$

and

$$\mathbf{T}_{MLk} = -(\mathbf{T}_{kj}\mathbf{T}_{MLj} + \mathbf{T}_{k\ell}) \qquad \text{(6-35b)}$$

which give $\mathbf{T}_{MLj}$ and $\mathbf{T}_{MLk}$ in terms of each other.

In summary, terms in the matrix $\mathbf{T}_{ML}$ can be obtained from member flexibilities and axis transformations by evaluating the formulas in Eq. (6-33). However, only one of the submatrices in Eq. (6-33) need be evaluated because the other can be found from equilibrium principles using either Eq. (6-35a) or Eq. (6-35b). This procedure can be applied to the members of any type of framed structure and is illustrated by the following beam example.

Example. Figure 6-13a shows a restrained beam member with actions $A_{\ell 1}$ (y force) and $A_{\ell 2}$ (z moment) applied at point ℓ. The submatrices of $\mathbf{S}_\mathrm{M}$ and $\mathbf{T}_\mathrm{ML}$ for this example will be found using member flexibilities and axis transformations. For the released structure in Fig. 6-13b, the flexibilities are

$$\mathbf{F}_{\mathrm{M}kk} = \frac{L}{6EI}\begin{bmatrix} 2L^2 & 3L \\ 3L & 6 \end{bmatrix}$$

Inversion of this matrix in accordance with Eq. (6-20) gives

$$\mathbf{S}_{\mathrm{M}kk} = \mathbf{F}_{\mathrm{M}kk}^{-1} = \frac{2EI}{L^3}\begin{bmatrix} 6 & -3L \\ -3L & 2L^2 \end{bmatrix}$$

From Eq. (6-23) the static equilibrants at j are

$$\mathbf{S}_{\mathrm{M}jk} = -\mathbf{T}_{jk}\mathbf{S}_{\mathrm{M}kk} = -\begin{bmatrix} 1 & 0 \\ L & 1 \end{bmatrix}\mathbf{S}_{\mathrm{M}kk} = \frac{2EI}{L^3}\begin{bmatrix} -6 & 3L \\ -3L & L^2 \end{bmatrix}$$

By transposition (see Eq. 6-24),

$$\mathbf{S}_{\mathrm{M}kj} = \mathbf{S}_{\mathrm{M}jk}^{\mathrm{T}} = \frac{2EI}{L^3}\begin{bmatrix} -6 & -3L \\ 3L & L^2 \end{bmatrix}$$

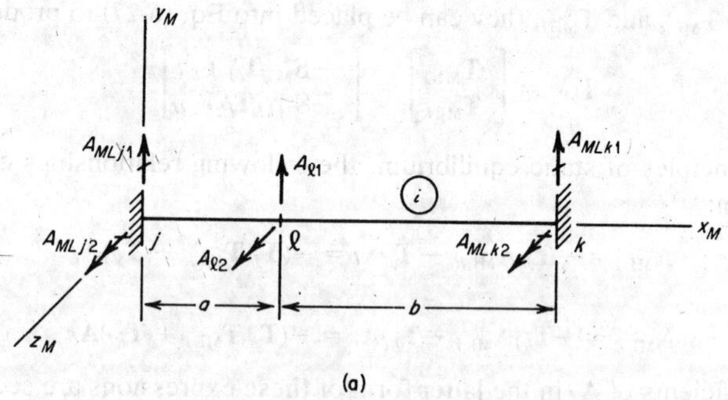

(a)

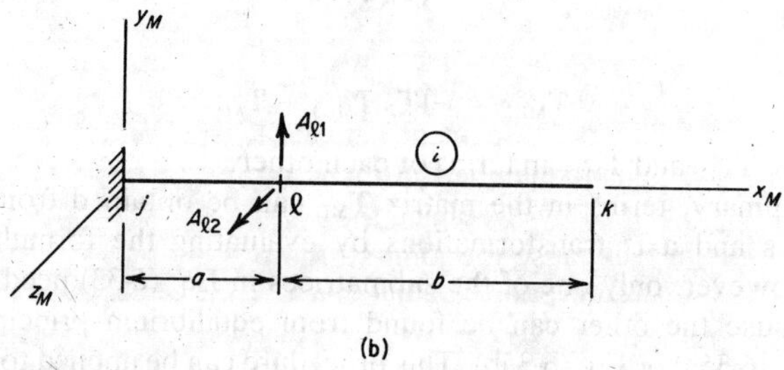

(b)

Fig. 6-13. Beam example.

Then Eq. (6-25) produces the equilibrants at j as

$$\mathbf{S}_{\mathrm{M}jj} = -\mathbf{T}_{jk}\mathbf{S}_{\mathrm{M}kj} = -\begin{bmatrix} 1 & 0 \\ L & 1 \end{bmatrix} \mathbf{S}_{\mathrm{M}kj} = \frac{2EI}{L^3}\begin{bmatrix} 6 & 3L \\ 3L & 2L^2 \end{bmatrix}$$

All of the terms in these stiffness submatrices are familiar expressions.

The less familiar submatrices of $\mathbf{T}_{\mathrm{ML}}$ will be obtained by working with the released structure in Fig. 6-13b. For unit values of $A_{\ell 1}$ and $A_{\ell 2}$, the flexibility matrix of the segment $j\ell$ is

$$\mathbf{F}_{a\ell\ell} = \frac{a}{6EI}\begin{bmatrix} 2a^2 & 3a \\ 3a & 6 \end{bmatrix}$$

and the transformation matrix for displacements at k due to unit displacements at ℓ is

$$\mathbf{T}_{\ell k}^{T} = \begin{bmatrix} 1 & b \\ 0 & 1 \end{bmatrix}$$

Therefore, the submatrix $\mathbf{T}_{\mathrm{ML}k}$ in Eq. (6-33) becomes

$$\mathbf{T}_{\mathrm{ML}k} = -\mathbf{S}_{\mathrm{M}kk}\mathbf{T}_{\ell k}^{T}\mathbf{F}_{a\ell\ell} = \frac{1}{L^3}\begin{bmatrix} -a^2(a+3b) & -6ab \\ a^2bL & aL(2b-a) \end{bmatrix}$$

Then the submatrix $\mathbf{T}_{\mathrm{ML}j}$ can be determined from equilibrium considerations (Eq. 6-35a), as follows:

$$\mathbf{T}_{\mathrm{ML}j} = -\mathbf{T}_{jk}\mathbf{T}_{\mathrm{ML}k} - \mathbf{T}_{j\ell} = -\begin{bmatrix} 1 & 0 \\ L & 1 \end{bmatrix}\mathbf{T}_{\mathrm{ML}k} - \begin{bmatrix} 1 & 0 \\ a & 1 \end{bmatrix}$$

$$= \frac{1}{L^3}\begin{bmatrix} -b^2(3a+b) & 6ab \\ -ab^2L & bL(2a-b) \end{bmatrix}$$

The submatrices of $\mathbf{T}_{\mathrm{ML}}$ found above are the same as those given previously in Eq. (6-4).

6.12 Nonprismatic Members. Members that do not have the same cross-sectional properties from one end to the other are called nonprismatic members. Examples are tapered members and members having reinforcements over parts of their lengths. In addition, any member that does not have a straight axis is nonprismatic. The topic of curved members is covered in Art. 6.13, and only nonprismatic members with straight axes are discussed in this article.

One feasible approach for analyzing a structure with nonprismatic members consists of assuming joints at intermediate points along the length of each nonprismatic member. A segment between two such joints may be approximated as a prismatic member (if it is not already prismatic) having section properties obtained by averaging those at the two ends of the segment. This approach will cause the joint stiffness matrix to become very large, but it does not require any changes in the computer programs of Chapter 5.

However, a more desirable approach consists of evaluating flexibilities for the nonprismatic member and then obtaining stiffnesses and fixed-end actions by the method described in the preceding article. The unit-load method in Appendix A provides a means of calculating flexibilities by direct integration, assuming that the variations in cross-sectional properties are expressed as continuous functions. Otherwise, the flexibilities can be obtained through an appropriate numerical integration procedure.

As an example, consider the tapered beam member shown in Fig. 6-14a. It will be assumed that this beam has a rectangular cross section of constant thickness t and that its height varies linearly from h_j at the j end to h_k at the k end. The j end of the member is fixed in preparation for calculating flexibilities at the k end. Also shown in the figure is an origin at point O, where the depth of the imaginary portion of the beam (shown by dashed lines) is zero. From this origin the auxiliary variable x_1 is measured and will be used for convenient integration. Figures 6-14b and 6-14c consist of moment diagrams M_{U1} and M_{U2} due to unit values of the actions A_{k1} and A_{k2}, respectively. These diagrams may be used to obtain the member flexibilities at the k end by the unit-load method. Thus, Eq. (A-31) in Appendix A-2 gives (for flexural deformations only)

$$(F_{Mkk})_{11} = \int_{L_1}^{L_1+L} \frac{M_{U1}^2}{EI_Z} dx_1 \tag{a}$$

In this equation the terms are

$$L_1 = \frac{h_k L}{h_j - h_k} \qquad M_{U1} = x_1 - L_1 \qquad I_Z = \frac{th^3}{12} \qquad h = \frac{h_k x_1}{L_1} \tag{b}$$

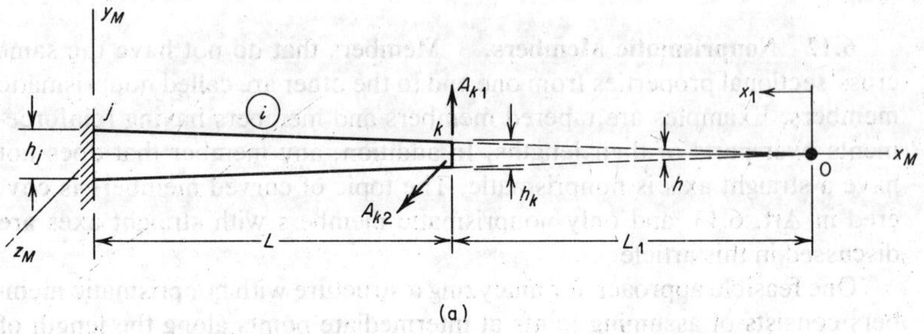

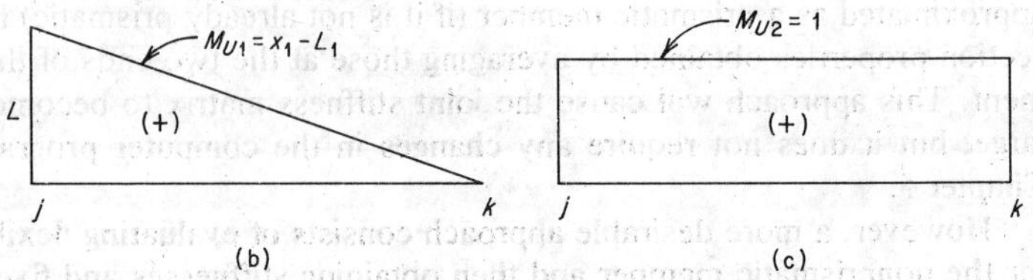

Fig. 6-14. Tapered beam member.

If $h_j = 2h_k$, the terms in expressions (b) simplify to

$$L_1 = L \qquad M_{U1} = x_1 - L \qquad I_Z = \frac{th_k^3 x_1^3}{12L^3} \tag{c}$$

Substitution of expressions (c) into (a) produces

$$(F_{Mkk})_{11} = \frac{12L^3}{th_k^3 E}\int_L^{2L} \frac{(x_1 - L)^2}{x_1^3}\, dx_1 \tag{d}$$

Intregration yields

$$(F_{Mkk})_{11} = \frac{12L^3}{th_k^3 E}\left(ln\, 2 - \frac{5}{8}\right) \tag{e}$$

Similarly,

$$(F_{Mkk})_{12} = \int_L^{2L} \frac{M_{U1}M_{U2}}{EI_Z}\, dx_1 = \frac{3L^2}{2th_k^3 E} \tag{f}$$

and

$$(F_{Mkk})_{22} = \int_L^{2L} \frac{M_{U2}^2}{EI_Z}\, dx_1 = \frac{9L}{2th_k^3 E} \tag{g}$$

From this point the member stiffness matrix $\mathbf{S}_{\mathrm{M}}$ and the transfer matrix $\mathbf{T}_{\mathrm{ML}}$ can be obtained as described in the preceding article.

Consider now a nonprismatic member composed of segments, each of which may be either prismatic or nonprismatic, as indicated in Figs. 6-15a and b. Assume that flexibilities, stiffnesses, and fixed-end actions for each segment have already been calculated in the manner described above. Two methods will be discussed for obtaining the matrices $\mathbf{S}_{\mathrm{M}}$ and $\mathbf{T}_{\mathrm{ML}}$ for this type of member. The first approach utilizes segment flexibilities and translation-of-axes transformations, whereas the second approach involves segment stiffnesses.

In the flexibility approach, let the j end of the segmented member be fixed (see Fig. 6-15a) and examine the influence of loads at the k end. Actions $\mathbf{A}_\ell$ at point ℓ that are statically equivalent to actions $\mathbf{A}_k$ at k are (from Eq. 6-13):

$$\mathbf{A}_\ell = \mathbf{T}_{\ell k}\mathbf{A}_k \tag{6-36}$$

Due to these actions, incremental displacements $\Delta\mathbf{D}_\ell$ at point ℓ, relative to those at point $\ell - 1$, become

$$\Delta\mathbf{D}_\ell = \mathbf{F}_{s\ell\ell}\mathbf{A}_\ell \tag{6-37}$$

where $\mathbf{F}_{s\ell\ell}$ represents the flexibility matrix of the ℓ end of the segment of length s. Substitute Eq. (6-36) into Eq. (6-37) to obtain

$$\Delta\mathbf{D}_\ell = \mathbf{F}_{s\ell\ell}\mathbf{T}_{\ell k}\mathbf{A}_k \tag{6-38}$$

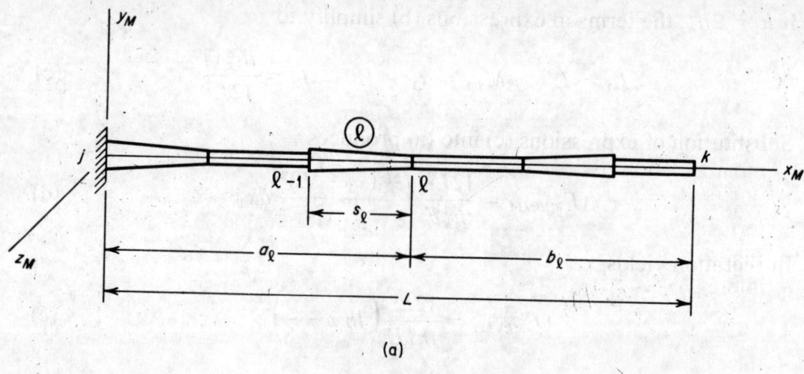

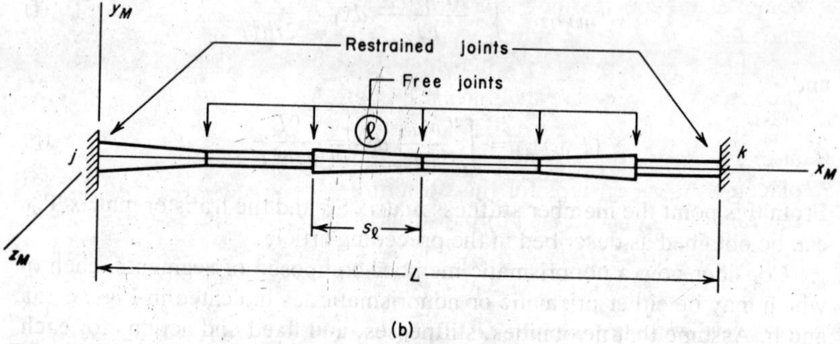

Fig. 6-15. Segmented member: (a) flexibility approach (b) stiffness approach.

Then calculate the incremental displacements at point k as (see Eq. 6-17)

$$\Delta \mathbf{D}_k = \mathbf{T}_{\ell k}^{\mathrm{T}} \Delta \mathbf{D}_\ell \tag{6-39}$$

Substitution of Eq. (6-38) into Eq. (6-39) gives

$$\Delta \mathbf{D}_k = \mathbf{T}_{\ell k}^{\mathrm{T}} \mathbf{F}_{s\ell\ell} \mathbf{T}_{\ell k} \mathbf{A}_k \tag{6-40}$$

This expression represents incremental displacements at k caused by actions $\mathbf{A}_k$ associated with the flexibilities of segment ℓ only. By superposition of the contributions of all segments, the total effect at k is

$$\mathbf{D}_k = \sum_{\ell=1}^{n_s} (\mathbf{T}_{\ell k}^{\mathrm{T}} \mathbf{F}_{s\ell\ell} \mathbf{T}_{\ell k}) \mathbf{A}_k \tag{6-41}$$

in which n_s is the number of segments. Hence,

$$\mathbf{F}_{Mkk} = \sum_{\ell=1}^{n_s} (\mathbf{T}_{\ell k}^{\mathrm{T}} \mathbf{F}_{s\ell\ell} \mathbf{T}_{\ell k}) \tag{6-42}$$

From this point the inversion of $\mathbf{F}_{\mathrm{M}kk}$ to obtain $\mathbf{S}_{\mathrm{M}kk}$, and so on, is the same as in the preceding article.

Similarly, expressions for the transfer matrix $\mathbf{T}_{\mathrm{ML}}$ can also be derived by the flexibility approach, as demonstrated previously for a prismatic member. Equation (6-33) provides the necessary formulas, but in this case the flexibility matrices $\mathbf{F}_{a\ell\ell}$ and $\mathbf{F}_{b\ell\ell}$ must be obtained by summation of terms from contributing segments to the left and to the right of point ℓ (see Fig. 6-15a). Furthermore, there is a different matrix $\mathbf{T}_{\mathrm{ML}\ell}$ for each point of interest, such that

$$\mathbf{T}_{\mathrm{ML}\ell} = \begin{bmatrix} \mathbf{T}_{\mathrm{ML}j\ell} \\ \mathbf{T}_{\mathrm{ML}k\ell} \end{bmatrix} \qquad (\ell = 1, 2, \ldots, n_s - 1) \tag{6-43}$$

Thus, the transfer matrix $\mathbf{T}_{\mathrm{ML}}$ for a point where two segments are joined is somewhat different from the one discussed in Art. 6.4, which was for fixed-end actions due to unit actions applied at any point on the axis of a member.

The stiffness approach will be developed using Fig. 6-15b, which shows the segmented member fixed at both ends. At each point where segments are joined there will be free displacements $\mathbf{D}_{\mathrm{F}\ell}$ ($\ell = 1, 2, \ldots, n_s - 1$), and at ends j and k there will be restrained displacements $\mathbf{D}_{\mathrm{R}}$. The action-displacement relationships for this structure (see Eq. 3-33) can be written in rearranged and partitioned form as follows:

$$\begin{bmatrix} \mathbf{S}_{\mathrm{FF}} & \mathbf{S}_{\mathrm{FR}} \\ \mathbf{S}_{\mathrm{RF}} & \mathbf{S}_{\mathrm{RR}} \end{bmatrix} \begin{bmatrix} \mathbf{D}_{\mathrm{F}} \\ \mathbf{D}_{\mathrm{R}} \end{bmatrix} = \begin{bmatrix} \mathbf{A}_{\mathrm{F}} \\ \mathbf{A}_{\mathrm{R}} \end{bmatrix} \tag{3-33}$$
repeated

Performing the indicated multiplications gives

$$\mathbf{S}_{\mathrm{FF}}\mathbf{D}_{\mathrm{F}} + \mathbf{S}_{\mathrm{FR}}\mathbf{D}_{\mathrm{R}} = \mathbf{A}_{\mathrm{F}} \tag{3-34a}$$
repeated

and

$$\mathbf{S}_{\mathrm{RF}}\mathbf{D}_{\mathrm{F}} + \mathbf{S}_{\mathrm{RR}}\mathbf{D}_{\mathrm{R}} = \mathbf{A}_{\mathrm{R}} \tag{3-34b}$$
repeated

Solving for $\mathbf{D}_{\mathrm{F}}$ in the first equation results in

$$\mathbf{D}_{\mathrm{F}} = \mathbf{S}_{\mathrm{FF}}^{-1}(\mathbf{A}_{\mathrm{F}} - \mathbf{S}_{\mathrm{FR}}\mathbf{D}_{\mathrm{R}}) \tag{3-35}$$
repeated

Then substituting this expression for $\mathbf{D}_{\mathrm{F}}$ into Eq. (3-34b) and rearranging terms yields

$$(\mathbf{S}_{\mathrm{RR}} - \mathbf{S}_{\mathrm{RF}}\mathbf{S}_{\mathrm{FF}}^{-1}\mathbf{S}_{\mathrm{FR}})\mathbf{D}_{\mathrm{R}} = \mathbf{A}_{\mathrm{R}} - \mathbf{S}_{\mathrm{RF}}\mathbf{S}_{\mathrm{FF}}^{-1}\mathbf{A}_{\mathrm{F}} \tag{6-44}$$

The right-hand side of Eq. (6-44) contains the actions of type R minus a term involving the effect of actions of type F. If the displacements $\mathbf{D}_{\mathrm{R}}$ are all equal to zero (fixed-end member), Eq. (6-44) gives the following relation between the actions of types R and F:

$$\mathbf{A}_{\mathrm{R}} = \mathbf{S}_{\mathrm{RF}}\mathbf{S}_{\mathrm{FF}}^{-1}\mathbf{A}_{\mathrm{F}} \tag{6-45}$$

Therefore, the transfer matrix $\mathbf{T}_{\mathrm{ML}}$ for fixed-end actions of type R due to unit actions of type F is

$$\mathbf{T}_{\mathrm{ML}} = \mathbf{S}_{\mathrm{RF}}\mathbf{S}_{\mathrm{FF}}^{-1} \tag{6-46}$$

On the other hand, if all actions of type F are set equal to zero, Eq. (6-44) becomes

$$(\mathbf{S}_{\mathrm{RR}} - \mathbf{S}_{\mathrm{RF}}\mathbf{S}_{\mathrm{FF}}^{-1}\,\mathbf{S}_{\mathrm{FR}})\mathbf{D}_{\mathrm{R}} = \mathbf{A}_{\mathrm{R}} \tag{6-47}$$

Since actions $\mathbf{A}_{\mathrm{R}}$ and displacements $\mathbf{D}_{\mathrm{R}}$ are corresponding, the expression in the parentheses in Eq. (6-47) is seen to be the stiffness matrix for unit displacements of the ends of the segmented member. Thus,

$$\mathbf{S}_{\mathrm{M}} = \mathbf{S}_{\mathrm{RR}} - \mathbf{S}_{\mathrm{RF}}\mathbf{S}_{\mathrm{FF}}^{-1}\mathbf{S}_{\mathrm{FR}} \tag{6-48}$$

Substitution of Eq. (6-46) into the last term of Eq. (6-48) yields

$$\mathbf{S}_{\mathrm{M}} = \mathbf{S}_{\mathrm{RR}} - \mathbf{T}_{\mathrm{ML}}\,\mathbf{S}_{\mathrm{FR}} \tag{6-49}$$

In summary, the member stiffness matrix $\mathbf{S}_{\mathrm{M}}$ and the transfer matrix $\mathbf{T}_{\mathrm{ML}}$ for a segmented member may be obtained using previous concepts applied to a single member. By this approach the over-all joint stiffness matrix is formed as shown in Eq. (3-33), giving the matrices $\mathbf{S}_{\mathrm{FF}}$, $\mathbf{S}_{\mathrm{FR}}$, $\mathbf{S}_{\mathrm{RF}}$, and $\mathbf{S}_{\mathrm{RR}}$. Inversion of $\mathbf{S}_{\mathrm{FF}}$ and substitution of the appropriate matrices into Eqs. (6-46) and (6-49) produce the transfer matrix $\mathbf{T}_{\mathrm{ML}}$ and the stiffness matrix $\mathbf{S}_{\mathrm{M}}$ for the segmented member.

The process of eliminating one or more matrices by partial solution of the problem (as was done in deriving Eq. 6-44) is sometimes referred to as *matrix condensation*. Thus, the matrices $\mathbf{S}_{\mathrm{M}}$ and $\mathbf{T}_{\mathrm{ML}}$ for the segmented nonprismatic member represent combinations of other matrices obtained by a condensation procedure. In this case the matrix $\mathbf{T}_{\mathrm{ML}}$ pertains to all the points where segments are joined. It is also possible to make a more refined analysis, in which the displacements $\mathbf{D}_{\mathrm{F}\ell}$ are eliminated one joint at a time.

Example. Assume that the two-segment beam member in Figs. 6-16a and b has the following properties;

$$L_1 = L_2 = L \qquad I_{Z1} = I \qquad I_{Z2} = 2I$$

Determine stiffnesses and fixed-end actions for this segmented member by the flexibility and stiffness approaches described above.

For the flexibility approach the following matrices are needed:

$$\mathbf{F}_{s11} = \frac{L}{6EI}\begin{bmatrix} 2L^2 & 3L \\ 3L & 6 \end{bmatrix} = 2\mathbf{F}_{s22} \qquad \mathbf{T}_{1k} = \begin{bmatrix} 1 & 0 \\ L & 1 \end{bmatrix} \qquad \mathbf{T}_{2k} = \begin{bmatrix} 1 & 0 \\ 0 & 1 \end{bmatrix}$$

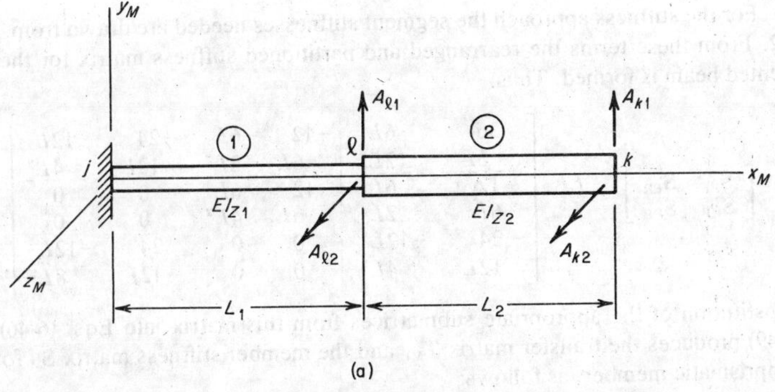

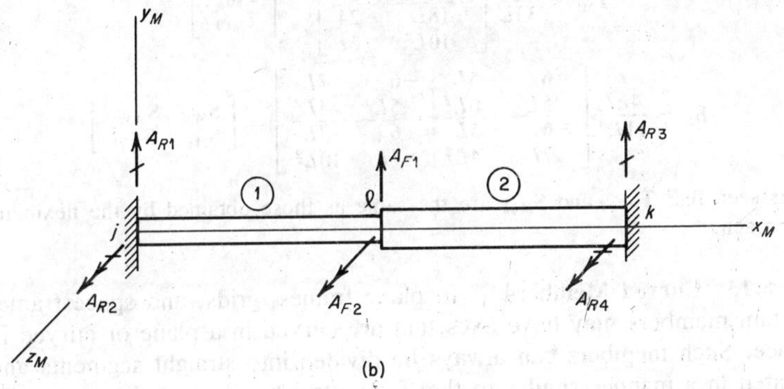

Fig. 6-16. Nonprismatic beam: (a) flexibility approach (b) stiffness approach.

From Eq. (6-42), the flexibilities of the k end are

$$\mathbf{F}_{\mathrm{M}kk} = \sum_{\ell=1}^{n_s} (\mathbf{T}_{\ell k}^{\mathrm{T}} \mathbf{F}_{s\ell\ell} \mathbf{T}_{\ell k}) = \frac{L}{6EI} \begin{bmatrix} 14L^2 & 9L \\ 9L & 12 \end{bmatrix} + \frac{L}{12EI} \begin{bmatrix} 2L^2 & 3L \\ 3L & 6 \end{bmatrix}$$

$$= \frac{L}{4EI} \begin{bmatrix} 10L^2 & 7L \\ 7L & 6 \end{bmatrix}$$

Inversion of this matrix yields

$$\mathbf{S}_{\mathrm{M}kk} = \mathbf{F}_{\mathrm{M}kk}^{-1} = \frac{4EI}{11L^3} \begin{bmatrix} 6 & -7L \\ -7L & 10L^2 \end{bmatrix}$$

From Eq. (6-33) the fixed-end actions $\mathbf{T}_{\mathrm{ML}k\ell}$ at the k end due to unit actions at point ℓ become

$$\mathbf{T}_{\mathrm{ML}k} = -\mathbf{S}_{\mathrm{M}kk} \mathbf{T}_{\ell k}^{\mathrm{T}} \mathbf{F}_{a\ell\ell}$$

$$= -\mathbf{S}_{\mathrm{M}kk} \begin{bmatrix} 1 & L \\ 0 & 1 \end{bmatrix} \begin{bmatrix} 2L^2 & 3L \\ 3L & 6 \end{bmatrix} \frac{L}{6EI} = \frac{2}{33L} \begin{bmatrix} -9L & -12 \\ 5L^2 & 3L \end{bmatrix}$$

For the stiffness approach the segment stiffnesses needed are drawn from Table 4-2. From these terms the rearranged and partitioned stiffness matrix for the segmented beam is formed. Thus,

$$\begin{bmatrix} \mathbf{S}_{FF} & \mathbf{S}_{FR} \\ \mathbf{S}_{RF} & \mathbf{S}_{RR} \end{bmatrix} = \frac{EI}{L^3} \left[\begin{array}{cc|cccc} 36 & 6L & -12 & -6L & -24 & 12L \\ 6L & 12L^2 & 6L & 2L^2 & -12L & 4L^2 \\ \hline -12 & 6L & 12 & 6L & 0 & 0 \\ -6L & 2L^2 & 6L & 4L^2 & 0 & 0 \\ -24 & -12L & 0 & 0 & 24 & -12L \\ 12L & 4L^2 & 0 & 0 & -12L & 8L^2 \end{array}\right]$$

Substitution of the appropriate submatrices from this matrix into Eqs. (6-46) and (6-49) produces the transfer matrix $\mathbf{T}_{ML}$ and the member stiffness matrix $\mathbf{S}_M$ for the nonprismatic member, as follows:

$$\mathbf{T}_{ML} = \frac{1}{33L} \left[\begin{array}{cc} -15L & 24 \\ -7L^2 & 9L \\ \hline -18L & -24 \\ 10L^2 & 6L \end{array}\right] = \begin{bmatrix} \mathbf{T}_{MLj\ell} \\ \mathbf{T}_{MLk\ell} \end{bmatrix}$$

$$\mathbf{S}_M = \frac{4EI}{11L^3} \left[\begin{array}{cc|cc} 6 & 5L & -6 & 7L \\ 5L & 6L^2 & -5L & 4L^2 \\ \hline -6 & -5L & 6 & -7L \\ 7L & 4L^2 & -7L & 10L^2 \end{array}\right] = \begin{bmatrix} \mathbf{S}_{Mjj} & \mathbf{S}_{Mjk} \\ \mathbf{S}_{Mkj} & \mathbf{S}_{Mkk} \end{bmatrix}$$

It is seen that $\mathbf{T}_{MLk\ell}$ and $\mathbf{S}_{Mkk}$ are the same as those obtained by the flexibility approach.

6.13 Curved Members. In plane frames, grids, and space frames certain members may have axes that are curved in a plane or curved in space. Such members can always be divided into straight segments and treated in a manner similar to that for a straight segmented member, as described in the preceding article. However, for this purpose rotation-of-axes as well as translation-of-axes transformation matrices are required. In the present article flexibilities, stiffnesses, and fixed-end actions are discussed for circular members in plane frames and grids. These cases occur rather commonly, and the exact integrals are not difficult to perform.

Figure 6-17 shows a plane frame member i in the shape of a circular arc with its center at point C. The member lies in the x-y plane, and its cross-sectional properties are assumed to be constant throughout its length. The j end is fixed, but the k end is free (in preparation for calculating flexibilities). Due to unit values of the actions A_{k1}, A_{k2}, and A_{k3} (two forces and a moment) the following bending moments exist in the member:

$$M_{U1} = -r(1 - \cos\theta) \qquad M_{U2} = r\sin\theta \qquad M_{U3} = 1 \tag{a}$$

In these expressions r is the radius of the member, and θ is the central angle measured from end k (see Fig. 6-17). The moments in expressions (a) can be used to determine the member flexibilities at the k end by the unit-load method. Thus, Eq. (A-31) in Appendix A-2 gives (for flexural deformations only)

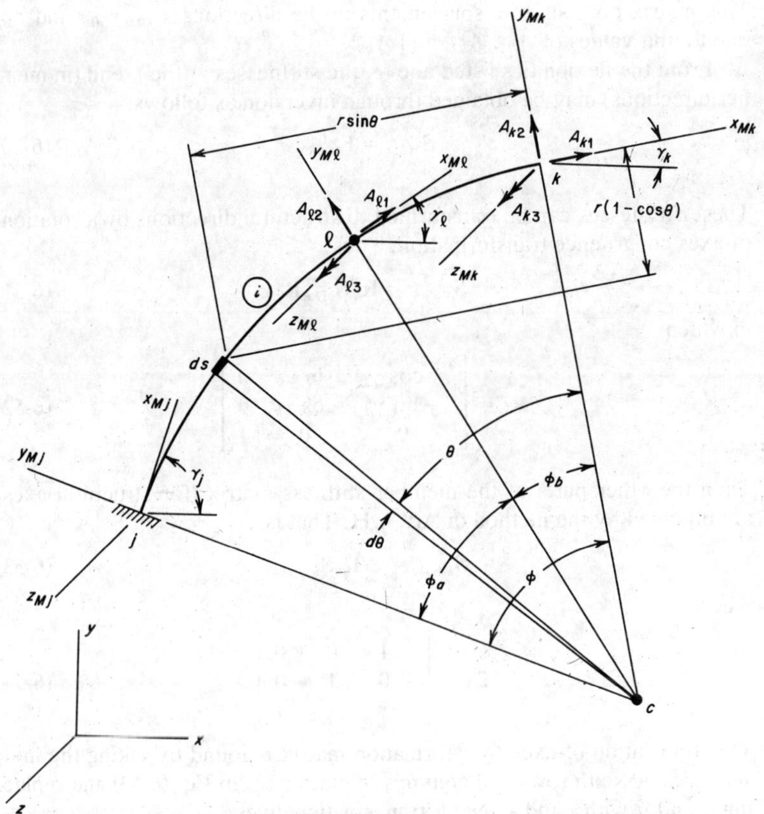

Fig. 6-17. Circular member in a plane frame.

$$(F_{Mkk})_{11} = \int \frac{M_{U1}^2}{EI_Z}\,ds = \int_0^{\phi} \frac{r^2(1-\cos\theta)^2}{EI_Z}\,r\,d\theta$$

$$= \frac{r^3}{2EI_Z}(3\phi - 4\sin\phi + \sin\phi\cos\phi) \tag{b}$$

After other flexibilities of this nature are developed at the k end, the member flexibility matrix is found to be

$$\mathbf{F}_{Mkk} = \frac{r}{2EI_Z}\begin{bmatrix} r^2(3\phi - 4\sin\phi + \sin\phi\cos\phi) & -r^2(2 - 2\cos\phi - \sin^2\phi) & -2r(\phi - \sin\phi) \\ & r^2(\phi - \sin\phi\cos\phi) & 2r(1-\cos\phi) \\ \text{Sym.} & & 2\phi \end{bmatrix} \tag{6-50}$$

This matrix consists of displacements in the directions x_{Mk}, y_{Mk}, and z_{Mk} due to unit values of A_{k1}, A_{k2}, and A_{k3}.

From the flexibilities listed above, the stiffnesses at the k end (in member directions) may be obtained through inversion as follows:

$$\mathbf{S}_{\mathrm{M}kk} = \mathbf{F}_{\mathrm{M}kk}^{-1} \tag{6-20}$$

repeated

These stiffnesses can be transformed to structural directions by a rotation-of-axes congruence transformation. Thus,

$$\mathbf{S}_{\mathrm{MS}kk} = \mathbf{R}_k^{\mathrm{T}}\mathbf{S}_{\mathrm{M}kk}\mathbf{R}_k \tag{6-51}$$

in which

$$\mathbf{R}_k = \begin{bmatrix} \cos\gamma_k & \sin\gamma_k & 0 \\ -\sin\gamma_k & \cos\gamma_k & 0 \\ 0 & 0 & 1 \end{bmatrix} \tag{6-52}$$

Then the other parts of the member stiffness matrix (for structural axes) are obtained by the method of Art. 6.11. That is,

$$\mathbf{S}_{\mathrm{MS}jk} = -\mathbf{T}_{jk}\mathbf{S}_{\mathrm{MS}kk} \tag{6-53}$$

where

$$\mathbf{T}_{jk} = \begin{bmatrix} 1 & 0 & 0 \\ 0 & 1 & 0 \\ -y_{jk} & x_{jk} & 1 \end{bmatrix} \tag{6-54}$$

This translation-of-axes transformation matrix is found by taking the first, second, and sixth rows and columns of matrix $\mathbf{T}_{pq}$ in Eq. (6-13) and replacing p and q with j and k. Next, transposition gives

$$\mathbf{S}_{\mathrm{MS}kj} = \mathbf{S}_{\mathrm{MS}jk}^{\mathrm{T}} = -\mathbf{S}_{\mathrm{MS}kk}\mathbf{T}_{jk}^{\mathrm{T}} \tag{6-55}$$

Finally,

$$\mathbf{S}_{\mathrm{MS}jj} = -\mathbf{T}_{jk}\mathbf{S}_{\mathrm{MS}kj} = \mathbf{T}_{jk}\mathbf{S}_{\mathrm{MS}kk}\mathbf{T}_{jk}^{\mathrm{T}} \tag{6-56}$$

These stiffness submatrices can be transformed to member directions, as follows:

$$\mathbf{S}_{\mathrm{M}} = \mathbf{R}_{\mathrm{T}}\mathbf{S}_{\mathrm{MS}}\mathbf{R}_{\mathrm{T}}^{\mathrm{T}} \tag{6-57}$$

in which

$$\mathbf{R}_{\mathrm{T}} = \begin{bmatrix} \mathbf{R}_j & \mathbf{O} \\ \mathbf{O} & \mathbf{R}_k \end{bmatrix} \tag{6-58}$$

In this operator the submatrix $\mathbf{R}_j$ is similar to $\mathbf{R}_k$ (see Eq. 6-52), but γ_j replaces γ_k.

To determine expressions for fixed-end actions, the concepts in Art.

6.11 can again be utilized. However, in the present case it is also necessary to transform certain actions and displacements by rotation-of-axes techniques. Consider the actions $A_{\ell 1}$, $A_{\ell 2}$, and $A_{\ell 3}$ applied at point ℓ in the directions $x_{M\ell}$, $y_{M\ell}$, and $z_{M\ell}$ (see Fig. 6-17). When these actions are premultiplied by a flexibility matrix $\mathbf{F}_{a\ell\ell}$, the displacements at ℓ in the member directions are

$$\mathbf{D}_\ell = \mathbf{F}_{a\ell\ell}\mathbf{A}_\ell \tag{c}$$

The matrix $\mathbf{F}_{a\ell\ell}$ in this expression is similar to $\mathbf{F}_{\mathrm{M}kk}$ in Eq. (6-50), but the central angle ϕ is replaced by ϕ_a (see Fig. 6-17). Rotation of axes to structural directions and translation of the results to point k gives

$$\mathbf{D}_k = \mathbf{T}_{\ell k}^{\mathrm{T}}\mathbf{R}_\ell^{\mathrm{T}}\mathbf{D}_\ell = \mathbf{T}_{\ell k}^{\mathrm{T}}\mathbf{R}_\ell^{\mathrm{T}}\mathbf{F}_{a\ell\ell}\mathbf{A}_\ell \tag{d}$$

In this expression the rotation matrix $\mathbf{R}_\ell$ is similar to $\mathbf{R}_k$ (see Eq. 6-52), but γ_ℓ replaces γ_k. Also, the translational operator $\mathbf{T}_{\ell k}$ is similar to $\mathbf{T}_{jk}$ (see Eq. 6-54), but $x_{\ell k}$ and $y_{\ell k}$ replace x_{jk} and y_{jk}. Then the flexibility equation (Eq. 2-9) can be applied to find the fixed-end actions at k, which are also transformed to member directions, as follows:

$$\mathbf{A}_{\mathrm{ML}k} = -\mathbf{R}_k\mathbf{S}_{\mathrm{MS}kk}\mathbf{D}_k = -\mathbf{R}_k\mathbf{S}_{\mathrm{MS}kk}\,\mathbf{T}_{\ell k}^{\mathrm{T}}\mathbf{R}_\ell^{\mathrm{T}}\mathbf{F}_{a\ell\ell}\mathbf{A}_\ell \tag{e}$$

In this formulation the coefficient of $\mathbf{A}_\ell$ is seen to be $\mathbf{T}_{\mathrm{ML}k}$.

Similarly, a matrix formula for $\mathbf{A}_{\mathrm{ML}j}$ can be obtained from the flexibility method by choosing the fixed-end actions at j as the redundants and releasing the j end instead of the k end. This approach leads to

$$\mathbf{A}_{\mathrm{ML}j} = -\mathbf{R}_j\mathbf{S}_{\mathrm{MS}jj}\mathbf{T}_{\ell j}^{\mathrm{T}}\mathbf{R}_\ell^{\mathrm{T}}\mathbf{F}_{b\ell\ell}\mathbf{A}_\ell = \mathbf{T}_{\mathrm{ML}j}\mathbf{A}_\ell \tag{f}$$

The matrix $\mathbf{F}_{b\ell\ell}$ in this expression is similar to $\mathbf{F}_{\mathrm{M}kk}$ in Eq. (6-50), but the central angle ϕ is replaced by ϕ_b (see Fig. 6-17) and the signs for elements 1,2 and 2,3 are reversed. Also, the translational operator $\mathbf{T}_{\ell j}$ is similar to $\mathbf{T}_{jk}$, but $x_{\ell j}$ and $y_{\ell j}$ replace x_{jk} and y_{jk}. Since the coefficients of $\mathbf{A}_\ell$ in Eqs. (e) and (f) are $\mathbf{T}_{\mathrm{ML}k}$ and $\mathbf{T}_{\mathrm{ML}j}$, they can be used in Eq. (6-27) to give

$$\mathbf{T}_{\mathrm{ML}} = \begin{bmatrix} \mathbf{T}_{\mathrm{ML}j} \\ \mathbf{T}_{\mathrm{ML}k} \end{bmatrix} = \begin{bmatrix} -\mathbf{R}_j\mathbf{S}_{\mathrm{MS}jj}\,\mathbf{T}_{\ell j}^{\mathrm{T}}\mathbf{R}_\ell^{\mathrm{T}}\mathbf{F}_{b\ell\ell} \\ -\mathbf{R}_k\mathbf{S}_{\mathrm{MS}kk}\mathbf{T}_{\ell k}^{\mathrm{T}}\mathbf{R}_\ell^{\mathrm{T}}\mathbf{F}_{a\ell\ell} \end{bmatrix} \tag{6-59}$$

Furthermore, principles of static equilibrium can be used to show that

$$\mathbf{T}_{\mathrm{ML}j} = -\mathbf{R}_j(\mathbf{T}_{jk}\mathbf{R}_k^{\mathrm{T}}\mathbf{T}_{\mathrm{ML}k} + \mathbf{T}_{j\ell}\mathbf{R}_\ell^{\mathrm{T}}) \tag{6-60a}$$

and

$$\mathbf{T}_{\mathrm{ML}k} = -\mathbf{R}_k(\mathbf{T}_{kj}\mathbf{R}_j^{\mathrm{T}}\mathbf{T}_{\mathrm{ML}j} + \mathbf{T}_{k\ell}\mathbf{R}_\ell^{\mathrm{T}}) \tag{6-60b}$$

which give $\mathbf{T}_{\mathrm{ML}j}$ and $\mathbf{T}_{\mathrm{ML}k}$ in terms of each other.

Consider now the grid type of structure, for which Fig. 6-18 shows a member i in the shape of a circular arc. Again the member lies in the x-y plane and has constant section properties. In this case, however, the

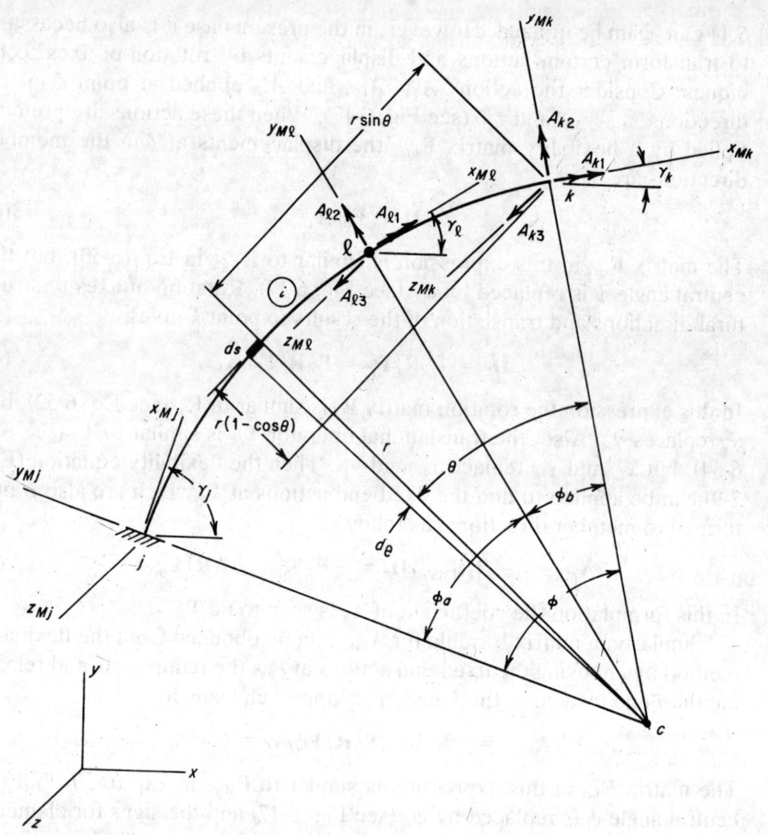

Fig. 6-18. Circular member in a grid.

actions A_{k1}, A_{k2}, and A_{k3} at the k end are two moments and a force. Due to unit values of these actions, the following bending moments and torques exist in the member:

$$\begin{array}{lll} M_{U1} = \sin\theta & M_{U2} = -\cos\theta & M_{U3} = r\sin\theta \\ T_{U1} = \cos\theta & T_{U2} = \sin\theta & T_{U3} = -r(1-\cos\theta) \end{array} \qquad \text{(g)}$$

In this case Eq. (A-31) in Appendix A-2 gives (due to flexural and torsional deformations):

$$(\mathbf{F}_{Mkk})_{11} = \int \frac{M_{U1}^2}{EI_Y}\,ds + \int \frac{T_{U1}^2}{GI_X}\,ds = \int_0^\phi \frac{\sin^2\theta}{EI_Y}\,r\,d\theta + \int_0^\phi \frac{\cos^2\theta}{GI_X}\,r\,d\theta$$

$$= \frac{r}{2EI_Y}(\phi - \sin\phi\cos\phi) + \frac{r}{2GI_X}(\phi + \sin\phi\cos\phi) \qquad \text{(h)}$$

After other flexibilities in $\mathbf{F}_{Mkk}$ are found, the matrix can be written as the sum of flexural and torsional terms, as follows:

$$\mathbf{F}_{Mkk} = (\mathbf{F}_{Mkk})_{\text{flex}} + (\mathbf{F}_{Mkk})_{\text{tors}} \tag{6-61}$$

where

$$(\mathbf{F}_{Mkk})_{\text{flex}} = \frac{r}{2EI_Y}\begin{bmatrix} \phi - \sin\phi\cos\phi & -\sin^2\phi & r(\phi - \sin\phi\cos\phi) \\ & \phi + \sin\phi\cos\phi & -r\sin^2\phi \\ \text{Sym.} & & r^2(\phi - \sin\phi\cos\phi) \end{bmatrix}$$

and (6-62a)

$$(\mathbf{F}_{Mkk})_{\text{tors}} = \frac{r}{2GI_X}\begin{bmatrix} \phi + \sin\phi\cos\phi & \sin^2\phi & r(\phi - 2\sin\phi + \sin\phi\cos\phi) \\ & \phi - \sin\phi\cos\phi & -r(2 - 2\cos\phi - \sin^2\phi) \\ \text{Sym.} & & r^2(3\phi - 4\sin\phi + \sin\phi\cos\phi) \end{bmatrix} \tag{6-62b}$$

Each of these arrays contains formulas for displacements in the directions x_{Mk}, y_{Mk}, and z_{Mk} due to unit values of A_{k1}, A_{k2}, and A_{k3}.

The determination of stiffnesses for the circular grid member is accomplished in the same manner as that described earlier for the plane frame member. However, for the grid member the translation-of-axes transformation matrix $\mathbf{T}_{jk}$ in Eq. (6-54) must be replaced by the following:

$$\mathbf{T}_{jk} = \begin{bmatrix} 1 & 0 & y_{jk} \\ 0 & 1 & -x_{jk} \\ 0 & 0 & 1 \end{bmatrix} \tag{6-63}$$

This matrix may be obtained from $\mathbf{T}_{pq}$ in Eq. (6-13) by taking the fourth, fifth, and third rows and columns (corresponding to x moment, y moment, and z force) in that order.

In addition, the matrix formulation of the fixed-end actions in $\mathbf{T}_{ML}$ for the circular grid member is symbolically the same as that for a plane frame member. In the case of the grid, however, the translational operators $\mathbf{T}_{\ell k}$ and $\mathbf{T}_{\ell j}$ are in the form of Eq. (6-63) instead of Eq. (6-54). Other comments made about the plane frame member apply also to the grid member, including the sign differences for terms in $\mathbf{F}_{a\ell\ell}$ and $\mathbf{F}_{b\ell\ell}$.

6.14 Discontinuities in Members. Figure 6-19 shows the types of partial discontinuities that can exist in the members of framed structures. The symbols in Figs. 6-19a, b, c, and d denote the inability to transmit

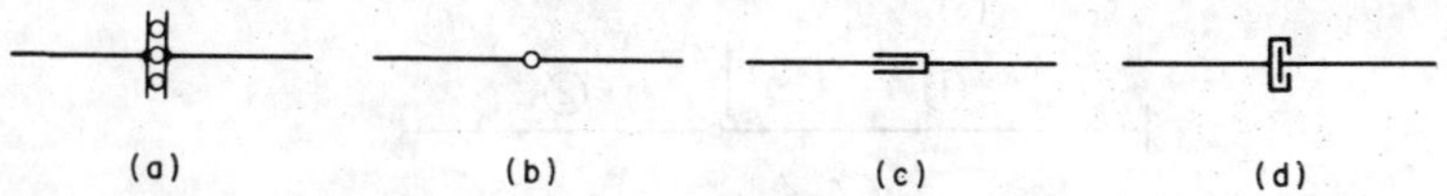

Fig. 6-19. Partial discontinuities.

shear, moment, thrust, and torque, respectively. These inabilities to transmit actions result in translational or rotational displacement discontinuities. Combinations of these discontinuities are also possible, and complete discontinuity takes the form of a free end. The general nature of a framed structure determines the types of discontinuities that are important. In beams, for example, only shear and moment discontinuities have significance. The member stiffness matrices and end-action transfer matrices for beams having such discontinuities are presented in this article, and the concepts may be extended to the members of other types of structures as well.

Consider first the possibility of a shear or moment discontinuity at one end of a prismatic beam. Figures 6-20a and 6-20b show such conditions at a small distance from the j end of a restrained member, and Figs. 6-20c and 6-20d indicate the same conditions at the k end. The member stiffness matrices $\mathbf{S}_M$ for these cases are easily obtained by analysis of the beams

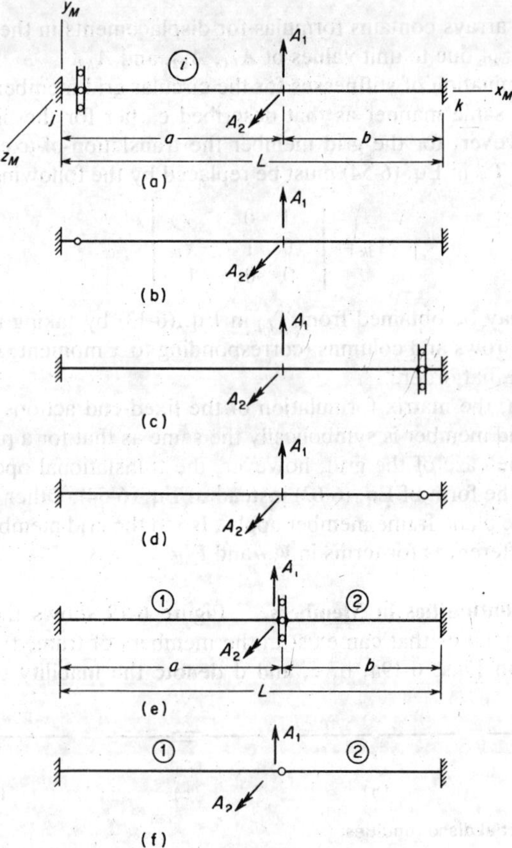

Fig. 6-20. Discontinuities in a beam member.

shown in the figure, and they are listed in Table 6-1 as matrices (a) through (d). Figures 6-20a through 6-20d also show actions A_1 and A_2 at the distance a from the j end of the member, and the transfer matrices $\mathbf{T}_{ML}$ for the resulting fixed-end actions are given in Table 6-2 as matrices (a) through (d). If the matrices $\mathbf{S}_M$ and $\mathbf{T}_{ML}$ shown in Tables 6-1 and 6-2 are used for a beam member having a shear or moment discontinuity at either end, the analysis may then proceed as before with no further modifications required.

Consider next the determination of the stiffness matrix $\mathbf{S}_M$ for a beam having a shear or moment discontinuity at an intermediate point. Figures 6-20e and 6-20f show such beams with the discontinuities at the distance a from the j end of the member. One method of analyzing these beams consists of treating them as two members by assuming a joint just to the left or right of the discontinuity. Then one of the submembers will have a discontinuity at one end (as in Figs. 6-20a through 6-20d), but the other submember will have no discontinuities. This approach has the advantage that no new matrices are required other than those discussed above. Alternatively, the stiffness matrices for beams having discontinuities at intermediate points always can be obtained directly from an elementary beam analysis. The results of such analyses for the beams in Figs. 6-20e and 6-20f are given in Table 6-1 as matrices (e) and (f).

Transfer matrices $\mathbf{T}_{ML}$ for fixed-end actions for beams with discontinuities at intermediate points may also be found by the procedures mentioned above. If actions A_1 and A_2 are applied just to the left of the discontinuities (see Figs. 6-20e and 6-20f), the transfer matrices have the form given in Table 6-2 as matrices (e) and (f). If the actions A_1 and A_2 are applied just to the right of the discontinuities, the matrices are as given in Table 6-2 as matrices (g) and (h).

Table 6-1

Member Stiffness Matrices for Beams of Fig. 6-20

$$\frac{EI_Z}{L}\begin{bmatrix} 0 & 0 & 0 & 0 \\ 0 & 1 & 0 & -1 \\ 0 & 0 & 0 & 0 \\ 0 & -1 & 0 & 1 \end{bmatrix} \qquad \frac{3EI_Z}{L^3}\begin{bmatrix} 1 & 0 & -1 & L \\ 0 & 0 & 0 & 0 \\ -1 & 0 & 1 & -L \\ L & 0 & -L & L^2 \end{bmatrix}$$

(a) (b)

$$\frac{EI_Z}{L}\begin{bmatrix} 0 & 0 & 0 & 0 \\ 0 & 1 & 0 & -1 \\ 0 & 0 & 0 & 0 \\ 0 & -1 & 0 & 1 \end{bmatrix} \qquad \frac{3EI_Z}{L^3}\begin{bmatrix} 1 & L & -1 & 0 \\ L & L^2 & -L & 0 \\ -1 & -L & 1 & 0 \\ 0 & 0 & 0 & 0 \end{bmatrix}$$

(c) (d)

$$\frac{EI_Z}{L}\begin{bmatrix} 0 & 0 & 0 & 0 \\ 0 & 1 & 0 & -1 \\ 0 & 0 & 0 & 0 \\ 0 & -1 & 0 & 1 \end{bmatrix} \qquad \frac{3EI_Z}{a^3+b^3}\begin{bmatrix} 1 & a & -1 & b \\ a & a^2 & -a & ab \\ -1 & -a & 1 & -b \\ b & ab & -b & b^2 \end{bmatrix}$$

(e) (f)

Table **6-2**

Transfer Matrices for Beams of Fig. 6-20

$$\frac{1}{2L}\begin{bmatrix} 0 & 0 \\ b^2 & -2b \\ -2L & 0 \\ b(L+a) & -2a \end{bmatrix} \qquad \frac{1}{2L^3}\begin{bmatrix} -b^2(2L+a) & 3b(L+a) \\ 0 & 0 \\ -a(3L^2-a^2) & -3b(L+a) \\ abL(L+a) & L(L^2-3a^2) \end{bmatrix}$$

(a) (b)

$$\frac{1}{2L}\begin{bmatrix} -2L & 0 \\ -a(L+b) & -2b \\ 0 & 0 \\ -a^2 & -2a \end{bmatrix} \qquad \frac{1}{2L^3}\begin{bmatrix} -b(3L^2-b^2) & 3a(L+b) \\ -abL(L+b) & L(L^2-3b^2) \\ -a^2(2L+b) & -3a(L+b) \\ 0 & 0 \end{bmatrix}$$

(c) (d)

$$\frac{1}{2L}\begin{bmatrix} -2L & 0 \\ -a(L+b) & -2b \\ 0 & 0 \\ -a^2 & -2a \end{bmatrix} \qquad \frac{1}{2(a^3+b^3)}\begin{bmatrix} -2b^3 & 3a^2 \\ -2ab^3 & a^3-2b^3 \\ 2a^3 & -3a^2 \\ 2a^3b & 3a^2b \end{bmatrix}$$

(e) (f)

$$\frac{1}{2L}\begin{bmatrix} 0 & 0 \\ b^2 & -2b \\ -2L & 0 \\ b(L+a) & -2a \end{bmatrix} \qquad \frac{1}{2(a^3+b^3)}\begin{bmatrix} -2b^3 & 3b^2 \\ -2ab^3 & 3ab^2 \\ -2a^3 & -3b^2 \\ 2a^3b & -2a^3+b^3 \end{bmatrix}$$

(g) (h)

To understand how terms in Tables 6-1 and 6-2 can be derived, consider the member in Fig. 6-20f with a hinge at an intermediate location. The method of matrix condensation (see Art. 6.12) may be applied to this problem to determine $\mathbf{S}_M$ and $\mathbf{T}_{ML}$. For that purpose, the joint stiffness matrix $\mathbf{S}_J$ is formed, using individual stiffness matrices for submembers 1 and 2 (see Fig. 6-20f), as follows:

Member 1

Member 2

$$\mathbf{S}_J = EI_Z \begin{bmatrix} \frac{12}{a^3} & \frac{6}{a^2} & -\frac{12}{a^3} & \frac{6}{a^2} & 0 & 0 \\ \frac{6}{a^2} & \frac{4}{a} & -\frac{6}{a^2} & \frac{2}{a} & 0 & 0 \\ -\frac{12}{a^3} & -\frac{6}{a^2} & \frac{12}{a^3}+\frac{3}{b^3} & -\frac{6}{a^2} & -\frac{3}{b^3} & \frac{3}{b^2} \\ \frac{6}{a^2} & \frac{2}{a} & -\frac{6}{a^2} & \frac{4}{a} & 0 & 0 \\ 0 & 0 & -\frac{3}{b^3} & 0 & \frac{3}{b^3} & -\frac{3}{b^2} \\ 0 & 0 & \frac{3}{b^2} & 0 & -\frac{3}{b^2} & \frac{3}{b} \end{bmatrix} \begin{matrix} R_1 \\ R_2 \\ F_1 \\ F_2 \\ R_3 \\ R_4 \end{matrix} \qquad \text{(a)}$$

$$\begin{matrix} R_1 & R_2 & F_1 & F_2 & R_3 & R_4 \end{matrix}$$

In this matrix the contributions from member 1 are drawn from Table 4-2, but those for member 2 are given by matrix (b) in Table 6-1. Rearrangement and partitioning of Eq. (a) yields

$$
\mathbf{S}_J = \begin{bmatrix} \mathbf{S}_{FF} & \mathbf{S}_{FR} \\ \mathbf{S}_{RF} & \mathbf{S}_{RR} \end{bmatrix} = EI_Z \left[\begin{array}{cc|cccc}
\frac{12}{a^3} + \frac{3}{b^3} & -\frac{6}{a^2} & -\frac{12}{a^3} & -\frac{6}{a^2} & -\frac{3}{b^3} & \frac{3}{b^2} \\
-\frac{6}{a^3} & \frac{4}{a} & \frac{6}{a^2} & \frac{2}{a} & 0 & 0 \\
\hline
-\frac{12}{a^3} & \frac{6}{a^2} & \frac{12}{a^3} & \frac{6}{a^2} & 0 & 0 \\
-\frac{6}{a^2} & \frac{2}{a} & \frac{6}{a^2} & \frac{4}{a} & 0 & 0 \\
-\frac{3}{b^3} & 0 & 0 & 0 & \frac{3}{b^3} & -\frac{3}{b^2} \\
\frac{3}{b^2} & 0 & 0 & 0 & -\frac{3}{b^2} & \frac{3}{b}
\end{array}\right] \qquad \text{(b)}
$$

From this form the matrices $\mathbf{T}_{ML}$ and $\mathbf{S}_M$ can be computed, using Eqs. (6-46) and (6-49) in Art. 6.12, to produce the arrays labeled (f) in Tables 6-1 and 6-2.

Note that the transfer matrices (e) through (h) in Table 6-2 are for end-actions at j and k due to unit actions applied adjacent to the discontinuities. If actions are applied at other points along the length, suitable transfer matrices can be developed for cases in which $x_{M1} \leqslant a$ and $x_{M2} \geqslant a$. One way of obtaining these transfer matrices is to make use of the matrices given in Tables 6-2. For this purpose, a temporary restraint is imposed at a point adjacent to the discontinuity, and matrices (a) through (d) are used to compute the restraint actions at that point. The negatives of these restraint actions constitute equivalent loads adjacent to the discontinuity, and matrices (e) through (h) may be used to transfer the effects to the ends of the member.

Matrices similar to those in Tables 6-1 and 6-2 can be developed for the members of other types of framed structures. Frames and grids may have such discontinuities; but the topic is of lesser importance in trusses, because the ends of all members are assumed to be pinned and any additional discontinuities are unlikely.

It is interesting to note that certain discontinuities in members may be taken into account by trivial changes in the analysis. For example, if a plane frame member has hinges at both ends, its member stiffness matrix will be the same as that for a plane truss member (expanded to size 6×6). Such a member may be handled in the computer program for a plane frame (Art. 5.8) simply by setting its cross-sectional moment of inertia I_Z equal to zero. Such "tricks" can often be used to good advantage without resorting to additional programming.

6.15 Elastic Connections. Joints of framed structures are usually idealized to be either pinned or completely rigid. However, the connections themselves may have a significant degree of flexibility that may be important in the analysis. If such connections are assumed to be linearly elastic, they may be incorporated into the stiffness properties of the individual members as modifications of the idealized cases.

Several types of elastic connections are theoretically possible, according to the relative translations and rotations that can occur at the joints of a structure. Connections for shear, bending moment, thrust, and torque may all possess a certain amount of flexibility, but the most important of these is the rotational type that transmits bending moment. This type of connection is discussed in this article in conjunction with the analysis of beams. The ideas may, of course, be extended to other types of elastic connections and other types of framed structures.

Figure 6-21a shows a prismatic beam with a rotational elastic connection at a small distance from each end. Let S_E be the stiffness constant for the elastic connection at each end of the member. A constant of this type is defined as the moment per unit relative rotation at the elastic connection. The stiffness constant S_E will be incorporated into the stiffness matrix $\mathbf{S_M}$ and the transfer matrix $\mathbf{T_{ML}}$ for the member. For convenience in writing these matrices, the following nondimensional parameters are defined:

$$e = \frac{EI_Z}{LS_E} \tag{6-64}$$

$$e_1 = e + 1 \qquad e_2 = 2e + 1 \qquad e_3 = 3e + 1$$
$$e_4 = 4e + 1 \qquad e_6 = 6e + 1$$

Terms in the modified member stiffness matrix $\mathbf{S_M}$ can be obtained using the flexibility approach explained in Art. 6.11. For this purpose the member is fixed at the j end and free at the k end, as shown in Figs. 6-21b through 6-21e. When a unit force is applied in the y_M direction at the k end (Fig. 6-21b), the flexibilities at that end become

$$F_{11} = \frac{L^3}{3EI_Z} + \frac{L^2}{S_E} = \frac{L^3}{3EI_Z} + \frac{L^3 e}{EI_Z} = \frac{L^3}{3EI_Z}(1 + 3e) = \frac{L^3}{3EI_Z} e_3 \tag{a}$$

and

$$F_{21} = \frac{L^2}{2EI_Z} + \frac{L}{S_E} = \frac{L^2}{2EI_Z} + \frac{L^2 e}{EI_Z} = \frac{L^2}{2EI_Z}(1 + 2e) = \frac{L^2}{2EI_Z} e_2 \tag{b}$$

Similarly, a unit moment in the y_M sense at the k end (Fig. 6-21c) produces

$$F_{12} = \frac{L^2}{2EI_Z} + \frac{L}{S_E} = \frac{L^2}{2EI_Z} e_2 \tag{c}$$

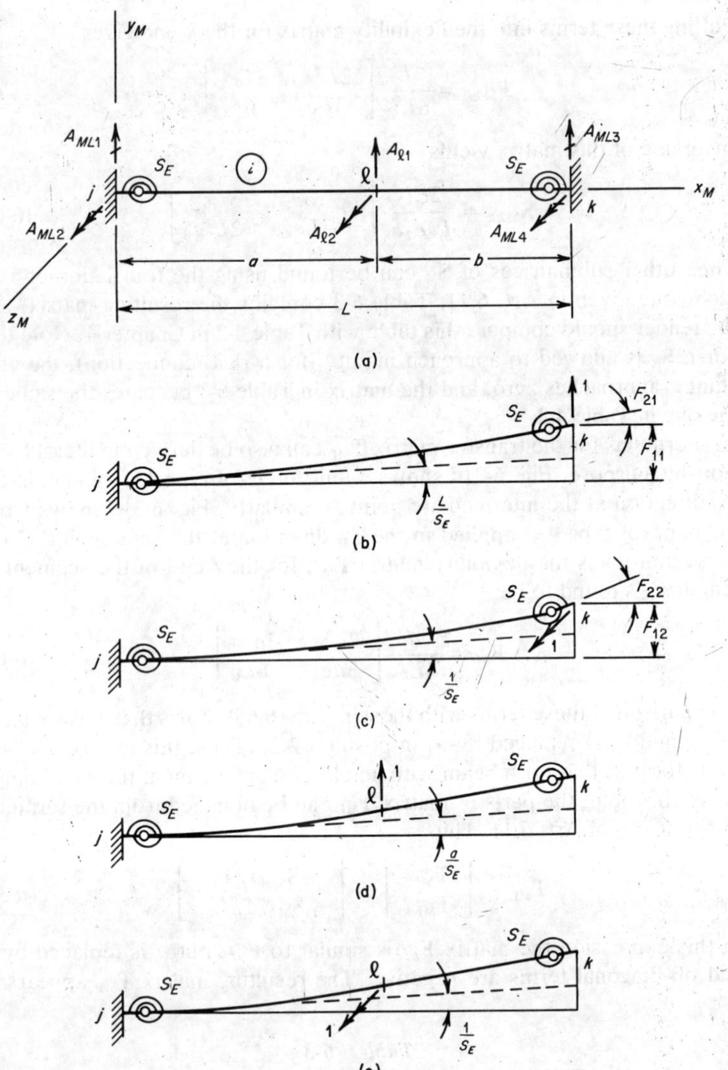

Fig. 6-21. Beam with elastic connections.

and

$$F_{22} = \frac{L}{EI_Z} + \frac{2}{S_E} = \frac{L}{EI_Z} + \frac{2Le}{EI_Z} = \frac{L}{EI_Z}\,(1 + 2e) = \frac{L}{EI_Z}\,e_2 \qquad \text{(d)}$$

Putting these terms into the flexibility matrix for the k end gives

$$\mathbf{F}_{\mathrm{M}kk} = \frac{L}{6EI_Z}\begin{bmatrix} 2L^2e_3 & 3Le_2 \\ 3Le_2 & 6e_2 \end{bmatrix} \tag{6-65}$$

Inversion of this matrix yields

$$\mathbf{S}_{\mathrm{M}kk} = \frac{2EI_Z}{L^3e_2e_6}\begin{bmatrix} 6e_2 & -3Le_2 \\ -3Le_2 & 2L^2e_3 \end{bmatrix} \tag{6-66}$$

Then other submatrices of $\mathbf{S}_{\mathrm{M}}$ can be found using the translation-of-axes technique given in Art. 6.11. Table 6-3 contains the resulting matrix, and the reader should compare this table with Table 4-2 in Chapter 4. Note that when S_{E} is allowed to approach infinity (for a rigid connection), the constant e approaches zero, and the matrix in Table 6-3 becomes the same as the one in Table 4-2.

Formulas for the transfer matrix $\mathbf{T}_{\mathrm{ML}}$ can also be derived in literal form. For this purpose, Fig. 6-21d shows a unit force (of type $A_{\ell 1}$) applied in the y_M direction at the intermediate point ℓ. Similarly, Fig. 6-21e shows a unit moment (of type $A_{\ell 2}$) applied in the z_M direction at the same point. From these unit loads the flexibility matrix $\mathbf{F}_{a\ell\ell}$, for the ℓ end of the segment of length a, is found to be:

$$\mathbf{F}_{a\ell\ell} = \frac{a}{6EI_Z}\begin{bmatrix} 2a^2e_3 & 3ae_2 \\ 3ae_2 & 6e_1 \end{bmatrix} \tag{6-67}$$

Comparison of these terms with those in Eq. (6-65) shows that L is replaced by a, and e_2 is replaced by e_1 in position 2,2. (Thus, this matrix is of the same form as that for a beam with an elastic connection at the j end only.) From this point the parts of matrix $\mathbf{T}_{\mathrm{ML}}$ can be obtained from the formulas in Eq. (6-33) of Art. 6.11. Thus,

$$\mathbf{T}_{\mathrm{ML}} = \begin{bmatrix} \mathbf{T}_{\mathrm{ML}j} \\ \mathbf{T}_{\mathrm{ML}k} \end{bmatrix} = \begin{bmatrix} -\mathbf{S}_{\mathrm{M}jj}\mathbf{T}_{\ell j}^{\mathrm{T}}\mathbf{F}_{b\ell\ell} \\ -\mathbf{S}_{\mathrm{M}kk}\mathbf{T}_{\ell k}^{\mathrm{T}}\mathbf{F}_{a\ell\ell} \end{bmatrix} \tag{6-33}$$

repeated

In this expression the matrix $\mathbf{F}_{b\ell\ell}$ is similar to $\mathbf{F}_{a\ell\ell}$, but a is replaced by b and off-diagonal terms are negative. The resulting matrix $\mathbf{T}_{\mathrm{ML}}$ appears in

Table 6-3

Member Stiffness Matrix for Beam with Elastic Connections

$$\mathbf{S}_{\mathrm{M}} = \frac{EI_Z}{e_2e_6}\left[\begin{array}{cc|cc} \frac{12}{L^3}e_2 & \frac{6}{L^2}e_2 & -\frac{12}{L^3}e_2 & \frac{6}{L^2}e_2 \\ \frac{6}{L^2}e_2 & \frac{4}{L}e_3 & -\frac{6}{L^2}e_2 & \frac{2}{L} \\ \hline -\frac{12}{L^3}e_2 & -\frac{6}{L^2}e_2 & \frac{12}{L^3}e^2 & -\frac{6}{L^2}e_2 \\ \frac{6}{L^2}e_2 & \frac{2}{L} & +\frac{6}{L^2}e_2 & \frac{4}{L}e_3 \end{array}\right]$$

Table 6-4, the parts of which are also related by equilibrium principles (see Eqs. 6-35a and b). When e is set equal to zero, elements in the table become the same as those in Eq. (6-4) for a beam with rigid joints (see Art. 6.4).

Matrices similar to those in Tables 6-3 and 6-4 may also be derived for the case of unequal elastic connections at ends j and k of the beam and for the case of an elastic connection at only one end of the beam. In addition, other types of elastic connections and other types of structural members may be considered.

6.16 Shearing Deformations. In beams, frames, and grids the deformations of members due to shear forces may be of significant magnitudes. If this is true, the effects of shearing deformations should be included in the analyses of such structures by making the appropriate modifications of the matrices $\mathbf{S}_M$ and $\mathbf{T}_{ML}$. The modified matrices for prismatic beams are presented in this article, and the changes made in the beam matrices may be easily extended to obtain the required changes in the matrices for frames and grids.

Elements of the modified stiffness matrix $\mathbf{S}_M$ for a beam member with shearing deformations can be derived in the same manner as that used in the preceding article for a beam member with elastic connections. Toward this end, Fig. 6-22a shows a cantilevered member with a unit force in the y_M direction applied at the k end. This force causes both flexural and shearing deformations in the member, as indicated in the figure. Thus, the translation at the k end may be written as

$$F_{11} = \frac{L^3}{3EI_Z} + \frac{fL}{GA_X} \tag{a}$$

where the first part is due to flexure and the second part is due to shear (see Appendix A). It is convenient to introduce a dimensionless shear constant g for the purpose of simplifying Eq. (a):

$$g = \frac{6fEI_Z}{GA_X L^2} \tag{6-68}$$

Table 6-4

Transfer Matrix for Beam with Elastic Connections

$$\mathbf{T}_{ML} = \frac{-1}{12e^2ab + e_4L^2}\begin{bmatrix} -\frac{b^2}{L}(3ae_2^2 + be_4) & \frac{6ab}{L}e_1e_2 \\ -\frac{ab^2}{L}(ae_2 + be_4) & \frac{b}{L^2}(2a^3e_1 + 3a^2be_2 - b^3e_4) \\ \hdashline -\frac{a^2}{L}(ae_4 + 3be_2^2) & -\frac{6ab}{L}e_1e_2 \\ \frac{a^2b}{L}(ae_4 + be_2) & \frac{a}{L^2}(-a^3e_4 + 3ab^2e_2 + 2b^3e_1) \end{bmatrix}$$

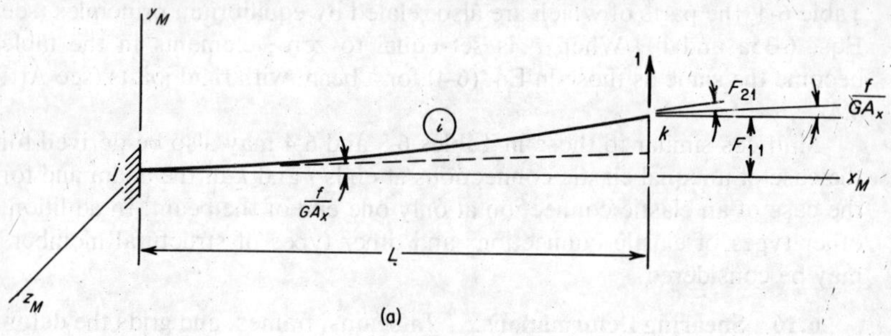

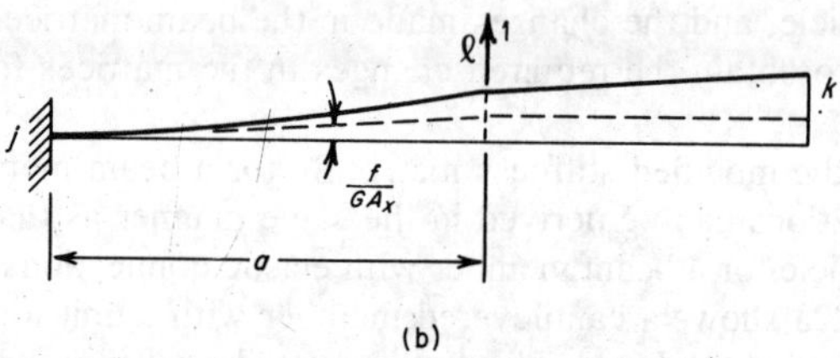

Fig. 6-22. Beam with shearing deformations.

Substitution of the form factor f (expressed in terms of g) into Eq. (a) gives

$$F_{11} = \frac{L^3}{3EI_Z} + \frac{L^3}{6EI_Z}g = \frac{L^3}{3EI_Z}\left(1 + \frac{g}{2}\right) \tag{b}$$

Due to shearing deformations, there is no rotation of the cross section at k (only translation). Therefore, the flexibility F_{21} (see Fig. 6-22a) is caused only by flexure and has the value

$$F_{21} = \frac{L^2}{2EI_Z} \tag{c}$$

By similar reasoning, a unit moment applied at k produces only flexural deformations; so the flexibility matrix becomes

$$\mathbf{F}_{Mkk} = \frac{L}{6EI_Z}\begin{bmatrix} 2L^2\left(1 + \frac{g}{2}\right) & 3L \\ 3L & 6 \end{bmatrix} \tag{6-69}$$

Inversion of $\mathbf{F}_{Mkk}$ yields

$$\mathbf{S}_{Mkk} = \frac{2EI_Z}{L^3(1 + 2g)}\begin{bmatrix} 6 & -3L \\ -3L & 2L^2\left(1 + \frac{g}{2}\right) \end{bmatrix} \tag{6-70}$$

Table 6-5
Member Stiffness Matrix for Beam with Shearing Deformations

$$\mathbf{S}_M = \frac{EI_Z}{1+2g}\left[\begin{array}{cc|cc} \frac{12}{L^3} & \frac{6}{L^2} & -\frac{12}{L^3} & \frac{6}{L^2} \\ \frac{6}{L^2} & \frac{4}{L}\left(1+\frac{g}{2}\right) & -\frac{6}{L^2} & \frac{2}{L}(1-g) \\ \hline -\frac{12}{L^3} & -\frac{6}{L^2} & \frac{12}{L^3} & -\frac{6}{L^2} \\ \frac{6}{L^2} & \frac{2}{L}(1-g) & -\frac{6}{L^2} & \frac{4}{L}\left(1+\frac{g}{2}\right) \end{array}\right]$$

Other parts of $\mathbf{S}_M$ are obtained as before, and the results are summarized in Table 6-5. Note that if the shear constant g is set equal to zero, the matrix in Table 6-5 becomes the same as that in Table 4-2.

The transfer matrix $\mathbf{T}_{ML}$ for a beam must also be modified for the effects of shearing deformations. This matrix can be developed using Fig. 6-22b, which shows the effect of a unit force (of type $A_{\ell 1}$) applied in the y_M direction at an intermediate point ℓ. Due to this force the translation at point ℓ is the same as that given by Eq. (b), except that L is replaced by a. In fact, the flexibility matrix for the ℓ end of the segment of length a is

$$\mathbf{F}_{a\ell\ell} = \frac{a}{6EI_Z}\begin{bmatrix} 2a^2\left(1+\frac{g}{2}\right) & 3a \\ 3a & 6 \end{bmatrix} \tag{6-71}$$

which is analogous to Eq. (6-69). From this point the matrix $\mathbf{T}_{ML}$ can be obtained using Eq. (6-33), and the results are listed in Table 6-6. If the constant g is set equal to zero, the elements in Table 6-6 become equal to those in Eq. (6-4).

Thus, the effects of shearing deformations in beams may be taken into account in convenient fashion by using the modified matrices $\mathbf{S}_M$ and $\mathbf{T}_{ML}$ described in this article.

6.17 Offset Connections. In structures with rigid joints the framing connections are often rather large in comparison with the lengths of the

Table 6-6
Transfer Matrix for Beam with Shearing Deformations

$$\mathbf{T}_{ML} = \frac{1}{1+2g}\left[\begin{array}{cc} -\frac{b^2}{L^3}[3a+b+2b(L/b)^2g] & \frac{6ab}{L^3} \\ -\frac{ab^2}{L^2}[1+(L/b)g] & \frac{b}{L^2}(2a-b-2gL) \\ \hline -\frac{a^2}{L^3}[a+3b+2a(L/a)^2g] & -\frac{6ab}{L^3} \\ \frac{a^2b}{L^2}[1+(L/a)g] & \frac{a}{L^2}(2b-a-2gL) \end{array}\right]$$

members. Steel building frames designed against lateral loads tend to have large moment-resisting connections, and reinforced concrete frames with deep members usually contain joints that are correspondingly bulky. If the finite dimensions of joints are ignored, significant errors can arise in the analysis and design of framed structures. Therefore, a method for including their effects is given in this article under the heading of offset connections.

Another type of offset connection is encountered when the cross section of a member is not doubly symmetric. In such a case flexural and torsional deflections may be coupled because the *shear-center axis** of the member does not coincide with the centroidal axis. Figure 6-23a shows a beam member with a channel-shaped cross section. Its shear-center axis through point O is eccentric with respect to its centroidal axis through point C. If the x_M axis of the member is taken to be the shear-center axis, forces in the x_M-y_M plane will cause the member to translate in the y_M direction without twisting. However, if applied forces are eccentric with respect to the shear-center axis, the cross section will deflect as indicated in Fig. 6-23b. In this figure it is assumed that the point O has translated in the negative y_M direction an amount D_1, and the cross section has also rotated about the x_M axis (shear-center axis) an amount D_2. This type of coupling can be avoided if the joints between members are located at the intersections of shear-center axes and if applied forces intersect those axes. Otherwise, the problem of shear-center eccentricities may be handled under the heading of offset connections.

When joints of finite size or other types of offset connections are taken to be infinitely stiff, they may be treated as rigid bodies within the structure.† Under this assumption the actions, displacements, and stiffnesses at

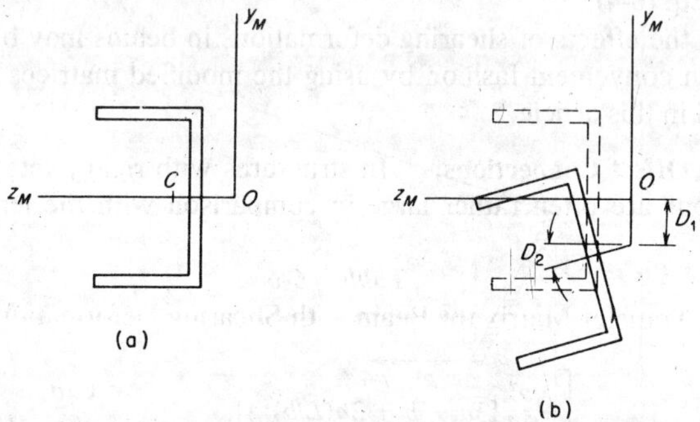

Fig. 6-23. Beam member with channel cross section.

*For a discussion of bending of unsymmetrical beams, see S. P. Timoshenko and J. M. Gere, *Mechanics of Materials*, D. Van Nostrand, New York, 1972, pp. 273–284.

†For a discussion of rigid bodies within framed structures, see W. Weaver, Jr., "Dynamics of Elastically Connected Rigid Bodies," *Developments in Theoretical and Applied Mechanics*, Vol. 3, edited by W. A. Shaw, Pergamon Press, New York, 1967, pp. 543–562.

the ends of members framing into such joints must be transformed to specified *working points* on the rigid bodies. Figure 6-24 shows a space frame member with such offset connections at both ends. Points p and q on the rigid bodies are taken as working points to which information about joints j and k will be referred. First, any actions $\mathbf{A}_j$ and $\mathbf{A}_k$ at the ends of the member may be transformed into statically equivalent actions $\mathbf{A}_p$ and $\mathbf{A}_q$ at the working points by the following generalized form of Eq. (6-13):

$$\begin{bmatrix} \mathbf{A}_p \\ \mathbf{A}_q \end{bmatrix} = \begin{bmatrix} \mathbf{T}_{pj} & \mathbf{O} \\ \mathbf{O} & \mathbf{T}_{qk} \end{bmatrix} \begin{bmatrix} \mathbf{A}_j \\ \mathbf{A}_k \end{bmatrix} \tag{6-72}$$

In the case of a space frame member, the vector $\mathbf{A}_j$ contains A_{j1}, $A_{j2}, \ldots, A_{j6}$, as indicated by the joint displacement indexes at joint j in Fig. 6-24. In addition, the vectors $\mathbf{A}_k$, $\mathbf{A}_p$, and $\mathbf{A}_q$ consist of similar actions for the points k, p, and q in the figure. The translation-of-axes transformation matrices $\mathbf{T}_{pj}$ and $\mathbf{T}_{qk}$ are both of the type given by Eq. (6-13) in Art. 6.11. Equation (6-72) may be expressed more concisely as

$$\mathbf{A}_{MB} = \mathbf{T}\mathbf{A}_{MS} \tag{6-73}$$

in which,

$$\mathbf{A}_{MS} = \{\mathbf{A}_j, \mathbf{A}_k\} \qquad \mathbf{A}_{MB} = \{\mathbf{A}_p, \mathbf{A}_q\} \tag{6-74}$$

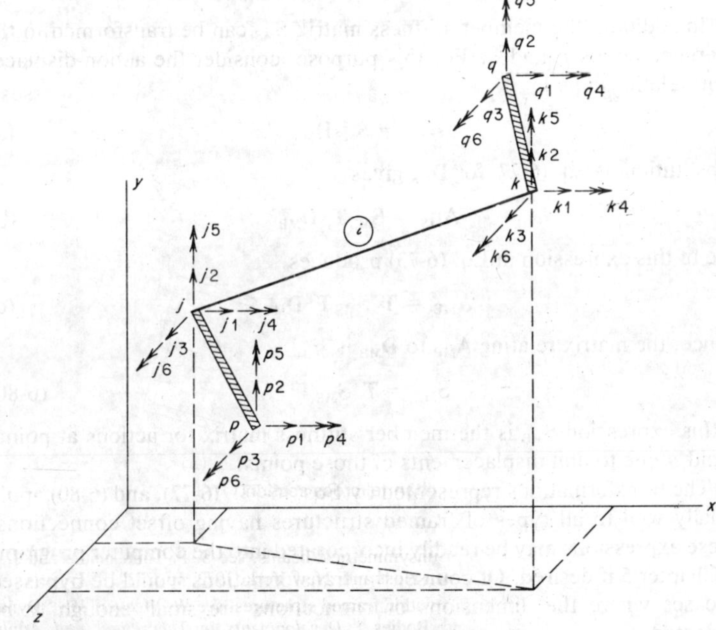

Fig. 6-24. Space frame member with offset connections.

and

$$\mathbf{T} = \begin{bmatrix} \mathbf{T}_{pj} & \mathbf{O} \\ \mathbf{O} & \mathbf{T}_{qk} \end{bmatrix} \tag{6-75}$$

The transformation matrix $\mathbf{T}$ is a combined operator that converts the actions in $\mathbf{A}_{\mathrm{MS}}$ into the statically equivalent actions in $\mathbf{A}_{\mathrm{MB}}$ for the rigid bodies. These actions are all in the directions of structural axes.

Displacements at joints j and k can also be expressed in terms of those at points p and q by an extended form of Eq. (6-17), as follows:

$$\begin{bmatrix} \mathbf{D}_j \\ \mathbf{D}_k \end{bmatrix} = \begin{bmatrix} \mathbf{T}_{pj}^{\mathrm{T}} & \mathbf{O} \\ \mathbf{O} & \mathbf{T}_{qk}^{\mathrm{T}} \end{bmatrix} \begin{bmatrix} \mathbf{D}_p \\ \mathbf{D}_q \end{bmatrix} \tag{6-76}$$

The displacement vectors in this equation correspond to the action vectors in Eq. (6-72). Equation (6-76) is expressed more succinctly as:

$$\mathbf{D}_{\mathrm{MS}} = \mathbf{T}^{\mathrm{T}}\mathbf{D}_{\mathrm{MB}} \tag{6-77}$$

in which,

$$\mathbf{D}_{\mathrm{MS}} = \{\mathbf{D}_j, \mathbf{D}_k\} \qquad \mathbf{D}_{\mathrm{MB}} = \{\mathbf{D}_p, \mathbf{D}_q\} \tag{6-78}$$

and

$$\mathbf{T}^{\mathrm{T}} = \begin{bmatrix} \mathbf{T}_{pj}^{\mathrm{T}} & \mathbf{O} \\ \mathbf{O} & \mathbf{T}_{qk}^{\mathrm{T}} \end{bmatrix} \tag{6-79}$$

In addition, the member stiffness matrix $\mathbf{S}_{\mathrm{MS}}$ can be transformed to the reference points p and q. For this purpose, consider the action-displacement relationships

$$\mathbf{A}_{\mathrm{MS}} = \mathbf{S}_{\mathrm{MS}}\mathbf{D}_{\mathrm{MS}} \tag{a}$$

Substitution of Eq. (6-77) for $\mathbf{D}_{\mathrm{MS}}$ gives

$$\mathbf{A}_{\mathrm{MS}} = \mathbf{S}_{\mathrm{MS}}\mathbf{T}^{\mathrm{T}}\mathbf{D}_{\mathrm{MB}} \tag{b}$$

Use of this expression in Eq. (6-73) produces

$$\mathbf{A}_{\mathrm{MB}} = \mathbf{T}\,\mathbf{S}_{\mathrm{MS}}\mathbf{T}^{\mathrm{T}}\mathbf{D}_{\mathrm{MB}} \tag{c}$$

Hence, the matrix relating $\mathbf{A}_{\mathrm{MB}}$ to $\mathbf{D}_{\mathrm{MB}}$ is

$$\mathbf{S}_{\mathrm{MB}} = \mathbf{T}\,\mathbf{S}_{\mathrm{MS}}\mathbf{T}^{\mathrm{T}} \tag{6-80}$$

In this expression $\mathbf{S}_{\mathrm{MB}}$ is the member stiffness matrix for actions at points p and q due to unit displacements of those points.

The transformations represented by Eqs. (6-73), (6-77), and (6-80) apply equally well to all types of framed structures having offset connections. These expressions may be readily incorporated into the computer programs of Chapter 5 if desired. Of course, the transformations would be bypassed in cases where the dimensions of connections are small enough to be neglected.

Example. A horizontal beam member within a rectangular plane frame (see Fig. 6-25) has offset connections at both ends. At the j end the offset distance is c, and at the k end the offset is d. For this example the translational operators are

$$\mathbf{T}_{pj} = \begin{bmatrix} 1 & 0 \\ c & 1 \end{bmatrix} \qquad \mathbf{T}_{qk} = \begin{bmatrix} 1 & 0 \\ -d & 1 \end{bmatrix} \tag{d}$$

Using these matrices in Eqs. (6-73), (6-77), and (6-80) produces

$$\mathbf{A}_{\mathrm{MB}} = \mathbf{TA}_{\mathrm{MS}} = \{A_{MS1},\ cA_{MS1} + A_{MS2},\ A_{MS3},\ -dA_{MS3} + A_{MS4}\} \tag{e}$$

$$\mathbf{D}_{\mathrm{MS}} = \mathbf{T}^{\mathrm{T}}\mathbf{D}_{\mathrm{MB}} = \{D_{MB1} + cD_{MB2},\ D_{MB2},\ D_{MB3} - dD_{MB4},\ D_{MB4}\} \tag{f}$$

$$\mathbf{S}_{\mathrm{MB}} = \mathbf{TS}_{\mathrm{MS}}\mathbf{T}^{\mathrm{T}}$$

$$= \frac{2EI_Z}{L^3}\begin{bmatrix} 6 & 3L + 6c & -6 & 3L + 6d \\ & 2L^2 + 6Lc + 6c^2 & -3L - 6c & L^2 + 3Lc + 3Ld + 6cd \\ & & 6 & -3L - 6d \\ & \text{Sym.} & & 2L^2 + 6Ld + 6d^2 \end{bmatrix} \tag{g}$$

6.18 Axial-Flexural Interaction. Throughout the earlier discussions in this book it was always assumed that there was no interaction between axial forces and bending moments in a structural member. This assumption is satisfactory in many ordinary situations; however, if the axial forces are large or if the member is slender, it may be necessary to make a more accurate analysis. Such an analysis must take into account the additional bending moments that are produced in a member by the axial forces when the member deflects laterally. This axial-flexural interaction, which is sometimes referred to as the *beam-column effect* or the *P-Δ effect,* results in changes in the member stiffness matrix $\mathbf{S}_{\mathrm{M}}$ and the transfer matrix $\mathbf{T}_{\mathrm{ML}}$.

Let us consider first the changes that are required in the member stiffness matrix $\mathbf{S}_{\mathrm{M}}$ for a beam member. Figure 6-26 shows a restrained beam which carries an axial compressive force P and which is subjected to a unit translation and a unit rotation at the j end (figures showing similar displacements at the k end are omitted). The axial force may be either tension or compression, but compression is usually of greater interest than tension because of the possibility of buckling. Because of the lateral deflections of

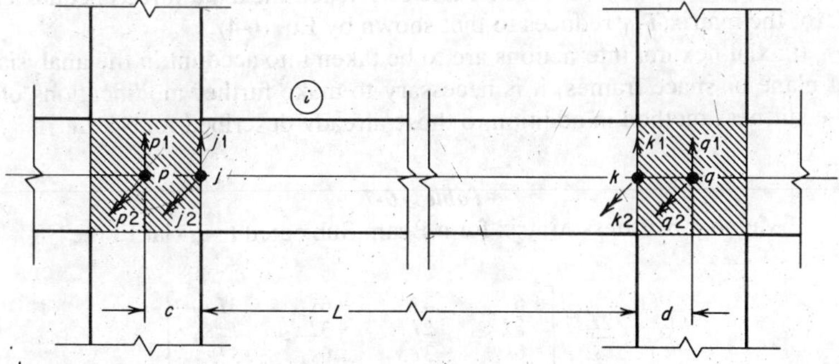

Fig. 6-25. Beam member with finite joint sizes.

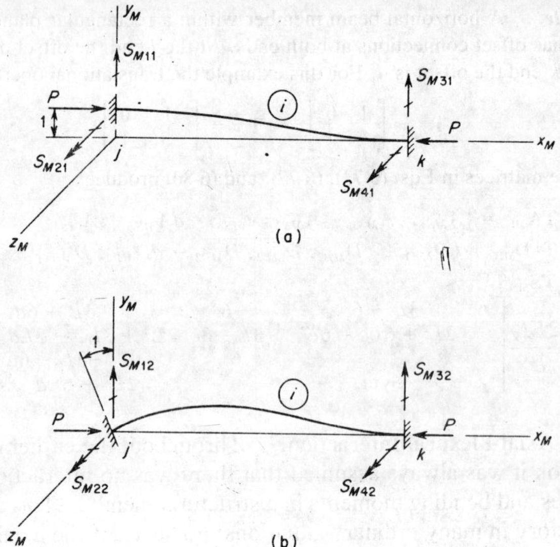

Fig. 6-26. Member stiffnesses for a beam subjected to an axial force.

the beam, as shown in the figure, the bending moments in the beam are modified by the presence of the axial force. The corresponding modified terms in the member stiffness matrix $\mathbf{S_M}$ are given in Table 6-7. Each term is expressed as the product of the stiffness without axial-flexural interaction (see Table 4-2) and a stiffness function s. The four stiffness functions are defined in Table 6-8 for axial forces which are compression, tension, and zero. All of the formulas given in this table can be derived by elementary beam analysis considering the presence of the axial force.

The elements of the transfer matrix $\mathbf{T_{ML}}$ are pictured in Fig. 6-27 for a beam subjected to an axial compressive force P. These elements may also be obtained by beam analysis, taking into account the axial-flexural interaction; the results are shown in Table 6-9. When the axial force is equal to zero, the matrix $\mathbf{T_{ML}}$ reduces to that shown by Eq. (6-4).

If axial-flexural interactions are to be taken into account in the analysis of plane or space frames, it is necessary to make further modifications of the stiffness method in addition to those already described for $\mathbf{S_M}$ and $\mathbf{T_{ML}}$.

Table 6-7

Member Stiffness Matrix for a Beam Subjected to Axial Force

$$\mathbf{S_M} = \frac{2EI_Z}{L^3}\begin{bmatrix} 6s_1 & 3Ls_2 & -6s_1 & 3Ls_2 \\ 3Ls_2 & 2L^2s_3 & -3Ls_2 & L^2s_4 \\ -6s_1 & -3Ls_2 & 6s_1 & -3Ls_2 \\ 3Ls_2 & L^2s_4 & -3Ls_2 & 2L^2s_3 \end{bmatrix}$$

Table 6-8
Stiffness Functions for a Beam Subjected to Axial Force

Function	Condition of Axial Force		
	Compression	*Zero*	*Tension*
s_1	$\frac{(kL)^3 \sin kL}{12\phi_C}$	1	$\frac{(kL)^3 \sinh kL}{12\phi_T}$
s_2	$\frac{(kL)^2(1 - \cos kL)}{6\phi_C}$	1	$\frac{(kL)^2 (\cosh kL - 1)}{6\phi_T}$
s_3	$\frac{kL(\sin kL - kL \cos kL)}{4\phi_C}$	1	$\frac{kL(kL \cosh kL - \sinh kL)}{4\phi_T}$
s_4	$\frac{kL(kL - \sin kL)}{2\phi_C}$	1	$\frac{kL(\sinh kL - kL)}{2\phi_T}$
$\phi_C = 2 - 2\cos kL - kL \sin kL$	$\phi_T = 2 - 2\cosh kL + kL \sinh kL$		$k = \sqrt{\frac{P}{EI_Z}}$

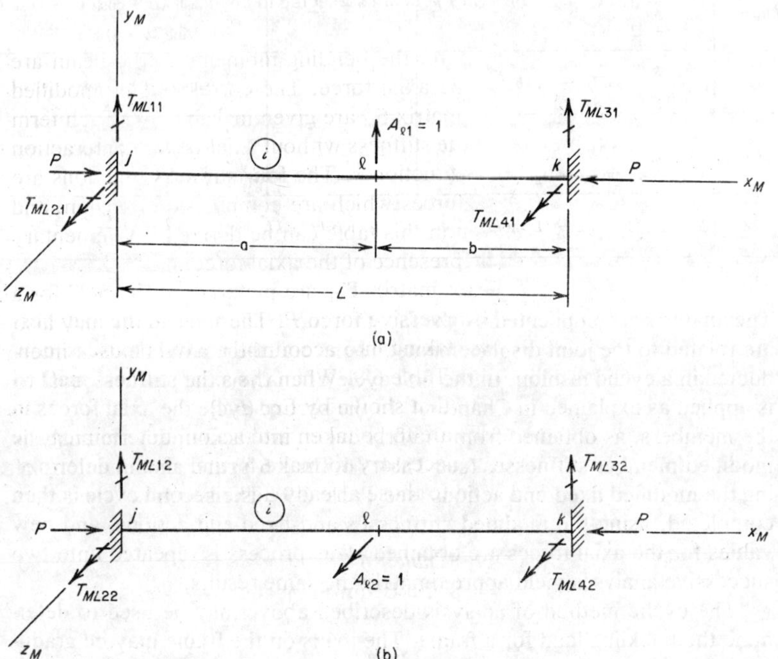

Fig. 6-27. Elements of transfer matrix T_{ML} for a beam subjected to an axial force.

Table 6-9
Elements of Transfer Matrix $\mathbf{T}_{ML}$ for a Beam Subjected to Axial Force

Element	Condition of Axial Force: Compression	Zero	Tension
T_{ML11}	$\frac{1}{\phi_C}(\cos kL - \cos ka + \cos kb + kb \sin kL - 1)$	$-\frac{b^2(3a+b)}{L^3}$	$\frac{1}{\phi_T}(\cosh kL - \cosh ka + \cosh kb\ kb \sinh kL - 1)$
T_{ML21}	$\frac{1}{k\phi_C}(\sin kL - \sin ka - \sin kb - kb \cos kL + kL \cos kb - ka)$	$-\frac{ab^2}{L^2}$	$\frac{1}{k\phi_T}(\sinh kL - \sinh ka - \sinh kb - kb \cosh kL + kL \cosh kb - ka)$
T_{ML31}	$\frac{1}{\phi_C}(\cos kL - \cos kb + \cos ka + ka \sin kL - 1)$	$-\frac{a^2(a+3b)}{L^3}$	$\frac{1}{\phi_T}(\cosh kL - \cosh kb + \cosh ka - ka \sinh kL - 1)$
T_{ML41}	$\frac{1}{k\phi_C}(-\sin kL + \sin kb + \sin ka + ka \cos kL - kL \cos ka + kb)$	$\frac{a^2b}{L^2}$	$\frac{1}{k\phi_T}(-\sinh kL + \sinh kb + \sinh ka + ka \cosh kL - kL \cosh ka + kb)$
T_{ML12}	$\frac{k}{\phi_C}(\sin ka + \sin kb - \sin kL)$	$\frac{6ab}{L^3}$	$\frac{k}{\phi_T}(-\sinh ka - \sinh kb + \sinh kL)$
T_{ML22}	$\frac{1}{\phi_C}(\cos kL + \cos kb - \cos ka + kL \sin kb - 1)$	$\frac{b(2a-b)}{L^2}$	$\frac{1}{\phi_T}(\cosh kL + \cosh kb - \cosh ka - kL \sinh kb - 1)$
T_{ML32}	$\frac{k}{\phi_C}(-\sin ka - \sin kb + \sin kL)$	$-\frac{6ab}{L^3}$	$\frac{k}{\phi_T}(\sinh ka + \sinh kb - \sinh kL)$
T_{ML42}	$\frac{1}{\phi_C}(\cos kL + \cos ka - \cos kb + kL \sin ka - 1)$	$\frac{a(2b-a)}{L^2}$	$\frac{1}{\phi_T}(\cosh kL + \cosh ka - \cosh kb - kL \sinh ka - 1)$
	$\phi_C = 2 - 2\cos kL - kL \sin kL$	$\phi_T = 2 - 2\cosh kL + kL \sinh kL$	$k = \sqrt{\frac{P}{EI_z}}$

The analysis is complicated by the fact that the axial forces in the members are related to the joint displacements. Therefore, the analysis must be conducted in a cyclic fashion. In the first cycle of analysis the stiffness method is applied as explained in Chapter 4. In the second cycle the axial forces in the members, as obtained from the first cycle, are used in determining the modified member stiffnesses (see Tables 6-7 and 6-8) and also in determining the modified fixed-end actions (see Table 6-9). The second cycle is then completed, using the modified stiffnesses and fixed-end actions, and new values for the axial forces are obtained. This process is repeated until two successive analyses yield approximately the same results.

The cyclic method of analysis described above may be used to determine the buckling load for a frame. The loads on the frame may be gradually increased until the stiffness matrix $\mathbf{S}_{FF}$ becomes singular. This singularity is the criterion for obtaining the magnitude of loading that causes elastic instability in the fundamental buckling mode.

Selected References

A. Elementary Structural Mechanics

1. **Laursen, H. I.,** *Structural Analysis,* 2nd ed., McGraw-Hill, New York, 1978.
2. **Norris, C. H., Wilbur, J. B.,** and **Utku, S.,** *Elementary Structural Analysis,* 3rd ed., McGraw-Hill, New York, 1976.

B. Matrix Analysis of Structures

1. **Beaufait, F. W., Rowan, W. H., Jr., Hoadley, P. G.,** and **Hackett, R. M.,** *Computer Methods of Structural Analysis,* Prentice-Hall, Englewood Cliffs, New Jersey, 1970.
2. **Hall, A. S.,** and **Woodhead, R. W.,** *Frame Analysis,* 2nd ed., Wiley, New York, 1967.
3. **Kardestuncer, H.,** *Elementary Matrix Analysis of Structures,* McGraw-Hill, New York, 1974.
4. **Laursen, H. I.,** *Matrix Analysis of Structures,* McGraw-Hill, New York, 1966.
5. **Martin, H. C.,** *Introduction to Matrix Methods of Structural Analysis,* McGraw-Hill, New York, 1966.
6. **McGuire, W.,** and **Gallagher, R.H.,** *Matrix Structural Analysis,* Wiley, New York, 1979.
7. **Meek, J. L.,** *Matrix Structural Analysis,* McGraw-Hill, New York, 1971.
8. **Rubinstein, M. F.,** *Matrix Computer Analysis of Structures,* Prentice-Hall, Englewood Cliffs, New Jersey, 1966.
9. **Rubinstein, M. F.,** *Structural Systems—Statics, Dynamics, and Stability,* Prentice-Hall, Englewood Cliffs, New Jersey, 1970.
10. **Vanderbilt, M. D.,** *Matrix Structural Analysis,* Quantum, New York, 1974.
11. **Wang, C. K.,** *Matrix Methods of Structural Analysis,* 2nd ed., International, Scranton, Pennsylvania, 1970.
12. **Willems, N.,** and **Lucas, W. M., Jr.,** *Matrix Analysis for Structural Engineers,* Prentice-Hall, Englewood Cliffs, New Jersey, 1968.

C. Finite Element Method

1. **Cook, R. D.,** *Concepts and Applications of Finite Element Analysis,* Wiley, New York, 1974.
2. **Desai, C. S.,** and **Abel, J. F.,** *Introduction to the Finite Element Method,* D. Van Nostrand, New York, 1972.
3. **Gallagher, R. H.,** *Finite Element Analysis Fundamentals,* Prentice-Hall, Englewood Cliffs, New Jersey, 1975.

4. **Martin, H. C.**, and **Carey, G. F.**, *Introduction to Finite Element Analysis*, McGraw-Hill, New York, 1973.

5. **Norrie, D. H.**, and **deVries, G.**, *The Finite Element Method*, Academic Press, New York, 1973.

6. **Oden, J. T.**, *Finite Elements of Nonlinear Continua*, McGraw-Hill, New York, 1972.

7. **Przemieniecki, J. S.**, *Theory of Matrix Structural Analysis*, McGraw-Hill, New York, 1968.

8. **Zienkiewicz, O. C.**, *The Finite Element Method*, 3rd ed., McGraw-Hill, London, 1977.

D. Numerical Methods

1. **Bathe, K. J.**, and **Wilson, E. L.**, *Numerical Methods in Finite Element Analysis*, Prentice-Hall, Englewood Cliffs, New Jersey, 1976.

2. **Fox, L.**, *An Introduction to Numerical Linear Algebra*, Oxford University Press, New York, 1965.

E. Computer Programming

1. **Hammond, R. H.**, **Rogers, W. B.**, and **Houck, B., Jr.**, *Introduction to FORTRAN IV*, 2nd ed., McGraw-Hill, New York, 1978.

2. **McCracken, D. D.**, *A Guide to FORTRAN IV Programming*, 2nd ed., Wiley, New York, 1972.

Notation

Table N-1
Matrices Used in the Flexibility Method (Arts. 2.1-2.5)

Matrix	*Order*	*Definition*
$\mathbf{Q}$	$q \times 1$	Unknown redundant actions (q = number of redundants)
$\mathbf{D}_{Q}$	$q \times 1$	Displacements in the actual structure corresponding to the redundants
$\mathbf{D}_{QL}$	$q \times 1$	Displacements in the released structure corresponding to the redundants and due to loads
$\mathbf{F}$	$q \times q$	Displacements in the released structure corresponding to the redundants and due to unit values of the redundants (flexibility coefficients)
$\mathbf{D}_{QT}$, $\mathbf{D}_{QP}$, $\mathbf{D}_{QR}$	$q \times 1$	Displacements in the released structure corresponding to the redundants and due to temperature, prestrain, and restraint displacements (other than those in $\mathbf{D}_{Q}$)
$\mathbf{D}_{QC}$	$q \times 1$	$\mathbf{D}_{QC} = \mathbf{D}_{QL} + \mathbf{D}_{QT} + \mathbf{D}_{QP} + \mathbf{D}_{QR}$
$\mathbf{D}_{J}$	$j \times 1$	Joint displacements in the actual structure (j = number of joint displacements)
$\mathbf{D}_{JL}$	$j \times 1$	Joint displacements in the released structure due to loads
$\mathbf{D}_{JQ}$	$j \times q$	Joint displacements in the released structure due to unit values of the redundants
$\mathbf{D}_{JT}$, $\mathbf{D}_{JP}$, $\mathbf{D}_{JR}$	$j \times 1$	Joint displacements in the released structure due to temperature, prestrain, and restraint displacements (other than those in $\mathbf{D}_{Q}$)
$\mathbf{D}_{JC}$	$j \times 1$	$\mathbf{D}_{JC} = \mathbf{D}_{JL} + \mathbf{D}_{JT} + \mathbf{D}_{JP} + \mathbf{D}_{JR}$
$\mathbf{A}_{M}$	$m \times 1$	Member end-actions in the actual structure (m = number of end-actions)
$\mathbf{A}_{ML}$	$m \times 1$	Member end-actions in the released structure due to loads
$\mathbf{A}_{MQ}$	$m \times q$	Member end-actions in the released structure due to unit values of the redundants
$\mathbf{A}_{R}$	$r \times 1$	Reactions in the actual structure (r = number of reactions)
$\mathbf{A}_{RL}$	$r \times 1$	Reactions in the released structure due to loads
$\mathbf{A}_{RQ}$	$r \times q$	Reactions in the released structure due to unit values of the redundants

Table N-2

Matrices Used in the Formalized Flexibility Method (Arts. 2.6-2.7)

Matrix	*Definition*
$\mathbf{F}_{Mi}$	Flexibilities at *k* end of member *i* (in member directions)
$\mathbf{A}_{Mi}$	Actions at *k* end of member *i* (in member directions)
$\mathbf{D}_{Mi}$	Displacements at *k* end of member *i* relative to *j* end (in member directions)
$\mathbf{F}_{M}$	Unassembled flexibility matrix (in member directions)
$\mathbf{A}_{M}$	Actions at *k* ends of all members (in member directions)
$\mathbf{D}_{M}$	Displacements at *k* ends of all members relative to *j* ends (in member directions)
$\mathbf{A}_{S}$	Actions in structural directions
$\mathbf{A}_{J}$	Actions at joints
$\mathbf{A}_{Q}$	Redundant actions
$\mathbf{B}_{MS}$	Actions $\mathbf{A}_{M}$ due to unit actions $\mathbf{A}_{S}$
$\mathbf{B}_{MJ}$	Actions $\mathbf{A}_{M}$ due to unit actions $\mathbf{A}_{J}$
$\mathbf{B}_{MQ}$	Actions $\mathbf{A}_{M}$ due to unit actions $\mathbf{A}_{Q}$
$\mathbf{D}_{S}$	Displacements in structural directions
$\mathbf{D}_{J}$	Displacements at joints
$\mathbf{D}_{Q}$	Displacements corresponding to redundants
$\mathbf{F}_{S}$	Assembled flexibility matrix
$\mathbf{F}_{JJ}$	Displacements $\mathbf{D}_{J}$ due to unit actions $\mathbf{A}_{J}$
$\mathbf{F}_{JQ}$	Displacements $\mathbf{D}_{J}$ due to unit actions $\mathbf{A}_{Q}$
$\mathbf{F}_{QJ}$	Displacements $\mathbf{D}_{Q}$ due to unit actions $\mathbf{A}_{J}$
$\mathbf{F}_{QQ}$	Displacements $\mathbf{D}_{Q}$ due to unit actions $\mathbf{A}_{Q}$
$\mathbf{A}_{MF}$	Fixed-end actions (in member directions)
$\mathbf{B}_{RS}$	Reactions $\mathbf{A}_{R}$ due to unit actions $\mathbf{A}_{S}$
$\mathbf{B}_{RJ}$	Reactions $\mathbf{A}_{R}$ due to unit actions $\mathbf{A}_{J}$
$\mathbf{B}_{RQ}$	Reactions $\mathbf{A}_{R}$ due to unit actions $\mathbf{A}_{Q}$
$\mathbf{A}_{RC}$	Combined loads applied at supports

Table N-3

Matrices Used in the Stiffness Method (Arts. 3.1-3.4)

Matrix	*Order*	*Definition*
$\mathbf{D}$	$d \times 1$	Unknown joint displacements (d = number of displacements)
$\mathbf{A}_D$	$d \times 1$	Actions in the actual structure corresponding to the unknown displacements
$\mathbf{A}_{DL}$	$d \times 1$	Actions in the restrained structure corresponding to the unknown displacements and due to all loads except those that correspond to the unknown displacements
$\mathbf{S}$	$d \times d$	Actions in the restrained structure corresponding to the unknown displacements and due to unit values of the displacements (stiffness coefficients)
$\mathbf{A}_{DT}$, $\mathbf{A}_{DP}$, $\mathbf{A}_{DR}$	$d \times 1$	Actions in the restrained structure corresponding to the unknown displacements and due to temperature, prestrain, and restraint displacement
$\mathbf{A}_{DC}$	$d \times 1$	$\mathbf{A}_{DC} = \mathbf{A}_{DL} + \mathbf{A}_{DT} + \mathbf{A}_{DP} + \mathbf{A}_{DR}$
$\mathbf{A}_M$	$m \times 1$	Member end-actions in the actual structure (m = number of end-actions)
$\mathbf{A}_{ML}$	$m \times 1$	Member end-actions in the restrained structure due to all loads except those that correspond to the unknown displacements
$\mathbf{A}_{MD}$	$m \times d$	Member end-actions in the restrained structure due to unit values of the displacements
$\mathbf{A}_{MT}$, $\mathbf{A}_{MP}$, $\mathbf{A}_{MR}$	$m \times 1$	Member end-actions in the restrained structure due to temperature, prestrain, and restraint displacement
$\mathbf{A}_{MC}$	$m \times 1$	$\mathbf{A}_{MC} = \mathbf{A}_{ML} + \mathbf{A}_{MT} + \mathbf{A}_{MP} + \mathbf{A}_{MR}$
$\mathbf{A}_R$	$r \times 1$	Reactions in the actual structure (r = number of reactions)
$\mathbf{A}_{RL}$	$r \times 1$	Reactions in the restrained structure due to all loads except those that correspond to the unknown displacements
$\mathbf{A}_{RD}$	$r \times d$	Reactions in the restrained structure due to unit values of the displacements
$\mathbf{A}_{RT}$, $\mathbf{A}_{RP}$, $\mathbf{A}_{RR}$	$r \times 1$	Reactions in the restrained structure due to temperature, prestrain, and restraint displacement
$\mathbf{A}_{RC}$	$r \times 1$	$\mathbf{A}_{RC} = \mathbf{A}_{RL} + \mathbf{A}_{RT} + \mathbf{A}_{RP} + \mathbf{A}_{RR}$

Table N-4

Matrices Used in the Formalized Stiffness Method (Arts. 3.5-3.6)

Matrix	*Definition*
$\mathbf{S}_{Mi}$	Stiffnesses at k end of member i (in member directions)
$\mathbf{A}_{Mi}$	Actions at k end of member i (in member directions)
$\mathbf{D}_{Mi}$	Displacements at k end of member i relative to j end (in member directions)
$\mathbf{S}_{M}$	Unassembled stiffness matrix (in member directions)
$\mathbf{A}_{M}$	Actions at k ends of all members (in member directions)
$\mathbf{D}_{M}$	Displacements at k ends of all members relative to j ends (in member directions)
$\mathbf{D}_{J}$	Displacements at all joints
$\mathbf{D}_{F}$	Free joint displacements
$\mathbf{D}_{R}$	Restrained joint displacements
$\mathbf{C}_{MJ}$	Displacements $\mathbf{D}_{M}$ due to unit displacements $\mathbf{D}_{J}$
$\mathbf{C}_{MF}$	Displacements $\mathbf{D}_{M}$ due to unit displacements $\mathbf{D}_{F}$
$\mathbf{C}_{MR}$	Displacements $\mathbf{D}_{M}$ due to unit displacements $\mathbf{D}_{R}$
$\mathbf{A}_{J}$	Actions at all joints
$\mathbf{A}_{F}$	Actions at free joints
$\mathbf{A}_{R}$	Reactions at restrained joints
$\mathbf{S}_{J}$	Assembled joint stiffness matrix
$\mathbf{S}_{FF}$	Actions $\mathbf{A}_{F}$ due to unit displacements $\mathbf{D}_{F}$
$\mathbf{S}_{FR}$	Actions $\mathbf{A}_{F}$ due to unit displacements $\mathbf{D}_{R}$
$\mathbf{S}_{RF}$	Reactions $\mathbf{A}_{R}$ due to unit displacements $\mathbf{D}_{F}$
$\mathbf{S}_{RR}$	Reactions $\mathbf{A}_{R}$ due to unit displacements $\mathbf{D}_{R}$
$\mathbf{A}_{ML}$	Member end-actions due to loads (in member directions)
$\mathbf{A}_{RC}$	Combined loads applied at supports

Table N-5

Matrices Used in the Computer-Oriented Direct Stiffness Method (Chapters 4, 5, and 6)

Matrix	*Definition*
$\mathbf{S}_{\mathrm{M}i}$	Member stiffnesses (for both ends of member *i*) in directions of member axes
$\mathbf{S}_{\mathrm{M}jj}$	Submatrix *jj* of $\mathbf{S}_{\mathrm{M}i}$
$\mathbf{S}_{\mathrm{M}jk}$	Submatrix *jk* of $\mathbf{S}_{\mathrm{M}i}$
$\mathbf{S}_{\mathrm{M}kj}$	Submatrix *kj* of $\mathbf{S}_{\mathrm{M}i}$
$\mathbf{S}_{\mathrm{M}kk}$	Submatrix *kk* of $\mathbf{S}_{\mathrm{M}i}$
$\mathbf{S}_{\mathrm{MS}i}$	Member stiffnesses (for both ends of member *i*) in directions of structural axes
$\mathbf{A}_{\mathrm{MS}i}$	Fixed-end actions (for both ends of member *i*) in directions of structural axes
$\mathbf{D}_{\mathrm{MS}i}$	Displacements (for both ends of member *i*) in directions of structural axes
$\mathbf{A}_{\mathrm{E}}$	Equivalent joint loads
$\mathbf{A}_{\mathrm{C}}$	Combined joint loads
$\mathbf{A}_{\mathrm{FC}}$	Combined joint loads corresponding to $\mathbf{D}_{\mathrm{F}}$
$\mathbf{A}_{\mathrm{RC}}$	Combined joint loads corresponding to $\mathbf{D}_{\mathrm{R}}$
$\mathbf{R}_{i}$	Rotation matrix for member *i*
$\mathbf{R}_{\mathrm{T}i}$	Rotation transformation matrix for member *i*
$\mathbf{D}_{\mathrm{J}i}$	Joint displacements at ends of member *i*
$\mathbf{A}_{\mathrm{MD}i}$	End-actions (for both ends of member *i*) in member directions, due to joint displacements
$\mathbf{A}_{\mathrm{RD}}$	Support reactions due to joint displacements
$\mathbf{T}_{\mathrm{ML}i}$	Transfer matrix for fixed-end actions due to unit values of concentrated loads
$\mathbf{A}_{\ell i}$	Concentrated loads at point ℓ between the ends of member *i*
$\mathbf{R}_{\mathrm{R}}$	Rotation transformation matrix for structure
$\mathbf{A}_{p}$	Actions at point *p*
$\mathbf{A}_{q}$	Actions at point *q*
$\mathbf{T}_{pq}$	Translation-of-axes transformation matrix
$\mathbf{D}_{p}$	Displacements at point *p*
$\mathbf{D}_{q}$	Displacements at point *q*
$\mathbf{T}_{jk}$	Specialization of $\mathbf{T}_{pq}$ to points *j* and *k*
$\mathbf{F}_{\mathrm{M}jj}$	Flexibilities for *j* end of member *i* (in member directions)
$\mathbf{F}_{\mathrm{M}kk}$	Flexibilities for *k* end of member *i* (in member directions)
$\mathbf{F}_{a\ell\ell}$	Flexibilities for ℓ end of segment $j\ell$ (in member directions)
$\mathbf{F}_{b\ell\ell}$	Flexibilities for ℓ end of segment ℓk (in member directions)
$\mathbf{A}_{\mathrm{MB}}$	Actions $\{\mathbf{A}_p, \mathbf{A}_q\}$ for rigid bodies
$\mathbf{D}_{\mathrm{MB}}$	Displacements $\{\mathbf{D}_p, \mathbf{D}_q\}$ for rigid bodies
$\mathbf{T}$	Combined translation-of-axes operator

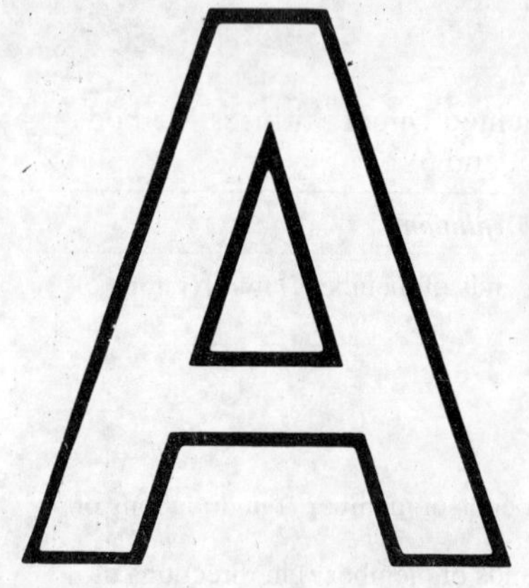

Displacements of Framed Structures

A.1 Stresses and Deformations in Slender Members. Whenever a load is applied to a structure, stresses will be developed within the material, and deformations will occur. Deformation means any change in the shape of some part of the structure, such as a change in shape of an element cut from a member, while stresses refer to the distributed actions that occur internally between adjoining elements of the structure. It is assumed in subsequent analyses that the deformations are very small and that the material is linearly elastic (Hooke's law). Under these conditions the stresses are proportional to the corresponding strains in the material, and the principle of superposition may be used for combining stresses, strains, and deformations due to various load systems.

The principal types of deformations to be considered are axial, flexural, torsional, and shearing deformations. These are caused by the corresponding stress resultants, which are axial forces, bending moments, torsional moments, and shearing forces, respectively. In each of these four cases the expressions for the stresses acting on the cross section, the strains in an element, and the deformation of an element are summarized in this article. In addition, deformations caused by temperature effects are described.

The calculation of displacements in structures is described in Articles A.2 and A.3. This subject is an important part of the flexibility method of analysis (see Chapter 2), and is presented in this Appendix for review purposes. Further information on the subject may be found in textbooks on mechanics of materials and elementary theory of structures.

Axial Deformations. The slender member shown in Fig. A-1a is assumed to be acted upon by a tensile force P at each end. The member will be in pure tension due to these forces, provided each force acts at the centroid of the cross-sectional area. At any distance x from the left end the tension stress σ_x on the cross section is

$$\sigma_x = \frac{P}{A} \tag{A-1}$$

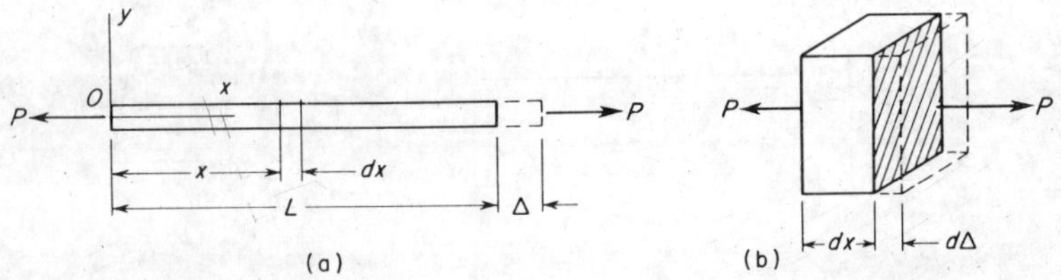

Fig. A-1. Axial deformations.

in which A is the cross-sectional area. The axial strain ϵ_x in the member is equal to the stress divided by the modulus of elasticity E of the material:

$$\epsilon_x = \frac{\sigma_x}{E} = \frac{P}{EA} \tag{A-2}$$

The quantity EA is called the *axial rigidity* of the member.

The change in length $d\Delta$ of an element of initial length dx is indicated in Fig. A-1b and is given by the formula

$$d\Delta = \epsilon_x\, dx = \frac{P}{EA}\, dx \tag{A-3}$$

The total elongation Δ of the member shown in Fig. A-1a is obtained by integration of $d\Delta$ over the length L, as follows:

$$\Delta = \int d\Delta = \int_0^L \frac{P}{EA}\, dx \tag{A-4}$$

If the member is prismatic and E is constant, the integration of Eq. (A-4) gives

$$\Delta = \frac{PL}{EA} \tag{A-5}$$

This equation can be used to calculate the change in length of a prismatic member subjected to a constant axial force.

If the axial force P varies along the length of the member, Eq. (A-4) can still be used. All that is necessary is to express P as a function of x and then perform the integration. If the member is tapered slightly, then A must be expressed as a function of x, after which the integration can be carried out.

Flexural Deformations. A member subjected to pure bending moment produced by couples M acting at each end is shown in Fig. A-2a. It is assumed that the plane of bending (the x-y plane) is a plane of symmetry of the beam, and hence the y axis is an axis of symmetry of the cross-sectional area (see Fig. A-2b). This requirement also means that the y and z axes are principal axes through point O, which is selected at the centroid of the cross section. With the bending moments M acting as shown in Fig. A-2, it

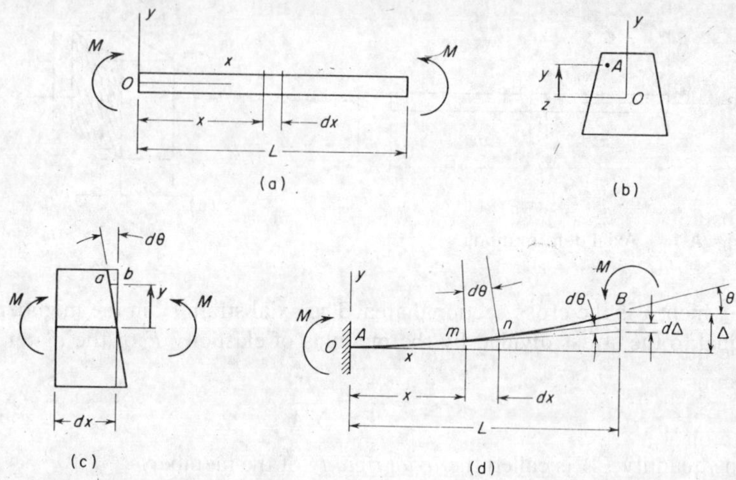

Fig. A-2. Flexural deformations.

follows that all displacements of the beam will be in the x-y plane. If the cross section of the beam is not symmetric about the y axis, the bending analysis becomes more complicated since bending will no longer occur in a single plane. It then becomes necessary to take the origin O at the shear center of the cross section and to take the y and z axes parallel to the principal centroidal axes. Then the beam is analyzed for bending in both principal planes, as well as for torsion; and the results are combined to give the final stresses and displacements.

At any cross section of the beam the flexural stress σ_x is given by the formula

$$\sigma_x = -\frac{My}{I_Z} \tag{A-6}$$

in which y is the distance from the neutral axis (the z axis) to any point A in the cross section (see Fig. A-2b), and I_Z is the moment of inertia of the cross-sectional area with respect to the z axis. The flexural strain ϵ_x at the same point is equal to the flexural stress divided by the modulus of elasticity; therefore,

$$\epsilon_x = \frac{\sigma_x}{E} = -\frac{My}{EI_Z} \tag{A-7}$$

The minus sign appears in Eqs. (A-6) and (A-7) because positive bending moment M produces negative stresses (compression) in the region where y is positive.

The relative angle of rotation $d\theta$ between two cross sections is shown in Fig. A-2c. For small angles of rotation this angle can be found by dividing the shortening *ab* of a fiber at distance y from the neutral axis by the distance y itself. Since the distance *ab* is equal to $-\epsilon_x dx$, the expression for $d\theta$ becomes

$$d\theta = \frac{-\epsilon_x dx}{y}$$

Substitution of Eq. (A-7) into this equation results in

$$d\theta = \frac{M}{EI_Z} dx \tag{A-8}$$

The quantity EI_Z in the denominator is called the *flexural rigidity* of the beam.

Expression (A-8) can sometimes be used to calculate angles of rotation and displacements of beams. An example of this kind is shown in Fig. A-2d, where it is assumed that the left end A of the beam in pure bending is fixed to a support and does not rotate. The angle of rotation θ of end B may be determined by integration of $d\theta$ (see Eq. A-8) along the length of the member. The expression for θ is

$$\theta = \int d\theta = \int_0^L \frac{M}{EI_Z} dx \tag{A-9}$$

in which dx is the length of the small element *mn* of the beam. If the member is prismatic and E is constant, integration of Eq. (A-9) gives the following expression for the angle at B for pure bending:

$$\theta = \frac{ML}{EI_Z} \tag{A-10}$$

However, Eq. (A-9) may also be used with good accuracy for cases in which the bending moment varies along the beam or in which the member is slightly tapered. The procedure is to substitute the appropriate expressions for M and I_Z into Eq. (A-9) before performing the integration.

The deflection Δ at end B of the beam (Fig. A-2d) is seen to consist of the summation of the small distances $d\Delta$, each of which is an intercept on the vertical through B of the tangent lines from points m and n. Thus, for small angles of rotation the intercept $d\Delta$ is

$$d\Delta = (L - x)d\theta$$

or, using Eq. (A-8),

$$d\Delta = (L - x)\frac{M}{EI_Z} dx \tag{A-11}$$

Integration over the length of the member gives the total displacement Δ for a prismatic beam, as follows:

$$\Delta = \int d\Delta = \int_0^L (L - x)\frac{M}{EI_Z}\,dx = \frac{ML^2}{2EI_Z} \tag{A-12}$$

This example for a beam in pure bending requires only very simple calculations in order to find the displacement. The same technique can be used if either M or the flexural rigidity EI_Z varies along the length. Under more general conditions it is necessary to use other methods for determining displacements of beams, such as the unit-load method described in Art. A.2.

Torsional Deformations. The deformations caused by pure torsion of a member having a circular cross section are illustrated in Fig. A-3. The member has a length L and is subjected to twisting moments T at its ends (see Fig. A-3a).* The deformation of an element located at distance x from one end of the member (Fig. A-3b) consists of a relative rotation about the x axis of one cross section with respect to another. The relative angle of rotation is denoted by $d\phi$ in the figure.

Associated with the deformation are shearing stresses τ and shearing strains γ. The torsional shearing stresses are directly proportional to the distance from the longitudinal axis; and, at a distance r from the axis (see Fig. A-3b), the stress intensity is given by the formula

$$\tau = \frac{Tr}{J} \tag{A-13}$$

The term J is the polar moment of inertia of the circular cross section; thus, J is equal to $\pi R^4/2$, where R is the radius of the member. The maximum shear stress occurs at the outer surface of the member and is obtained from the formula

$$\tau_{\max} = \frac{TR}{J} \tag{A-14}$$

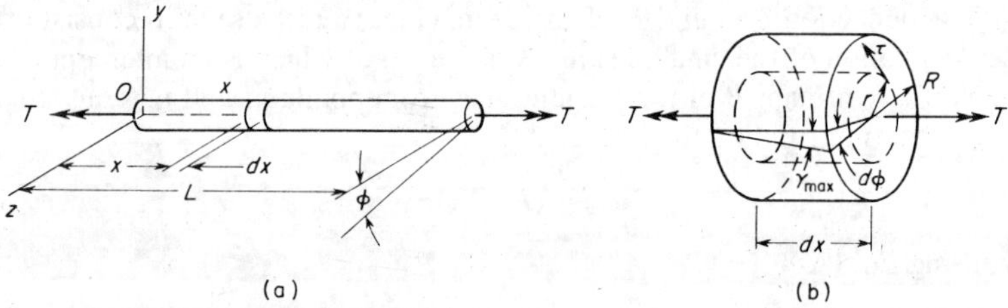

Fig. A-3. Torsional deformations.

*In this book a double-headed arrow (or vector) is used to represent a moment. The sense of the moment is according to the right-hand rule of signs.

The shear stresses acting on the circular cross sections are always in a tangential direction (normal to the radius) and in the same sense as the torque T.

The shear strain γ at the radius r is equal to the shear stress divided by the shearing modulus of elasticity G of the material; therefore,

$$\gamma = \frac{\tau}{G} = \frac{Tr}{GJ} \tag{A-15}$$

The expression for the maximum shear strain (see Fig. A-3b) is

$$\gamma_{max} = \frac{TR}{GJ} \tag{A-16}$$

The quantity GJ appearing in this formula is called the *torsional rigidity* of the member.

The relative angle of rotation $d\phi$ between the cross sections of the element in Fig. A-3b is

$$d\phi = \frac{\gamma_{max}}{R}\,dx$$

as can be seen from the geometry of the figure. This expression takes the following form when Eq. (A-16) is substituted:

$$d\phi = \frac{T}{GJ}\,dx \tag{A-17}$$

From Eq. (A-17) the total angle of twist ϕ (see Fig. A-3a) can be found by integration of $d\phi$ over the length of the member. The result is

$$\phi = \int d\phi = \int_0^L \frac{T}{GJ}\,dx \tag{A-18}$$

which for a cylindrical member with constant torque T becomes

$$\phi = \frac{TL}{GJ} \tag{A-19}$$

All of the formulas given above may be used for either a solid or a tubular circular member. Of course, in the latter case J must be taken equal to the polar moment of inertia of the annular cross section.

It should be noted that Eq. (A-18) may be used for a member that is subjected to a torque T which varies along its length. It may also be used when J varies, provided the variation is gradual over the range of integration. In either of these cases the expression for T or J as a function of x is substituted into Eq. (A-18) before the integration is performed.

If the cross section of the member is not circular or annular, the torsion analysis is more complicated than the one described above. However, for pure torsion in which the twisting moment T is constant along the length, the formula for the angle of twist (see Eq. A-19) can still be used with good accuracy, provided that J is taken as the appropriate torsion constant for the particular cross section. Torsion constants for several shapes of cross sections are tabulated in Appendix C.

If the cross section of the bar is not circular, there will be warping of the cross sections. Warping refers to the longitudinal displacement of points in the cross section, so that the cross section is no longer planar. Warping occurs in the case of I beams and channel beams, as well as most other sections, and a more complicated analysis can be made. However, in such cases it is usually found that the analysis based upon pure torsion alone, with the warping effects neglected, gives acceptable results. Torsion in which warping occurs is called *nonuniform torsion*.*

Shearing Deformations. There are usually shearing forces as well as bending moments acting on the cross sections of a beam. For example, at distance x from the fixed support of the cantilever beam shown in Fig. A-4a there will be a bending moment M (assumed positive when the top of the beam is in compression) given by the equation

$$M = -P(L - x) \tag{A-20}$$

The positive direction for M is shown in Fig. A-4b, which shows an element of length dx from the beam. The shearing force V is constant throughout the length of the beam and is obtained in this example from the expression

$$V = P \tag{A-21}$$

It is assumed that positive shearing force is downward on the right-hand side of the element and upward on the left-hand side (see Fig. A-4b). In a more general case, the shearing force V as well as the bending moment M will vary along the length of the beam.

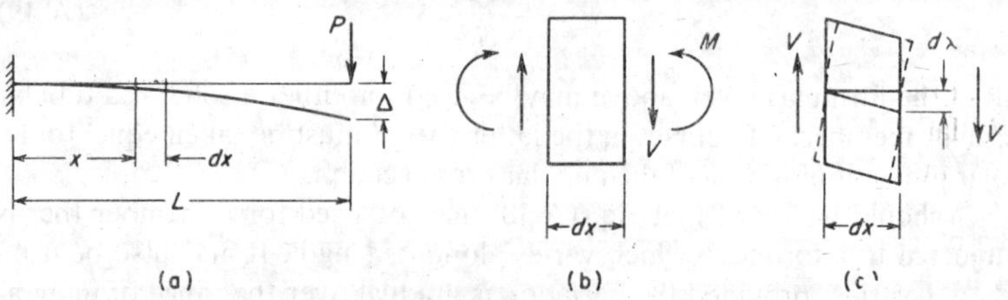

Fig. A-4. Shearing deformations.

*For the theory of nonuniform torsion of beams, see Oden, J. T., *Mechanics of Elastic Structures*, McGraw-Hill, New York, 1967, Chapter 7.

The shearing stresses on the cross section of a beam with rectangular cross section, due to a shear force V, can be found from the formula

$$\tau = \frac{VQ}{I_Z b} \tag{A-22}$$

in which Q is the first moment (or static moment) about the neutral axis of the portion of the cross-sectional area which is outside of the section where the shear stress is to be determined; I_Z is the moment of inertia of the cross-sectional area about the neutral axis; and b is the width of the rectangular beam. Equation (A-22) can be used to find shear stresses in a few other shapes of beams; for example, it can be used to calculate the shear stresses in the web of an I beam, provided that b is taken as the thickness of the web of the beam. On the other hand, it cannot be used to calculate stresses in a beam of circular cross section. The shearing strain γ can be found by dividing the shear stress τ by the shearing modulus of elasticity G.

Previously, the deformation of an element of a beam due solely to the action of bending moments was considered (see Fig. A-2c). In the present discussion, only the deformations caused by the shearing forces V will be taken into account. These deformations consist of a relative displacement $d\lambda$ of one face of the element with respect to the other (see Fig. A-4c). The displacement $d\lambda$ is given by the expression

$$d\lambda = f\frac{V dx}{GA} \tag{A-23}$$

in which A is the area of the cross section and f is a form factor* that is dependent upon the shape of the cross section. Values of the form factor for several shapes of cross section are given in Appendix C. The quantity GA/f is called the *shearing rigidity* of the bar.

The presence of shearing deformations $d\lambda$ in the elements of a beam means that the total displacement at any other point along the beam will be influenced by both flexural and shearing deformations. Usually the effects of shear are small compared to the effects of bending and can be neglected; however, if it is desired to include the shearing deformations in the calculations of displacements, it is possible to do so by using the unit-load method, as described in Art. A.2.

In a few elementary cases the deflections due to shearing deformations can be calculated by a direct application of Eq. (A-23). This equation can be used, for example, in calculating the deflection Δ at the end of the cantilever beam in Fig. A-4a. The portion Δ_s of the total deflection that is due

*See S. P. Timoshenko and J. M. Gere, *Mechanics of Materials*, D. Van Nostrand, New York, 1972, pp. 370–372.

solely to the effect of shearing deformations is equal to (see Eqs. A-21 and A-23)

$$\Delta_s = \int d\lambda = \frac{fP}{GA}\int_0^L dx = \frac{fPL}{GA} \tag{A-24}$$

The remaining part Δ_b of the deflection is due to bending and can be found by integrating Eq. (A-11). However, Eq. (A-11) was derived on the basis that upward deflection Δ was positive, whereas in Fig. A-4a the deflection Δ is downward. Thus, it is necessary to change the sign of the expression appearing in Eq. (A-11); the result is

$$\Delta_b = \int_0^L -(L-x)\frac{M}{EI_Z}dx$$

The bending moment M for the beam is given by Eq. (A-20) and, when this expression is substituted into the above equation, the following result is obtained:

$$\Delta_b = \int_0^L \frac{(L-x)^2P}{EI_Z}dx = \frac{PL^3}{3EI_Z} \tag{A-25}$$

Summing the deflections due to both bending and shearing deformations gives the total deflection Δ, as follows:

$$\Delta = \Delta_b + \Delta_s = \frac{PL^3}{3EI_Z} + \frac{fPL}{GA} \tag{A-26}$$

From this equation it is found that the ratio of the shear deflection to the bending deflection is $3fEI_Z/GAL^2$. This ratio is very small compared to unity except in the case of short, deep beams; hence, in most cases it can be neglected.

Temperature Deformations. When the temperature of a structure varies, there is a tendency to produce changes in the shape of the structure. The resulting deformations and displacements may be of considerable importance in the analysis of the structure. In order to obtain formulas for the deformation due to temperature changes, consider the member shown in Fig. A-5a. A uniform temperature change throughout the member results in an increase in its length by the amount

$$\Delta = \alpha LT \tag{A-27}$$

in which Δ is the change in length (positive sign denotes elongation); α is the coefficient of thermal expansion; L is the length of the member; and T is the change in temperature (positive sign means increase in temperature). In addition, all other dimensions of the member will be changed proportionately, but only the change in length will be of importance in the analysis of framed structures.

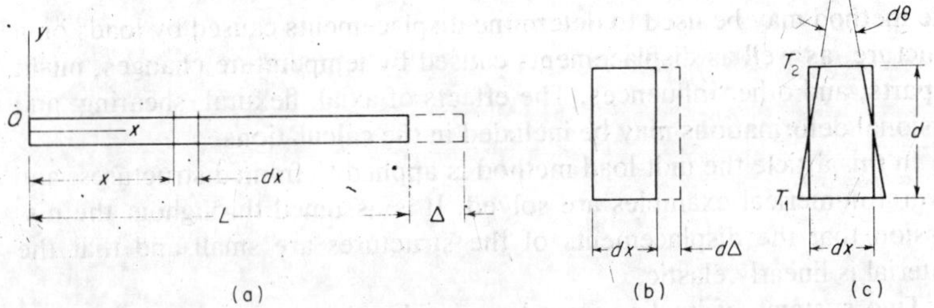

Fig. A-5. Temperature deformations.

The deformation of an element of the member of length dx (Fig. A-5a) will be analogous to that for the entire member. The longitudinal deformation of the element is shown in Fig. A-5b and is seen to be of the same type as that caused by an axial force (see Fig. A-1b). This deformation is given by the expression

$$d\Delta = \alpha T\, dx \tag{A-28}$$

Equation (A-28) can be used in calculating displacements of structures due to uniform temperature changes, as described in the next article.

A temperature differential between two sides of a beam causes each element of the member to deform as shown in Fig. A-5c. If the temperature varies linearly between T_2 at the top of the beam and a higher temperature T_1 at the bottom of the beam, the cross sections of the member will remain plane, as illustrated in the figure. The relative angle of rotation $d\theta$ between the sides of the element is

$$d\theta = \frac{\alpha(T_1 - T_2)\,dx}{d} \tag{A-29}$$

in which d is the depth of the beam. The deformation represented by the angle $d\theta$ in Eq. (A-29) is similar to that caused by bending moments acting on the member (see Fig. A-2c). The use of Eq. (A-29) in finding beam displacements will be shown also in Art. A.2.

The formulas in the preceding paragraphs have been presented for reference use in the solution of problems and examples throughout this book. For a more complete treatment of stresses and deformations, a textbook on mechanics of materials should be consulted.

A.2 Displacements by the Unit-Load Method. The unit-load method derived in Art. 1.14 is a very general and versatile technique for calculating displacements of structures. It may be used (in theory) for either determinate or indeterminate structures, although for practical calculations it is applied almost exclusively to statically determinate structures because its use requires that the stress resultants be known throughout the structure.

The method may be used to determine displacements caused by loads on a structure, as well as displacements caused by temperature changes, misfit of parts, and other influences. The effects of axial, flexural, shearing, and torsional deformations may be included in the calculations.

In this article the unit-load method is applied to framed structures, and several numerical examples are solved. It is assumed throughout the discussion that the displacements of the structures are small and that the material is linearly elastic.

Two systems of loading must be considered when using the unit-load method. The first system consists of the structure in its actual condition; that is, subjected to the actual loads, temperature changes, or other effects. The second system consists of the same structure subjected to a unit load corresponding to the desired displacement in the actual structure. The unit load is a fictitious or dummy load and is introduced solely for purposes of analysis. By a unit load corresponding to the displacement is meant a load at the particular point of the structure where the displacement is to be determined and acting in the positive direction of that displacement. The term "displacement" is used here in the generalized sense, as discussed in Art. 1.4. That is, a displacement may be the translation of a point on the structure, the angle of rotation of the axis of a member, or a combination of translations and rotations. If the displacement to be calculated is a translation, then the unit load is a concentrated force at the point where the displacement occurs. In addition, the unit load must be in the same direction as the displacement and have the same positive sense. If the displacement to be calculated is a rotation, then the unit load is a moment at the point where the rotation occurs and is assumed positive in the same sense as a positive rotation. Similarly, if the displacement is the relative translation of two points along the line joining them, the unit load consists of two collinear and oppositely directed forces acting at the two points; and if the displacement is a relative rotation between two points, the unit load consists of two equal and oppositely directed moments at the points.

In Art. 1.14 the principle of complementary virtual work was specialized to the unit-load method (see Eq. 1-43), which may be stated in words as

$$\text{(unit virtual load)(unknown displacement)} = \int_V \text{(virtual stresses)(real strains)}\, dV$$

where the left-hand side is external work and the right-hand side is internal work. For the slender members of framed structures, however, integration over volume may be replaced by integration over length by working with virtual stress resultants and the corresponding internal displacements. The virtual stress resultants caused by the unit load will be represented by the symbols N_U, M_U, T_U, and V_U, denoting the axial force, bending moment, twisting moment, and shearing force, respectively, at any cross section in the members of the structure.

The corresponding incremental displacements will be denoted as $d\Delta$ for axial deformation (see Fig. A-1b), $d\theta$ for flexural deformation (Fig. A-2c), $d\phi$ for torsional deformation (Fig. A-3b), and $d\lambda$ for shearing deformation (Fig. A-4c). Thus, the internal work of the virtual stress resultants and the corresponding incremental displacements for an infinitesimal element may be written as

$$N_U d\Delta + M_U d\theta + T_U d\phi + V_U d\lambda$$

The first term in this expression is the work done by the axial force N_U (produced by the unit load) when the displacement $d\Delta$ (due to the actual loads or other effects) is imposed on the element. A similar statement can be made about each of the other terms. Then the total work of the virtual internal actions is

$$\int N_U d\Delta + \int M_U d\theta + \int T_U d\phi + \int V_U d\lambda$$

where the integrations are carried out over the lengths of all members of the structure.

The external work done by the unit load is

$$(1)\ \Delta$$

in which Δ represents the desired unknown displacement. Equating the work of the external and internal actions gives the equation of the unit-load method for framed structures as

$$\Delta = \int N_U d\Delta + \int M_U d\theta + \int T_U d\phi + \int V_U d\lambda \qquad \text{(A-30)}$$

Because the unit load has been divided from the left-hand side of Eq. (A-30), leaving only the term Δ, it is necessary to consider the quantities N_U, M_U, T_U, and V_U as having the dimensions of force or moment per unit of the applied unit load.

The quantities $d\Delta$, $d\theta$, $d\phi$, and $d\lambda$ appearing in Eq. (A-30) can be expressed in terms of the properties of the structure. The expression for $d\Delta$ when the axial deformations are caused by loads only is (compare with Eq. A-3)

$$d\Delta = \frac{N_L dx}{EA}$$

in which N_L represents the axial force in the member due to the actual loads on the structure. Similarly, if the deformations are caused by a uniform temperature increase, the expression for $d\Delta$ is (see Eq. A-28)

$$d\Delta = \alpha T\, dx$$

in which α is the coefficient of thermal expansion and T is the temperature change. The expressions for the remaining deformation quantities due to loads (compare with Eqs. A-8, A-17, and A-23) are

$$d\theta = \frac{M_L dx}{EI} \qquad d\phi = \frac{T_L dx}{GJ} \qquad d\lambda = \frac{fV_L dx}{GA}$$

The quantities M_L, T_L, and V_L represent the bending moment, twisting moment, and shearing force caused by the loads. If there is a temperature differential across the beam, Eq. (A-29) can be used for the deformation $d\theta$.

When the relations given above for the deformations due to loads only are substituted into Eq. (A-30), the equation for the displacement takes the form

$$\Delta = \int \frac{N_U N_L dx}{EA} + \int \frac{M_U M_L dx}{EI} + \int \frac{T_U T_L dx}{GJ} + \int \frac{fV_U V_L dx}{GA} \qquad \text{(A-31)}$$

Each term in this equation represents the effect of one type of deformation on the total displacement Δ that is to be found. In other words, the first term represents the displacement caused by axial deformations; the second term represents the displacement caused by bending deformations; and so forth for the remaining terms. The sign conventions used for the quantities appearing in Eq. (A-31) must be consistent with one another. Thus, the axial forces N_U and N_L must be obtained according to the same convention; for example, tension is positive. Similarly, the bending moments M_U and M_L must have the same sign convention, as also must T_U and T_L, and V_U and V_L. Only if the sign conventions are consistent will the displacement Δ have the same positive sense as the unit load.

The procedure for calculating a displacement by means of Eq. (A-31) is as follows: (1) determine forces and moments in the structure due to the loads (that is, obtain N_L, M_L, T_L, and V_L); (2) place a unit load on the structure corresponding to the displacement Δ that is to be found; (3) determine forces and moments in the structure due to the unit load (that is, find N_U, M_U, T_U, and V_U); (4) form the products shown in Eq. (A-31) and integrate each term for the entire structure; and (5) sum the results to obtain the total displacement.

Usually not all of the terms given in Eq. (A-31) are required for the calculation of displacements. In a truss with hinged joints and with loads acting only at the joints, there will be no bending, torsional, or shearing deformations. Furthermore, if each member of the truss is prismatic, the cross-sectional area A will be a constant for each member. In such a case the equation for Δ can be written as

$$\Delta = \sum \frac{N_U N_L L}{EA} \qquad \text{(A-32)}$$

in which L represents the length of a member. The summation is carried out for all members of the truss.

In a beam it is quite likely that only bending deformations are important Therefore, the equation for the displacement simplifies to

$$\Delta = \int \frac{M_U M_L dx}{EI} \tag{A-33}$$

In an analogous manner it is possible to calculate displacements by using any appropriate combination of terms from Eq. (A-31), depending upon the nature of the structure and the degree of refinement required for the analysis. Other terms can be used when displacements due to temperature changes, prestrains, etc., are to be found. All that is necessary is to substitute into Eq. (A-30) the appropriate expressions for the deformations. Some examples of the use of the unit-load method will now be given.

Example 1. The truss shown in Fig. A-6a is subjected to loads P and $2P$ at joint A. All members of the truss are assumed to have the same axial rigidity EA. The horizontal displacement Δ_1 of joint B (positive to the right) is to be calculated.

The calculations for the displacement Δ_1 by the unit-load method are given in Table A-1. The first two columns in the table identify the members of the truss and their lengths. The axial forces N_L, which are determined by static equilibrium for the truss shown in Fig. A-6a, are listed in column (3) of the table. The unit load corresponding to the displacement Δ_1 is shown in Fig. A-6b, and the resulting axial forces N_{U1} are given in column (4). Finally, the products $N_{U1}N_L L$ are obtained for each member (column 5), summed, and divided by EA. Thus, the displacement Δ_1 is (see Eq. A-32):

$$\Delta_1 = -3.828 \frac{PL}{EA}$$

The negative sign in this result means that Δ_1 is in the direction opposite to the unit load (i.e., to the left).

A similar procedure can be used to find any other displacement of the truss. For example, suppose that it is desired to determine the relative displacement Δ_2 of joints A and D (see Fig. A-6a) along the line joining them. The corresponding unit

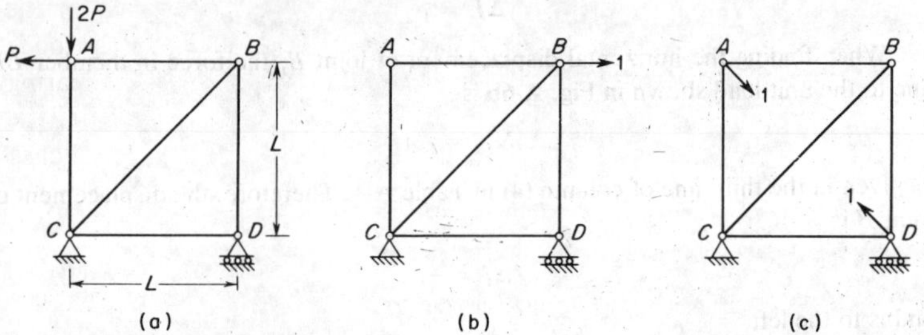

Fig. A-6. Examples 1 and 2.

Table A-1

Member	*Length*	N_L	N_{U1}	$N_{U1}N_LL$	N_{U2}	$N_{U2}N_LL$
(1)	(2)	(3)	(4)	(5)	(6)	(7)
AB	L	P	0	0	$-1/\sqrt{2}$	$-PL/\sqrt{2}$
AC	L	$-2P$	0	0	$-1/\sqrt{2}$	$2PL/\sqrt{2}$
BD	L	P	-1	$-PL$	$-1/\sqrt{2}$	$-PL/\sqrt{2}$
CD	L	0	0	0	$-1/\sqrt{2}$	0
CB	$\sqrt{2}L$	$-\sqrt{2}P$	$\sqrt{2}$	$-2\sqrt{2}PL$	1	$-2PL$
				$-3.828PL$		$-2PL$

load consists of two unit forces, as shown in Fig. A-6c. The resulting axial forces N_{U2} in the truss are listed in column (6) of Table A-1, and the products $N_{U2}N_LL$ are given in column (7). Thus, the relative displacement of joints A and D is

$$\Delta_2 = -2\frac{PL}{EA}$$

in which the minus sign indicates that the distance between points A and D has increased (that is, it is opposite to the sense of the unit loads).

Example 2. Consider again the truss shown in Fig. A-6a, and assume now that member BD has been fabricated with a length which is greater by an amount e than the theoretical length L. The horizontal displacement Δ_1 of joint B and the relative displacement Δ_2 between points A and D are to be determined (see Figs. A-6b and A-6c for the corresponding unit loads).

Any displacement of the truss caused by the increased length of member BD can be found by using Eq. (A-30) and retaining only the first term on the right-hand side. For a truss the equation may be expressed in the form

$$\Delta = \sum N_U(\Delta L)$$

in which the summation is carried out for all members of the truss and ΔL represents the change in length of any member. In this example the only member in the truss of Fig. A-6a which has a change in length is member BD itself, and hence there is only one term in the summation. For member BD, the term ΔL is

$$\Delta L = e$$

When finding the horizontal displacement of joint B, the force in member BD due to the unit load shown in Fig. A-6b is

$$N_{U1} = -1$$

as given in the third line of column (4) in Table A-1. Therefore, the displacement of joint B is

$$\Delta_1 = -e$$

and is to the left.

When the decrease in distance between joints A and D due to the lengthening of member BD is to be found, the value for N_{U2} becomes

$$N_{U2} = -\frac{1}{\sqrt{2}}$$

as given in column (6) of Table A-1; therefore,

$$\Delta_2 = -\frac{e}{\sqrt{2}}$$

The negative sign for Δ_2 shows that joints A and D move apart from one another.

A uniform temperature change in one or more members of the truss is handled in the same manner as a change in length. The only difference is that the change in length ΔL now is given by Eq. (A-27). Thus, the horizontal displacement of joint B due to a temperature increase of T degrees in member BD becomes

$$\Delta_1 = -\alpha LT$$

and the change in distance between points A and D becomes

$$\Delta_2 = -\frac{\alpha LT}{\sqrt{2}}$$

Example 3. The cantilever beam AE shown in Fig. A-7a is subjected to loads at points B, C, D and E. The translational displacements Δ_1 and Δ_2 of the beam (positive upward) at points C and E, respectively, are to be determined.

The displacements Δ_1 and Δ_2 can be found from Eq. (A-33), which is expressed in terms of the bending moments M_L and M_U. The former moments are due to the actual loads on the beam, and the latter are due to unit loads corresponding to the desired displacements. Expressions for M_L and M_U must be obtained for each segment of the beam between applied loads, and then these expressions are substituted into Eq. (A-33) to obtain the displacements.

The required calculations are shown in Table A-2. The first two columns of the

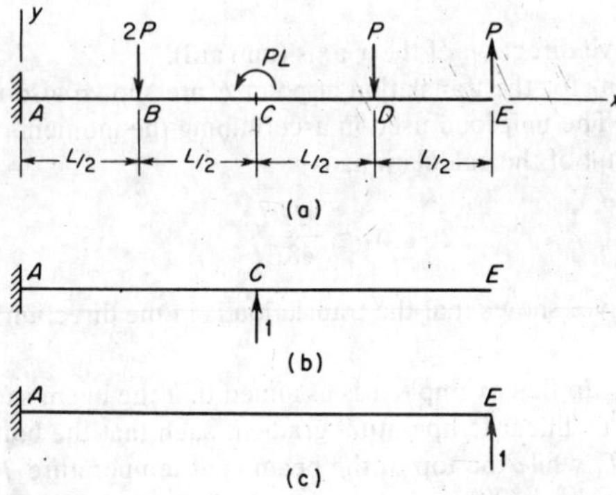

Fig. A-7. Example 3.

Table A-2

Segment	Limits for x	M_L	Unit Load at C		Unit Load at E	
			M_{U1}	$\int M_{U1} M_L dx$	M_{U2}	$\int M_{U2} M_L dx$
(1)	*(2)*	*(3)*	*(4)*	*(5)*	*(6)*	*(7)*
AB	0 to $\frac{L}{2}$	$\frac{P}{2}(4x+L)$	$L-x$	$\frac{17PL^3}{48}$	$2L-x$	$\frac{41PL^3}{48}$
BC	$\frac{L}{2}$ to L	$\frac{3PL}{2}$	$L-x$	$\frac{3PL^3}{16}$	$2L-x$	$\frac{15PL^3}{16}$
CD	L to $\frac{3L}{2}$	$\frac{PL}{2}$	0	0	$2L-x$	$\frac{3PL^3}{16}$
DE	$\frac{3L}{2}$ to $2L$	$P(2L-x)$	0	0	$2L-x$	$\frac{PL^3}{24}$
				$\frac{13PL^3}{24}$		$\frac{97PL^3}{48}$

table list the segments of the beam and the limits for the distance x, which is measured from the fixed support. Column (3) gives the expressions for the bending moments in the beam due to the actual loads, assuming that compression on top of the beam corresponds to positive bending moment. The moments M_{U1} in column (4) are those caused by a unit load at point C (Fig. A-7b). These moments are evaluated according to the same sign convention that was used in determining the moments M_L. Next, column (5) shows the results of evaluating the integral given in Eq. (A-33), except that the factor EI has been omitted, since it is assumed to be the same for all segments of the beam. When the expressions in column (5) are summed, and the total is divided by EI, the result is the displacement corresponding to the unit load. Therefore, the displacement at point C in the y direction is

$$\Delta_1 = \frac{13PL^3}{24EI}$$

and is in the positive direction of the y axis (upward).

The calculations for the translation at point E are shown also in Table A-2 (see columns 6 and 7). The unit load used in ascertaining the moments M_{U2} is shown in Fig. A-7c. The result of the calculations is

$$\Delta_2 = \frac{97PL^3}{48EI}$$

which, being positive, shows that the translation is in the direction of the y-axis.

Example 4. In this example it is assumed that the beam shown in Fig. A-7a is now subjected to a linear temperature gradient such that the bottom of the beam is at temperature T_1 while the top of the beam is at temperature T_2 (see Fig. A-8). The formula for the displacements is obtained by using only the second term in the

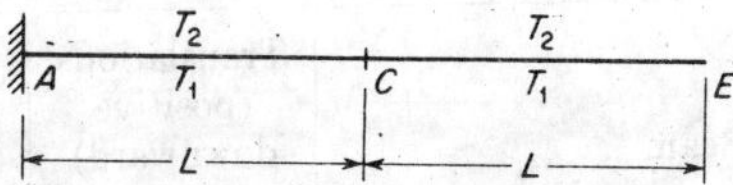

Fig. A-8. Example 4.

right-hand side of Eq. (A-30) and substituting for $d\theta$ the expression given in Eq. (A-29). Thus,

$$\Delta = \int M_U \frac{\alpha(T_1 - T_2)dx}{d}$$

The expressions for M_U that are to be substituted into this equation are given in columns (4) and (6) of Table A-2, assuming that the vertical translations Δ_1 and Δ_2 at points C and E are to be found. The calculations become as follows:

$$\Delta_1 = \frac{\alpha(T_1 - T_2)}{d}\int_0^L (L - x)dx = \frac{\alpha(T_1 - T_2)L^2}{2d}$$

$$\Delta_2 = \frac{\alpha(T_1 - T_2)}{d}\int_0^{2L} (2L - x)dx = \frac{2\alpha(T_1 - T_2)L^2}{d}$$

These results show that if T_1 is greater than T_2, the beam deflects upward.

The preceding examples illustrate the determination of translational displacements in trusses and beams due to various causes. Other types of structures can be analyzed in an analogous manner. Also, techniques that are similar to those illustrated can be used to find rotations at a point (the unit load corresponding to a rotation is a unit moment), as well as to obtain displacements due to shearing and torsional deformations, all of which are included in Eq. (A-30).

A.3 Displacements of Beams. In many of the problems and examples given in Chapters 1 and 2, it is necessary to determine displacements of beams. Such displacements can be found in all cases by the unit-load method, although other standard methods (including integration of the differential equation for displacements of a beam, and the moment-area method) may be suitable also. In most of the examples, however, the desired displacements can be obtained with the aid of the formulas given in Table A-3 for prismatic beams.

As an illustration of the use of the formulas, consider a cantilever beam with constant EI that is subjected to a concentrated load P at its midpoint (Fig. A-9). The displacement Δ at the end of the beam can be obtained readily by making the following observation: the displacement Δ is equal to the displacement at B plus the rotation at B times the distance from B to C. Thus, from Case 7 in Table A-3, the following expression is obtained:

$$\Delta = \Delta_B + \theta_B \frac{L}{2} = P\left(\frac{L}{2}\right)^3 \frac{1}{3EI} + P\left(\frac{L}{2}\right)^2 \frac{1}{2EI}\frac{L}{2} = \frac{5PL^3}{48EI}$$

Table A-3
Displacements of Prismatic Beams

Beam	Translations (positive downward)	Rotations (positive clockwise)
1	$\Delta_C = \dfrac{5wL^4}{384EI}$	$\theta_A = -\theta_B = \dfrac{wL^3}{24EI}$
2	$\Delta_C = \dfrac{PL^3}{48EI}$	$\theta_A = -\theta_B = \dfrac{PL^2}{16EI}$
3	$\Delta_C = \dfrac{23PL^3}{648EI}$	$\theta_A = -\theta_B = \dfrac{PL^2}{9EI}$
4	$\Delta_C = 0$	$\theta_A = \theta_B = \dfrac{ML}{24EI}$
5	$\Delta_C = \dfrac{ML^2}{16EI}$	$\theta_A = \dfrac{ML}{6EI}$ $\theta_B = -\dfrac{ML}{3EI}$
6	$\Delta_B = \dfrac{wL^4}{8EI}$	$\theta_B = \dfrac{wL^3}{6EI}$
7	$\Delta_B = \dfrac{PL^3}{3EI}$	$\theta_B = \dfrac{PL^2}{2EI}$
8	$\Delta_B = \dfrac{ML^2}{2EI}$	$\theta_B = \dfrac{ML}{EI}$

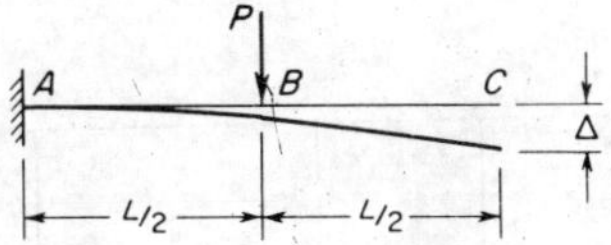

Fig. A-9.

Techniques of this kind can be very useful for finding displacements in beams and plane frames.

A.4 Integrals of Products for Computing Displacements. Equation (A-31) in Art. A.2 provides a useful tool for calculating displacements in framed structures by the unit-load method. Evaluation of the integrals of products in that equation can become repetitious, however (see Example 3 in Art. A.2), and it is possible to avoid much of the work by using a table of *product integrals*.

The members of framed structures are usually prismatic and have constant material properties along their lengths. In all such cases the rigidities may be taken outside of the integral signs, as follows:

$$\Delta = \frac{1}{EA}\int N_U N_L\,dx + \frac{1}{EI}\int M_U M_L\,dx + \frac{1}{GJ}\int T_U T_L\,dx + \frac{f}{GA}\int V_U V_L\,dx \qquad \text{(A-34)}$$

The product integrals in this expression must be evaluated over the length of each member and then added for all members. For a structure in which only flexural deformations are considered, Eq. (A-34) reduces to

$$\Delta = \frac{1}{EI}\int M_U M_L\,dx \qquad \text{(A-35)}$$

Table A-4 contains product integrals for the most commonly encountered functions (constant, linear, and quadratic). Although the results in the table are in terms of M_U and M_L (see Eq. A-35), these functions can be replaced by any others, such as N_U and N_L (see Eq. A-34).

To demonstrate the use of Table A-4, consider the simple beam in Fig. A-10a, subjected to a uniformly distributed load of intensity w. The rotation θ_B (taken as positive clockwise) at end B will be determined, assuming that EI is constant over the length. For this purpose, a unit load in the form of a moment is applied at point B (see Fig. A-10c). It is seen from Figs. A-10b and A-10d that the functions M_L and M_U are quadratic and linear, respectively. For this example, Table A-4 yields

$$\theta_B = \frac{1}{EI}\int M_U M_L\,dx = \frac{1}{EI}\left[\frac{L}{3}M_1 M_3\right] = \frac{1}{EI}\left[\frac{L}{3}(-1)\left(\frac{wL^2}{8}\right)\right] = -\frac{wL^3}{24EI}$$

The result is the same as that given in Case 1 of Table A-3.

Table A-4

Product Integrals $\int_0^L M_U \cdot M_L \, dx$

M_L \ M_U	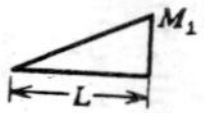 Triangle (M_1, length L)	 Trapezoid (M_1, M_2, length L)	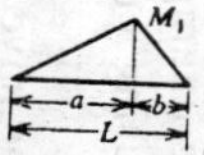 Triangle with peak M_1 (a, b, L)	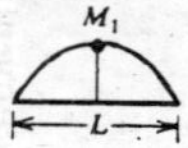 Parabola (M_1, length L)
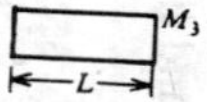 Rectangle (M_3, L)	$\frac{L}{2} M_1M_3$	$\frac{L}{2}(M_1+M_2)M_3$	$\frac{L}{2} M_1M_3$	$\frac{2L}{3} M_1M_3$
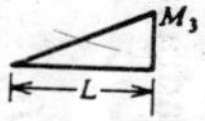 Triangle (M_3 at right, L)	$\frac{L}{3} M_1M_3$	$\frac{L}{6}(M_1+2M_2)M_3$	$\frac{L}{6}\left(1+\frac{a}{L}\right)M_1M_3$	$\frac{L}{3} M_1M_3$
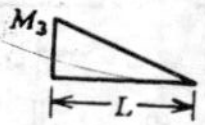 Triangle (M_3 at left, L)	$\frac{L}{6} M_1M_3$	$\frac{L}{6}(2M_1+M_2)M_3$	$\frac{L}{6}\left(1+\frac{b}{L}\right)M_1M_3$	$\frac{L}{3} M_1M_3$
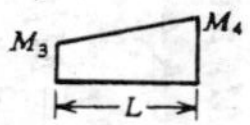 Trapezoid (M_3, M_4, L)	$\frac{L}{6} M_1(M_3+2M_4)$	$\frac{L}{6} M_1(2M_3+M_4)$ $+\frac{L}{6} M_2(M_3+2M_4)$	$\frac{L}{6}\left(1+\frac{b}{L}\right)M_1M_3$ $+\frac{L}{6}\left(1+\frac{a}{L}\right)M_1M_4$	$\frac{L}{3} M_1(M_3+M_4)$
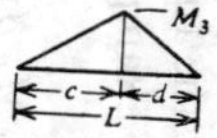 Triangle with peak M_3 (c, d, L)	$\frac{L}{6}\left(1+\frac{c}{L}\right)M_1M_3$	$\frac{L}{6}\left(1+\frac{d}{L}\right)M_1M_3$ $+\frac{L}{6}\left(1+\frac{c}{L}\right)M_2M_3$	For $c \leq a$: $\frac{L}{3} M_1M_3$ $-\frac{L(a-c)^2}{6ad} M_1M_3$	$\frac{L}{3}\left(1+\frac{cd}{L^2}\right)M_1M_3$
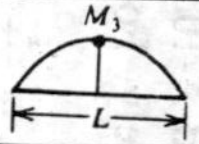 Parabola (M_3, L)	$\frac{L}{3} M_1M_3$	$\frac{L}{3}(M_1+M_2)M_3$	$\frac{L}{3}\left(1+\frac{ab}{L^2}\right)M_1M_3$	$\frac{8L}{15} M_1M_3$
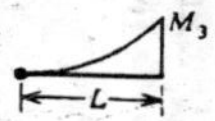 Parabolic spandrel (M_3, L)	$\frac{L}{4} M_1M_3$	$\frac{L}{12}(M_1+3M_2)M_3$	$\frac{L}{12}\left(1+\frac{a}{L}+\frac{a^2}{L^2}\right)M_1M_3$	$\frac{L}{5} M_1M_3$

Note: All curved lines are second-degree parabolas with vertices shown by black dots.

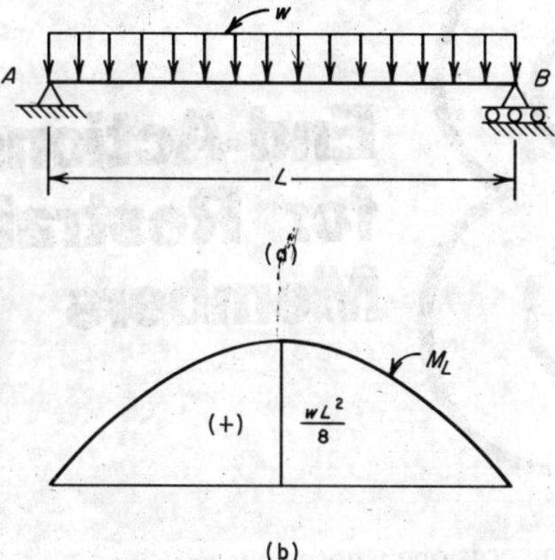

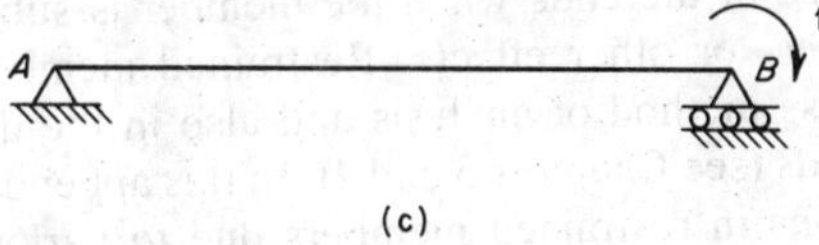

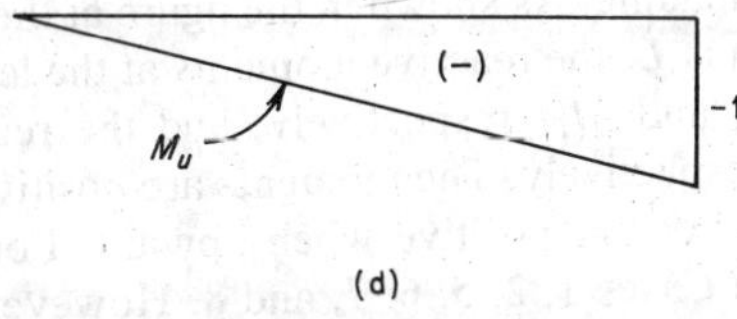

Fig. A-10.

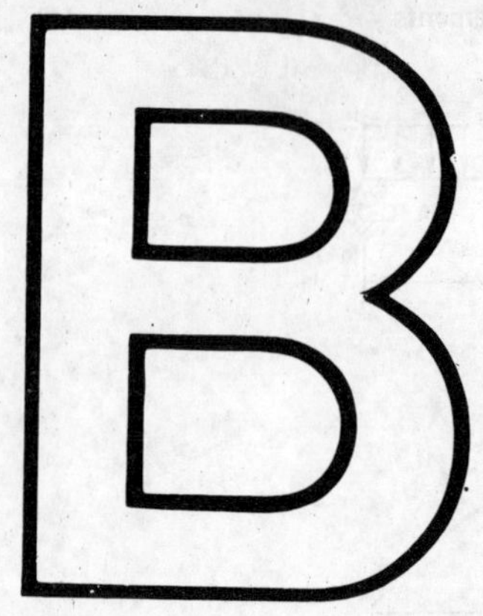

End-Actions for Restrained Members

A restrained member is one whose ends are restrained against displacements (translation and rotation), as in the case of a fixed-end beam. The end-actions for a restrained member are the reactive actions (forces and moments) developed at the ends when the member is subjected to loads, temperature changes, or other effects. Restrained members are encountered in the stiffness method of analysis and also in the determination of equivalent joint loads (see Chapters 3 and 4). In this appendix, formulas are given for end-actions in restrained members due to various causes. It is assumed in each case that the member is prismatic.

Table B-1 gives end-actions in fixed-end beams that are subjected to various conditions of loading. As shown in the figure at the top of the table, the length of the beam is L, the reactive moments at the left and right-hand ends are denoted M_A and M_B, respectively, and the reactive forces are denoted R_A and R_B, respectively. The moments are positive when counterclockwise, and the forces are positive when upward. Formulas for these quantities are given in Cases 1, 2, 5, 6, 7, and 8. However, Cases 3 and 4 differ slightly because of the special nature of the loads. In Case 3 the load is an axial force P, and therefore the only reactions are the two axial forces shown in the figure. In Case 4 the load is a twisting moment T, which produces reactions in the form of twisting moments only.

All of the formulas given in Table B-1 can be derived by standard methods of mechanics of materials. For instance, many of the formulas for beams can be obtained by integration of the differential equation for bending of a beam. The flexibility method, as described in Chapter 2, can also be used to obtain the formulas. Furthermore, the more complicated cases of loading frequently can be obtained from the simpler cases by using the principle of superposition.

Fixed-end actions due to temperature changes are listed in Table B-2. Case 1 of this table is for a beam subjected to a uniform temperature increase of T degrees. The resulting end-actions consist of axial compressive forces that are independent of the length of the member. The second case is a beam subjected to a linear temperature gradient such that the top

Table B-1

Fixed-End Actions Caused by Loads

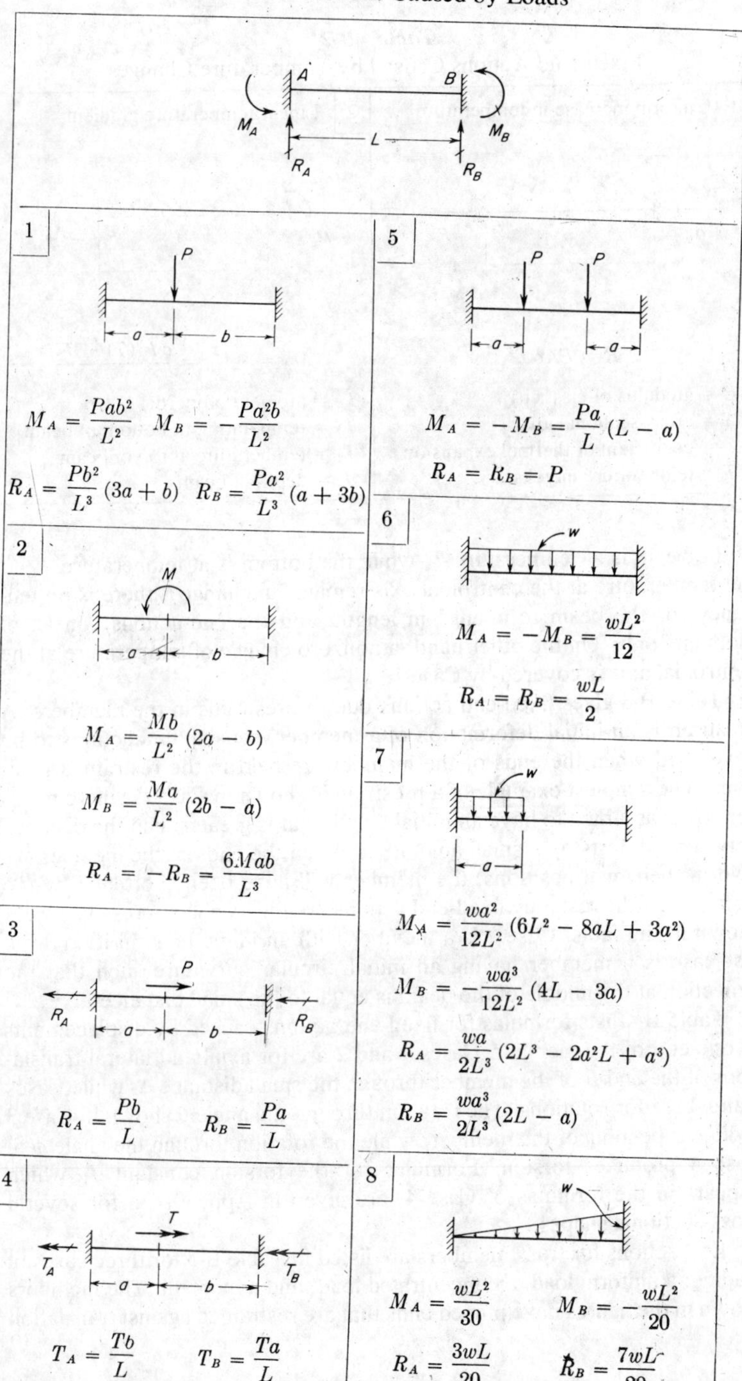

1

$$M_A = \frac{Pab^2}{L^2} \quad M_B = -\frac{Pa^2b}{L^2}$$

$$R_A = \frac{Pb^2}{L^3}(3a + b) \quad R_B = \frac{Pa^2}{L^3}(a + 3b)$$

2

$$M_A = \frac{Mb}{L^2}(2a - b)$$

$$M_B = \frac{Ma}{L^2}(2b - a)$$

$$R_A = -R_B = \frac{6Mab}{L^3}$$

3

$$R_A = \frac{Pb}{L} \qquad R_B = \frac{Pa}{L}$$

4

$$T_A = \frac{Tb}{L} \qquad T_B = \frac{Ta}{L}$$

5

$$M_A = -M_B = \frac{Pa}{L}(L - a)$$

$$R_A = R_B = P$$

6

$$M_A = -M_B = \frac{wL^2}{12}$$

$$R_A = R_B = \frac{wL}{2}$$

7

$$M_A = \frac{wa^2}{12L^2}(6L^2 - 8aL + 3a^2)$$

$$M_B = -\frac{wa^3}{12L^2}(4L - 3a)$$

$$R_A = \frac{wa}{2L^3}(2L^3 - 2a^2L + a^3)$$

$$R_B = \frac{wa^3}{2L^3}(2L - a)$$

8

$$M_A = \frac{wL^2}{30} \qquad M_B = -\frac{wL^2}{20}$$

$$R_A = \frac{3wL}{20} \qquad R_B = \frac{7wL}{20}$$

Table B-2
Fixed-End Actions Caused by Temperature Changes

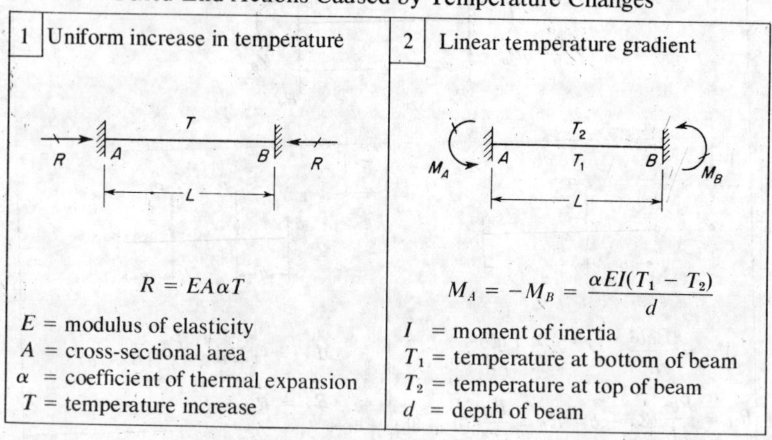

1 Uniform increase in temperature	2 Linear temperature gradient
$R = EA\alpha T$	$M_A = -M_B = \dfrac{\alpha EI(T_1 - T_2)}{d}$
E = modulus of elasticity A = cross-sectional area α = coefficient of thermal expansion T = temperature increase	I = moment of inertia T_1 = temperature at bottom of beam T_2 = temperature at top of beam d = depth of beam

of the beam is at temperature T_2 while the bottom is at temperature T_1. If the temperature at the centroidal axis remains unchanged, there is no tendency for the beam to change in length; and the end-actions consist of moments only. On the other hand, a nonzero change of temperature at the centroidal axis is covered by Case 1.

Table B-3 gives fixed-end actions due to prestrains in the members. A prestrain is an initial deformation of a member, causing end-actions to be developed when the ends of the member are held in the restrained positions. The simplest example of a prestrain is shown in Case 1, where member AB is assumed to have an initial length that is greater than the distance between supports by a small amount e. When the ends of the member are held in their final positions, the member will have been shortened by the distance e. The resulting fixed-end actions are the axial compressive forces shown in the table. Case 2 is a member with an initial bend in it, and the last case is a member having an initial circular curvature such that the deflection at the middle of the beam is equal to the small distance e.

Table B-4 lists formulas for fixed-end actions caused by displacements of one end of the member. Cases 1 and 2 are for axial and lateral translations of the end B of the member through the small distance Δ, while Cases 3 and 4 are for rotations. The rotation through the angle θ shown in Case 3 produces bending of the member, while the rotation through the angle ϕ in Case 4 produces torsion. Formulas for the torsion constant J, which appears in the formulas of Case 4, are given in Appendix C for several cross-sectional shapes.

End-actions for truss members are listed in Table B-5 for three cases of loading: a uniform load, a concentrated load, and a moment. The members shown in the figures have pinned ends that are restrained against translation

Table B-3
Fixed-End Actions Caused by Prestrains

1 Bar with excess length

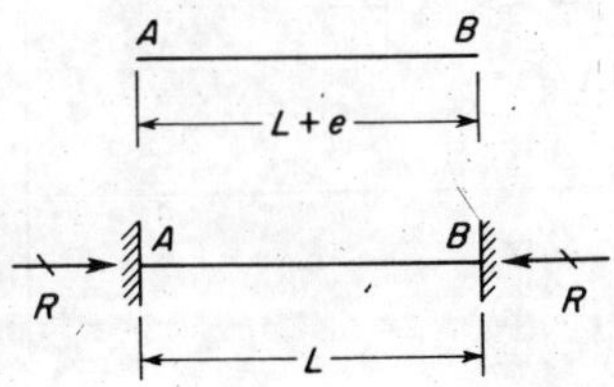

$$R = \frac{EAe}{L}$$

E = modulus of elasticity

A = cross-sectional area

e = excess length

2 Bar with a bend

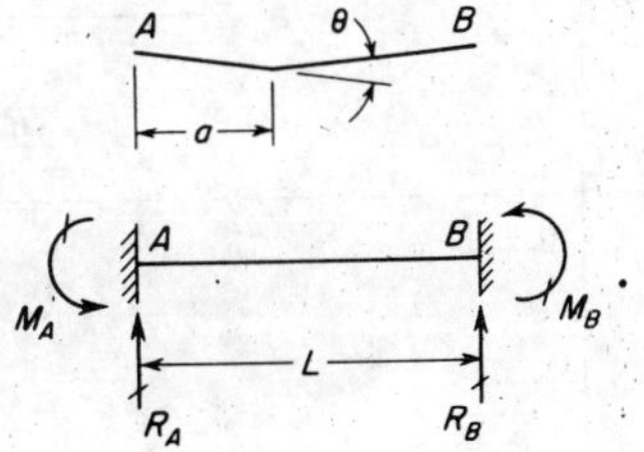

$$M_A = \frac{2EI\theta}{L^2}(2L - 3a)$$

$$M_B = \frac{2EI\theta}{L^2}(L - 3a)$$

$$R_A = -R_B = \frac{6EI\theta}{L^3}(L - 2a)$$

I = moment of inertia

θ = angle of bend

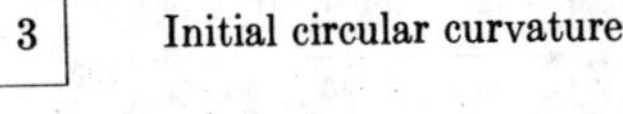

3 Initial circular curvature

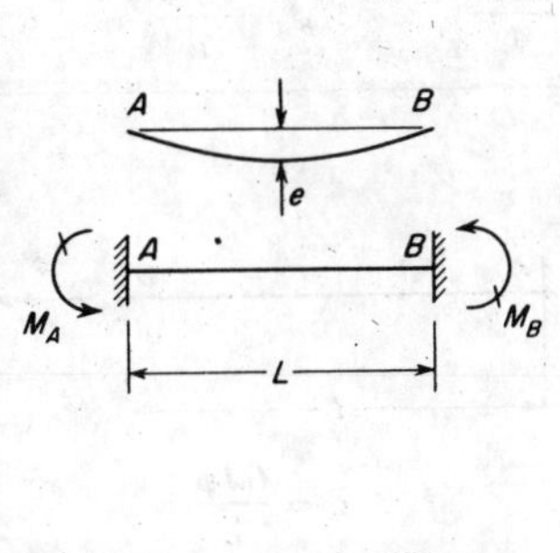

$$M_A = -M_B = \frac{8EIe}{L^2}$$

e = initial deflection at middle of bar

Table B-4

Fixed-End Actions Caused by End-Displacements

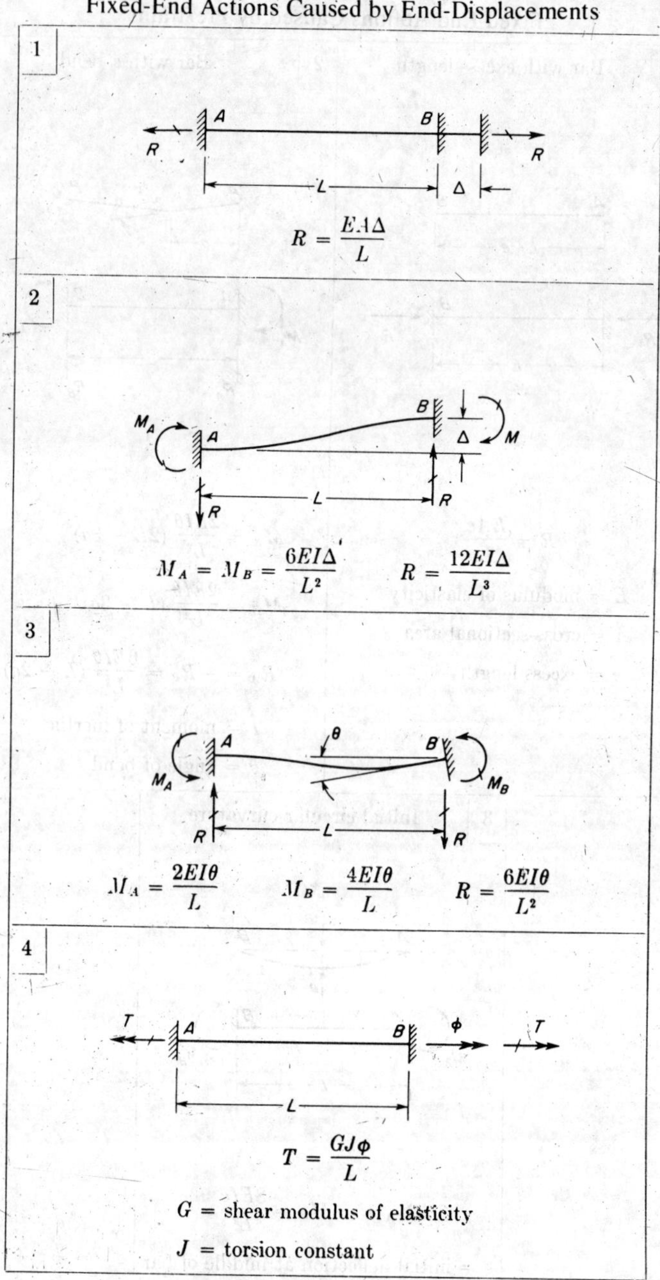

Table B-5
End-Actions for Truss Members

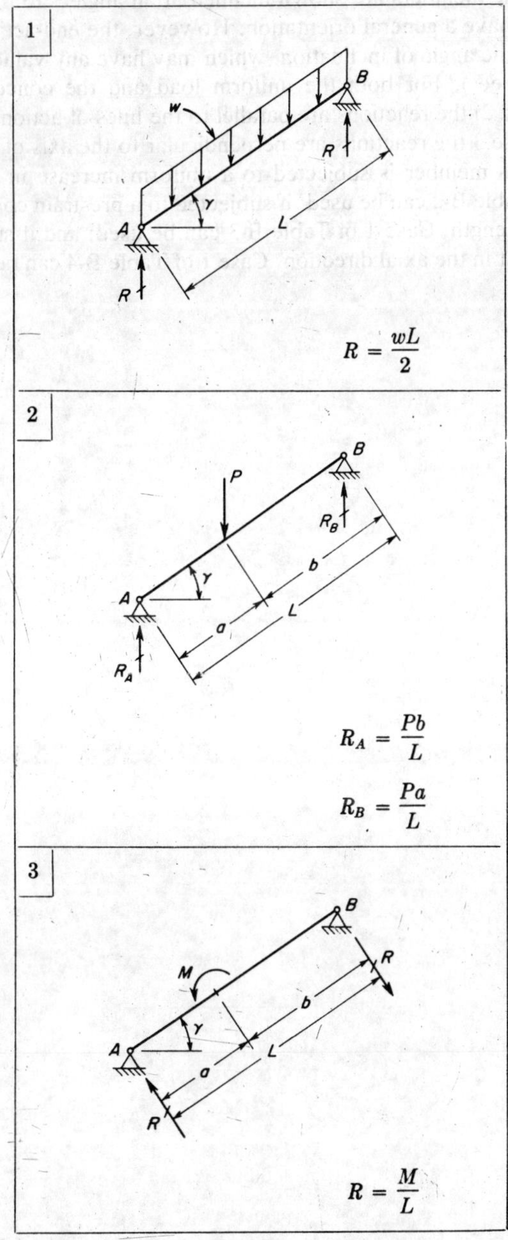

but not rotation, because only joint translations are of interest in a truss analysis. The members are shown inclined at an angle γ to the horizontal, in order to have a general orientation. However, the end-actions are independent of the angle of inclination, which may have any value (including 0 and 90 degrees). For both the uniform load and the concentrated load (Cases 1 and 2) the reactions are parallel to the lines of action of the loads, while in Case 3 the reactions are perpendicular to the axis of the member.

If a truss member is subjected to a uniform increase in temperature, Case 1 of Table B-2 can be used; if subjected to a prestrain consisting of an increase in length, Case 1 of Table B-3 can be used; and if subjected to a displacement in the axial direction, Case 1 of Table B-4 can be used.

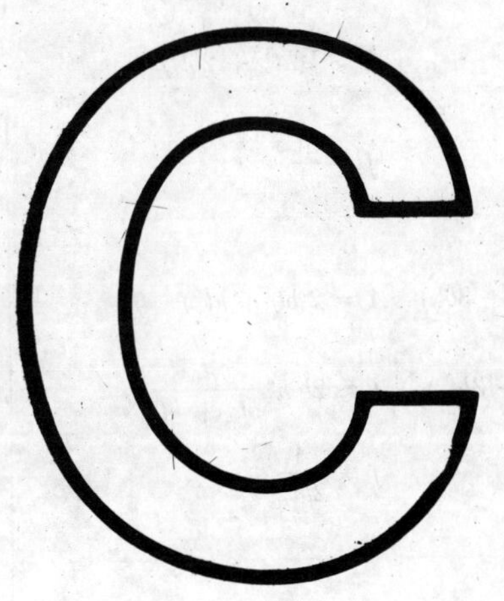

Properties of Sections

Appendix C Properties of Sections

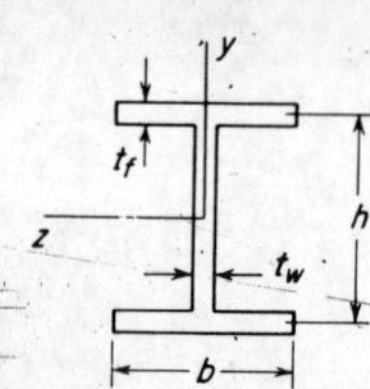

$$I_Z \approx \frac{h^2}{12}(ht_w + 6bt_f) \qquad A \approx ht_w + 2bt_f$$

$$I_Y \approx \frac{b^3 t_f}{6} \qquad J \approx \tfrac{1}{3}(ht_w^3 + 2bt_f^3)$$

$$f = \frac{A}{ht_w}$$

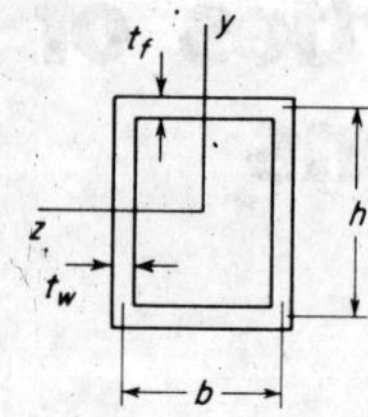

$$I_Z \approx \frac{h^2}{6}(ht_w + 3bt_f) \qquad A = 2(bt_f + ht_w)$$

$$I_Y \approx \frac{b^2}{6}(bt_f + 3ht_w) \qquad J \approx 2b^2h^2 \frac{t_f t_w}{bt_w + ht_f}$$

$$f = \frac{A}{2ht_w}$$

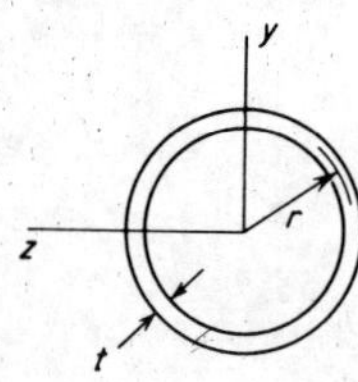

$$I_Z = I_Y \approx \pi r^3 t \qquad J \approx 2\pi r^3 t$$

$$A \approx 2\pi rt \qquad f = 2$$

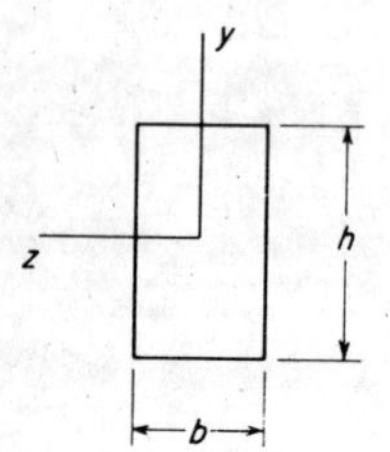

$$I_Z = \frac{bh^3}{12} \qquad J = \beta\, hb^3$$

For $h \geqslant b$,

$$I_Y = \frac{hb^3}{12} \qquad \beta \approx \frac{1}{3} - 0.21\,\frac{b}{h}\left(1 - \frac{b^4}{12h^4}\right)$$

$$A = bh \qquad f = \frac{6}{5}$$

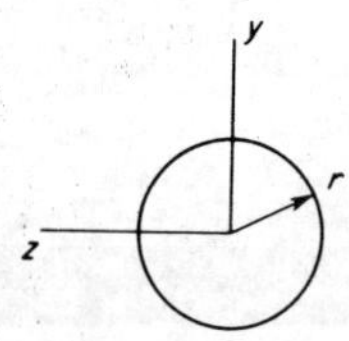

$$I_Z = I_Y = \frac{\pi r^4}{4} \qquad J = \frac{\pi r^4}{2}$$

$$A = \pi r^2 \qquad f = \frac{10}{9}$$

I_Z = moment of inertia of cross section about z axis
I_Y = moment of inertia of cross section about y axis
A = area of cross section
J = torsion constant
f = form factor for shear

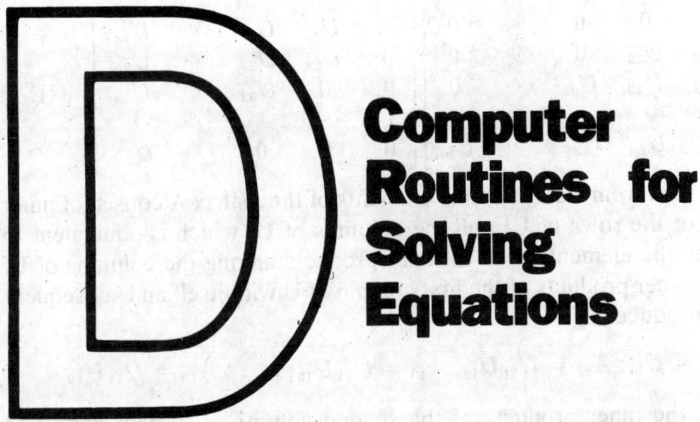

D Computer Routines for Solving Equations

D.1 Factorization Method for Symmetric Matrices. The primary mathematical task associated with matrix analysis of framed structures consists of solving a set of n simultaneous linear algebraic equations for n unknowns. There are many methods for computing the unknowns in such equations,* one of which is called the *factorization method* (also referred to as the *method of decomposition*). This approach is particularly well suited for matrix analysis of structures because it provides the efficiency of the well-known Gaussian elimination process within a matrix format. Since the stiffness and flexibility matrices of linearly elastic structures are always symmetric, a specialized type of factorization known as the *Cholesky method* will be developed for symmetric matrices. Recurrence equations derived in this article will be extended to banded matrices and implemented as computer subprograms in subsequent articles.

To begin the discussion, let the symbol **A** represent a symmetric matrix of size $n \times n$. If this matrix is positive definite† as well as symmetric, it can be factored into the product of a lower triangular matrix and an upper triangular matrix, each of which is the transpose of the other. Thus, the factorization of **A** may be stated as

$$\mathbf{A} = \mathbf{U}^{\mathrm{T}}\mathbf{U} \tag{D-1}$$

The symbol **U** in this expression denotes the upper triangular matrix, and $\mathbf{U}^{\mathrm{T}}$ is its transpose. Equation (D-1) in expanded form becomes

$$\begin{bmatrix} A_{11} & A_{12} & A_{13} & \cdots & A_{1n} \\ A_{21} & A_{22} & A_{23} & \cdots & A_{2n} \\ A_{31} & A_{32} & A_{33} & \cdots & A_{3n} \\ \cdots & \cdots & \cdots & \cdots & \cdots \\ A_{n1} & A_{n2} & A_{n3} & \cdots & A_{nn} \end{bmatrix} =$$

*For detailed descriptions of various calculation techniques, see Bathe, K.J., and Wilson, E.L., *Numerical Methods in Finite Element Analysis,* Prentice-Hall, Englewood Cliffs, N.J., 1976.

†Gere, J.M., and Weaver, W., Jr., *Matrix Algebra for Engineers,* D. Van Nostrand, 1965.

$$= \begin{bmatrix} U_{11} & 0 & 0 & \cdots & 0 \\ U_{12} & U_{22} & 0 & \cdots & 0 \\ U_{13} & U_{23} & U_{33} & \cdots & 0 \\ \cdots & \cdots & \cdots & \cdots & \cdots \\ U_{1n} & U_{2n} & U_{3n} & \cdots & U_{nn} \end{bmatrix} \begin{bmatrix} U_{11} & U_{12} & U_{13} & \cdots & U_{1n} \\ 0 & U_{22} & U_{23} & \cdots & U_{2n} \\ 0 & 0 & U_{33} & \cdots & U_{3n} \\ \cdots & \cdots & \cdots & \cdots & \cdots \\ 0 & 0 & 0 & \cdots & U_{nn} \end{bmatrix} \quad \text{(D-2)}$$

It can be seen from Eq. (D-2) that elements of the matrix **A** consist of inner products of the rows of $\mathbf{U}^{\mathrm{T}}$ and the columns of **U**, which is equivalent to calculating the elements of **A** as inner products among the columns of **U**. Thus, the inner products of the first column of **U** with itself and subsequent columns produce

$$A_{11} = U_{11}^2;\ A_{12} = U_{11}U_{12};\ A_{13} = U_{11}U_{13};\ \ldots;\ A_{1n} = U_{11}U_{1n}$$

Similarly, the inner products of the second column of **U** with itself and subsequent columns are

$$A_{22} = U_{12}^2 + U_{22}^2;\ A_{23} = U_{12}U_{13} + U_{22}U_{23};\ \ldots;\ A_{2n} = U_{12}U_{1n} + U_{22}U_{2n}$$

and for the third column of **U** the inner products are

$$A_{33} = U_{13}^2 + U_{23}^2 + U_{33}^2;\ \ldots;\ A_{3n} = U_{13}U_{1n} + U_{23}U_{2n} + U_{33}U_{3n}$$

In general, a diagonal term A_{ii} in matrix **A** can be written as

$$A_{ii} = U_{1i}^2 + U_{2i}^2 + U_{3i}^2 + \ldots + U_{ii}^2$$

or

$$A_{ii} = \sum_{k=1}^{i} U_{ki}^2 \qquad (i = j) \qquad \text{(a)}$$

In a similar manner, the off-diagonal term A_{ij} in an upper triangular position is seen to be

$$A_{ij} = U_{1i}U_{1j} + U_{2i}U_{2j} + U_{3i}U_{3j} + \ldots + U_{ii}U_{ij}$$

or

$$A_{ij} = \sum_{k=1}^{i} U_{ki}U_{kj} \qquad (i < j) \qquad \text{(b)}$$

Then the elements of **U** may be determined by rearranging Eqs. (a) and (b) as follows:

$$U_{ii} = \sqrt{A_{ii} - \sum_{k=1}^{i-1} U_{ki}^2} \qquad (1 < i = j) \qquad \text{(D-3)}$$

$$U_{ij} = \frac{1}{U_{ii}}\left(A_{ij} - \sum_{k=1}^{i-1} U_{ki}U_{kj}\right) \qquad (1 < i < j) \qquad \text{(D-4)}$$

$$U_{ij} = 0 \qquad (i > j) \qquad \text{(D-5)}$$

Equations (D-3), (D-4), and (D-5) constitute recurrence formulas in an algorithm for factoring the matrix **A** into the form given by Eq. (D-1). Because it was developed by Cholesky and a square root term appears in Eq. (D-3), this technique is known as the *Cholesky square root method*. It applies only to symmetric, positive definite matrices, for which the term within the square root sign in Eq. (D-3) is always a positive number.

If the symmetric matrix **A** were not positive definite, it could still be factored into the following triple product:

$$\mathbf{A} = \overline{\mathbf{U}}^{\mathrm{T}}\mathbf{D}\overline{\mathbf{U}} \tag{D-6}$$

In this expression the symbol **D** represents a diagonal matrix containing the squares of terms factored from the rows of **U**. If such factored terms are chosen to be U_{ii}, then the typical diagonal term in **D** is

$$D_{ii} = U_{ii}^2 \qquad (i = 1, 2, \ldots, n) \tag{D-7}$$

Since this term is the square of that in Eq. (D-3), it becomes possible to avoid taking square roots by factoring **A** as indicated by Eq. (D-6) instead of Eq. (D-1). For this purpose the recurrence formulas given by Eqs. (D-3) and (D-4) must be modified, and the resulting technique is known as the *modified Cholesky method*. In expanded form, the factorization represented by Eq. (D-6) is

$$\mathrm{A} = \begin{bmatrix} 1 & 0 & 0 & \cdots & 0 \\ \overline{U}_{12} & 1 & 0 & \cdots & 0 \\ \overline{U}_{13} & \overline{U}_{23} & 1 & \cdots & 0 \\ \cdots & \cdots & \cdots & \cdots & \cdots \\ \overline{U}_{1n} & \overline{U}_{2n} & \overline{U}_{3n} & \cdots & 1 \end{bmatrix} \times$$

$$\begin{bmatrix} D_{11} & 0 & 0 & \cdots & 0 \\ 0 & D_{22} & 0 & \cdots & 0 \\ 0 & 0 & D_{33} & \cdots & 0 \\ \cdots & \cdots & \cdots & \cdots & \cdots \\ 0 & 0 & 0 & \cdots & D_{nn} \end{bmatrix} \begin{bmatrix} 1 & \overline{U}_{12} & \overline{U}_{13} & \cdots & \overline{U}_{1n} \\ 0 & 1 & \overline{U}_{23} & \cdots & \overline{U}_{2n} \\ 0 & 0 & 1 & \cdots & \overline{U}_{3n} \\ \cdots & \cdots & \cdots & \cdots & \cdots \\ 0 & 0 & 0 & \cdots & 1 \end{bmatrix} \tag{D-8}$$

From this form it is seen that $A_{11} = D_{11}$ and that other diagonal terms in **A** can be written as

$$A_{ii} = D_{11}\overline{U}_{1i}^2 + D_{22}\overline{U}_{2i}^2 + D_{33}\overline{U}_{3i}^2 + \ldots + D_{ii}$$

or

$$A_{ii} = D_{ii} + \sum_{k=1}^{i-1} D_{kk}\overline{U}_{ki}^2 \qquad (1 < i = j) \tag{c}$$

Similarly, off-diagonal terms in the first row of **A** are $A_{1j} = D_{11}\overline{U}_{1j}$ and in other rows are

$$A_{ij} = D_{11}\overline{U}_{1i}\overline{U}_{1j} + D_{22}\overline{U}_{2i}\overline{U}_{2j} + D_{33}\overline{U}_{3i}\overline{U}_{3j} + \ldots + D_{ii}\overline{U}_{ij}$$

or

$$A_{ij} = D_{ii}\overline{U}_{ij} + \sum_{k=1}^{i-1} D_{kk}\overline{U}_{ki}\overline{U}_{kj} \qquad (1 < i < j) \qquad \text{(d)}$$

Then the elements of **D** and $\overline{\mathbf{U}}$ can be found by rearranging Eqs. (c) and (d), as follows:

$$D_{ii} = A_{ii} - \sum_{k=1}^{i-1} D_{kk}\overline{U}_{ki}^2 \qquad (1 < i = j) \qquad \text{(D-9)}$$

$$\overline{U}_{ij} = \frac{1}{D_{ii}}\left(A_{ij} - \sum_{k=1}^{i-1} D_{kk}\overline{U}_{ki}\overline{U}_{kj}\right) \qquad (1 < i < j) \qquad \text{(D-10)}$$

$$\overline{U}_{ij} = 0 \qquad (i > j) \qquad \text{(D-11)}$$

These recurrence formulas imply twice as many multiplications as those in Eqs. (D-3) and (D-4). However, this increase in the number of operations can be avoided, as shown in the subsequent discussion.

The recurrence equations (D-9) and (D-10) imply that the diagonal term D_{ii} is computed first, followed by the calculation of the terms in the i-th row of $\overline{U}$. This row-wise generation of terms can be changed to the column-wise sequence:

$$\overline{U}_{ij} = \frac{1}{D_{ii}}\left(A_{ij} - \sum_{k=1}^{i-1} D_{kk}\overline{U}_{ki}\overline{U}_{kj}\right) \qquad (1 < i < j) \qquad \text{(e)}$$

$$D_{jj} = A_{jj} - \sum_{k=1}^{j-1} D_{kk}\overline{U}_{kj}^2 \qquad (1 < i = j) \qquad \text{(f)}$$

Note that the product $D_{kk}\overline{U}_{kj}$ appears in both Eq. (e) and Eq. (f). Let this product be

$$\overline{U}^*_{kj} = D_{kk}\overline{U}_{kj} \qquad \text{(g)}$$

and perform the calculations for $\overline{U}_{ij}$ and D_{jj} in the following manner (for $j = 2, 3, \ldots, n$):

$$\overline{U}^*_{ij} = A_{ij} - \sum_{k=1}^{i-1} \overline{U}_{ki}\overline{U}^*_{kj} \qquad (1 < i < j) \qquad \text{(D-12)}$$

$$D_{jj} = A_{jj} - \sum_{k=1}^{j-1} \overline{U}_{kj}\overline{U}^*_{kj} \qquad (1 < i = j) \qquad \text{(D-13)}$$

where

$$\overline{U}_{kj} = \frac{1}{D_{kk}}\overline{U}^*_{kj} \qquad \text{(D-14)}$$

Thus, for column j the intermediate product $\overline{U}^*_{ij}$ is obtained for each off-diagonal term after the first (see Eq. D-12). Then the diagonal term D_{jj} is computed (Eq. D-13), during which calculation the final value of each off-diagonal term is also found (Eq. D-14). By this sequence of operations, the number of multiplications is reduced to that for the Cholesky square root method, and the calculation of square roots is avoided.

Assume that the following system of linear algebraic equations is to be solved:

$$\mathbf{AX} = \mathbf{B} \tag{D-15}$$

in which **X** is a column vector of n unknowns and **B** is a column vector of constant terms. As a preliminary step, substitute Eq. (D-6) into Eq. (D-15) to obtain

$$\overline{\mathbf{U}}^{\mathrm{T}}\mathbf{D}\overline{\mathbf{U}}\mathbf{X} = \mathbf{B} \tag{D-16}$$

Then define the vector **Y** to be

$$\overline{\mathbf{U}}\mathbf{X} = \mathbf{Y} \tag{D-17}$$

In expanded form this expression is

$$\begin{bmatrix} 1 & \overline{U}_{12} & \overline{U}_{13} & \cdots & \overline{U}_{1n} \\ 0 & 1 & \overline{U}_{23} & \cdots & \overline{U}_{2n} \\ 0 & 0 & 1 & \cdots & \overline{U}_{3n} \\ \cdots & \cdots & \cdots & \cdots & \cdots \\ 0 & 0 & 0 & \cdots & 1 \end{bmatrix} \begin{bmatrix} X_1 \\ X_2 \\ X_3 \\ \cdots \\ X_n \end{bmatrix} = \begin{bmatrix} Y_1 \\ Y_2 \\ Y_3 \\ \cdots \\ Y_n \end{bmatrix} \tag{D-18}$$

In addition, define the vector **Z** to be

$$\mathbf{DY} = \mathbf{Z} \tag{D-19}$$

for which the expanded form is

$$\begin{bmatrix} D_{11} & 0 & 0 & \cdots & 0 \\ 0 & D_{22} & 0 & \cdots & 0 \\ 0 & 0 & D_{33} & \cdots & 0 \\ \cdots & \cdots & \cdots & \cdots & \cdots \\ 0 & 0 & 0 & \cdots & D_{nn} \end{bmatrix} \begin{bmatrix} Y_1 \\ Y_2 \\ Y_3 \\ \cdots \\ Y_n \end{bmatrix} = \begin{bmatrix} Z_1 \\ Z_2 \\ Z_3 \\ \cdots \\ Z_n \end{bmatrix} \tag{D-20}$$

Substitution of Eq. (D-17) into Eq. (D-19) and then the latter into Eq. (D-16) yields

$$\overline{\mathbf{U}}^{\mathrm{T}}\mathbf{Z} = \mathbf{B} \tag{D-21}$$

Or, in expanded form:

$$\begin{bmatrix} 1 & 0 & 0 & \cdots & 0 \\ \overline{U}_{12} & 1 & 0 & \cdots & 0 \\ \overline{U}_{13} & \overline{U}_{23} & 1 & \cdots & 0 \\ \cdots & \cdots & \cdots & \cdots & \cdots \\ \overline{U}_{1n} & \overline{U}_{2n} & \overline{U}_{3n} & \cdots & 1 \end{bmatrix} \begin{bmatrix} Z_1 \\ Z_2 \\ Z_3 \\ \cdots \\ Z_n \end{bmatrix} = \begin{bmatrix} B_1 \\ B_2 \\ B_3 \\ \cdots \\ B_n \end{bmatrix} \tag{D-22}$$

The original vector of unknowns **X** may now be obtained in three steps using Eqs. (D-21), (D-19), and (D-17). In the first step Eq. (D-21) is solved for the vector **Z.** Since $\overline{\mathbf{U}}^T$ is a lower triangular matrix (see Eq. D-22), the elements of **Z** can be calculated in a series of forward substitutions. For example, the first term in **Z** is

$$Z_1 = B_1 \tag{h}$$

The second term in **Z** is found to be

$$Z_2 = B_2 - \overline{U}_{12}Z_1 \tag{i}$$

and the third term is

$$Z_3 = B_3 - \overline{U}_{13}Z_1 - \overline{U}_{23}Z_2 \tag{j}$$

In general, the recurrence formula for elements of **Z** becomes

$$Z_i = B_i - \sum_{k=1}^{i-1} \overline{U}_{ki}Z_k \qquad (1 < i) \tag{D-23}$$

The second step consists of solving for the vector **Y** in Eq. (D-19). Since **D** is a diagonal matrix (see Eq. D-20), the elements of **Y** can be found by dividing terms in **Z** by corresponding diagonals of **D,** as follows:

$$Y_i = \frac{Z_i}{D_{ii}} \qquad (i = 1, 2, \ldots, n) \tag{D-24}$$

This recurrence formula can be applied in either a forward or a backward sequence.

In the third step the vector **X** is found from Eq. (D-17). Since $\overline{\mathbf{U}}$ is an upper triangular matrix (see Eq. D-18), the elements of **X** are determined in a backward substitution procedure. The last term in **X** is

$$X_n = Y_n \tag{k}$$

The next-to-last term is

$$X_{n-1} = Y_{n-1} - \overline{U}_{n-1,n}X_n \tag{l}$$

and so on. In general, the elements of **X** (other than the last) may be calculated from the recurrence formula:

$$X_i = Y_i - \sum_{k=i+1}^{n} \overline{U}_{ik}X_k \qquad (i < n) \tag{D-25}$$

This step completes the solution of the original equations (Eq. D-15) for the unknown quantities.

Numbers of arithmetic operations for the modified Cholesky method are summarized in Table D-1 for factoring an $n \times n$ symmetric coefficient matrix and solving for n unknowns. Also shown in the table are the corre-

Table D-1
Numbers of Arithmetic Operations

Method		*Multiplications*	*Divisions*	*Additions*
Modified Cholesky	Factor	$\frac{n^3}{6} - \frac{n}{6}$	$\frac{n^2}{2} - \frac{n}{2}$	$\frac{n^3}{6} - \frac{n}{6}$
	Solve	n^2	n	$n^2 - n$
	Totals	$\frac{n^3}{6} + n^2 - \frac{n}{6}$	$\frac{n^2}{2} + \frac{n}{2}$	$\frac{n^3}{6} + n^2 - \frac{7n}{6}$
Compact Gaussian elimination		$\frac{n^3}{6} + \frac{3n^2}{2} - \frac{2n}{3}$	n	$\frac{n^3}{6} + n^2 - \frac{7n}{6}$

sponding numbers for the method of compact Gaussian elimination*, which is known to require the least number of operations. It is seen that the sum of the number of multiplications and divisions is the same for both methods, as is the number of additions.

The factorization and solution procedures described above apply also to banded symmetric matrices. For a banded matrix the upper triangular component U (and hence $\overline{U}$) has the same semi-band width as the original matrix **A**. Therefore, fewer calculations are required in the recurrence formulas for both factorization and solution.

D.2 Subprogram FACTOR. In this article the factorization of a symmetric matrix by the modified Cholesky approach, as described in the preceding article, is cast into the form of a computer subprogram. The name of this subprogram is

FACTOR(N,A,*)

The first argument in the parentheses is the integer number N, which denotes the size of the matrix to be factored. The second identifier represents a symmetric matrix **A** of real numbers, and the third symbol (an asterisk) signifies a *nonstandard RETURN* to an error message in the main program if **A** is found not to be positive definite. In addition to this notation, the integer numbers I, J, K, I1, and J1 serve as local indexes in the body of the subprogram; and the real variables SUM and TEMP are used for temporary storage.

Subprogram FACTOR appears in Flow Chart D-1, which implements Eqs. (D-12), (D-13), and (D-14) from the preceding article. Elements of the upper triangular matrix $\overline{\mathbf{U}}$ are generated column-wise in the storage locations originally occupied by the upper triangular part of the matrix **A**. Thus, the identifier *A* remains in use throughout the flow chart. In addition, the

*Fox, L., *An Introduction to Numerical Linear Algebra,* Oxford University Press, New York, 1965, p. 184.

Flow Chart D-1: Subprogram FACTOR (N, A, *)

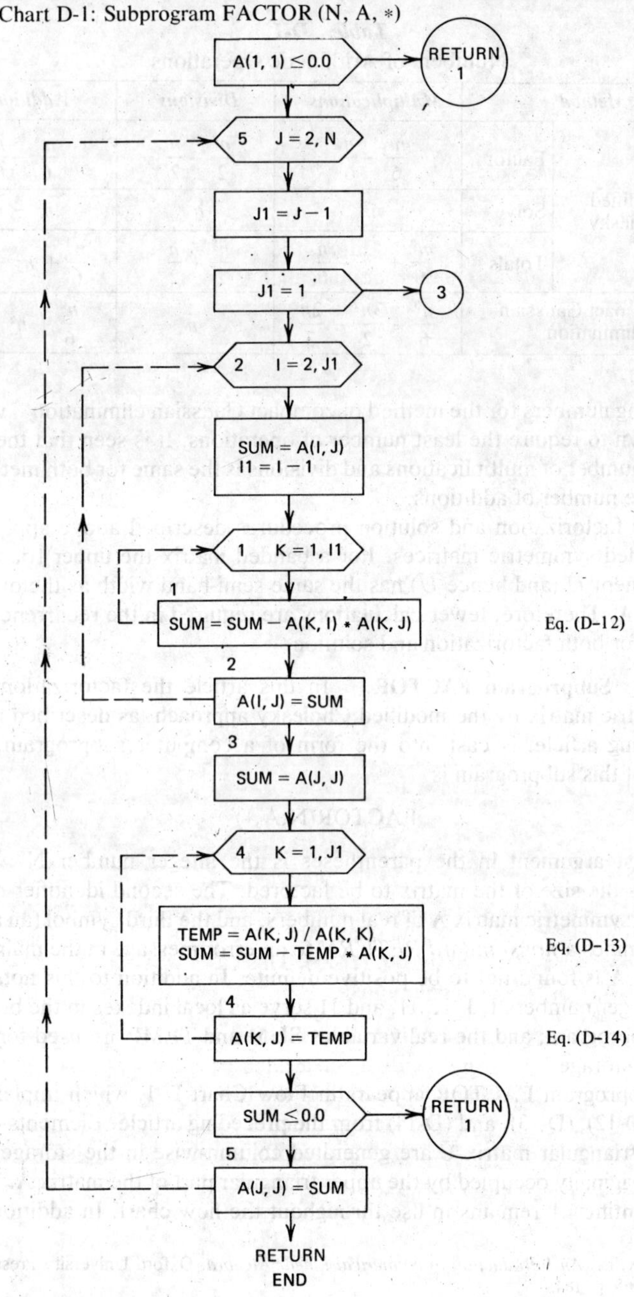

diagonal elements D_{ii} are stored in the diagonal positions A_{ii} of matrix **A**. If a zero or negative value of D_{ii} is detected, control is transferred (by means of the nonstandard RETURN) to an error message in the main program. Elements of **A** below the main diagonal are left undisturbed by this subprogram.

Neither Subprogram FACTOR nor Subprogram SOLVER (described in the next article) are utilized by the structural analysis programs in this book. However, they are included in the series of subprograms because of their general usefulness. In addition, they serve as guides to understanding the more complicated subprograms called BANFAC and BANSOL, which are given in subsequent articles and used in the structural analysis programs.

D.3 Subprogram SOLVER. The second subprogram in this series accepts the factored matrix from Subprogram FACTOR and solves for the unknowns in the original system of equations. The name of this subprogram is

SOLVER(N,U,B,X)

The argument N has the same meaning as previously, and the symbol U denotes the matrix from Subprogram FACTOR. The identifiers B and X represent real vectors of constant terms and unknowns, respectively (see Eq. D-15).

Flow Chart D-2 shows the logic for Subprogram SOLVER. In the first portion of the flow chart the intermediate vector **Z** is computed by forward substitutions, according to Eq. (D-23). Note that the vector **X** is used as temporary storage for **Z** in this part of the subprogram.

The second portion of the chart involves finding the vector **Y** by dividing each value of Z_i by the corresponding diagonal term D_{ii} (see Eq. D-24). In this instance the vector **X** is used as temporary storage for **Y**, and the terms D_{ii} are known to be in the diagonal positions U_{ii} (see Subprogram FACTOR).

In the last portion of the chart, the final values of the elements in the vector **X** are calculated by Eq. (D-25). This backward sweep completes the solution of the original equations.

D.4 Subprogram BANFAC. The factorization method is more efficient for a banded matrix than for a filled array because no calculations need be made for elements outside of the band. Figure D-1a illustrates the general appearance of a banded symmetric matrix. The symbol NB shown in the figure denotes the semi-band width, and N is the size of the matrix. Only the upper portion of the band (including the diagonal elements) has to be stored, as indicated by the small squares in Fig. D-1a. A more efficient pattern for storing the upper band portion of the matrix appears in Fig. D-1b. In this arrangement the required elements are stored as a rectangular

Flow Chart D-2: Subprogram SOLVER (N, U, B, X)

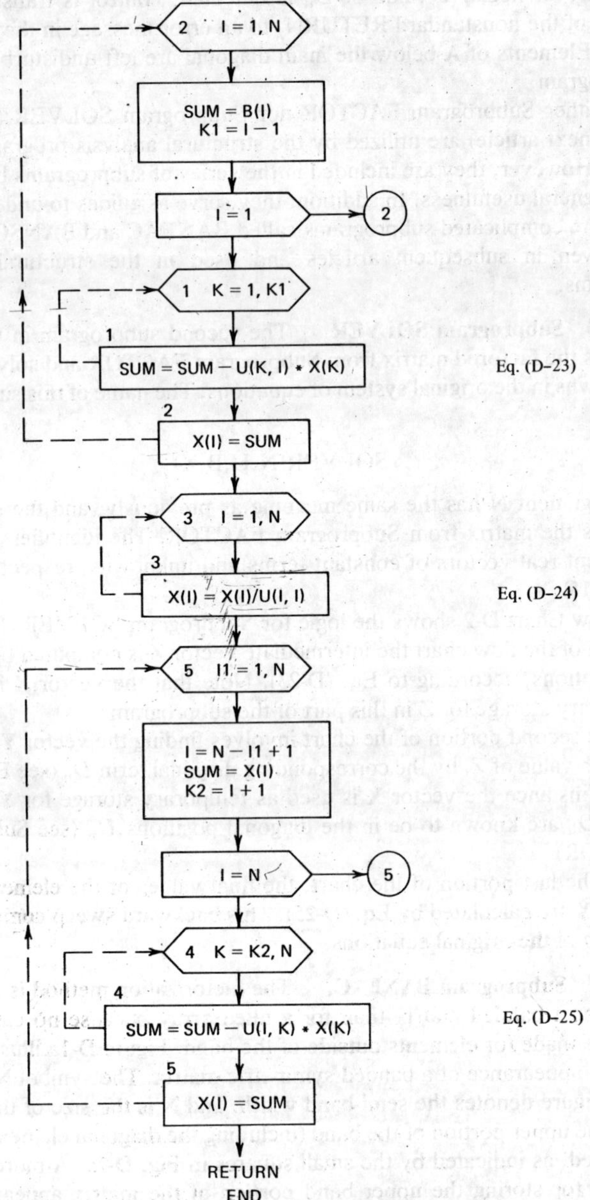

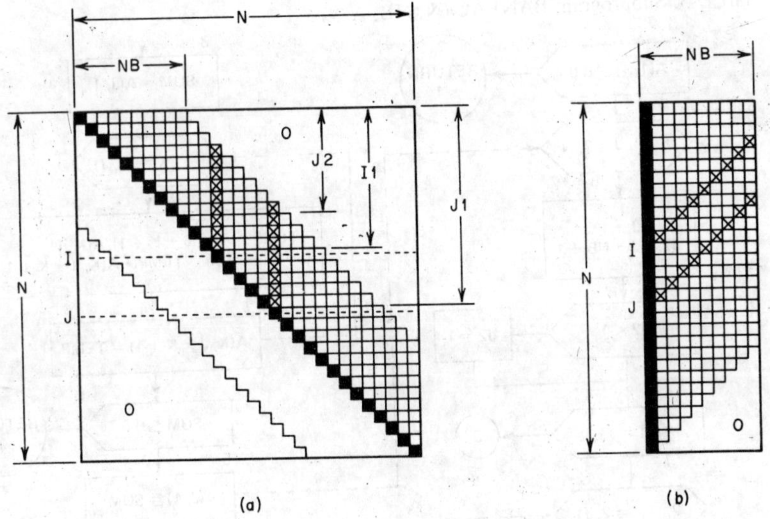

Fig. D-1. Banded matrix: (a) usual form of storage and (b) upper band stored as a rectangular array.

array with the diagonal elements (shown shaded) in the first column. Comparison of Fig. D-1a with Fig. D-1b shows that the rows of the matrix have been shifted to the left, and most of the excess terms have been removed.

Flow Chart D-3 contains the steps for a subprogram that factors the upper band of a symmetric matrix stored as a rectangular array. The name of this subprogram is

BANFAC(N,NB,A,*)

Arguments in the parentheses have all been defined, and most of the other identifiers in the body of this subprogram were used before. However, a new index J2 is introduced for the purpose of limiting calculations to nonzero elements. When the column number J exceeds the semi-band width NB in Fig. D-1a, the first nonzero term in that column has the row index

$$J2 = J - NB + 1 \qquad (NB < J \leq N) \tag{a}$$

Otherwise, for the first NB columns (except column 1),

$$J2 = 1 \qquad (1 < J \leq NB) \tag{b}$$

The sequence of operations in Subprogram BANFAC follows that in Subprogram FACTOR (see Flow Chart D-1), except that additional statements are required to determine the index J2. Furthermore, the column subscripts of terms in Eqs. (D-12), (D-13), and (D-14) are modified due to

Flow Chart D-3: Subprogram BANFAC̸ (N, NB, A, *)

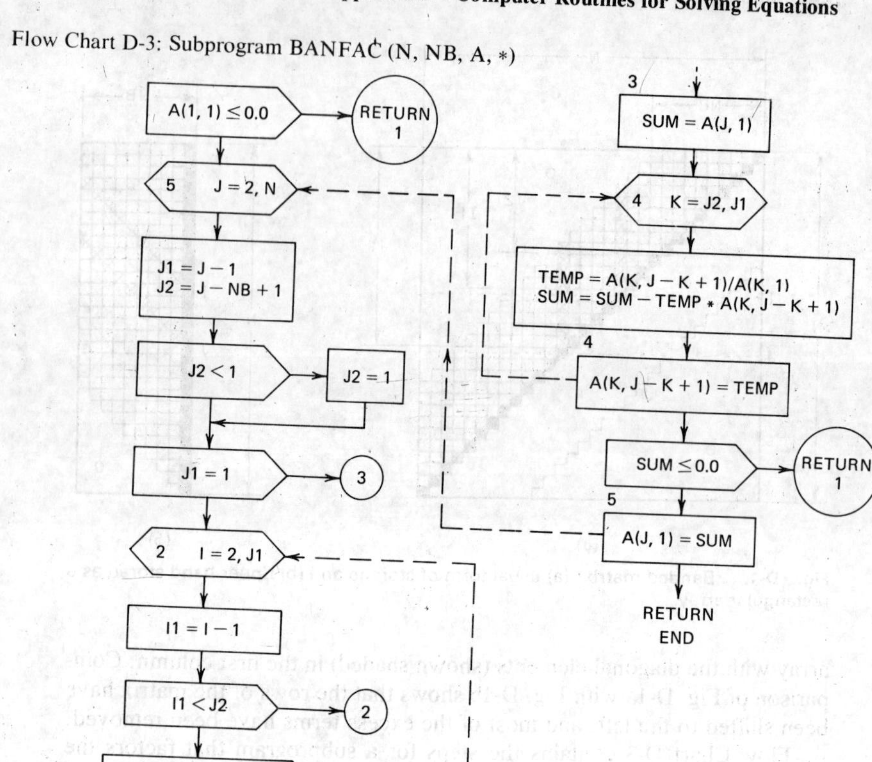

the fact that the upper band portion of matrix **A** is stored as a rectangular array. Whereas the recurrence equations pertain to columns I and J in Fig. D-1a, they involve staircase patterns of elements in the matrix of Fig. D-1b.

As in the earlier subprogram for factorization, the matrix $\overline{\mathbf{U}}$ is generated and placed in the storage locations originally occupied by the matrix **A**, but the identifier A remains in use. In addition, the diagonal elements D_{ii} are stored in the first column of **A** for convenience in later calculations. If D_{ii} is found to be zero or negative, control is transferred (by means of the nonstandard RETURN) to an error message in the main program.

Subprogram BANSOL, given in the next article, is intended to be used in conjunction with Subprogram BANFAC. They are both used in the structural analysis programs in Chapter 5, which take advantage of the band width of the stiffness matrix.

D.5 Subprogram BANSOL. The last subprogram in the series is analogous to Subprogram SOLVER (see Flow Chart D-2), except that it applies to a banded matrix. This subprogram accepts the upper band of the matrix $\overline{\mathbf{U}}$ from Subprogram BANFAC and solves for the unknowns in the original system of equations. The name of the subprogram is

BANSOL(N,NB,U,B,X)

All identifiers in the parentheses are familiar terms that have been used before.

Flow Chart D-4 for Subprogram BANSOL bears much similarity to that for Subprogram SOLVER in Flow Chart D-2. However, it is complicated by the fact that the upper band of $\overline{\mathbf{U}}$ is stored in rectangular form. In both the forward and backward substitutions (see Eqs. D-23 and D-25) the index J is used to delineate nonzero terms to be included in the calculations. In the forward sweep, the row index for the first nonzero item in column I is

$$\mathrm{J} = \mathrm{I} - \mathrm{NB} + 1 \qquad (\mathrm{NB} < \mathrm{I} \leq \mathrm{N}) \tag{a}$$

Otherwise, for the first NB columns (except column 1),

$$\mathrm{J} = 1 \qquad (1 < \mathrm{I} \leq \mathrm{NB}) \tag{b}$$

Similarly, in the backward sweep, the column index for the last nonzero element in row I of matrix $\overline{\mathbf{U}}$ is

$$\mathrm{J} = \mathrm{I} + \mathrm{NB} - 1 \qquad [1 \leq \mathrm{I} < (\mathrm{N} - \mathrm{NB})] \tag{c}$$

Otherwise, for the last NB rows (except row N),

$$\mathrm{J} = \mathrm{N} \qquad [(\mathrm{N} - \mathrm{NB}) \leq \mathrm{I} < \mathrm{N}] \tag{d}$$

In addition, the column indexes of the elements in $\overline{\mathbf{U}}$ are modified according to their actual locations in the rectangular array.

As in the earlier solution subprogram, the intermediate vectors **Z** and **Y** are generated in the vector **X**, and final values of **X** are calculated in the backward sweep. The elements of $\overline{\mathbf{U}}$ and **B** are left unaltered by this solution routine; so it can be used repeatedly for the same matrix $\overline{\mathbf{U}}$ but for different vectors of constant terms.

Flow Chart D-4: Subprogram BANSOL (N, NB, U, B, X)

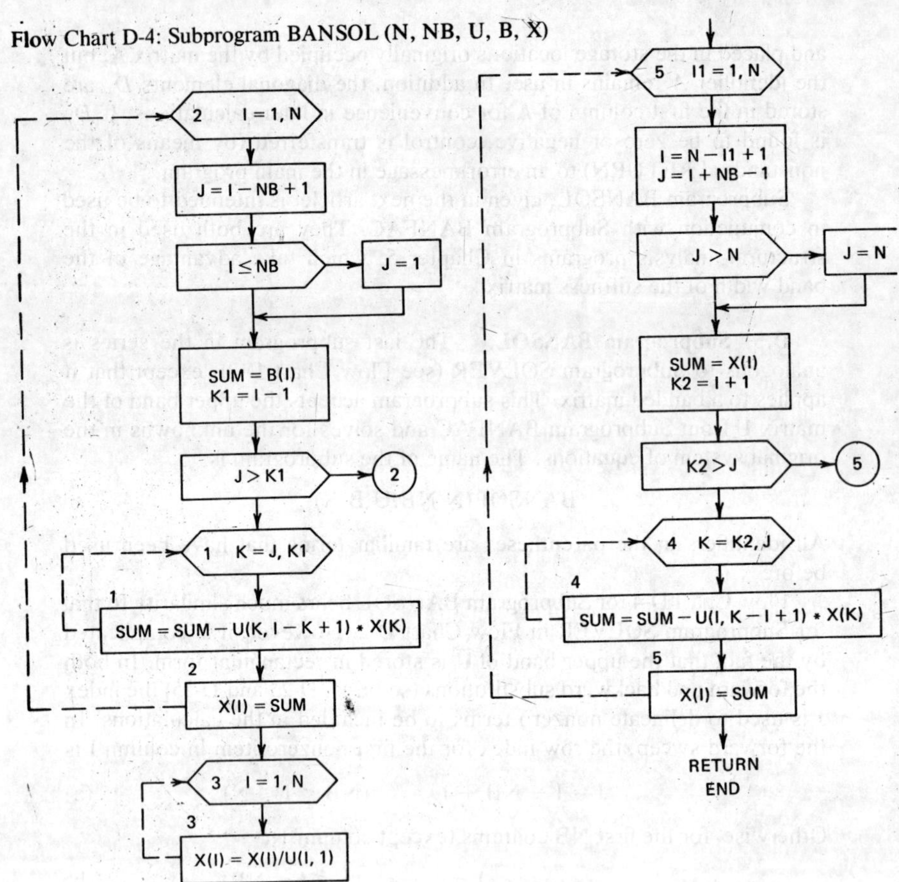

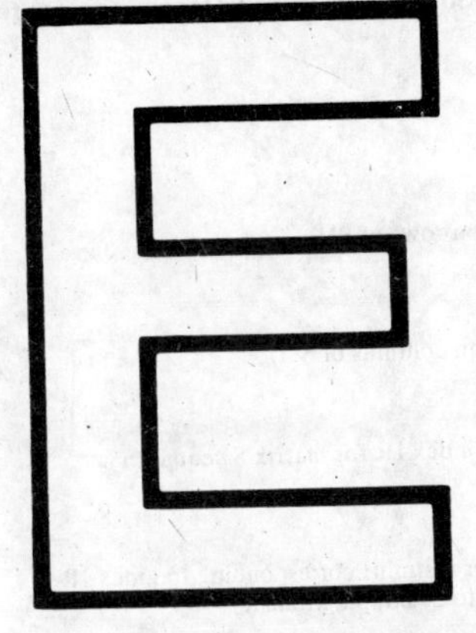

Solution Without Rearrangement

In the stiffness method of analysis it is possible to solve the joint equilibrium equations in place (without rearrangement). This can be accomplished by modifying the stiffness and load matrices to convert the equations for support reactions into trivial displacement equations embedded within the complete set of equations. Then the whole set can be solved for the unknown joint displacements as well as the known support displacements without having to rearrange and partition the matrices.

To show the technique, a small example will suffice. Suppose that a hypothetical structure has only four possible joint displacements, as indicated by the following joint equilibrium equations:

$$\begin{bmatrix} S_{J11} & S_{J12} & S_{J13} & S_{J14} \\ S_{J21} & S_{J22} & S_{J23} & S_{J24} \\ S_{J31} & S_{J32} & S_{J33} & S_{J34} \\ S_{J41} & S_{J42} & S_{J43} & S_{J44} \end{bmatrix} \begin{bmatrix} D_{J1} \\ D_{J2} \\ D_{J3} \\ D_{J4} \end{bmatrix} = \begin{bmatrix} A_{J1} \\ A_{J2} \\ A_{J3} \\ A_{J4} \end{bmatrix} \tag{E-1}$$

In addition, suppose that the third displacement is specified to be a nonzero support displacement $D_{J3} \neq 0$. Then the terms involving D_{J3} can be subtracted from both sides of Eq. (E-1), and the third equation can be replaced by the trivial expression $D_{J3} = D_{J3}$ to obtain:

$$\begin{bmatrix} S_{J11} & S_{J12} & 0 & S_{J14} \\ S_{J21} & S_{J22} & 0 & S_{J24} \\ 0 & 0 & 1 & 0 \\ S_{J41} & S_{J42} & 0 & S_{J44} \end{bmatrix} \begin{bmatrix} D_{J1} \\ D_{J2} \\ D_{J3} \\ D_{J4} \end{bmatrix} = \begin{bmatrix} A_{J1} - S_{J13}D_{J3} \\ A_{J2} - S_{J23}D_{J3} \\ D_{J3} \\ A_{J4} - S_{J43}D_{J3} \end{bmatrix} \tag{E-2}$$

These equations may now be solved for the four joint displacements, including D_{J3}.

Negative terms on the right-hand side of Eq. (E-2) represent equivalent joint loads due to the specified support displacement D_{J3}. Of course, if $D_{J3} = 0$ these equivalent joint loads are also zero. Any number of specified support displacements can be handled in this manner. This technique precludes the calculation of support reactions by the matrix multiplication approach described in Art. 6.7. Instead, it is necessary to obtain the reac-

Flow Chart E-1: Alternative Method

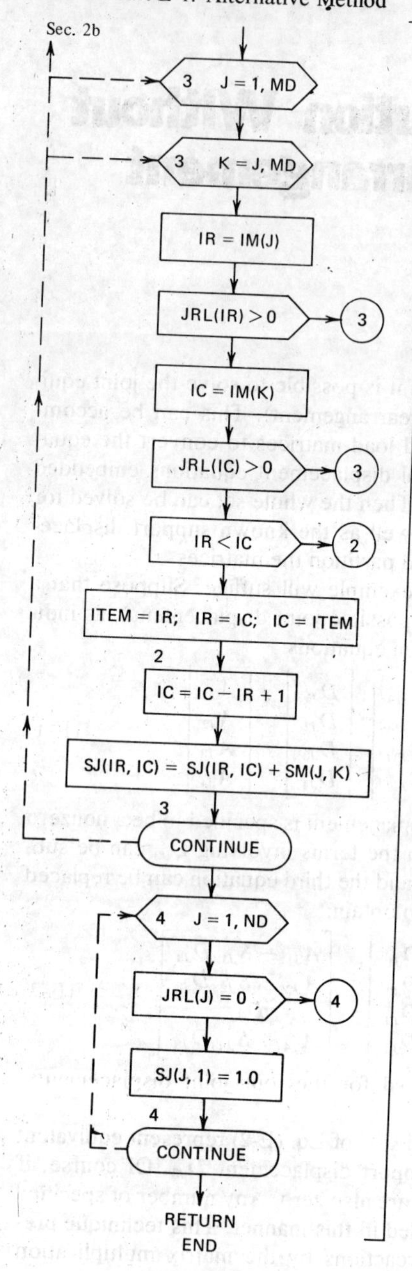

(Index on rows of SM)

(Index on columns of SM)

Set row index IR for matrix SJ equal to IM(J).

Check for restraint corresponding to index IR. If so, go to 3, skipping transfer.

Set column index IC for matrix SJ equal to IM(K).

Check for restraint corresponding to index IC. If so, go to 3, skipping transfer.

Check for upper triangular position. If so, go to 2, keeping subscripts as they are.

For lower triangular position, switch subscripts, thereby changing to upper triangular position.

Reset column index for storing upper band of matrix SJ as a rectangular array.

Transfer an element of SM to the position in matrix SJ given by subscripts IR and IC

(Index on joint displacements)

Check for restraint corresponding to index J. If none, go to 4, skipping next statement.

Set SJ(J, 1) equal to unity in preparation for solving in place without rearrangement.

tions from end-actions of members framing into the supports, as in the programs of Chapter 5.

To implement the technique described above, it is necessary to revise the flow charts in Chapter 5. Flow Chart E-1 shows the changes required in generating the stiffness matrix for any type of framed structure. By this routine no transfers of stiffness terms are made to rows or columns corresponding to support restraints. In addition, the value 1.0 is placed in the diagonal position (first column of SJ) wherever a support restraint exists.

Other less important changes must also be made at various places in all of the programs of Chapter 5, as follows:

1. Because there is no rearrangement, the identifiers SFF and DF may be replaced by SJ and DJ, respectively.
2. Section 1e on joint displacement indexes may be omitted.
3. When calling Subprograms BANFAC and BANSOL, use ND instead of N.
4. In Sec. 4b no rearrangement of the load vector is required.
5. Before calling BANSOL, set the loads at supports equal to zero.
6. In Sec. 5a omit the expansion of the displacement vector.

Although the coded programs are somewhat simplified by avoiding rearrangement, they can be wasteful of computer time and storage if there are numerous restraints. Therefore, the method involving rearrangement was chosen for the programs in this book because of its greater efficiency.

Answers to Problems

Chapter 1

1.4-1 The increase in length of the bar.

1.4-2 A horizontal force acting to the right at joint C, and a clockwise couple acting at joint C.

1.4-3 $D_{11} = \dfrac{A_1L^3}{48EI} \quad D_{12} = \dfrac{A_2L^2}{16EI} \quad D_1 = D_{11} + D_{12}$

1.4-4 $D_{11} = D_{31} = \dfrac{A_1L}{2EI} \quad D_{21} = \dfrac{3A_1L^2}{8EI}$

1.4-5 $D_{11} = \dfrac{A_1L}{3EI} \quad D_{23} = -\dfrac{A_3L^3}{32EI} \quad D_{33} = \dfrac{A_3L^3}{8EI}$

1.7-1 (a) 3
(b) 2
(c) 3
(d) 2

1.7-2 (a) 2
(b) 7
(c) 4

1.7-3 (a) 1
(b) 5

1.7-4 10

1.7-5 (a) 3
(b) 9

1.7-6 (a) 3
(b) 6
(c) 3

1.7-7 (a) 1
(b) 8
(c) 5

1.7-8 (a) 8
(b) 16
(c) 8

1.7-9 (a) 0
(b) 9

1.7-10 (a) 3
(b) 6

1.7-11 (a) 21
(b) 12

1.7-12 (a) 30
(b) 36
(c) 25

1.14-1 (a) $\theta_A = (8A_1 - A_2)L/24EI$ (b) $\Delta_C = A_1L^2/16EI$
(c) $\theta_C = (-A_1 + 2A_2)L/24EI$

1.14-2 $F_{11} = L/EI \quad F_{12} = F_{21} = -L^2/8EI \quad F_{22} = L^3/24EI$

1.14-3 $F_{11} = L^3/8EI \quad F_{12} = F_{21} = L^2/48EI \quad F_{22} = L/12EI$

1.14-4 $D_1 = (4.83A_1 + 3.41A_2)L/EA \quad D_2 = 3.41(A_1 + A_2)L/EA$

1.14-5 $F_{11} = 2(1 + \sqrt{2})L/EA \quad F_{12} = F_{21} = -L/EA \quad F_{22} = L/EA$

1.14-6 $F_{12} = 125L/18EA \quad F_{22} = 42L/EA \quad F_{23} = 32L/3EA$

Chapter 2

2.3-1 $Q_1 = \dfrac{PL}{8} - \dfrac{M}{4} \quad Q_2 = \dfrac{PL}{8} + \dfrac{M}{4}$

2.3-2 $Q_1 = -\frac{31}{56}PL \quad Q_2 = \frac{5PL}{14}$ **2.3-3** $Q_1 = -\frac{96sEI}{7L^3} \quad Q_2 = \frac{30sEI}{7L^3}$

2.3-4 $Q_1 = \frac{3PL}{28} \quad Q_2 = \frac{17P}{14}$ **2.3-5** $Q_1 = -\frac{5wL^2}{608} \quad Q_2 = -\frac{33wL^2}{152}$

2.3-6 $Q_1 = -\frac{wL^2}{24} \quad Q_2 = -\frac{wL^2}{12}$ **2.3-7** $Q_1 = -0.243P \quad Q_2 = 0$

2.3-8 $Q_1 = -0.243P \quad Q_2 = 0.172P$

2.3-9 $Q_1 = 0.0267P \quad Q_2 = 0.4698P$

2.3-10 $Q_1 = 3.517P \quad Q_2 = 0.753P$

2.3-11 $Q_1 = -wL/16 \quad Q_2 = 7wL/16 \quad Q_3 = -wL^2/24$

2.3-12 $F_{11} = \frac{L}{EA} + \frac{H^3}{3EI} + \frac{fH}{GA} \quad F_{22} = \frac{L^3}{3EI} + \frac{fL}{GA} + \frac{H}{EA}$

Other terms in **F** are the same as in Eq. (a) of Example 4, Art. 2.3.

2.3-13 (a) $F_{11} = \frac{2L^3}{3EI} \quad F_{12} = F_{23} = 0 \quad F_{13} = -\frac{L^2}{EI} \quad F_{22} = \frac{8L^3}{3EI}$

$F_{33} = \frac{4L}{EI}$ (b) $F_{11} = \frac{2L^3}{3EI} + \frac{2L}{EA} + \frac{2fL}{GA} \quad F_{12} = F_{23} = 0$

$F_{13} = -\frac{L^2}{EI} \quad F_{22} = \frac{8L^3}{3EI} + \frac{2L}{EA} + \frac{2fL}{GA} \quad F_{33} = \frac{4L}{EI}$

2.3-14 $F_{11} = \frac{8L^3}{3EI} \quad F_{12} = \frac{3L^3}{EI} \quad F_{13} = \frac{3L^2}{EI} \quad F_{22} = \frac{20L^3}{3EI} \quad F_{23} = \frac{4L^2}{EI}$

$F_{33} = \frac{4L}{EI}$ **2.3-15** $Q_1 = -\frac{P}{16} \quad Q_2 = \frac{11P}{16}$

2.3-16 $Q_1 = -0.156P \quad Q_2 = 0.290P$

2.3-17 $F_{11} = \frac{2L^3}{3EI} + \frac{L^3}{GJ} \quad F_{12} = \frac{L^2}{2EI} + \frac{L^2}{GJ} \quad F_{13} = \frac{L^2}{2EI}$

$F_{22} = F_{33} = \frac{L}{EI} + \frac{L}{GJ} \quad F_{23} = 0$

2.3-18 $Q_1 = \frac{3P}{16}\frac{1}{2+3\rho}; \quad \rho = \frac{EI}{GJ}$

2.3-19 $\mathbf{Q} = \{-77, 9\}PL/336$

2.3-20 $F_{11} = F_{33} = \frac{2L^3}{3EI} + \frac{2L}{EA} \quad F_{12} = 0 \quad F_{13} = -\frac{L}{EA}$

$F_{22} = \frac{4L^3}{3EI} + \frac{2L}{EA} \quad F_{23} = \frac{2L^3}{3EI} + \frac{L}{EA}$

2.3-21 $D_{QL1} = -\frac{PL^3}{2EI} \quad D_{QL2} = -\frac{PL^3}{3EI} \quad D_{QL3} = 0 \quad F_{11} = \frac{8L^3}{3EI} + \frac{L^3}{GJ}$

$F_{12} = -F_{13} = \frac{L^3}{2EI} \quad F_{22} = \frac{2L^3}{3EI} + \frac{L^3}{GJ} \quad F_{23} = \frac{L^3}{2EI} + \frac{L^3}{GJ}$

$F_{33} = \frac{5L^3}{3EI} + \frac{L^3}{GJ}$ **2.4-1** $Q_1 = 0 \quad Q_2 = \frac{\alpha EI(T_1 - T_2)}{d}$

4-2 $D_{QT2} = 2D_{QT1} = \frac{\alpha L(T_1 - T_2)}{d}$ **2.4-3** $D_{QT1} = \alpha LT \quad D_{QT2} = 0$

2.4-4 $Q_1 = 0 \quad Q_2 = -\dfrac{\beta EI}{L}$ **2.4-5** $D_{QP1} = 0 \quad D_{QP2} = -e\sqrt{2}$

2.4-6 $Q_1 = -\dfrac{12EIs}{L^3} + \dfrac{6EI\beta}{L^2} \quad Q_2 = -\dfrac{6EIs}{L^2} + \dfrac{2EI\beta}{L}$

2.4-7 $D_{QR1} = -s \quad D_{QR2} = 0$

2.4-8 $D_{QR1} = -\dfrac{2s_1}{L} + \dfrac{s_2}{L} \quad D_{QR2} = \dfrac{s_1}{L} - \dfrac{2s_2}{L}$

2.4-9 $D_{QC1} = \alpha LT - \beta H \quad D_{QC2} = -\dfrac{5PL^3}{48EI} - \alpha HT - s$

$D_{QC3} = -\dfrac{PL^2}{8EI} + \beta$

2.5-1 $\mathbf{D}_J = \dfrac{PL^2}{720EI}\begin{bmatrix} -16 \\ 2 \\ -7 \\ 26 \end{bmatrix} \quad \mathbf{A}_R = \dfrac{P}{120}\begin{bmatrix} 46 \\ 129 \\ 144 \\ 41 \end{bmatrix}$

2.5-2 $\mathbf{D}_J = \dfrac{PL}{EA}\begin{bmatrix} -1.172 \\ -1.828 \end{bmatrix} \quad \mathbf{A}_M = P\begin{bmatrix} 1.172 \\ -1.828 \\ 0 \end{bmatrix} \quad \mathbf{A}_R = P\begin{bmatrix} -0.1716 \\ 1.828 \end{bmatrix}$

2.5-3 $\mathbf{D}_J = \{0, 0, 1\}PL^2/64EI \quad \mathbf{A}_R = \{-3, 13, L\}P/32$

Chapter 3

3.3-1 $A_{R1} = \dfrac{5wL}{8} \quad A_{R2} = \dfrac{wL^2}{8} \quad A_{R3} = \dfrac{3wL}{8}$

3.3-2 $A_{M1} = \dfrac{11P}{16} \quad A_{M2} = \dfrac{3PL}{16} \quad A_{M3} = \dfrac{5P}{16}$

3.3-3 $A_{R1} = 2P \quad A_{R2} = \dfrac{13PL}{18} \quad A_{R3} = \dfrac{5PL}{18}$

3.3-4 $A_{M1} = \dfrac{13P}{8} \quad A_{M2} = \dfrac{PL}{2} \quad A_{M3} = \dfrac{P}{4} \quad A_{M4} = \dfrac{3PL}{8}$

$A_{R1} = -\dfrac{3P}{8} \quad A_{R2} = -\dfrac{P}{4}$

3.3-5 $A_{M1} = \dfrac{P}{3} \quad A_{M2} = \dfrac{5PL}{72} \quad A_{R1} = P \quad A_{R2} = \dfrac{7PL}{72}$

3.3-6 $A_{M1} = \dfrac{37P}{60} \quad A_{M2} = \dfrac{59PL}{360} \quad A_{M3} = \dfrac{23P}{60} \quad A_{M4} = -\dfrac{17PL}{360}$

$A_{R1} = \dfrac{49P}{120} \quad A_{R2} = \dfrac{23P}{60}$

3.3-7 $A_{M1} = \dfrac{PL}{72} \quad A_{M2} = -A_{M3} = -\dfrac{25PL}{72} \quad A_{M4} = \dfrac{5PL}{18}$

$A_{M5} = \dfrac{13PL}{18} \quad A_{M6} = \dfrac{17PL}{72} \quad A_{R1} = \dfrac{127P}{48} \quad A_{R2} = \dfrac{79P}{48}$

3.3-8 $S_{11} = S_{44} = \frac{4EI}{L}$ $S_{12} = S_{34} = \frac{2EI}{L}$ $S_{22} = S_{33} = \frac{6EI}{L}$

$S_{23} = \frac{EI}{L}$ $S_{13} = S_{14} = S_{24} = 0$

3.3-9 $S_{22} = 2S_{11} = \frac{8EI}{L}$ $S_{12} = S_{23} = \frac{2EI}{L}$ $S_{33} = 5S_{34} = \frac{20EI}{3L}$

$S_{44} = \frac{8EI}{3L}$ $S_{13} = S_{14} = S_{24} = 0$

3.3-10 $S_{11} = S_{66} = \frac{4EI}{L}$ $S_{22} = S_{33} = S_{44} = S_{55} = \frac{8EI}{L}$

$S_{12} = S_{23} = S_{34} = S_{45} = S_{56} = \frac{2EI}{L}$

3.3-11 $S_{11} = S_{44} = \frac{8EI}{L}$ $S_{24} = 2S_{12} = \frac{4EI}{L}$ $S_{22} = \frac{12EI}{L}$

$S_{23} = S_{34} = -\frac{24EI}{L^2}$ $S_{33} = \frac{96EI}{L^3}$

3.3-12 $S_{11} = S_{44} = \frac{12EI}{L^3}$ $S_{22} = S_{33} = \frac{12EI}{L}$ $S_{12} = -S_{34} = \frac{6EI}{L^2}$

$S_{23} = \frac{4EI}{L}$

3.3-13 $S_{11} = S_{55} = 2S_{12} = 2S_{45} = 4EI/L$ $S_{22} = S_{44} = 28EI/3L$

$S_{33} = 4S_{23} = 4S_{34} = 32EI/3L$

3.3-14 $A_{R1} = \frac{19P}{22}$ $A_{R2} = \frac{79PL}{264}$ $A_{R3} = \frac{3P}{22}$ $A_{R4} = -\frac{19PL}{264}$

3.3-15 $A_{M1} = \frac{P}{2}$ $A_{M2} = 0.433P$ $A_{M3} = \frac{P}{4}$

3.3-16 $A_{M1} = A_{M3} = 0.293P$ $A_{M2} = 2A_{M1}$

3.3-17 $A_{M1} = A_{M3} = 0.293P - 0.060wL$ $A_{M2} = 0.586P - 0.621wL$

3.3-18 $A_{M1} = P$ $A_{M2} = 0.586P$ $A_{M3} = -0.414P$

3.3-19 $A_{M1} = -A_{M5} = 0.236P$ $A_{M2} = 5A_{M4} = 0.417P$ $A_{M3} = 0.354P$

3.3-20 $A_{R1} = 0.446wL$ $A_{R2} = 2.153wL$ $A_{R3} = 2.968wL$

$A_{R4} = -0.446wL$ $A_{R5} = 0$ $A_{R6} = 1.707wL$

3.3-21 $S_{11} = S_{22} = S_{33} = S_{44} = 1.354\frac{EA}{L}$ $S_{12} = -S_{34} = -0.354\frac{EA}{L}$

$S_{24} = -\frac{EA}{L}$

3.3-22 $S_{11} = S_{22} = S_{33} = S_{44} = 1.354\frac{EA}{L}$ $S_{14} = -S_{13} = -S_{34} = 0.354\frac{EA}{L}$

3.3-23 $S_{11} = S_{33} = S_{44} = S_{66} = S_{88} = \frac{253}{250}\frac{EA}{L}$ $S_{22} = S_{77} = \frac{233}{375}\frac{EA}{L}$

$S_{55} = S_{99} = \frac{179}{375}\frac{EA}{L}$ $S_{13} = S_{46} = S_{68} = -\frac{EA}{4L}$

$S_{14} = S_{18} = S_{36} = -\frac{32}{125}\frac{EA}{L}$

$$S_{15} = -S_{19} = S_{24} = -S_{28} = -S_{37} = -S_{45} = S_{89} = -\frac{24}{125}\frac{EA}{L}$$

$$S_{25} = S_{29} = -\frac{18}{125}\frac{EA}{L} \qquad S_{27} = -\frac{EA}{3L}$$

3.3-24 $\mathbf{A}_M = \frac{M}{4L}\begin{bmatrix} 3 \\ 3 \\ -2L \end{bmatrix} \qquad \mathbf{A}_R = \frac{M}{4L}\begin{bmatrix} -3 \\ -3 \\ -L \end{bmatrix}$

3.3-25 $A_{R1} = \frac{27P}{448} \qquad A_{R2} = \frac{31P}{56} \qquad A_{R3} = \frac{PL}{7} \qquad A_{R4} = -\frac{27P}{224}$

$$A_{R5} = \frac{3P}{2} \qquad A_{R6} = \frac{3PL}{112}$$

3.3-26 $\mathbf{A}_M = \frac{wL}{28}\begin{bmatrix} -45 \\ 40 \\ 26L \end{bmatrix} \qquad \mathbf{A}_R = \frac{3wL}{28}\begin{bmatrix} 15 \\ \\ -4 \end{bmatrix}$

3.3-27 (a) $S_{11} = \frac{24EI}{L^3} \qquad S_{12} = S_{13} = \frac{6EI}{L^2} \qquad S_{22} = S_{33} = \frac{6EI}{L} \qquad S_{23} = \frac{EI}{L}$

(b) $S_{11} = S_{4} = \frac{12EI}{L^3} + \frac{EA}{2L} \qquad S_{22} = S_{55} = \frac{3EI}{2L^3} + \frac{EA}{L}$

$$S_{13} = S_{46} = \frac{6EI}{L^2} \qquad S_{14} = -\frac{EA}{2L}$$

$$S_{23} = S_{26} = -S_{35} = -S_{56} = \frac{3EI}{2L^2} \qquad S_{25} = -\frac{3EI}{2L^3}$$

$$S_{33} = S_{66} = 6S_{36} = \frac{6EI}{L}$$

3.3-28 $S_{11} = S_{33} = S_{44} = \frac{4EI}{L} \qquad S_{12} = S_{23} = S_{24} = \frac{2EI}{L} \qquad S_{22} = \frac{12EI}{L}$

3.3-29 $S_{22} = 2S_{11} = -2S_{12} = \frac{48EI_1}{H^3} \qquad S_{33} = S_{44} = \frac{4EI_1}{H} + \frac{4EI_2}{L}$

$$S_{55} = S_{66} = \frac{8EI_1}{H} + \frac{4EI_2}{L} \qquad S_{13} = S_{14} = S_{15} = S_{16} =$$

$$= -S_{23} = -S_{24} = \frac{6EI_1}{H^2} \qquad S_{34} = S_{56} = \frac{2EI_2}{L} \qquad S_{35} = S_{46} = \frac{2EI_1}{H}$$

3.3-30 $A_{R1} = \frac{3P}{8}\frac{9 + 2\eta}{4 + \eta} \qquad A_{R2} = -\frac{PL}{16}\frac{\eta(5 + 2\eta)}{(1 + \eta)(4 + \eta)}$

$$A_{R3} = \frac{PL}{16}\frac{4\eta^2 + 23\eta + 22}{(1 + \eta)(4 + \eta)} \qquad \eta = \frac{GJ}{EI}$$

3.3-31 $A_{M1} = \frac{P}{8}\frac{5 + 2\eta}{4 + \eta} \qquad A_{M2} = -\frac{PL}{16}\frac{(2 + \eta)(5 + 2\eta)}{(1 + \eta)(4 + \eta)}$

$$A_{M3} = \frac{3PL}{16}\frac{\eta}{(1 + \eta)(4 + \eta)}$$

3.3-32 $\mathbf{D} = \frac{wL^3}{24EI(1 + \eta)}\begin{bmatrix} -L(3 + \eta) \\ 4 \\ -4 \end{bmatrix}$

3.3-33 $A_{R1} = \frac{P}{4}\frac{11 + 2\eta}{4 + \eta} \qquad A_{R2} = 0 \qquad A_{R3} = \frac{PL}{8}\frac{6 + \eta}{4 + \eta}$

3.3-34 $S_{11} = S_{77} = \frac{4EI}{L}$ $S_{13} = S_{46} = S_{57} = \frac{2EI}{L}$

$S_{22} = -S_{24} = -S_{35} = -S_{68} = S_{88} = \frac{GJ}{L}$

$S_{33} = S_{44} = S_{55} = S_{66} = \frac{4EI}{L} + \frac{GJ}{L}$

3.3-35 $S_{11} = -S_{14} = -S_{46} = S_{66} = \frac{GJ}{L}$ $S_{22} = S_{77} = \frac{4EI}{L}$

$S_{23} = -S_{34} = -S_{37} = -\frac{6EI}{L^2}$ $S_{25} = S_{57} = \frac{2EI}{L}$ $S_{33} = \frac{36EI}{L^3}$

$S_{44} = \frac{4EI}{L} + \frac{2GJ}{L}$ $S_{55} = \frac{8EI}{L} + \frac{GJ}{L}$

3.4-1 $\mathbf{A}_R = \frac{3\alpha EI(T_1 - T_2)}{7dL}\begin{bmatrix} 2 \\ 3L \\ -1 \\ -1 \end{bmatrix}$ **3.4-2** $\mathbf{A}_R = \frac{6EI}{7L^3}\begin{bmatrix} 11s_1 - 3s_2 \\ 6s_1L - s_2L \\ -16s_1 + 5s_2 \\ 5s_1 - 2s_2 \end{bmatrix}$

3.4-3 $A_{M1} = \frac{EI\beta}{4L^2}$ $A_{M2} = \frac{EI\beta}{12L}$ $A_{M3} = \frac{4EI\beta}{9L^2}$ $A_{M4} = -\frac{EI\beta}{6L}$

$A_{R1} = \frac{7EI\beta}{36L^2}$ $A_{R2} = -\frac{61EI\beta}{36L^2}$

3.4-4 $A_{M1} = 0$ $A_{M2} = -\frac{\alpha EI(T_1 - T_2)}{d}$ $A_{R1} = A_{R3} = 0$

$A_{R2} = -A_{R4} = -A_{M2}$ **3.4-5** $A_{R1} = 0$ $A_{R2} = -A_{R3} = -\frac{EI\beta}{L}$

3.4-6 $A_{DC1} = -\frac{PL}{8} - \frac{6EIs}{L^2}$ $A_{DC2} = \frac{PL}{12} + \frac{2EI\beta}{L}$ $A_{RC1} = \frac{P}{2} + \frac{12EIs}{L^3}$

$A_{RC2} = \frac{P}{2} + \frac{6EI\beta}{L^2}$

3.4-7 $A_{M1} = A_{M4} = -0.1745EA\alpha T$ $A_{M2} = A_{M3} = 0.1277EA\alpha T$

3.4-8 $A_{M1} = 2A_{M3} = \frac{\sqrt{3}EAe}{4L}$ $A_{M2} = -\frac{5EAe}{8L}$

3.4-9 $A_{DC} = -EA\left(\frac{s}{L} + \sqrt{2}\alpha T\right)$ $A_{MC1} = A_{MC2} = -EA\alpha T$

$A_{MC2} = -\frac{EAs}{L}$

3.4-10 $A_{DC1} = EA\left(\frac{s}{L} - 1.707\alpha T\right)$ $A_{DC2} = EA\left(\frac{e}{L} + 1.707\alpha T\right)$

$A_{DC3} = EA\left(0.354\frac{s}{L} - 1.707\alpha T\right)$ $A_{DC4} = A_{DC3} - \frac{EAe}{L}$

3.4-11 $A_{DC1} = A_{MC1} = -EA\alpha T$ $A_{DC2} = \frac{EAs_1}{L} + \frac{12EIs_2}{L^3}$

$A_{DC3} = \frac{\alpha EI(T_1 - T_2)}{d} - \frac{6EIs_2}{L^2}$ $A_{MC2} = \frac{12EIs_2}{L^3}$ $A_{MC3} = -\frac{6EIs_2}{L^2}$

Chapter 4

4.9-1 $\mathbf{D} = \dfrac{PL^2}{384EI}\{7, -53\}$ $\quad \mathbf{A}_{\mathrm{R}} = \dfrac{P}{576}\{351, 93L, 1049, 427, 765, -207L\}$

$\mathbf{A}_{\mathrm{M1}} = \dfrac{P}{192}\{117, 31L, 75, -10L\}$ $\quad \mathbf{A}_{\mathrm{M2}} = \dfrac{P}{288}\{124, 15L, 308, -153L\}$

$\mathbf{A}_{\mathrm{M3}} = \dfrac{P}{192}\{-63, -90L, 255, -69L\}$

4.9-2 $\mathbf{D} = -\dfrac{PL^2}{240EI}\{6, 13L\}$ $\quad \mathbf{A}_{\mathrm{R}} = \dfrac{P}{240}\{204, 48L, 516, 36L\}$

$\mathbf{A}_{\mathrm{M1}} = \dfrac{P}{240}\{204, 48L, 276, -84L\}$ $\quad \mathbf{A}_{\mathrm{M2}} = \dfrac{P}{240}\{240, 84L, 0, 36L\}$

4.9-3 $\mathbf{D} = \dfrac{PL^2}{144EI}\{-8, 23\}$ $\quad \mathbf{A}_{\mathrm{R}} = \dfrac{P}{144}\{24, 2L, 381, 237, -66, 34L\}$

$\mathbf{A}_{\mathrm{M1}} = \dfrac{P}{72}\{12, L, 60, -25L\}$ $\quad \mathbf{A}_{\mathrm{M2}} = \dfrac{P}{144}\{117, 50L, 27, 40L\}$

$\mathbf{A}_{\mathrm{M3}} = \dfrac{P}{72}\{105, 52L, -33, 17L\}$

4.9-4 $\mathbf{D} = \dfrac{PL^2}{528EI}\{-7L, 2\}$ $\quad \mathbf{A}_{\mathrm{R}} = \dfrac{P}{264}\{228, 79L, 36, -19L\}$

$\mathbf{A}_{\mathrm{M1}} = \dfrac{P}{264}\{228, 79L, 36, 17L\}$ $\quad \mathbf{A}_{\mathrm{M2}} = \dfrac{P}{264}\{-36, -17L, 36, -19L\}$

4.9-5 $\mathbf{D} = -\dfrac{wL^3}{384EI}\{16, 15L, 36\}$ $\quad \mathbf{A}_{\mathrm{R}} = \dfrac{wL}{4}\{1, 0, 6\}$

$\mathbf{A}_{\mathrm{M1}} = \dfrac{wL}{4}\{1, 0, 3, -L\}$ $\quad \mathbf{A}_{\mathrm{M2}} = \dfrac{wL}{4}\{3, L, -1, 0\}$

4.9-6 $\mathbf{D} = \dfrac{PL^2}{416EI}\{109, -36, 9\}$ $\quad \mathbf{A}_{\mathrm{R}} = \dfrac{P}{416}\{646, 24, 1464, 362, -86L\}$

$\mathbf{A}_{\mathrm{M1}} = \dfrac{P}{416}\{646, 416L, -230, 22L\}$

$\mathbf{A}_{\mathrm{M2}} = \dfrac{P}{416}\{254, -22L, 578, -140L\}$

$\mathbf{A}_{\mathrm{M3}} = \dfrac{P}{416}\{470, 140L, 362, -86L\}$

4.9-7 $\mathbf{D} = \dfrac{PL^2}{2448EI}\{387L, 279, -819\}$

$\mathbf{A}_{\mathrm{R}} = \dfrac{P}{1224}\{702, -1152L, -3159, 3681, -1125L\}$

$\mathbf{A}_{\mathrm{M1}} = \dfrac{P}{612}\{351, -576L, -351, -297L\}$

$\mathbf{A}_{\mathrm{M2}} = \dfrac{P}{612}\{-261, 297L, -963, 54L\}$

$\mathbf{A}_{\mathrm{M3}} = \dfrac{P}{1224}\{-1233, -1332L, 3681, -1125L\}$

4.9-8 Nonzero elements of rearranged $\mathbf{S}_J$ after factoring $\dfrac{EI}{2L^3}$:

$S_{J11} = 2S_{J12} = \frac{2}{3}S_{J22} = 4S_{J23} = \frac{2}{3}S_{J33} = 2S_{J34} = S_{J44} = 8L^2$

$S_{J15} = -S_{J16} = S_{J25} = -\frac{4}{3}S_{J26}$

$= -4S_{J27} = 4S_{J36} = \frac{4}{3}S_{J37} = -S_{J38} = S_{J47} = -S_{J48} = 12L$

$S_{J55} = -S_{J56} = \frac{8}{9}S_{J66} = -8S_{J67} = \frac{8}{9}S_{J77} = -S_{J78} = S_{J88} = 24$

4.9-9 Nonzero elements of rearranged $\mathbf{S}_J$ after factoring $\dfrac{EI}{L^3}$:

$S_{J11} = 4S_{J12} = 4S_{J16} = \frac{2}{3}S_{J22} = 2S_{J24} = S_{J44} = 2S_{J66} = 8L^2$

$S_{J15} = -S_{J18} = -\frac{1}{4}S_{J23} = S_{J27} = \frac{1}{3}S_{J28}$

$= -\frac{1}{4}S_{J34} = \frac{1}{4}S_{J48} = S_{J56} = -S_{J67} = 6L$

$\frac{1}{8}S_{J33} = -\frac{1}{8}S_{J38} = S_{J55} = -S_{J57} = \frac{1}{2}S_{J77} = -S_{J78} = \frac{1}{9}S_{J88} = 12$

4.9-10 Nonzero elements of rearranged $\mathbf{S}_J$ after factoring $\dfrac{EI}{L^3}$:

$S_{J11} = -S_{J16} = S_{J44} = -S_{J47} = \frac{1}{3}S_{J66} = -\frac{1}{2}S_{J67} = \frac{1}{3}S_{J77} = 12$

$S_{J12} = S_{J15} = S_{J26}$

$= -\frac{1}{2}S_{J27} = -S_{J34} = \frac{1}{2}S_{J36} = -S_{J37} = -S_{J48} = -S_{J56} = S_{J78} = 6L$

$S_{J22} = 3S_{J23} = 6S_{J25} = S_{J33} = 6S_{J38} = 3S_{J55} = 3S_{J88} = 12L^2$

4.9-11 Nonzero elements of rearranged $\mathbf{S}_J$ after factoring $\dfrac{EI}{L^3}$:

$S_{J11} = 2S_{J12} = \frac{1}{2}S_{J22} = 2S_{J23} = \frac{1}{2}S_{J33} = 2S_{J34} = \frac{1}{2}S_{J44} = 2S_{J45} = S_{J55}$

$= 4L^2 \quad S_{J16} = -S_{J17} = S_{J26}$

$= -S_{J28} = S_{J37} = -S_{J39} = S_{J48}$

$= -S_{J410} = S_{J59} = -S_{J510} = 6L$

$S_{J66} = -S_{J67} = \frac{1}{2}S_{J77} = -S_{J78}$

$= \frac{1}{2}S_{J88} = -S_{J89} = \frac{1}{2}S_{J99}$

$= -S_{J910} = S_{J1010} = 12$

4.9-12 Nonzero elements of rearranged $\mathbf{S}_J$ after factoring $\dfrac{2EI}{L^3}$:

$S_{J11} = 4S_{J13} = 4S_{J18} = \frac{4}{3}S_{J33}$

$= 8S_{J34} = 2S_{J44} = 8S_{J46} = 4S_{J66} = 2S_{J88} = 8L^2$

$S_{J12} = S_{J17} = -2S_{J23}$

$= 2S_{J24} = S_{J39} = -2S_{J310} = -2S_{J45} = -2S_{J56} = 2S_{J610} = S_{J78} = -S_{J89}$

$= 6L$

$S_{J22} = -\frac{3}{2}S_{J29} = -3S_{J210}$

$= 3S_{J55} = -3S_{J510} = \frac{3}{2}S_{J77} = -\frac{3}{2}S_{J79} = \frac{3}{4}S_{J99} = \frac{3}{2}S_{J1010} = 18$

4.12-1 $\mathbf{D} = \dfrac{\sqrt{2}PL}{(1+\sqrt{2})EA_X}\{1 + \sqrt{2}, 1\}$

$\mathbf{A}_R = \dfrac{P}{2(1+\sqrt{2})}\{-(2 + \sqrt{2}), -(2 + \sqrt{2}), -2(1 + \sqrt{2}), -2\sqrt{2},$
$-\sqrt{2}, \sqrt{2}\}$

$\mathbf{A}_{M1} = P\{-1, 0, 1, 0\}$

$\mathbf{A}_{M2} = \dfrac{P}{1+\sqrt{2}}\{-\sqrt{2}, -(1 + \sqrt{2}), \sqrt{2}, -(1 + \sqrt{2})\}$

$\mathbf{A}_{M3} = \dfrac{P}{1+\sqrt{2}}\{1, 0, -1, 0\}$

4.12-2 $\mathbf{D} = \dfrac{PL}{EA_X}\{1.30, 1.60\}$

$\mathbf{A}_R = P\{-3.80, 1.50, -1.00, -0.600, -0.200, 1.10\}$

$\mathbf{A}_{M1} = P\{1.30, -2.00, -1.30, 2.00\}$ $\mathbf{A}_{M2} = P\{0, 1.00, 0, 1.00\}$

$\mathbf{A}_{M3} = P\{-1.60, 1.00, 1.60, 1.00\}$

$\mathbf{A}_{M4} = P\{0.500, 0.500, 0.500, 0.500\}$ $\mathbf{A}_{M5} = P\{-0.500, 0, 0.500, 0\}$

$\mathbf{A}_{M6} = P\{0, 0, 0, 0\}$

4.12-3 $\mathbf{D} = \dfrac{wL^2}{EA_X}\{-1.26, 0\}$ $\mathbf{A}_R = wL\{0, 1.71, -0.446, 0.446, 2.15, 2.97\}$

$\mathbf{A}_{M1} = wL\{0, 0.500, 0, 0.500\}$ $\mathbf{A}_{M2} = wL\{0, 0.500, 0, 0.500\}$

$\mathbf{A}_{M3} = wL\{0.500, 0, 0.500, 0\}$ $\mathbf{A}_{M4} = wL\{1.76, 0, -0.762, 0\}$

$\mathbf{A}_{M5} = wL\{-0.500, 0.500, -0.500, 0.500\}$

$\mathbf{A}_{M6} = wL\{1.131, 0.500, 0.131, 0.500\}$

4.12-4 $\mathbf{D} = \dfrac{wL^2}{EA_X}\{-0.288, -0.815, -0.731\}$

$\mathbf{A}_R = wL\{-0.0630, 0.351, 2.48, -0.288, 2.32\}$

$\mathbf{A}_{M1} = wL\{-0.288, 0.300, 0.288, 0.300\}$

$\mathbf{A}_{M2} = wL\{0, 0.300, 0, 0.300\}$ $\mathbf{A}_{M3} = wL\{1.22, 0, -0.416, 0\}$

$\mathbf{A}_{M4} = wL\{1.13, 0, -0.332, 0\}$

$\mathbf{A}_{M5} = wL\{0.080, 0.300, -0.880, 0.300\}$

$\mathbf{A}_{M6} = wL\{0.985, 0.300, -0.185, 0.300\}$

4.12-5 $\mathbf{D} = \dfrac{PL}{EA_X}\{0.667, -2.00, -4.00\}$

$\mathbf{A}_R = P\{-2.40, 0.700, 0.21, 0.700, 4.79\}$

$\mathbf{A}_{M1} = P\{0.667, 1.00, -0.667, 2.00\}$ $\mathbf{A}_{M2} = P\{0, -1.00, 0, -1.00\}$

$\mathbf{A}_{M3} = P\{1.40, 0, -1.40, 0\}$ $\mathbf{A}_{M4} = P\{2.07, 0, -2.07, 0\}$

$\mathbf{A}_{M5} = P\{4.33, 0.500, -2.60, 0.500\}$

4.12-6 $\mathbf{D} = \dfrac{wL^2}{EA_X}\{-2.92, -2.92\}$

$\mathbf{A}_R = wL\{-1.40, 1.40, -1.40, 2.25, 1.40, 2.25\}$

$\mathbf{A}_{M1} = wL\{0, 0.400, 0, 0.400\}$ $\mathbf{A}_{M2} = wL\{0, 0.400, 0, 0.400\}$

$\mathbf{A}_{M3} = wL\{0.300, 0, 0.300, 0\}$ $\mathbf{A}_{M4} = wL\{0.150, 0, 0.150, 0\}$

$\mathbf{A}_{M5} = wL\{-2.05, 0.400, 1.45, 0.400\}$

$\mathbf{A}_{M6} = wL\{2.05, 0.400, -1.45, 0.400\}$

4.12-7 Nonzero elements of $\mathbf{S}_J$ after factoring $\dfrac{EA_X}{L}$:

$S_{J11} = -2S_{J15} = -2S_{J17} = -\tfrac{2}{3}S_{J26} = -\tfrac{2}{3}S_{J28} = -\tfrac{2}{3}S_{J36} = \tfrac{2}{3}S_{J38}$
$= -\tfrac{2}{3}S_{J45} = \tfrac{2}{3}S_{J47} = 0.500$

$S_{J16} = -S_{J18} = S_{J25} = -S_{J27} = S_{J46} = S_{J48} = -0.433$ $S_{J22} = 3.23$

$S_{J24} = -1.73$ $S_{J33} = -2S_{J35} = -2S_{J37} = S_{J44} = 2.60$

$S_{J55} = S_{J77} = 2.55$ $S_{J56} = S_{J66} = -S_{J78} = S_{J88} = 1.18$ $S_{J57} = -1.00$

4.12-8 Nonzero elements of $\mathbf{S}_J$ after factoring $\dfrac{EA_X}{L}$:

$S_{J11} = S_{J33} = 1.38$

$S_{J12} = -\tfrac{1}{2}S_{J16} = S_{J18} = -\tfrac{1}{2}S_{J25} = S_{J27} = -S_{J34} = \tfrac{1}{2}S_{J36} = -S_{J310}$
$= \tfrac{1}{2}S_{J45} = -S_{J49} = \tfrac{1}{2}S_{J57} = \tfrac{1}{2}S_{J59} = \tfrac{3}{2}S_{J68} = \tfrac{3}{2}S_{J610} = -0.217$

$S_{J13} = 4S_{J15} = 8S_{J17} = -\frac{8}{9}S_{J22} = \frac{4}{3}S_{J26} = \frac{8}{3}S_{J28} = 4S_{J35} = 8S_{J39}$
$= -\frac{8}{9}S_{J44} = \frac{4}{3}S_{J46} = \frac{8}{3}S_{J410} = 4S_{J58} = -4S_{J510} = 4S_{J67}$
$= -4S_{J69} = -1.00$
$S_{J55} = 1.368 \quad S_{J66} = 1.79 \quad S_{J77} = S_{J99} = 0.559$
$S_{J78} = -S_{J910} = 0.467 \quad S_{J88} = S_{J1010} = 0.520$

4.12-9 Nonzero elements of rearranged $\mathbf{S}_J$ after factoring $\frac{EA_X}{L}$:

$S_{J11} = S_{J33} = S_{J77} = S_{J99} = 1.71$
$S_{J12} = S_{J27} = S_{J35} = S_{J49} = -1.00 \quad S_{J22} = 2.71$
$S_{J14} = S_{J15} = -S_{J16} = -S_{J23} = S_{J29} = S_{J210} = S_{J39} = S_{J310} = S_{J46}$
$= S_{J47} = -S_{J48} = S_{J56} = -S_{J57} = S_{J58} = -S_{J66} = S_{J78}$
$= -S_{J88} = -S_{J910} = -S_{J1010} = -0.707$
$S_{J44} = S_{J55} = 2.42$

4.12-10 Nonzero elements of $\mathbf{S}_J$ after factoring $\frac{EA_X}{L}$:

$S_{J11} = S_{J33} = \frac{1}{3}S_{J55} = \frac{1}{3}S_{J77} = \frac{1}{2}S_{J99} = \frac{1}{2}S_{J1111} = 2.03$
$S_{J12} = \frac{4}{3}S_{J17} = -S_{J18} = -S_{J27} = \frac{3}{4}S_{J28} = -S_{J34} = \frac{4}{3}S_{J35} = S_{J36}$
$= S_{J45} = \frac{3}{4}S_{J46} = \frac{2}{3}S_{J511} = -\frac{1}{2}S_{J512} = -\frac{1}{2}S_{J611} = \frac{3}{8}S_{J612}$
$= \frac{2}{3}S_{J79} = \frac{1}{2}S_{J710} = \frac{1}{2}S_{J89} = \frac{3}{8}S_{J810} = -\frac{1}{2}S_{J910} = -\frac{3}{8}S_{J1010}$
$= \frac{1}{2}S_{J1112} = -\frac{3}{8}S_{J1212} = -0.480$
$S_{J15} = S_{J37} = \frac{1}{2}S_{J59} = \frac{1}{2}S_{J711} = -1.67 \quad S_{J22} = S_{J44} = 1.89$
$S_{J24} = \frac{1}{2}S_{J68} = -1.25 \quad S_{J56} = -S_{J78} = -0.48$
$S_{J66} = S_{J88} = 4.42$

Answers to problems for Art. 4.18 in US units are kips, inches, and radians, those in SI units are kilonewtons, meters, and radians.

4.18-1 $\mathbf{D} = \{-0.000438, -0.00193, 0.000532\}$
$\mathbf{A}_R = \{0.913, 5.97, 227.7, -0.913, 4.03, 43.59\}$
$\mathbf{A}_{M1} = \{0.913, 5.97, 227.7, -0.913, 4.03, -87.94\}$
$\mathbf{A}_{M2} = \{4.03, 0.913, 87.94, -4.03, -0.913, 43.59\}$

4.18-2 $\mathbf{D} = \{-0.000216, -0.00358, 0.000424\}$
$\mathbf{A}_R = \{0.360, 3.34, 249.6, 0.360, -0.303, 28.19, -0.720, 8.96, 46.00\}$
$\mathbf{A}_{M1} = \{0.360, 3.34, 249.6, -0.360, 2.66, -151.5\}$
$\mathbf{A}_{M2} = \{-0.360, 0.303, 59.09, 0.360, -0.303, 28.19\}$
$\mathbf{A}_{M3} = \{8.96, 0.721, 92.36, -8.96, -0.721, 46.0\}$

4.18-3 $\mathbf{D} = \{0.7491, -0.1650, 1.980\} \times 10^{-4}$
$\mathbf{A}_R = \{-24.97, 39.00, -62.29, -25.03, 11.00, 44.49\}$
$\mathbf{A}_{M1} = \{24.97, 11.00, 44.31, -24.97, 39.00, -62.29\}$
$\mathbf{A}_{M2} = \{11.00, 25.03, 44.49, -11.00, 24.97, -14.31\}$

4.18-4 $\mathbf{D} = \{0.5934, -2.225, -1.466\} \times 10^{-4}$
$\mathbf{A}_R = \{-20.77. 19.54, 20.36, -20.77, 62.58, -66.02, -38.47, 77.89, 38.21\}$
$\mathbf{A}_{M1} = \{-20.77, 19.54, 20.36, 20.77, 20.46, -22.21\}$
$\mathbf{A}_{M2} = \{20.77, 57.42, 50.56, -20.77, 62.58, -66.02\}$
$\mathbf{A}_{M3} = \{77.89, 38.47, 38.21, -77.89, 41.54, -44.35\}$

4.18-5 $\mathbf{D} = \{0, -0.000620, 0\}$
$\mathbf{A}_R = \{-0.323, 0, -8.55, 0, 0.434, 0, 0.323, 1.57, 8.77\}$
$\mathbf{A}_{M1} = \{-0.434, 0, 0, 0.434, 0, 0\}$
$\mathbf{A}_{M2} = \{1.52, 0.503, 8.77, 0.215, 0.497, -8.55\}$

4.18-6 $\mathbf{D} = \{0.5302, -1.457, -0.02613\} \times 10^{-4}$
$\mathbf{A}_R = \{-141.40, -63.40, 5.444, 51.40, 93.40, 40.75\}$
$\mathbf{A}_{M1} = \{141.40, 63.40, 4.747, -141.40, -63.40, 5.444\}$
$\mathbf{A}_{M2} = \{102.39, 29.70, 40.75, -144.81, 12.73, -4.747\}$

4.18-7 $\mathbf{D} = \{0.4160, -0.9615, 0.1880\} \times 10^{-4}$
$\mathbf{A}_R = \{-13.87, 24.77, 33.35, -20.80, 13.29, -18.00, 34.66, 61.94, 17.23\}$
$\mathbf{A}_{M1} = \{-13.87, 24.77, 33.35, 13.87, 5.235, -4.765\}$
$\mathbf{A}_{M2} = \{20.80, 6.707, 4.824, -20.80, 13.29, -18.00\}$
$\mathbf{A}_{M3} = \{70.35, 9.435, 17.23, -46.35, 8.565, -0.059\}$

4.18-8 Nonzero elements of $\mathbf{S}_J$:
$S_{J11} = S_{J44} = 1541.7$
$S_{J13} = S_{J19} = -S_{J37} = S_{J46} = S_{J412} = -S_{J610} = -S_{J79}$
$= -S_{J1012} = 5000.0$
$S_{J14} = \frac{1}{2}S_{J28} = \frac{1}{2}S_{J511} = -\frac{1}{2}S_{J88} = -\frac{1}{2}S_{J1111} = -1500.0$
$S_{J17} = S_{J410} = -S_{J77} = -S_{J1010} = -41.7 \qquad S_{J22} = S_{J55} = 3005.2$
$S_{J23} = S_{J26} = -S_{J35} = -S_{J56} = 1250.0 \qquad S_{J25} = -5.22$
$\frac{1}{3}S_{J33} = 2S_{J36} = S_{J39} = \frac{1}{3}S_{J66} = S_{J612} = \frac{1}{2}S_{J99} = \frac{1}{2}S_{J1212} = 4.0 \times 10^5$

4.18-9 Nonzero elements of $\mathbf{S}_J$:
$S_{J11} = S_{J22} = 3041.7$
$S_{J13} = S_{J19} = S_{J23} = S_{J26} = -S_{J35} = -S_{J37} = -S_{J79} = 5000.0$
$S_{J14} = S_{J28} = -S_{J88} = -3000.0 \qquad S_{J17} = S_{J25} = -S_{J77} = -41.7$
$\frac{1}{4}S_{J33} = S_{J36} = S_{J39} = \frac{5}{18}S_{J66} = \frac{5}{4}S_{J612} = \frac{1}{2}S_{J99} = \frac{5}{8}S_{J1212} = 4.0 \times 10^5$
$S_{J44} = 3877.7 \qquad S_{J45} = -S_{J411} = -S_{J510} = S_{J1011} = -1143.0$
$S_{J46} = S_{J412} = -S_{J610} = -S_{J1012} = 2560.0 \qquad S_{J410} = -S_{J1010} = -877.7$
$S_{J55} = 1585.0 \qquad S_{J56} = -3080.0 \qquad S_{J511} = -S_{J1111} = -1543.7$
$S_{J512} = -S_{J611} = -S_{J1112} = 1920.0$

Index